TRANSPORT MODELING FOR ENVIRONMENTAL
ENGINEERS AND SCIENTISTS

ENVIRONMENTAL SCIENCE AND TECHNOLOGY

A Wiley-Interscience Series of Texts and Monographs

Edited by JERALD L. SCHNOOR, *University of Iowa*
  ALEXANDER ZEHNDER, *Swiss Federal Institute for Water Resources and Water Pollution Control*

PHYSIOCHEMICAL PROCESSES FOR WATER QUALITY CONTROL
  Walter J. Weber, Jr., Editor
pH AND pION CONTROL IN PROCESS AND WASTE STREAMS
  F. G. Shinskey
AQUATIC POLLUTION: An Introductory Text
  Edward A. Laws
INDOOR AIR POLLUTION: Characterization, Prediction, and Control
  Richard A. Wadden and Peter A. Scheff
PRINCIPLES OF ANIMAL EXTRAPOLATION
  Edward J. Calabrese
SYSTEMS ECOLOGY: An Introduction
  Howard T. Odum
INTEGRATED MANAGEMENT OF INSECT PESTS OF POME AND STONE FRUITS
  B. A. Croft and S. C. Hoyt, Editors
WATER RESOURCES: Distribution, Use and Management
  John R. Mather
ECOGENETICS: Genetic Variation in Susceptibility to Environmental Agents
  Edward J. Calabrese
GROUNDWATER POLLUTION MICROBIOLOGY
  Gabriel Bitton and Charles P. Gerba, Editors
CHEMISTRY AND ECOTOXICOLOGY OF POLLUTION
  Des W. Connell and Gregory J. Miller
SALINITY TOLERANCE IN PLANTS: Strategies for Crop Improvement
  Richard C. Staples and Gary H. Toenniessen, Editors
ECOLOGY, IMPACT ASSESSMENT, AND ENVIRONMENTAL PLANNING
  Walter E. Westman
CHEMICAL PROCESSES IN LAKES
  Werner Stumm, Editor
INTEGRATED PEST MANAGEMENT IN PINE-BARK BEETLE ECOSYSTEMS
  William E. Waters, Ronald W. Stark, and David L. Wood, Editors
PALEOCLIMATE ANALYSIS AND MODELING
  Alan D. Hecht, Editor
BLACK CARBON IN THE ENVIRONMENT: Properties and Distribution
  E. D. Goldberg
GROUND WATER QUALITY
  C. H. Ward, W. Giger, and P. L. McCarty, Editors
TOXIC SUSCEPTIBILITY: Male/Female Differences
  Edward J. Calabrese
ENERGY AND RESOURCE QUALITY: The Ecology of the Economic Process
  Charles A. S. Hall, Cutler J. Cleveland, and Robert Kaufmann
AGE AND SUSCEPTIBILITY TO TOXIC SUBSTANCES
  Edward J. Calabrese

The list of the titles in this series is continued at the end of this volume.

TRANSPORT MODELING FOR ENVIRONMENTAL ENGINEERS AND SCIENTISTS

MARK M. CLARK

Department of Civil and Environmental Engineering
University of Illinois at Urbana-Champaign
Urbana, Illinois

A WILEY-INTERSCIENCE PUBLICATION
JOHN WILEY & SONS
New York · Chichester · Brisbane · Toronto · Singapore

Library of Congress Cataloging in Publication Data:

Clark, Mark M.
    Transport modeling for environmental engineers and scientists /
Mark M. Clark.
        p.   cm. — (Environmental science and technology)
    "A Wiley-Interscience publication."
    Includes Index.
    ISBN 0-471-12348-X (cloth; alk. paper)
    1. Environmental chemistry — Mathematical models.   2. Transport
theory — Mathematical models.   I. Title.   II. Series.
TD193.C55   1996
628.5′2 — dc20   96-4814

Printed in the United States of America

10 9 8 7 6 5 4 3 2 1

This book is dedicated in loving memory to Dan William Clark

CONTENTS

SERIES PREFACE
Environmental Science and Technology

We are in the third decade of the Wiley-Interscience Series of texts and monographs in Environmental Science and Technology. It has a distinguished record of publishing outstanding reference texts on topics in the environmental sciences and engineering technology. Classic books have been published here, graduate students have benefited from the textbooks in this series, and the series has also provided for monographs on new developments in various environmental areas.

As new editors of this Series, we wish to continue the tradition of excellence and to emphasize the interdisciplinary nature of the field of environmental science. We publish texts and monographs in environmental science and technology as it is broadly defined from basic science (biology, chemistry, physics, toxicology) of the environment (air, water, soil) to engineering technology (water and wastewater treatment, air pollution control, solid, soil, and hazardous wastes). The series is dedicated to a scientific description of environmental processes, the prevention of environmental problems, and to preservation and remediation technology.

There is a new clarion for the environment. No longer are our pollution problems only local. Rather, the scale has grown to the global level. There is no such place as "upwind" any longer; we are all "downwind" from somebody else in the global environment. We must take care to preserve our resources as never before and to learn how to internalize the cost to prevent environmental degradation into the product that we make. A new "industrial ecology" is emerging that will lessen the impact our way of life has on our surroundings.

In the next 50 years, our population will come close to doubling, and if the developing countries are to improve their standard of living as is needed, we will reuire a gross world product several times what we currently have. This will create new pressures on the environment, both locally and globally. But there are new opportunities also. The world's people are recognizing the need for sustainable development and leaving a legacy of resources for future generations at least equal to what we had. The goal of this series is to help understand the environment, its functioning, and how problems can be overcome; the series will also provide new insights and new sustainable technologies that will allow us to preserve and hand down an intact environment to future generations.

JERALD L. SCHNOOR
ALEXANDER J. B. ZEHNDER

PREFACE

Transport Modeling for Environmental Engineers and Scientists was written for students as an introduction to the modeling of mass- and momentum-transport processes in the environment. This book presents a new approach to education in environmental engineering. First, it is based on the philosophy that mass-, momentum-, and energy-transport processes can be taught in the same course. About 30 years ago, this philosophy was introduced in Bird, Stewart, and Lightfoot's *Transport Phenomena*, which changed forever chemical-engineering education by demonstrating the underlying unity of transport processes. This book adopts an approach similar to the integrated heat-, mass-, and momentum-transport courses of chemical and mechanical engineering curricula. Second, this book provides examples of transport processes relevant to the three main environmental media — air, water, and soil. In the early stages of environmental study, much can be learned from approaches developed in the allied fields of air- and water-pollution control, ecology, and soil and groundwater transport. Hence, this book stresses the common features underlying pollutant transport and pollution-control processes. Although neither a design nor "unit processes" presentation is used, the principles introduced here clarify existing processes, and will, I hope, form the basis of new and improved models and designs in environmental engineering.

The main prerequisite a student should possess before reading this book is a background in calculus through differential equations and college-level physics (mechanics). Courses in thermodynamics, physical chemistry, and fluid mechanics would be useful, but are not required. The book can be used as a textbook in third- or fourth-year undergraduate or first-year graduate courses in environmental and chemical engineering. At the University of Illinois, the material in this book forms the basis of a required first-semester course in the graduate-level environmental engineering and science program. Although this course is composed mostly of students with undergraduate Civil or Chemical Engineering degrees, students from the chemistry and biology departments also find the course quite accessible so long as the mathematics prerequisites are honored.

Although the mathematics in this book is at about the sophomore undergraduate engineering level, there is a good deal of algebra and first-order differential equations. As an undergraduate student, I often glossed over equation derivations, probably because I was not often tested on this material. But I encourage readers to think about the math and algebraic steps, and to really ask themselves, "Do I understand this?" Many of the mathematical steps, algebraic simplifications, and approximations are traditional techniques in applied mathematics, engineering, and physical chemistry. In a sense, they form some of the underlying culture of our field. Learning these techniques is a part of the maturation of environmental engineers and scientists.

I have tried to use SI units in the book, but have also used other systems of units when they are more traditionally used in the literature. Many approximations such as $\gg$ or $\ll$ and $\ggg$ or $\lll$, are also utilized. In this book, these symbols have the following meanings:

$<$ or $>$	Less than or greater than.
$\ll$ or $\gg$	Much less than, or much greater than. The compared quantities differ by *at least* a factor of 10.
$\lll$ or $\ggg$	Much, much less than, or much, much greater than. The compared quantities differ by *at least* a factor of 100.

There is also a fair amount of work with tabular data and curve fitting in the exercises. Thus it is extremely helpful for students to have access to spreadsheet and graphics programs.

I sincerely hope you find environmental modeling as exciting and rewarding as the many scientists who have come before you.

Urbana, Illinois   MARK M. CLARK
February 1996

ACKNOWLEDGMENTS

I am indebted to many colleagues and University of Illinois students and staff for their help with this project. Marcelo Garcia, Paul Newton, Gary Peyton, Massoud Rostam-Abadi, and Albert Valocchi kindly reviewed and critiqued selected chapters of the book. Jean-Luc Bersillon, Jeffrey Collett, Edwin Herricks, Wayland Eheart, David Freedman, Susan Larson, Valentina Lazarova, Bruce Rittmann, Mark Rood, and Vernon Snoeyink contributed valuable references and important feedback on various parts of the book. A number of students commented on errors and made valuable suggestions and other contributions. They include Samer Adham, Betsy Andrews, Richard Bernard, Fred Cannon, Joel Ducoste, Mary Jo Kirisits, Detlef Knappe, Catherine Jucker, Timothy Kramer, and Mark Rhodes. Mesenia Atenas worked the majority of exercises and made numerous corrections and suggestions. Ron Winburn expertly redrew approximately 200 figures, and Claudia Cook and Helen Maris provided much assistance in reformatting and typing parts of the manuscript. My wife, Joan Stolz, proofread and edited numerous drafts of the book; she also nobly bore my preoccupations and the permanent loss of our dining room table. I also acknowledge my mentors at the Johns Hopkins University, Charles O'Melia and the late Stanley Corrsin, who guided me during my early studies of environmental transport and fluid mechanics.

Finally, I wish to thank Joël Mallevialle, former director of the Centre International de Recherche sur l'Eau et l'Environnement of Lyonnaise des Eaux, for supporting me during my sabbatical in Paris. Writing the middle of my book in an apartment overlooking the Seine River was one of the most rewarding experiences of my academic career.

LIST OF SYMBOLS

a	particle radius, L
$\mathbf{a}$	initial position in Lagrangian coordinate system, L
a_L	longitudinal dispersivity coefficient, L
a_T	transverse dispersivity coefficient, L
A	Hamaker constant, J
A	surface area of projected area, L^2
A_F	constant, dimensionless
A_p	surface area of prticle, L^2
A_i	total interfacial area of system of particles, L^2
b	ratio of rate constants in Langmuir isotherm, dimensionless
Bi	Biot number, dimensionless
c_A	concentration of component A, mol/L^3
$\tilde{c}_{A,R}$	dimensionless concentration
$c_{A,s}$	concentration of A at surface, mol/L^3
$c_{A,\infty}$	concentration of component A in solution phase at equilibrium, mol/L^3
$c_{A,\infty}$	concentration of component A at $r \to \infty$ (or very far from surface), mol/L^3
$c_{A,ave}$	average concentration of A, mol/L^3
c_i	concentration of ith component, mol/L^3
c'_A	concentration of component A in interstitial fluid, mol/L^3
$\tilde{c}_A$	dimensionless concentration
c'_A	fluctuating component of A, mol/L^3
c_A^*	equilibrium concentration of A, mol/L^3
C	concentration, mol/L^3
$\bar{C}_A$	mean concentration of A, mol/L^3
$c_{A,i}$	inlet concentration of A, mol/L^3
c_A^*	characteristic concentration scale or difference, mol/L^3
C_c	Cunningham slip coefficient, dimensionless
C_c	carbon concentration, M/L^3
C_D	drag coefficient, dimensionless

C_{fL} average skin friction coefficient, dimensionless
C_{fx} local skin friction coefficient, dimensionless
C_n infinite series of constants, dimensionless
C_p specific heat, energy/$(M\text{-}°C)$
C_p particulate-associated pesticide concentration, M/L^3
d diameter, L
d decay term, mol-L^{-3}-T^{-1}
d_c cylindrical collector diameter, L
dw differential work, $M\text{-}L^2\text{-}T^{-2}$
D system or box depth, L
D_{AB} diffusion coefficient of component A in component B, L^2/T
D_{AB}^{aer} diffusion coefficient in aerobic layer, L^2/T
D_{AB}^{an} diffusion coefficient in anaerobic layer, L^2/T
D_A^{II} second Damköhler number, dimensionless
D_{dis} dispersion coefficient, L^2/T
D_{eff} effective diffusion coefficient, L^2/T
$D_{i,j}$ relative diffusion coefficient of class i and j particles, L^2/T
$D_{i,j}$ mechanical dispersion tensor, L/T
D_j diffusion coefficient of class j particles, L^2/T
$D_{i,j}^{(h)}$ hydrodynamic dispersion tensor, L^2/T
$\bar{D}_{i,j}^{(h)}$ principle component form of hydrodynamic dispersion tensor, L^2/T
D_{dis}^t turbulent dispersion coefficient, L^2/T
$\tilde{D}$ depth of uncleared fluid, L
$\mathbf{e}_r$ unit vector in radial direction, dimensionless
$\mathbf{e}_\theta$ unit vector in θ direction, dimensionless
E electrical field strength, V/L
f friction factor, dimensionless
f function of
$f(t)$ residence time density function, T^{-1}
$f(\tau)$ dimensionless residence time density
$f_{data}(t_i)$ residence time density function based on tracer data, T^{-1}
$f_{model}(t_i)$ residence time density function based on model, T^{-1}
f_f fanning friction factor, dimensionless
F force, $M\text{-}L\text{-}T^{-2}$
F Lagrangian integral time scale, T
F_e force on particle due to electrical field, $M\text{-}L\text{-}T^{-2}$
F_D drag force, $M\text{-}L\text{-}T^{-2}$
$F(t)$ cumulative residence time distribution function, dimensionless
$\mathbf{F}_g$ net force of gravity on particle, $M\text{-}L\text{-}T^{-2}$
$\mathbf{F}_i$ ith component of force acting on particle, $M\text{-}L\text{-}T^{-2}$

$\mathbf{F}_n$	nonsteady particle drag force, $M\text{-}L\text{-}T^{-2}$
g	acceleration of gravity, L/T^2
g	dimensionless function
g_θ	θ-component of acceleration of gravity, L/T^2
G	Gibbs function, $M\text{-}L^2\text{-}T^{-2}$
G	potential energy, L^2/T^2
G_n	infinite series of constants, dimensionless
Gr	Grashof number, dimensionless
Gv	gravitational parameter, dimensionless
h	heat transfer coefficient, energy$/(L^2\text{-}T\text{-}°C)$
H	Henry's Law constant, atm/mole fraction or $M\text{-}L^2\text{-}mol^{-1}\text{-}T^{-2}$
H_A	attraction or adhesion group, dimensionless
$\mathbf{i}$	unit vector along x-axis, dimensionless
I	current, A
$\mathbf{j}$	unit vector along y-axis, dimensionless
$\mathbf{j}_A$	mass flux of component A relative to moving coordinates, $M\text{-}L^{-2}\text{-}T^{-1}$
$\mathbf{j}_i$	mass flux of ith component relative to moving coordinates, $M\text{-}L^{-2}\text{-}T^{-1}$
J	mass flux, $M\text{-}L^{-2}\text{-}T^{-1}$ or $mol\text{-}L^{-2}\text{-}T^{-1}$
$\tilde{J}_A$	flux of component A in transformed coordinate systems, $mol\text{-}L^{-2}\text{-}T^{-1}$
$J_{A,\text{ave}}$	average flux over plate of length L, $mol\text{-}L^{-2}\text{-}T^{-1}$
$\mathbf{J}_A^*$	relative molar flux of component A, $mol\text{-}L^{-2}\text{-}T^{-1}$
$\mathbf{J}_i^*$	molar flux of ith component relative to moving coordinates, $mol\text{-}L^{-2}\text{-}T^{-1}$
k	reaction rate constant, $mol^{1-n}\text{-}L^{3-3n}\text{-}T$ (where n is reaction order)
k	Boltzman constant, $J\text{-}K^{-1}$
k	permeability, L^2
k	thermal conductivity, energy$/(L\text{-}T\text{-}°C)$
k_a	adsorption rate constant, $L^3\text{-}T^1\text{-}M^{-1}$
k_c	mass transfer coefficient, L/T
k_d	partition coefficient, L^3/M
k_d	desorption rate constant, T^{-1}
k_e	constant in Ergun equation, L
k_s	sedimentation rate constant, T^{-1}
k_o	constant, dimensionless
k_A	first-order homogeneous rate constant, T^{-1}
$k_{A,s}$	first-order heterogeneous reaction rate constant, L/T
k_A^0	zero-order homogeneous rate constant, $mol\text{-}L^{-3}\text{-}T^{-1}$

k_G	gas-phase mass transfer resistance, mol-T-M^{-1}-L^{-1}
k_L	liquid phase mass transfer resistance, L/T
k_*	sum of k_1 and k_2, T^{-1}
k	unit vector along z-axis, dimensionless
K	Kozeny constant, dimensionless
K_{AW}	air-water partition coefficient, dimensionless
K_{eq}	equilibrium constant for reversible first-order reaction, dimensionless
K_f	coefficient in Freundlich equation, $(M/M)(L/M)^{1/n}$ where n is Freundlich parameter
K_G	(overall) gas mass transfer coefficient, mol-T-M^{-1}-L^{-1}
K_L	(overall) liquid mass transfer coefficient, L/T
$K_L a$	liquid mass transfer coefficient, L/T
K_{ow}	octanol-water partition coefficient, dimensionless
Kn	Knudson number, dimensionless
l	system or box length, L
l	mixing length, L
L	characteristic length scale
L_A	Avogadro's number, mol^{-1}
L_c	length required for fully developed concentration boundary layer, L
L_e	tube length in ideal porous medium, L
L_e	length required for fully developed flow, L
L_f	fiber length, L
L	length dimension
L	liter, L^3
Lo	van der Waals force number, dimensionless
m	hydraulic radius, L
m	source strength, L^2/T
m_a	molar mass, M/mol
m_A	molar mass of gas A, M/mol
M	mass dimension
M	mass tracer injected
$M_{A,t}$	mass component A adsorbed at time t, M
$M_{A,\infty}$	mass component A adsorbed at equilibrium, M
M_f	mass of fluid displaced by particle, M
M_p	particle mass, M
M_s	suspended solids concentration, M/L^3
M_T	total mass in multiple system, M
M_T	total mass of tracer
M	molarity, mol/L

n	coefficient in Freundlich equation, dimensionless
$n_{A,x}$	x-component of flux of component A, $M\text{-}L^{-2}\text{-}T^{-1}$
n_i	amount of ith type ions, mol
n_i	number concentration of ith type ions or particles, L^{-3}
n_m	number moles in mobil phase, dimensionless
n_s	number moles in stationary phase, dimensionless
n_∞	particle concentration outside control volume, L^{-3}
$\mathbf{n}$	outward unit normal vector, dimensionless
$\mathbf{n}_A$	mass flux, $M\text{-}L^{-2}\text{-}T^{-1}$
N	number per unit volume, L^{-3}
N_{ij}	rate of collisions between particles, $T^{-1}\text{-}L^{-3}$
N_L	particle concentration at bed exit, L^{-3}
N_0	particle concentration at bed entrance, L^{-3}
N_∞	total particle number concentration, L^{-3}
$\mathbf{N}_A$	molar flux of component A, $\text{mol-}L^{-2}\text{-}T^{-1}$
$\mathbf{N}_{A,z}$	molar flux component A at position z, $\text{mol-}L^{-2}\text{-}T^{-1}$
$\mathbf{N}_i$	molar flux of ith component, $\text{mol-}L^{-2}\text{-}T^{-1}$
Nu	Nusselt number, dimensionless
p	modified pressure, $M\text{-}L^{-1}\text{-}T^{-2}$
p	production term, $\text{mol-}L^{-3}\text{-}T^{-1}$
p	partial pressure, $M\text{-}L^{-1}\text{-}T^{-2}$
$p_{B,lm}$	log mean concentration, dimensionless
p_∞	reference pressure, $M\text{-}L^{-1}\text{-}T^{-2}$
$\bar{P}$	average pressure, $M\text{-}L^{-1}\text{-}T^{-2}$
$p_A(x, t)$	probability density function at position x and time t, L^{-1}
p_A^*	equilibrium partial pressure of A, $M\text{-}L^{-1}\text{-}T^{-2}$
P	pressure, $M\text{-}L^{-1}\text{-}T^{-2}$
P_A	density of constituent A in porous medium fluid phase, M/L^3
$\bar{P}_r$	reaction probability, dimensionless
Pe	Peclet number, dimensionless
Pr	Prandtl number, dimensionless
q	particle charge, C
q	recycle flow, L^3/T
q	heat flux, $\text{energy}/(L^2\text{-}T)$
q_e	equilibrium adsorbent capacity, mol/M
q_n	roots of generating function, dimensionless
q_r	unit flow in r-direction, L^2/T
Q	flow, L^3/T
Q_p	permeate flow, L^3/T
Q^0	ultimate adsorbent capacity, mol/M
r	radial distance, L

r	reaction rate, $M\text{-}L^{-3}\text{-}T^{-1}$ or $\text{mol-}L^{-3}\text{-}T^{-1}$
r_A	reaction rate of component A, mol/T
r_a	adsorption rate, $M\text{-}L^{-3}\text{-}T^{-1}$
r_B	reaction rate of component B, mol/T
r_b	boundary layer resistance, T/L
r_c	surface reaction resistance, T/L
$r_a(z)$	aerodynamic resistance, T/L
$r_{A,z}$	reaction rate of component A at position z, $\text{mol-}L^{-3}\text{-}T^{-1}$
r_d	desorption rate, $M\text{-}L^{-3}T^{-1}$
r_f	forward reaction rate, mol/T
r_i	radius of ith class particle, L
r_l	radial coordinate of leading edge, L
r_r	reverse reaction rate, mol/T
r_s	sedimentation rate, $M\text{-}L^{-3}\text{-}T^{-1}$
r_t	radial coordinate of trailing edge, L
$\bar{r}$	dimensionless radial coordinate
R	interception parameter, dimensionless
R	recycle ratio, dimensionless
R	retardation coefficient, dimensionless
R	fractional removal, dimensionless
R	universal gas constant, $\text{J-mol}^{-1}\text{-K}^{-1}$
R	ventilation rate, T^{-1}
R_b	biofilm mass transfer resistance, T/L
R_d	diffusion mass transfer resistance, T/L
R_f	liquid film mass transfer resistance, T/L
R_{fpb}	mass transfer resistance for fully penetrated biofilm, T/L
R_G	gas side resistance, $M\text{-}L\text{-mol}^{-1}\text{-}T^{-1}$
R_i	inner dimension (radius) of couette device, L
R_L	liquid side resistance, $M\text{-}L\text{-mol}^{-1}\text{-}T^{-1}$
R_n	infinite series of constants, dimensionless
R_o	outer dimension (radius) of couette device, L
R_{mn}	reaction mass transfer resistance, T/L
R_s	surface reaction resistance, T/L
R_τ	Lagrangian autocorrelation coefficient, dimensionless
Re	Rynolds number, dimensionless
Re_L	boundary layer Reynolds number at x = L, dimensionless
Re_p	particle Reynolds number, dimensionless
Re_{pb}	Reynolds number for packed bed, dimensionless
Re_x	boundary layer Reynolds number, dimensionless
$\mathscr{R}$	retention, dimensionless
s	standard deviation, L

s	solid fraction, dimensionless
S	entropy (J/K)
S	total number of adsorption sites, dimensionless
S	source of material, $M\text{-}L^{-3}\text{-}T^{-1}$
S	stopping distance, L
S_A	mass of A adsorbed per unit mass solids, dimensionless
$S_{Aj\infty}$	equilibrium concentration of A in sphere, mol/L^3
S_d	sedimentation coefficient, T
S_0	specific surface, L^{-1}
S_1	grain surface area per unit of grain volume, L^{-1}
S_1	number of adsorption sites occupied, dimensionless
Sc	Schmidt number, dimensionless
Sh	Sherwood number, dimensionless
St	Stokes number, dimensionless
$\bar{t}$	mean residence or detention time, T
t_{arr}	arrival time, T
t_R	retention time, T
t_v	viscous disturbance time, T
t_θ	exit time of nonadsorbing solutes, T
t_0	starting time, T
$\tilde{t}$	dimensionless time
$\bar{t}_p$	mean residence time for plug flow reactor, T
$\bar{t}_{RTD}$	mean residence time computed from the residence time density, T
$\bar{t}_s$	mean residence time for perfect mixer, T
t_*	time parameter in recycle model, T
T	time dimension
T	averaging period, T
T	temperature, K or °C
T_s	temperature at surface, K or °C
T_∞	temperature far from surface, K or °C
$\tilde{T}$	dimensionless temperature
$\bar{T}_N$	time parameter for recycle model, T
u_s	velocity at surface, L/T
u_x	x-component of velocity, L/T
u_z	streamwise velocity, L/T
$u_{z,ave}$	average velocity in tube, L/T
$u_{\tilde{z}}$	velocity in transformed coordinate system, L/T
u_θ	θ-component of velocity, L/T
$\mathbf{u}$	vector velocity, L/T
u_i	component i of velocity, L/T
u_1	x_1-component of velocity, L/T

u_2	x_2-component of velocity, L/T
u_3	x_3-component of velocity, L/T
u_1'	x_1-component of fluctuating velocity, L/T
u_2'	x_2-component of fluctuating velocity, L/T
u_3'	x_3-component of fluctuating velocity, L/T
u_θ	θ-component of velocity, L/T
u_*	friction velocity, L/T
$\mathbf{u^*}$	molar average velocity, L/T
$\tilde{u}$	dimensionless velocity
U_{ap}	approach velocity, L/T
U_{max}	boundary velocity, L/T
$\bar{U}_{max}$	centerline turbulent average velocity, L/T
U_{sup}	superficial bed velocity, L/T
U_0	characteristic velocity scale, L/T
$\bar{U}_1$	x_1-component of average velocity, L/T
$\bar{U}_2$	x_2-component of average velocity, L/T
$\bar{U}_3$	x_3-component of average velocity, L/T
U^+	dimensionless boundary layer velocity
U	characteristic velocity, L/T
U_i	component i of velocity, L/T
v	particle velocity, L/T
v	average linear velocity in porous medium, L/T
v	y-component of velocity, L/T
v_i	settling velocity of the ith size class particle, L/T
v_i	ith component of average linear velocity in porous medium, L/T
v_r	r-component of particle velocity, L/T
v_t	terminal sedimentation velocity, L/T
v_t	Lagrangian velocity at time t, L/T
v_x	x-component of particle velocity, L/T
v_i'	fluctuation in ith component of average linear velocity in porous medium, L/T
v_t'	Lagrangian velocity fluctuation at time t, L/T
$\mathbf{v}$	velocity in Lagrangian coordinate system, L/T
$\mathbf{v}$	vector form of average linear velocity in porous medium, L/T
v_s	average migration velocity of adsorbing species, L/T
v_0	average migration velocity of non-adsorbing species, L/T
V	volume, L^3
V	voltage, V
V	volume, L^3
V_A	molar volume of gas A, L^3/mol
V_i	volume of ith size class particle, L^3

V_i	ith component of linear velocity in porous medium, L/T
V_m	volume of a monolayer, L^3
V_M	mobile phase volume, L^3
V_R	retention volume, L^3
V_s	stationary phase volume, L^3
V_s	porous medium solid volume, L^3
V_t	total volume of system of particles, L^3
$\bar{V}_t$	Lagrangian mean velocity at time t, L/T
V_T	porous medium total volume, L^3
V_v	porous medium void volume, L^3
V_o	overflow velocity, L/T
w	system or box width, L
w	z-component of velocity, L/T
W	inverse of stability ratio, dimensionless
W_A	mass transfer of component A, M/T
$W(t)$	washout function, dimensionless
W'	unit width mass transfer rate, $M\text{-}L^{-1}\text{-}T^{-1}$
x	liquid phase mole fraction, dimensionless
x	amount of unreacted reactant, mol/L^3
x	rectangular coordinate, L
x_A	mole fraction component A, dimensionless
x_c	position of peak in concentration, L
x'_1	x coordinate in moving coordinate system, L
$\tilde{x}$	dimensionless distance
$\mathbf{x}$	position in Lagrangian coordinate system, L
y	gas phase mole fraction, dimensionless
y	rectangular coordinate, L
y^+	dimensionless boundary layer distance
z	rectangular coordinate, L
z_i	valence ith type ion, dimensionless
z_o	roughness length, L
$\tilde{z}$	transformed coordinate system, L
$\tilde{z}$	dimensionless axial coordinate

Greek

α	thermal diffusivity, L^2/T
α	equilibrium ratio of solute in solution phase to mass of solute in sphere, dimensionless
α	collision efficiency function, dimensionless
α_d	attachment efficiency, dimensionless
β_n	roots of equation, dimensionless

$\beta(r_i, r_j)$	collision frequency for class i and j particles, L^{-3}
Γ_i	surface excess of ith type ion, mol/L^2
γ	parameter in recycle model, T
γ	potential parameter in Gouy-Chapman theory, dimensionless
Δ	del operator
ΔH_{ads}	heat of adsorption, J/mol
ΔH_{vap}	heat of vaporization, J/mol
$\Delta m_{i,i+1}$	incremental mass of tracer, M
δ	boundary layer thickness, L
δ	Dirac delta function
δ_d	concentration boundary layer thickness, L
δ_f	film thickness
ϵ	relative permittivity, dimensionless
ϵ	porosity, dimensionless
ϵ	parameter in recycle model, dimensionless
ε	unit mass energy dissipation rate, L^2/T^3
ϵ_{ij}	component of eddy diffusivity, L^2/T
ϵ_0	permittivity of vacuum, $C^2\text{-}L^{-1}\text{-}J^{-1}$
ζ	dummy integration variable, dimensionless
η	dimensionless variable
η	Kolmogoroff microscale, L
η	dimensionless independent variable in boundary layer
η	single fiber efficiency, dimensionless
$\eta_{F,cd}$	cylinder (fiber) collection efficiency for convective diffusion, dimensionless
$\eta_{F,di}$	cylinder (fiber) collection efficiency for direct interception, dimensionless
$\eta_{F,ii}$	cylinder (fiber) collection efficiency for inertial impaction, dimensionless
$\eta_{S,cd}$	sphere collection efficiency for convective diffusion, dimensionless
$\eta_{S,di}$	sphere collection efficiency for direct interception, dimensionless
$\eta_{S,gd}$	sphere collection efficiency for gravitational deposition, dimensionless
θ	angle in cylindrical or spherical coordinates, radians
θ	fractional surface coverage, dimensionless
κ	inverse of double layer thickness, L^{-1}
κ	von Kármán constant, dimensionless
λ	mean free path, L
λ_n	infinite series of eigenvalues, dimensionless
μ	absolute or dynamic viscosity, $M\text{-}L^{-1}\text{-}T^{-1}$
μ_i	chemical potential species i, J/mol
μ_n	nth moment about the origin, L^n
μ'_n	nth moment about the mean, L^n
μ_t	eddy viscosity, $M\text{-}L^{-1}\text{-}T^{-1}$

ν	kinematic viscosity, L^2/T
ν_t	eddy diffusivity of momentum, L^2/T
ξ	dimensionless independent variable
O	order of
ρ	charge density, C/m^3
ρ	density, M/L^3
ρ_A	density of component A, M/L^3
ρ'_A	fluctuation of density of constituent A in porous medium fluid phase, M/L^3
ρ_b	bulk density, M/L^3
ρ_f	fluid density, M/L^3
ρ_i	density ith component, M/L^3
ρ_p	particle density, M/L^3
σ	interfacial tension, $M\text{-}T^{-2}$
σ	surface charge density, C/m^2
σ^2	variance, L^2
σ_τ^2	dimensionless variance
τ	characteristic time, T
τ	diffusion time scale, T
τ	relaxation time, T
τ'	tortuosity, dimensionless
τ_d	diffusion time scale, T
τ_v	viscous diffusion time scale, T
τ_{ij}	stress tensor, $M\text{-}L^{-1}\text{-}T^{-2}$
τ_p	dimensionless mean residence time for plug flow reactor
τ_r	reactime time scale, T
$\tau_{r\theta}$	r-θ component of stress (cylindrical coordinates), $M\text{-}L^{-1}\text{-}T^{-2}$
τ_δ	dimensionless mean residence time for perfect mixer
τ_w	wall stress, $M\text{-}L^{-1}\text{-}T^2$
τ_{ij}^t	Reynolds stress tensor, $M\text{-}L^{-1}\text{-}T^{-2}$
ϕ	potential function, L^2/T
Φ_R	repulsive energy of interaction, $M\text{-}L^2\text{-}T^{-2}$
χ	shape factor, dimensionless
ψ	electric field potential, V
ψ	stream function, L^2/T
ω	angular velocity, T^{-1}
ω	vorticity, T^{-1}
ω_A	mass fraction, dimensionless
ω_0	frequency of seasonal system input, T^{-1}

Math

∇	gradient operator, L^{-1}
$\tilde{\nabla}$	dimensionless gradient operator

Subscripts

batch	batch reactor
e	equilibrium
eq	equilibrium
sys	system
in	inlet
out	outlet
0	time = 0
∞	time = ∞
∞	reference value of variable

Superscripts

(s)	surface
s	skin
f	form

1

CONSERVATION LAWS
AND CONTINUA

1.1. INTRODUCTION

The laws of conservation of mass and energy are the most important laws of environmental modeling. In this chapter, several different forms of the conservation of mass law are developed. By way of introduction, we can say that conservation of mass is a little like a no-interest bank account. When the amount of deposits made during a certain period equals the amount of withdrawals, the amount of money in the bank stays the same. Or in the terminology of this chapter, we might say the accumulation term is zero. However, if the amount of withdrawals made during the period exceeds the deposits, the money in the bank must decrease. Essentially the same principles apply to environmental systems that exchange mass (or energy) with the surrounding environment. In general, for a fixed system, the mass into the system (per unit time) will only equal the mass out of the system (per unit time) when there are no sources, losses, or accumulations of mass within the system. The conservation of mass idea usually boils down to an equation. If the equation can be solved, we may be able to make some important inferences about the state or the performance of the system. In fact, we will find that the laws of conservation of mass work themselves into almost every analysis covered in this book.

1.2. CONSERVATION LAWS—SYSTEMS APPROACH

The law of *conservation of mass (or energy)* can be stated as follows: *Mass (or energy) can be neither created nor destroyed.* This law and several extensions of it can be used to solve numerous problems in environmental science and engineering. The conservation laws are often applied to a *system*, which here is a *representation of a region of space, whose boundaries may or may not correspond exactly to a real physical space.* An example of a system exchanging mass with its surroundings is shown in Figure 1.1.

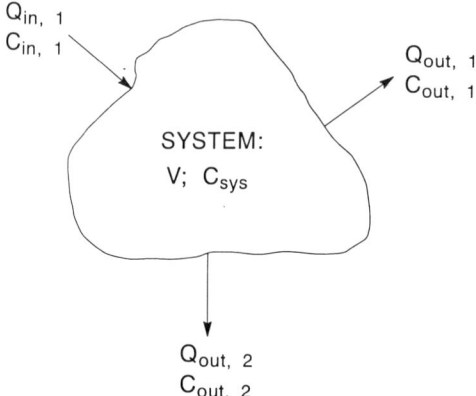

Figure 1.1. System with a single input, two outputs, and no reactions, sources, or sinks.

In Figure 1.1, the Q values are flows (L^3/T) going into or out of the system; the C values are mass concentrations (M/L^3) of some component of interest; V is the volume of the system, which can be variable; and C_{sys} is the concentration of the component of interest in the system. For a system with no reactions (i.e., *conservative* mass constituents), sources, or sinks, and no molecular diffusion across the boundaries (see Chapter 6), a differential equation based on the conservation of mass is

$$\frac{d}{dt}(VC_{sys}) = \sum_{i=1}^{N} Q_{in,i}C_{in,i} - \sum_{j=1}^{M} Q_{out,j}C_{out,j} \tag{1.1}$$

The subscripts "in" and "out" refer to inputs and outputs, respectively. The reader should verify that the units of this equation are mass per time (M/T). Equation 1.1 simply states that *the rate of change or accumulation of mass of some component within a system is equal to the difference between the rate of mass input and mass output.*

Note the number of unknowns in Eq. 1.1 — V, C_{sys}, and the Q and C values. These unknowns may also vary in time. One common simplification is to assume a constant volume system. Equation 1.1 can then be expressed somewhat more compactly as

$$\frac{dC_{sys}}{dt} = \frac{1}{V}\sum_{i=1}^{N} Q_{in,i}C_{in,i} - \frac{1}{V}\sum_{j=1}^{M} Q_{out,j}C_{out,j} \tag{1.2}$$

For *steady-state* conditions, the time derivative in Eq. 1.1 is zero and all the Q and C values are constants. Hence, a new equation results:

$$\sum_{i=1}^{N} Q_{in,i}C_{in,i} = \sum_{j=1}^{M} Q_{out,j}C_{out,j} \tag{1.3}$$

Another important equation results if we let concentration in Eq. 1.1 actually refer to the mass concentration of fluid in the system, that is, the density, (ρ). Then Eq. 1.1 becomes

$$V\frac{d\rho_{sys}}{dt} + \rho_{sys}\frac{dV}{dt} = \sum_{i=1}^{N} Q_{in,i}\rho_{in,i} - \sum_{j=1}^{M} Q_{out,j}\rho_{out,j} \tag{1.4}$$

If the density is assumed to be constant (an *incompressible* fluid), Eq. 1.4 yields

$$\frac{dV}{dt} = \sum_{i=1}^{N} Q_{in,i} - \sum_{j=1}^{N} Q_{out,j} \tag{1.5}$$

Finally, for a constant density *and* constant volume system, Eq. 1.5 becomes the famous *continuity equation* of hydraulics:

$$\sum_{i=1}^{N} Q_{in,i} = \sum_{j=1}^{M} Q_{out,j} \tag{1.6}$$

Example 1.1. A certain pollution-control device is processing a waste stream, which has a flow rate of 3.00 liters per second (L/s) and a contaminant concentration of 102 milligrams per liter (mg/L). We want to produce a treated product flow of 2.80 L/s with a contaminant concentration of 9.0 mg/L. Determine the flow and concentration of the residual stream.

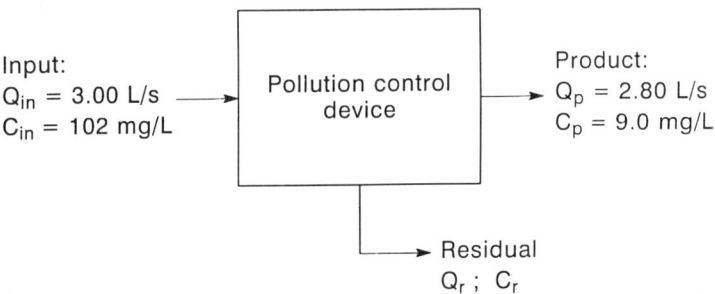

Input:
Q_{in} = 3.00 L/s
C_{in} = 102 mg/L

Pollution control device

Product:
Q_p = 2.80 L/s
C_p = 9.0 mg/L

Residual
Q_r ; C_r

Example 1.1

SOLUTION. Since there is no reason to believe this is not a steady-flow problem, Eq. 1.3 can be used:

$$Q_{in}C_{in} = Q_p C_p + Q_r C_r$$

There are two unknowns in the equation, Q_r and C_r. However, if a constant-density and constant-volume system is assumed, Eq. 1.6 gives the flow of the residual stream:

$$Q_r = Q_{in} - Q_p = 3.00 - 2.80 = 0.20 \, \text{L/s}$$

Therefore, from Eq. 1.3, the concentration of contaminant in the residual stream is

$$C_r = \frac{Q_{in} C_{in} - Q_p C_p}{Q_r} = \frac{3.00(102) - 2.80(9.0)}{0.20} = 1.4 \times 10^3 \, \text{mg/L}$$

Note that a strength and weakness of the systems approach is that the system is essentially treated as a "black box." Although this approach is satisfactory for many problems, limitations become apparent when there are transformations of mass in the system and we need to forecast a system output knowing only the input. The approach in this section is then only useful if some rather heroic (but common) assumptions are made about the system. (For example, see the "box model" discussion of Section 1.6.)

1.3.   CONSERVATION LAWS—CONTROL VOLUME APPROACH

For our purposes, a *control volume* is *a region of space corresponding to an actual physical space, in which we can describe the conditions of mass, momentum, and/or energy exchange across the surface of the control volume (i.e., the control surface)*. Thus, we need to develop the ability to draw control volumes. Figure 1.2 shows several examples; note that control volumes are often drawn immediately inside an actual physical boundary, as in Figures 1.2b and c. We can imagine that the control volume is an infinitesimally thin boundary between the physical boundary and any internal fluid. It is also occasionally advantageous to draw control volumes outside physical boundaries.

We will develop the integral form of the equation of conservation of mass with reference to the control volume shown in Figure 1.3. Note that vector notation is used in the derivation. (A review of relevant aspects of vector algebra is provided in Appendix II.) The vector $\mathbf{u}$ is the local velocity at the infinitesimal area dA. The vector $\mathbf{n}$ is called the *outward unit normal vector*, which is normal (perpendicular) to the surface dA, points away from the control volume, and has a magnitude of unity. Using this notation, we can derive the three terms of the equation of conservation of mass:

$$\begin{bmatrix} \text{I: rate of} \\ \text{mass inflow} \end{bmatrix} - \begin{bmatrix} \text{II: rate of} \\ \text{mass outflow} \end{bmatrix} = \begin{bmatrix} \text{III. rate of} \\ \text{accumulation of mass} \end{bmatrix} \tag{1.7}$$

Note in Figure 1.3 that the local velocity component normal (perpendicular) to the surface element dA is the simple dot product, $\mathbf{n} \cdot \mathbf{u} \, (L/T)$. The scalar *mass*

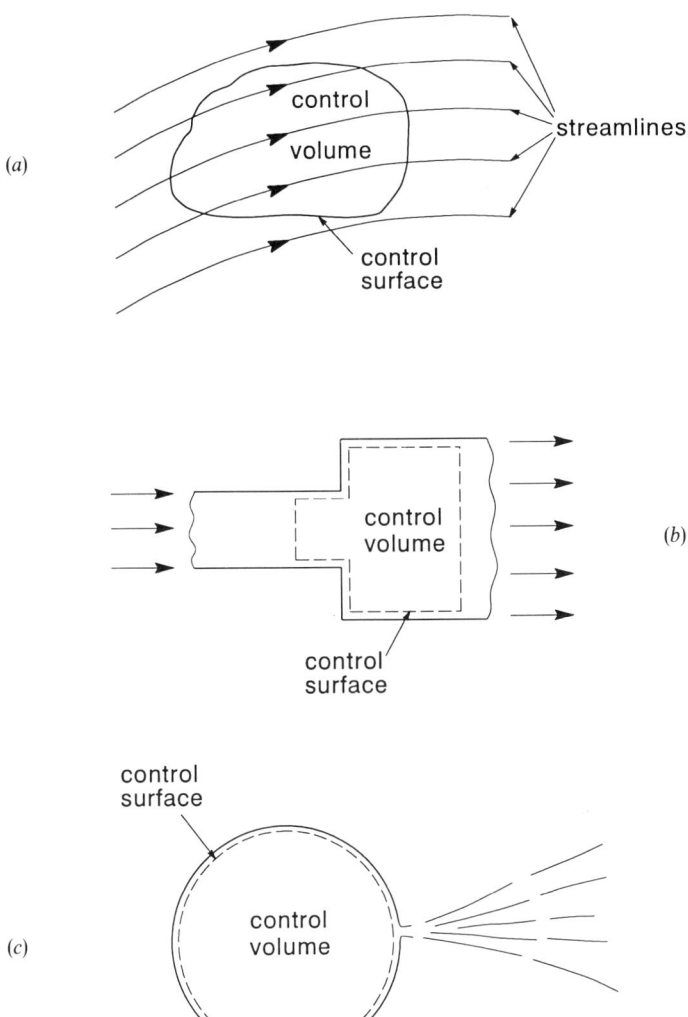

Figure 1.2. Examples of control volumes: (*a*) general control volume in a flowing fluid; (*b*) control volume drawn inside a pipe expansion; (*c*) control volume for a balloon or ruptured cell.

flux across dA is simply $\rho(\mathbf{n} \cdot \mathbf{u})$ $(M\text{-}L^{-2}\text{-}T^{-1})$. Then the rate of mass flowing out through dA is

$$\begin{bmatrix} \text{rate of mass} \\ \text{outflow at } dA \end{bmatrix} = \rho(\mathbf{n} \cdot \mathbf{u})dA \quad (M/T) \tag{1.8}$$

We might now move the differential element around and sum all contributions

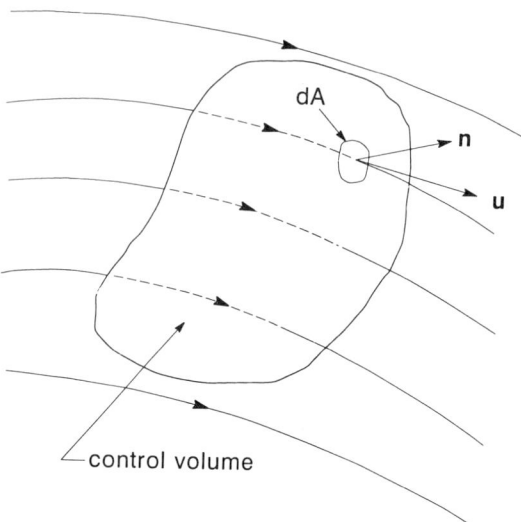

Figure 1.3. General control volume for derivation of integral relation for conservation of mass.

to term II in Eq. 1.7; however, we find that a surface integral in Eq. 1.7 does the trick more elegantly than summations. In fact, the surface integral accounts for terms I and II in Eq. 1.7, since the dot product accounts for the sign changes in mass transport at different points on the control surface:

$$\begin{bmatrix} \text{rate of} \\ \text{mass inflow} \end{bmatrix} - \begin{bmatrix} \text{rate of} \\ \text{mass outflow} \end{bmatrix} = - \iint_{\text{control surface}} \rho(\mathbf{n} \cdot \mathbf{u}) dA \quad (1.9)$$

The negative sign in Eq. 1.9 is added to ensure that if mass outflows are larger than mass inflows, the integral of all inflows and outflows will indeed be negative.

The rate of accumulation of mass in the control volume is equally elegant in vector notation:

$$\begin{bmatrix} \text{rate of accumulation} \\ \text{of mass} \end{bmatrix} = \frac{\partial}{\partial t} \iiint_{\text{control volume}} \rho \, dV \quad (1.10)$$

Equations 1.7, 1.9, and 1.10 can now be combined to yield the *integral form of the equation of conservation of mass:*

$$\iint_{\text{control surface}} \rho \mathbf{n} \cdot \mathbf{u} \, dA + \frac{\partial}{\partial t} \iiint_{\text{control volume}} \rho \, dV = 0 \quad (1.11)$$

Several things are worth noting about Eq. 1.11:

1. The parentheses around **n·u** have been dropped in the final form. Since ρ and dA are scalers, the form shown in Eq. 1.11 is strictly unambiguous. (The parentheses were used earlier for better visualization.)

2. The scalar density has been retained inside the double and triple integral terms, allowing for the possibility of a spatial variation in ρ.

3. A partial derivative with respect to time is used in the second term on the left-hand side of Eq. 1.11. This is because density can be a function of space and time.

Some useful simplifications of Eq. 1.11 are noted. For *steady flows*, all time derivatives are zero; hence, Eq. 1.11 yields

$$\iint\limits_{\text{control surface}} \rho\mathbf{n}\cdot\mathbf{u}\,dA = 0 \qquad (1.12)$$

For steady flows and *incompressible fluids* (i.e., ρ = costant), Eq. 1.12 simplifies to

$$\iint\limits_{\text{control surface}} \mathbf{n}\cdot\mathbf{u}\,dA = 0 \qquad (1.13)$$

Example 1.2. The velocity at the inlet to the pipe constriction below has a parabolic distribution, given by $u_1 = 3(1 - r^2/9)$. Find the average velocity U_2 at the pipe outlet.

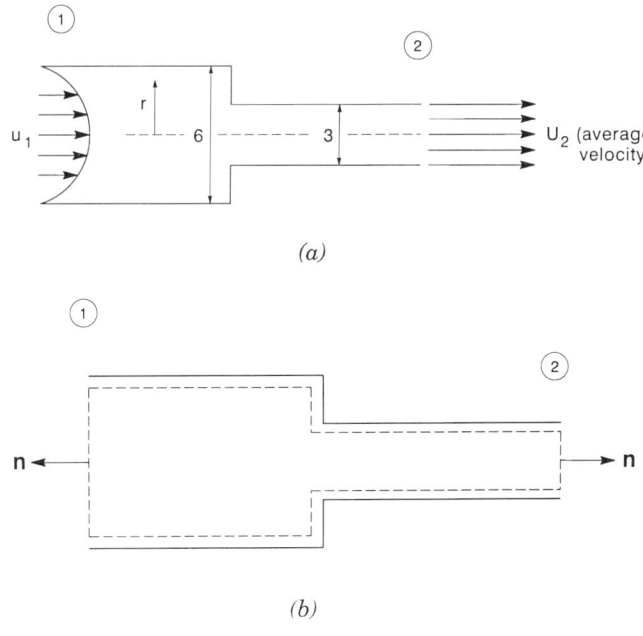

(a)

(b)

Example 1.2

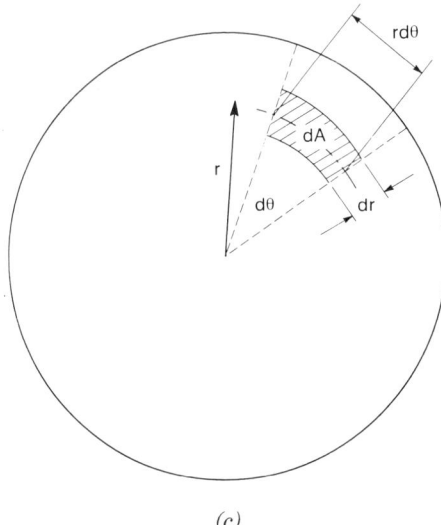

(c)

Example 1.2. (*Continued*)

SOLUTION. The first step is to draw a control volume, as shown in Example 1.2b. Note that with this selection of the control volume, there is mass flux into and out of the control volume at sections 1 and 2 only. Note also that at section 1, the outward unit normal vector is in a sense opposite to velocity, whereas at section 2, the outward unit normal and velocity vectors are colinear. This will cause these two contributions to the mass flux to have opposite signs. Since we have not stated otherwise, assume that this is a steady flow and an incompressible fluid; then, from Eq. 1.13:

$$\iint_{\text{control surface}} \mathbf{n} \cdot \mathbf{u}\, dA = \iint_{\text{section 1}} \mathbf{n} \cdot \mathbf{u}\, dA + \iint_{\text{section 2}} \mathbf{n} \cdot \mathbf{u}\, dA = 0$$

Realizing that the velocities must be integrated over circular end sections, we find:

$$-\iint_{\text{section 1}} \mathbf{n} \cdot \mathbf{u}\, dA = \iint_{\text{section 2}} \mathbf{n} \cdot \mathbf{u}\, dA$$

Example 1.2c may be of use in setting up the integral at section 1. Therefore, dA is equal to $r\, d\theta\, dr$; hence

$$\int_0^{2\pi} \int_0^3 3\left(1 - \frac{r^2}{9}\right) r\, dr\, d\theta = U_2 \pi (3/2)^2$$

or

$$U_2 = 6$$

The idea of *mass flux* introduced in this section, is one of the most powerful tools of environmental modeling, because we are so often interested in *mass transport* into or out of certain regions. Although the equations thus far concern the mass transport of the fluid medium itself, they can also describe the mass transport of any conservative mass constituent dissolved or transported with the fluid, as long as this mass constituent can be characterized by a well-defined concentration.* In other words, the approach in this section can be used to consider the conservation of mass of almost any constituent (e.g., contaminants, nutrients, or particles, in either liquid or gaseous phases). Referring to Figure 1.3 and thinking about the mass transport of some component with concentration C, we can write an equation analogous to Eq. 1.11:

$$\iint_{control\,surface} C\mathbf{n}\cdot\mathbf{u}\,dA + \frac{\partial}{\partial t}\iiint_{control\,volume} C\,dV = 0 \qquad (1.14)$$

Note that the only difference between Eqs. 1.11 and 1.14 is that ρ has been replaced by C, which represents the concentration of some mass component of interest. The equation says that any nonzero net mass transport across the control surface (i.e., a positive or negative value of the surface integral) results in a positive or negative accumulation of mass in the system. This is an extremely powerful tool, since it allows us, to a certain degree, to discover what is going on within a system knowing only mass transfer at the boundaries of the system or control volume (the control surface).

Calculation of the mass transport of C across the control surface is accomplished with the first term on the left-hand side of Eq. 1.14. The local scalar mass flux at a point on a surface or control volume is given by the term within the surface integral:

$$\mathbf{C}\mathbf{n}\cdot\mathbf{u} \qquad (1.15)$$

From this expression, the units of mass flux are $M\text{-}L^{-2}\text{-}T^{-1}$. In other words, flux is like the amount of mass crossing a unit surface (per surface) per unit time. Integrating over the control surface eliminates the units of surface (L^2), resulting in *mass transport*, which in this book always has the units M/T or mol/T. Note also from Eq. 1.15 that mass flux only involves the component of velocity *normal* (perpendicular) to the surface (i.e., the dot product of $\mathbf{n}$ and $\mathbf{u}$). This of course makes sense since the component of velocity *parallel* to the

---

*As in Section 1.2, we do not allow molecular diffusion across any control surface. This is taken up in Chapter 6.

surface cannot transport mass across the surface. So the dot product adds mathematical elegance to a simple but important idea. Mass fluxes are discussed extensively in this book.

Finally, note that the product $C\mathbf{u}$ has the same units as flux $(M\text{-}L^{-2}\text{-}T^{-1})$. This term is also a flux, but it is the *vector flux*, since the product of a scalar (concentration) and velocity (vector) is also a vector.

1.4. CONSERVATION LAWS—DIFFERENTIAL ELEMENT APPROACH

Our final method for deriving the conservation of mass equation requires that we imagine a differential element fixed in the flow region of interest. This is the so-called Eulerian coordinate system. Figure 1.4 shows a small-volume element with sides Δx, Δy, and Δz. As in Section 1.3, we will evaluate the terms in Eq. 1.7, but in this case, three components of terms I and II must be evaluated separately. For example, considering the x direction in Figure 1.4, the mass transport through the left face of the volume element is simply $\rho u_x \Delta y \Delta z$. But how do we express the mass leaving the right face of the element? From elementary calculus, the *Taylor theorem* allows us to approximate the value of a function at one point in terms of its known value at a nearby point. The Taylor theorem states that the value of $f(x)$ at b can be estimated from its values at a, provided Δx is sufficiently small (Figure 1.5):

$$f(b) \cong f(a) + \frac{df(a)}{dx}(b - a) + \frac{1}{2!}\frac{d^2f(a)}{dx^2}(b - a)^2 + \cdots \qquad (1.16)$$

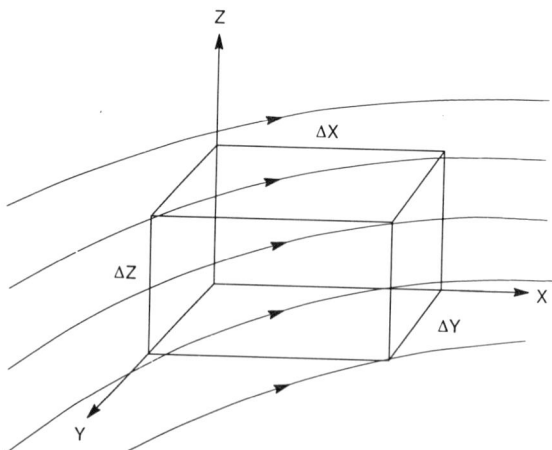

Figure 1.4. Volume element for derivation of the differential relation for conservation of mass.

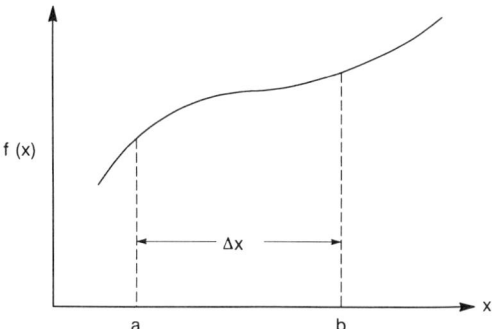

Figure 1.5. Function used to demonstrate a Taylor series.

Equation 1.16 can be called a *Taylor series* expansion or approximation. Using this idea, we can express the rate of mass transport out of the right-hand side of the control element as

$$\left(\rho + \frac{\partial \rho}{\partial x}\Delta x + \frac{1}{2!}\frac{\partial^2 \rho}{\partial x^2}\Delta x^2 + \cdots\right)\left(u_x + \frac{\partial u_x}{\partial x}\Delta x + \frac{1}{2!}\frac{\partial^2 u_x}{\partial x^2}\Delta x^2 + \cdots\right)\Delta y\,\Delta z$$

(1.17)

Next we observe that as Δx becomes sufficiently close to zero, the terms of higher order in Δx approach zero much "faster" than lower-order terms. In other words, as we let Δx approach zero, we can cancel all the higher-order terms. This is a classic maneuver in applied math and engineering, which makes sense after studying Figure 1.5. As Δx becomes smaller, a and b become closer; therefore, fewer terms should be required to estimate $f(b)$ based on $f(a)$. Retaining only the two lowest order terms in Eq. 1.17, the difference in mass transport between left and right faces is

$$\left[\rho u_x - \left(\rho + \frac{\partial \rho}{\partial x}\Delta x\right)\left(u_x + \frac{\partial u_x}{\partial x}\Delta x\right)\right]\Delta y\,\Delta z$$

(1.18)

Expanding the term in square brackets we find:

$$\left(\rho u_x - \rho u_x - \rho\frac{\partial u_x}{\partial x}\Delta x - u_x\frac{\partial \rho}{\partial x}\Delta x - \frac{\partial \rho}{\partial x}\frac{\partial u_x}{\partial x}\Delta x^2\right)\Delta y\,\Delta z$$

(1.19)

The first terms cancel each other, while we argue that as Δx approaches zero, the last term approaches zero faster than the previous lower-order terms;

hence, Eq. 1.19 reduces to

$$-\left(\rho\frac{\partial u_x}{\partial x} + u_x\frac{\partial\rho}{\partial x}\right)\Delta x\,\Delta y\,\Delta z \tag{1.20}$$

The rate of mass accumulation in the volume element is simply

$$\frac{\partial\rho}{\partial t}\Delta x\,\Delta y\,\Delta z \tag{1.21}$$

Finally, we can write expressions similar to Eq. 1.20 for the y and z components of mass transport across the various faces of the volume element. These expressions and Eq. 1.21 can be substituted into Eq. 1.7, and after simplifying, the *differential form of the equation of conservation of mass* is found to be

$$\frac{\partial\rho}{\partial t} + \rho\left(\frac{\partial u_x}{\partial x} + \frac{\partial u_y}{\partial y} + \frac{\partial u_z}{\partial z}\right) + \mu_x\frac{\partial\rho}{\partial x} + u_y\frac{\partial\rho}{\partial y} + u_z\frac{\partial\rho}{\partial z} = 0 \tag{1.22}$$

Referring to Appendix II, we recognize that some of the terms in Eq. 1.22 can be expressed in more compact vector notation:

$$\frac{\partial\rho}{\partial t} + \rho\nabla\cdot\mathbf{u} + \mathbf{u}\cdot\nabla\rho = 0 \tag{1.23}$$

An interesting conclusion is made from Eq. 1.23 if we assume that density is constant (i.e., ρ = constant — an incompressible fluid). Then, since the density is constant, the first and third terms on the left-hand side are zero and

$$\nabla\cdot\mathbf{u} = 0 \tag{1.24}$$

Equation 1.24 is another form of the *continuity equation* of fluid mechanics (cf. Eq. 1.6).

As in Section 1.3, an equation similar to Eq. 1.23 can be derived in terms of the mass concentration of some nonreactive ("conservative") component of local concentration C as long as the component is not diffusing:

$$\frac{\partial C}{\partial t} + C\nabla\cdot\mathbf{u} + \mathbf{u}\cdot\nabla C = 0 \tag{1.25}$$

For an incompressible fluid, this equation simplifies to

$$\frac{\partial C}{\partial t} + \mathbf{u}\cdot\nabla C = 0 \tag{1.26}$$

The left-hand side of Eq. 1.26 appears in Chapter 7 during the development of the famous *convective-diffusion* equation.

1.5. CONTINUA

In the preceding sections, we imagined scalars like density and concentration and vectors like velocity to vary smoothly and continuously in space. But is this always a good assumption? This is not a trivial question, since many concepts of integral and differential calculus rely on this assumption. In particular, application of some of the previous equations would be hard to imagine in the absence of a continuum.

One way to examine this problem is to complete a thought experiment. Assume that we have an instrument that precisely measures fluid density. The instrument consists of a probe, density meter, and strip chart recorder (Figure 1.6). From the probe tip detail shown in the figure, it appears that the probe senses the local density in a region characterized by dimension L. We can presume that the probe senses an average value of density in the sensitive region.

Continuing the thought experiment, assume that the manufacturer will sell us probe tips of any dimension L we wish, even down to molecular dimensions. The output of the meter is shown in Figure 1.7 for three choices of probe tip dimension L. Note that as L increases relative to the spacing of the molecules of the fluid, the measured density function becomes smoother. Since we require that density have a well-defined value in a continuum, we could say that a continuum can be assumed for the following condition:

$$\frac{1}{N} \lll L^3 \tag{1.27}$$

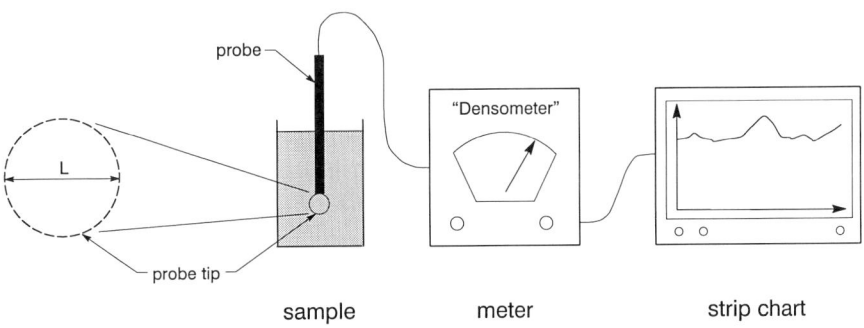

Figure 1.6. Imaginary density-measuring equipment.

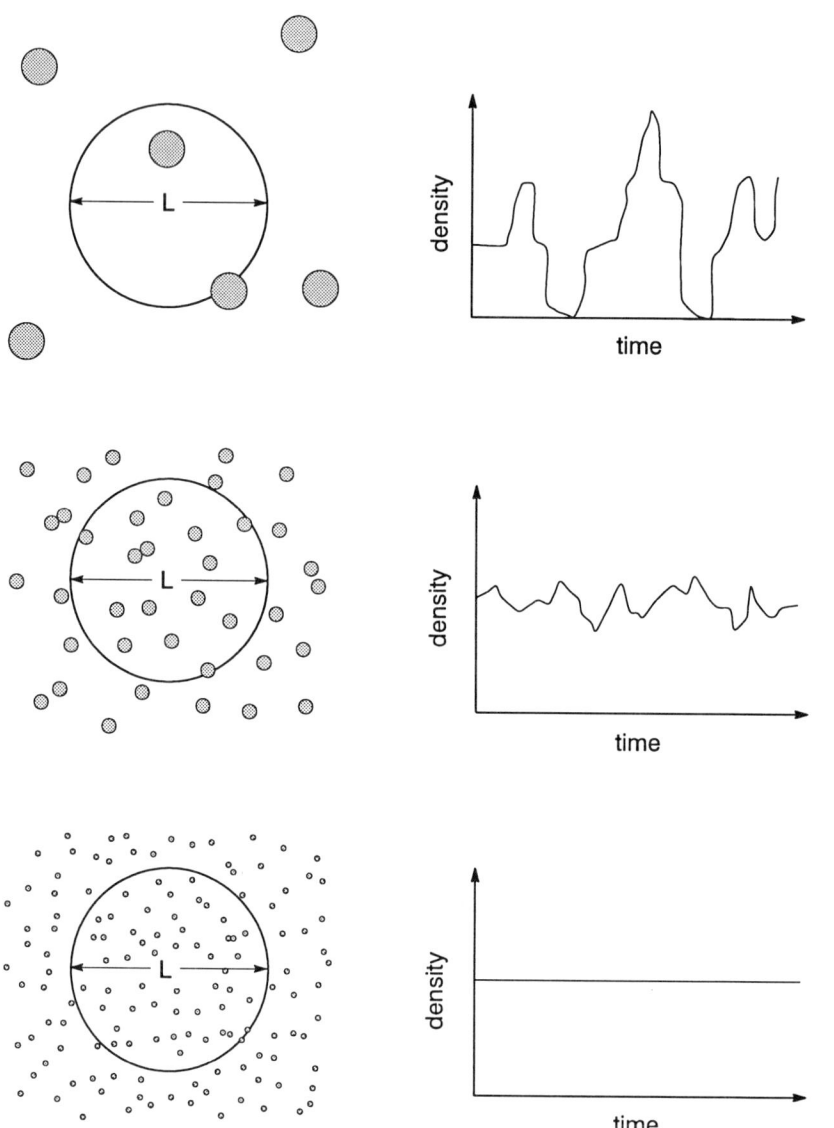

Figure 1.7. Density measurements for three choices of the sensor-characteristic dimension L.

where N is the number of molecules per unit volume. Now terminating our successful thought experiment, what relevance does the experiment have to the real world? The answer is known as the *continuum hypothesis* of fluid mechanics: *A continuum is a region of space where characteristic flow scales L are large enough that properties like density and velocity can be assumed to vary smoothly and therefore have "point" values.* In water, N is about $3.3 \times 10^{28}/m^3$; in air, N

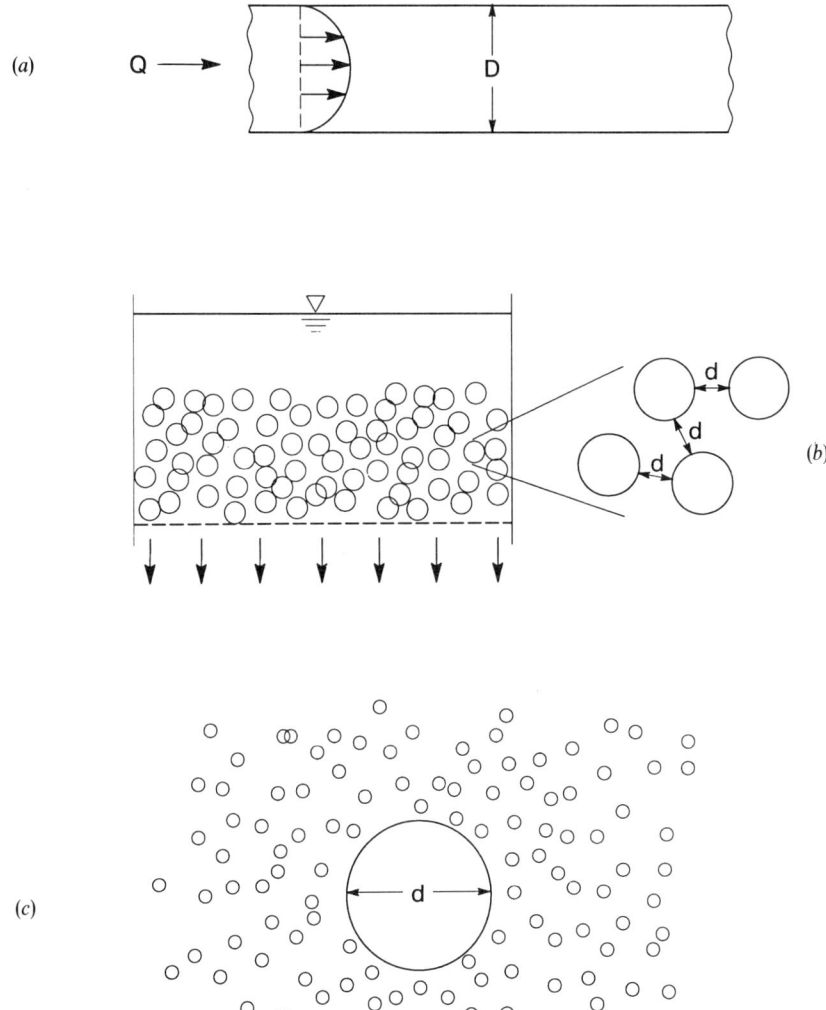

Figure 1.8. Examples of characteristic flow scales: (a) pipe diameter in laminar pipe flow; (b) pore "diameter" in porous media flow; (c) aerosol particle diameter in diffusion or sedimentation.

is about $2.5 \times 10^{25}/m^3$. It is almost safe to assume a continuum in liquids; however, gases cannot always be presumed to be continua.

To see these problems more clearly, we need to know a little more about characteristic flow scales. The best way to introduce the idea of flow scales is through some examples. Figure 1.8a shows the classic case of fully developed laminar flow in a smooth pipe. The only characteristic flow scale is the pipe diameter, which is usually large enough that it is safe to asssume a continuum. (An important exception may be flow through some membrane pores.) Porous media flow, such as filter or groundwater flow, is somewhat more complicated (Figure 1.8b). We would agree that the characteristic length scale is related to the distance separating the grains of the porous media. But since the grains are arranged randomly in the media, what is the correct length scale? Here we rely on our intuition and consider that the *typical* or *characteristic* length scale of the flow is the same order of magnitude as the grains themselves. Although typical natural porous media have a variety of grain sizes, all are generally large enough to support the continuum assumption.

Another interesting case is suggested in Figure 1.8c. Here a particle is diffusing or sedimenting in air (the air molecules are the little circles). As long as the aerosol particle senses a continuum, the characteristic flow length is the particle diameter. But does the aerosol particle sense a continuum, or a series of discrete bombardments by the air molecules? This question arises so often in air physics and air-pollution-control engineering that scientists frequently make use of a dimensionless parameter known as the *Knudsen number*:

$$\text{Kn} = \frac{2\lambda}{d} \tag{1.28}$$

Here d is the aerosol particle diameter and λ is the *mean free path* in the fluid — the average distance a fluid molecule travels before striking another molecule. The mean free path in air is given by the following formula from Flagan and Seinfeld (1988):

$$\lambda = \frac{\mu}{0.499\, P \left(\dfrac{8\, m_a}{\pi R T}\right)^{1/2}} \tag{1.29}$$

Here μ is the fluid (gas) viscosity, P is the pressure, m_a is the molar mass of the gas, R is the universal gas constant, and T is the absolute temperature. In air at 25°C and 1 atm pressure, P is 1.013×10^5 Pa and μ is 1.8×10^{-5} Pa-s. Taking the molar mass of air as 0.0289 kg/mol, we find from Eq. 1.28 that the mean free path is 0.0653 μm.

When the Knudsen number is much less than unity ($\text{Kn} \ll 1$), a continuum can be assumed. When $\text{Kn} \gg 1$, a continuum does not exist (the "free-molecule" range); even so, it is still possible to solve problems of environmental interest (see Chapter 2 for examples).

1.6. SOURCES, SINKS, REACTIONS, AND BOX MODELS

Note that in this chapter we have assumed that nothing "happens" to a particular species while in the system volume, control volume, or differential element. However, we know that this assumption will eventually have to be relaxed: Some of the most important problems in environmental modeling are those in which there exist chemical reactions, adsorption, sedimentation, biological degradation, volatilization, and other sources and sinks for mass constituents. Although these problems are a little more complicated, it is not difficult to see how some of the equations in this chapter must be modified. For example, a new version of Eq. 1.1 that takes into consideration reaction within the system could be

$$\frac{d}{dt}(VC_{sys}) = \sum_{i=1}^{N} Q_{in,i}C_{in,i} - \sum_{j=1}^{M} Q_{out,j}C_{out,j} + \sum_{k=1}^{L} Vr_k \tag{1.30}$$

Here r_k are any reactions, sources, or sinks of material in the system volume. The units of r_k are apparently $M\text{-}L^{-3}\text{-}T^{-1}$. Reactions can have either positive or negative magnitudes, depending on the reaction and constituent (reactant or product) being considered.

For the control volume approach, the equation corresponding to Eq. 1.14, which includes a single reaction, is

$$\iint_{control\ surface} C\mathbf{n}\cdot\mathbf{u}\,dA + \frac{\partial}{\partial t}\iiint_{control\ volume} C\,dV = \iiint_{control\ volume} r\,dV \tag{1.31}$$

and the equation corresponding to Eq. 1.26 for one reaction is

$$\frac{\partial C}{\partial t} + \mathbf{u}\cdot\nabla C = r \tag{1.32}$$

Example 1.3. A house has a ventilation rate of $100\,m^3/h$. A kerosene heater is being used in the house, which emits $1000\,g/h$ of CO_2. What is the average

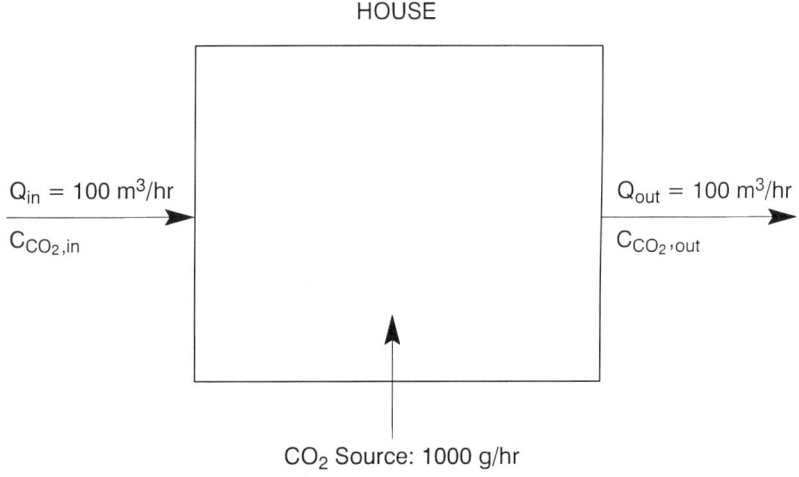

CO$_2$ Source: 1000 g/hr

Example 1.3

steady-state CO_2 concentration in the house? Note: the concentration of CO_2 in dry air at 20°C and 1 atm pressure is 0.031% (by volume).

SOLUTION It is always a good idea to draw a picture of the physical system, as in Section 1.2. We model the input of CO_2 from the heater as a source of CO_2. (See the figure on p. 17.)
    Substituting into Eq. 1.30 we find, at steady state:

$$0 = Q_{in} C_{CO_2, \, in} - Q_{out} C_{CO_2, \, out} + V r_{CO_2}$$

We know $Q_{in} = Q_{out} = 100 \, m^3/h$. $C_{CO_2, \, in}$ is just the normal background concentration of CO_2. If the volumetric concentration of CO_2 in the atmosphere is 0.031%, and if 1 mol of ideal gas occupies 24.0 L at 20°C and 1 atm pressure, the concentration of CO_2 in the air coming into the house is

$$C_{CO_2, \, in} = 0.00031 \frac{1 \, mol}{24.0 \, L} \frac{44 \, g}{mol} \frac{1000 \, L}{m^3} = 0.568 \frac{g \, CO_2}{m^3}$$

where the molar mass of carbon dioxide is 44 g. Finally, note that the source term, $V r_{CO_2}$, must be 1000 g/h, as shown in the figure. Substituting into the equation above, we find that $C_{CO_2, \, out} = 10.6 \, g/m^3$. The concentration at specific points in the house may be greater or less than this value, but the average concentration in the house must be 10.6 g/m³. This is about 17 times the normal background concentration of CO_2.

    Example 1.3 and Eq. 1.30 are essentially the systems approach to mass conservation modified to account for reactions, sources, and sinks of mass within the system. Equation 1.30 is used so often in environmental modeling that it goes by a special name — the *box model*. This indicates that the environmental system under consideration is treated as a box (often a "black box"). Almost always, a special type of internal condition called *perfect mixing* is assumed (Chapter 10). In perfect mixing, any molecule entering the system is instantaneously mixed with all other molecules in the system. This is a convenient assumption because any internal gradients in concentration are ignored. The box model is used at several junctures in this book, as well as in several exercises at the end of this chapter. Perfect mixing is not a critical assumption for simple mass balances with sources and sinks (like Example 1.3). However, when reactions are present, the common perfect mixing assumption of box modeling must be considered carefully. We return to box modeling at several junctures in this book, and the reader's understanding of the critical assumptions will be increased.*

---

*Chapter 10 includes a thorough introduction to several ideal mixing models (e.g., perfect mixing). The reader is welcome to jump ahead to this material at any time; however, a rigorous understanding of mixing is not needed to understand the material covered in the first nine chapters of this book.

1.7. SUMMARY

In this chapter the conservation of mass law was derived in three different forms. Some extensions and simplifications of this law were shown that result from assumptions like steady state and incompressibility, and which are useful in problem solving. The reader may wonder why three forms of the conservation equation are necessary. The simple answer is that engineering (and environmental) mass conservation problems are usually more easily cast and solved in one of the basic forms.

Box modeling techniques were introduced, and some of the power of this technique was demonstrated in a simple example. The idea of the continuum was also introduced. The continuum assumption is important in work with fluids (liquids and gases), and has particular importance in some air-pollution-control studies.

Completing the exercises at the end of this chapter should provide the student more confidence in applying the laws of conservation of mass (and energy). The surprising utility of the simple concept of conservation should become evident to the reader.

EXERCISES

1.1. A power plant delivers 2000 MW of power to the distribution grid:

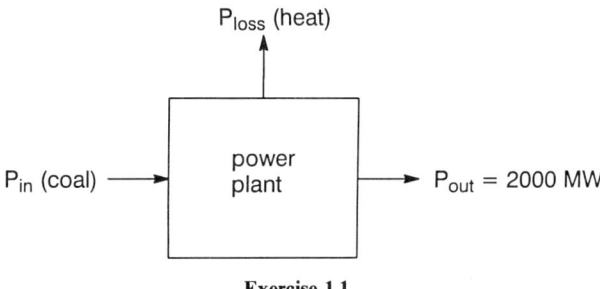

Exercise 1.1

If the plant is 45% efficient (i.e., $P_{out} = 0.45 P_{in}$), what is the plant's consumption of coal if the coal has an energy content of 40 kW-h/lb? Hint: In analogy to Section 1.2,

$$\begin{bmatrix} \text{rate of accumulation} \\ \text{of energy} \end{bmatrix} = \begin{bmatrix} \text{rate of} \\ \text{energy input} \end{bmatrix} - \begin{bmatrix} \text{rate of} \\ \text{energy output} \end{bmatrix}$$

What is the rate of heat loss to the environment?

1.2. A 500-m^3 detention pond initially contains 200 m^3 of water. At the 200-m^3 level or below), there is no discharge from the pond, but discharge will begin for any level above 200 m^3.

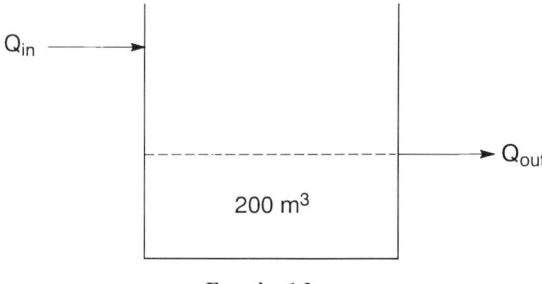

Exercise 1.2

An input flow of 0.2 m^3/s begins, and at the same time, a valve controls the output flow according to the following expression:

$$Q_{out} = 0.001[V - 200]\left(\frac{m^3}{s}\right) \quad \text{for } V > 200 \, m^3$$

In the equation, V is the detention pond volume. Determine the pond volume over time, as well as the steady-state pond volume.

1.3. This problem uses the same detention pond described in Exercise 1.2. A thunderstorm upstream of the pond produces a short flood described in the following sketch:

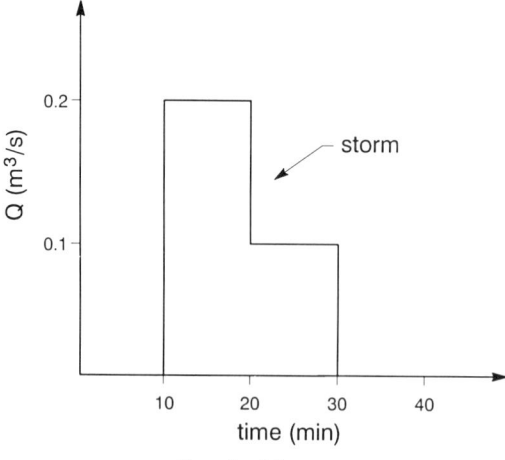

Exercise 1.3

Using the same outflow function of Exercise 1.2, determine the pond volume and outflow over time (include sketches of these functions). From the point of view of someone downstream of the retention pond, what effect do retention ponds have on floods? (Express answer in words.)

1.4. Consider a steady-state system with constant volume, a single constant inflow, and a single constant outflow, but with spatially varying fluid density within the system. Use Eq. 1.4 to derive a simple formula for ρ_{out} if Q_{in}, Q_{out}, and ρ_{in} are known. Why might fluid density change within a system?

1.5. Consider gas flow in a tube. Pressure and temperature are measured at points 1 and 2 as shown:

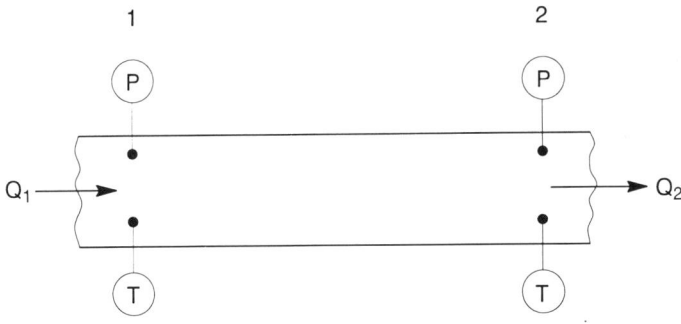

Exercise 1.5

(a) Use the ideal gas law to show how gas density is related to pressure and temperature. Recall: $PV = nRT$, where P is pressure, V is volume, n is the number of moles of gas, R is the universal gas constant, and T is temperature.

(b) If the gas flow is known at point 1, develop a steady-state equation to compute the gas flow at point 2 in terms of Q_1, P_1, P_2, T_1, and T_2. Hint: The results of Exercise 1.4 should be helpful. Assume steady state.

1.6. A fish that is exchanging mass with the surrounding water is shown below. Typically speaking, what types of mass would be exchanged between the water and the fish?

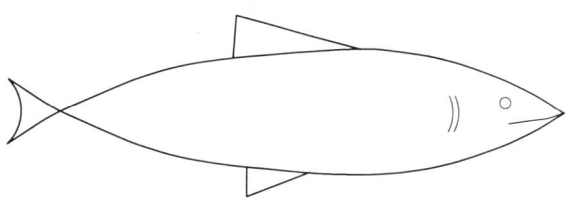

Exercise 1.6a

For modeling purposes, consider a cylindrical fish in which all oxygen is transferred to the fish from water which is brought into the fish through the mouth, and which leaves the fish through the gills:

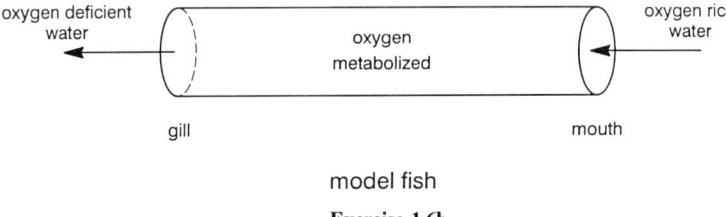

model fish

Exercise 1.6b

(a) If the oxygen content of the water is 7.5×10^{-3} kg/m³, and the fish is 25% efficient in transferring oxygen from the water, at what rate is oxygen metabolized by the fish if the fish processes 1 L of water every minute? Give answer in kilograms of O_2 per hour. Hint: Can the metabolism of O_2 in the fish be considered a sink for O_2 in the fish? If so, the approach in Section 1.6 may be useful.

(b) In the described analysis above, does it really matter what shape we have assumed for the fish?

1.7. A 5000-m stretch of a certain river can be approximated for modeling purposes as:

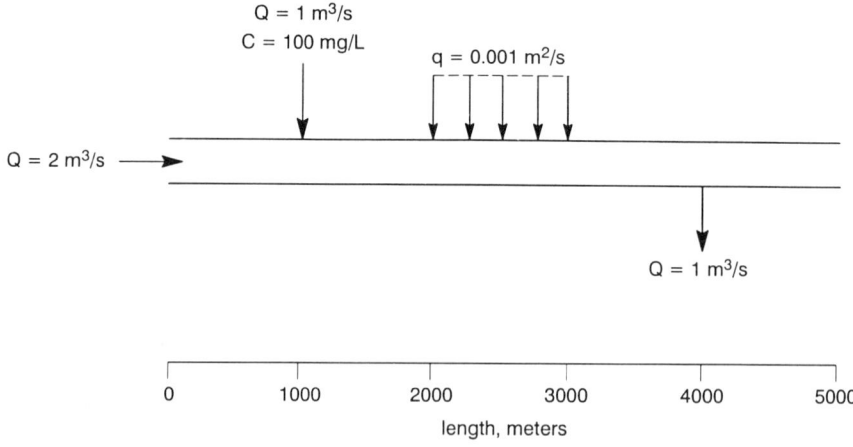

length, meters

Exercise 1.7

In the figure, q is a distributed "nonpoint source" flow input given in units of $L^3/(L\text{-}T)$. As shown in the figure, at 1000 m, a flow enters the river with a contaminant concentration of 100 mg/L. This is a conservative contaminant, which means that it undergoes no reactions or degradation (e.g., NaCl). Plot the concentration of the contaminant along the 5000-m stretch of river. You may assume perfect mixing of the contaminant in the river cross section, but consider only transport downstream due to the average flow velocity.

1.8. A countercurrent gas absorption tower is shown in the figure below. It is called countercurrent because the liquid and gas phases move through the tower in opposite directions on average. The tower can be optimized to transfer dissolved gases from the liquid to the gas phase ("stripping"), or to transfer gas from the gas to the liquid phase ("scrubbing"). Using the approach in Section 1.2, write a steady-state mass balance on a gaseous contaminant which is also soluble in the water phase.

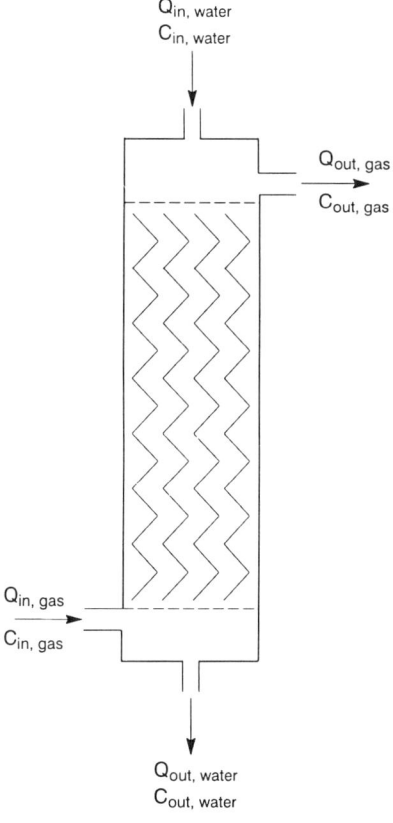

Exercise 1.8

Comment: This type of problem could be called a "multiphase mass balance."

1.9. The outflow and inflow to a $10^6\,\text{m}^3$ lake are $1\,\text{m}^3/\text{s}$. At a certain point in time, contamination of the inflow to the lake starts. Thereafter, the contaminant concentration in the lake inflow is $25\,\text{mg/L}$.

  (a) If there is no prior contaminant in the lake inflow, determine the concentration of contaminant in the lake over time. Assume that once the contaminant enters the lake, it is efficiently mixed with the rest of the lake water; that is, in the terminology of Section 1.2, $C_{sys} = C_{out}$.

  (b) What is the concentration in the lake effluent at a very long time after the contamination begins $(t \rightarrow \infty)$? Hint: This is called the steady-state concentration in the lake.

1.10. A refractory organic compound (a compound not easily degraded biologically) enters a lake with a seasonally varying concentration profile which can be modeled as a sinusoidal function.

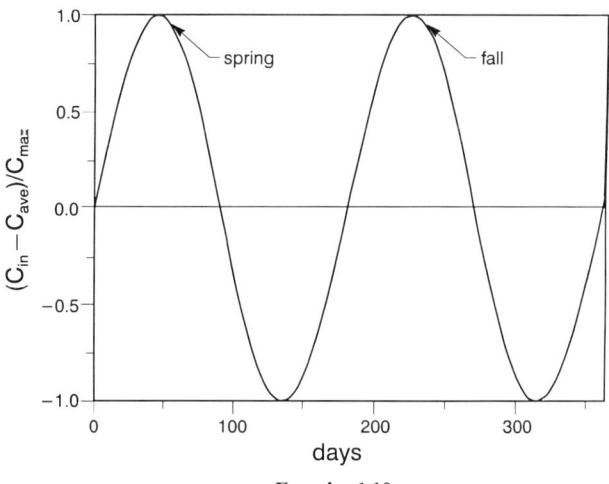

Exercise 1.10

The sinusoidal input can be described by the following function:

$$C_{in} = C_{ave} + C_{max}\sin(\omega_0 t)$$

where ω_0 is the frequency of the sinusoidal function:

$$\omega_0 = \frac{4\pi}{365}\frac{\text{rad}}{\text{day}}$$

(a) If there is no decay or other reaction of the compound in the lake (i.e., a conservative compound), use the systems approach (with the perfect mixing assumption) to show that the appropriate differential equation describing the system is

$$\frac{dC}{dt} + \frac{1}{\bar{t}}C = \frac{1}{\bar{t}}[C_{ave} + C_{max}\sin(\omega_0 t)]$$

where $\bar{t}$ is the mean hydraulic detention time in the lake:

$$\bar{t} = \frac{V}{Q_{in}} = \frac{V}{Q_{out}}$$

(b) Show that for the initial condition $C = C_{ave}$ at $t = 0$, the solution to the differential equation given above is

$$C = C_{ave} + \frac{C_{max}}{\frac{1}{\bar{t}} + \omega_0^2 \bar{t}}\left[\frac{1}{\bar{t}}\sin(\omega_0 t) - \omega_0 \cos(\omega_0 t) + \omega_0 e^{-t/\bar{t}}\right]$$

Hint: The integrating factor method of solving differential equations may be of some use.

(c) For lake volume $V = 10^6\,\text{m}^3$ and $Q_{in} = Q_{out} = 0.35\,\text{m}^3/\text{s}$, plot the solution on the same figure with the input function shown above.

(d) For $t \to \infty$ and $1/\bar{t} \ggg \omega_0$, show that the solution to (b) simplifies to

$$C = C_{ave} + C_{max}\sin(\omega_0 t)$$

This is the same function as the input function, C_{in}. Next show that for $t \to \infty$ and $1/\bar{t} \lll \omega_0$, the solution to (b) simplifies to

$$C = C_{ave}$$

Try to explain why these two simplified solutions make sense (i.e., explain the effect of the lake on the concentration of organic matter leaving the lake).

1.11. An air flow enters the impaction device shown in the figure on p. 26. What is the average velocity leaving the circular section? Hint: Use the approach in Section 1.3.

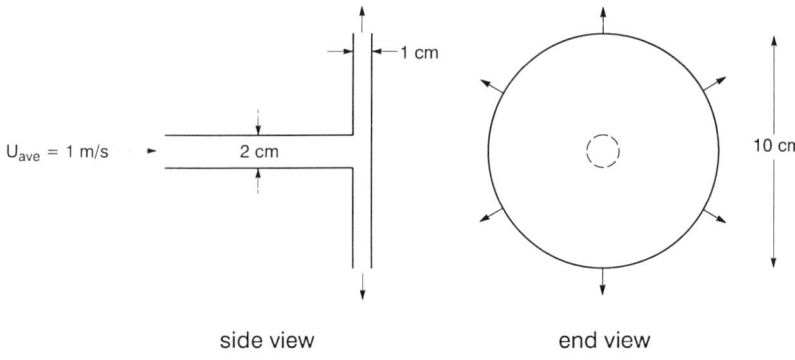

side view   end view

Exercise 1.11

1.12. Ultrafiltration membranes can be used to separate macromolecules and small particles from liquid streams. If the membrane pores are much smaller than the particle or molecular diameters, the membrane is thought to pass the suspending liquid but reject or retain the particles or molecules. A section of membrane-filtering particles from a liquid stream is shown below.

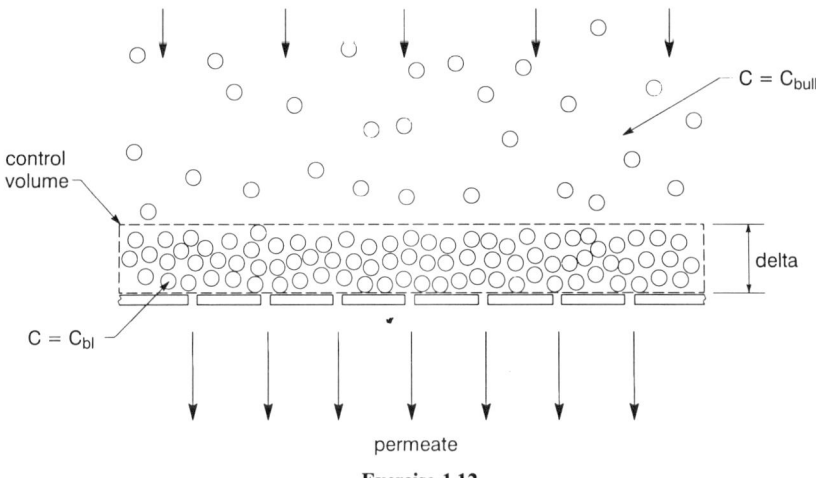

permeate

Exercise 1.12

As shown in the diagram, the thickness of a concentrated particle layer above the membrane surface increases as more fluid is filtered. Use the control volume approach discussed in Section 1.3 to calculate the thickness (delta) of the particle layer in terms of the membrane area (A), the volume of "permeate" passing through the membrane (V_p), the particle concentration in the concentrated layer (C_{bl}), and the particle concentration in the liquid approaching the membrane (C_{bulk}). Hint: Assume that the permeate flow is constant over time; hence, the control volume is increasing uniformly in size over time.

1.13. Equation 1.25 has a compact form because of vector notation. However, when practical problems are solved, a vector equation often has to be expanded into its cartesian component form (i.e., x, y, and z components). Expand Eq. 1.25 into component form.

1.14. A certain microbe has a spherical shape with a diameter of $1\,\mu m$. Assuming the interior of the microbe can be approximated as water, is the interior of the microbe a fluid continuum? Suppose you are told that the interior of the microbe has a pH of 7.5. How many H^+ atoms are expected on average to be inside the microbe? Is the interior of the microbe a continuum with respect to the H^+ concentration?

1.15. In air, are particles of 0.005-, 0.05-, 0.5-, and 5.0-μm diameter in the continuum range?

1.16. Ryan et al. (1983) studied the generation of nitrous oxide (NO_2) by portable kerosene heaters and the effect of this pollutant on indoor air quality. The authors' model of this nonconservative compound imagines a house as a relatively closed system with an input of NO_2 from the heater, loss of NO_2 due to unspecified sink reactions, and exchange of NO_2 with the atmosphere outside the house due to ventilation. (Ventilation can be from an actual ventilation system and/or from leaks in walls, ceilings, and windows.)

(a) Using the box model approach with an assumption of perfect mixing inside the house, show that the appropriate differential equation governing the concentration of NO_2 is

$$\frac{dC}{dt} = R(C_e - C) - kC + S$$

where C is the concentration of NO_2 in the house (M/L^3), C_e is the concentration of NO_2 in the external atmosphere (M/L^3), S is the NO_2 source ($M\text{-}L^{-3}\text{-}T^{-1}$), and R is the so-called ventilation rate (which is not the same as the ventilation rate used in Example 1.3—the units are number of house volumes exchanged per unit time, T^{-1}). Here k is the first-order decay constant for the unspecified sink, which is connected to r in Eq. 1.30,

$$r = -kC$$

(First-order reactions are studied in more detail in subsequent chapters, including a fundamental definition in Section 9.2.)

(b) Using the initial conditions $C = C_0$ at $t = 0$, show that the equation above integrates to

$$C = \frac{S'}{R'}(1 - e^{-R't}) + C_0 e^{-R't}$$

where $R' = R + k$ and $S' = S + RC_e$.

(c) Plot the indoor concentration of NO_2 over time for $k = 0.5\,h^{-1}$, $S = 200\,\mu g\text{-m}^{-3}\text{-h}^{-1}$, and $R = 0.3$, 1.0, and $3.0\,h^{-1}$. Assume $C_0 = C_e = 0$.

(d) Solve the integrated equation for steady-state conditions (i.e., $t \to \infty$). Does the equation predict the same steady-state values as shown in your plot?

REFERENCES

Flagan, R. C. and Seinfeld, J. H., *Fundamentals of Air Pollution Engineering*, Prentice Hall, Englewood Cliffs, NJ, 1988.

Ryan, B. P., Spengler, J. D., and Letz, R., "The Effects of Kerosene Heaters on Indoor Pollutant Concentrations: A Monitoring and Modeling Study," *Atmos. Environ.*, **17**, 7:1339–1345 (1983).

BIBLIOGRAPHY

Batchelor, G. K., *An Introduction to Fluid Dynamics*, Cambridge University Press, Cambridge, UK, 1967 (available in paperback).

Felder, R. M. and Rousseau, R. W., *Elementary Principles of Chemical Processes*, Wiley, New York, 1986.

Mackay, D., *Multimedia Environmental Models: The Fugacity Approach*, Lewis Publishers, Chelsea, MI, 1991.

Schnoor, J. L. and McAvoy, D. C., "Pesticide Transport and Bioconcentration Model," *J. Environ. Eng.*, **107**, EE6:1229–1246 (1981).

Sorenson, R., "Thought Experiments," *Am. Sci.*, **79**, 3:250–263 (1991).

Van den Berg, G. B. and Smolders, C. A., "Flux Decline in Ultrafiltration Processes," *Desalination*, **77**, 101–133 (1990).

Welty, J. R., Wicks, C. E., and Wilson, R. E., *Fundamentals of Momentum, Heat, and Mass Transport*, Wiley, New York, 1976.

Wylie, C. R., *Advanced Engineering Mathematics*, McGraw-Hill, New York, 1975.

2

LOW-CONCENTRATION
PARTICLE SUSPENSIONS
AND FLOWS

2.1. INTRODUCTION

Many environmental mass transport processes are inextricably connected with
the transport of small particles (on the order of $10\,\mu m$ and smaller). The
pollutant or material of interest may be the particle itself, for example, soot, or
ash in the air, or organic particles from wastewater treatment plants. However,
in many other cases, our concern is with a compound that is transported with
the particle, because it is adsorbed or partitioned into the particle. For
example, clay particles suspended in natural water streams may not be
considered pollutants, but organic compounds, like pesticides adsorbed to the
surface of the particle, could be a major concern in the overall transport of the
pesticide in the environment. In either case, we need to understand the basic
equation of motion of small particles in liquid or gaseous environments, which
is the focus of this chapter. We will answer the following question: What are
the main forces controlling the motion of small particles in the environment
and in pollution-control equipment?

2.2 DRAG ON A SPHERE

Imagine a small, smooth, spherical particle fixed in a flow field, which is
passing by the particle at uniform and steady velocity U (Figure 2.1). The
particle Reynolds number for this situation is defined as

$$\mathrm{Re}_p = \frac{dU\rho_f}{\mu} \tag{2.1}$$

where d is the particle diameter, ρ_f is the fluid density, and μ is the absolute
viscosity of the fluid. Here U is called the *free-stream* velocity; it is the uniform
velocity of the flow field at a distance far enough from the particle that the

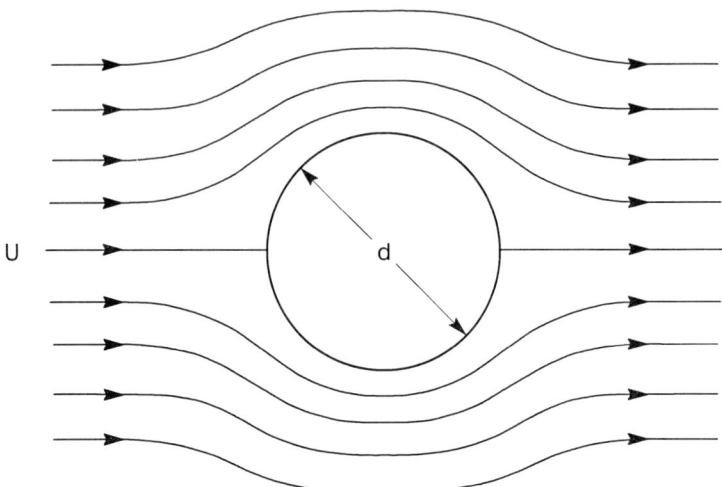

Figure 2.1. Particle fixed in low-Reynolds-number flow.

velocity is unaffected by the presence of the particle. This is an important "boundary condition" for this flow situation. For values of the particle Reynolds number much less than unity ($Re_p \ll 1$), the special type of flow field shown in Figure 2.1 exists in the region of the particle (see Chapter 5). This "Stokes" flow is characterized by fore and aft flow symmetry, that is, no "separation" of the flow on the back end of the particle. The drag on the particle is considered to be the sum of the "form drag" and the "skin drag." (The concepts of form and skin drag are developed more fully in Chapter 5.) The total drag of the fluid on the particle (sum of form and skin drag) is given exactly by

$$F_D = 3\pi\mu U d \tag{2.2}$$

The Reynolds number can be interpreted as the ratio of viscous and convective time scales (Chapter 5). In particular, for a Reynolds number much less than unity, the viscous time scale is much smaller than the convective time scale. In the situation shown in Figure 2.1, this means that viscous forces on the particle are much greater than inertial forces. However, as the Reynolds number increases above unity, the flow field shown in Figure 2.1 begins to be disturbed, and inertial forces become more and more important until they reach the point at which they completely dominate viscous forces.

Although Eq. 2.2 gives an exact value for the drag on a particle at low Reynolds number, drag at Reynolds numbers greater than unity cannot generally be predicted very accurately. Therefore, engineers have resorted to experimental determination of drag forces. The general expression used to

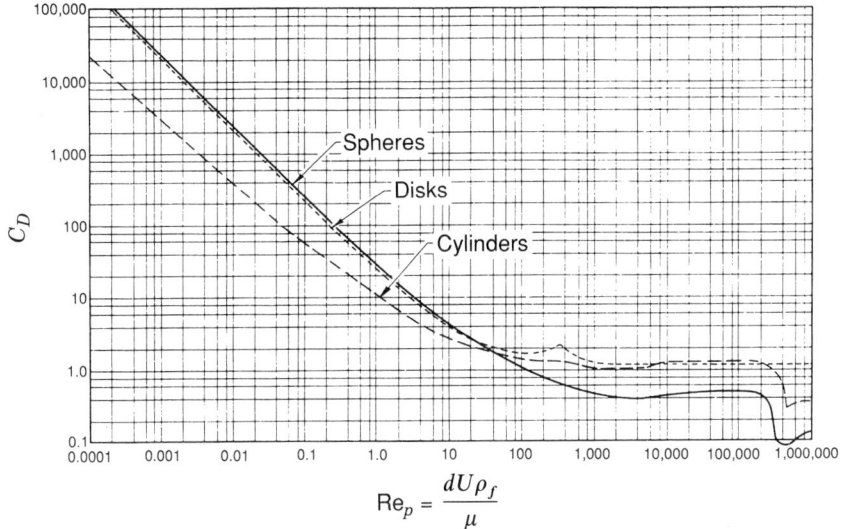

Figure 2.2. Steady drag coefficient for spheres, disks, and cylinders. Disks have flat side normal to flow. The axis of the infinite length cylinder is normal to flow direction. For disks and cylinders, d in Eq. 2.1 is the disk or cylinder diameter (respectively). *Source:* Reprinted with permission from Lapple and Shepard, *Ind. Eng. Chem.* **32**, 605 (1940). Copyright 1940, American Chemical Society.

"correlate" experimental drag data was proposed by Newton:

$$F_D = \tfrac{1}{2} C_D \rho_f U^2 A \tag{2.3}$$

Here F_D is the steady drag force, C_D is the steady drag coefficient, ρ_f is the density of the fluid, and A is the projected area of the particle. Figure 2.2 provides a plot of C_D versus Re_p for spheres, cylinders, and disks over a wide range of Reynolds numbers. Note that in the Stokes flow region, the drag coefficient for spheres is inversely proportional to the Reynolds number. (This inverse relationship between C_D and Re_p in the low Reynolds number range can be found by manipulating Eqs. 2.1–2.3; see Eq. 2.2-4 below.)

It is not always convenient to pick a C_D value off a plot like Figure 2.2; therefore, equations have been developed to approximate C_D over wide ranges of Re_p. For spheres, these relationships are

$$C_D = \frac{24}{\mathrm{Re}_p} \qquad 0 \leqslant \mathrm{Re}_p \leqslant 2 \tag{2.4}$$

$$C_D = \frac{18.5}{\mathrm{Re}_p^{0.6}} \qquad 2 \leqslant \mathrm{Re}_p \leqslant 500 \tag{2.5}$$

$$C_D = 0.44 \qquad 500 \leqslant \mathrm{Re}_p \leqslant 2 \times 10^5 \tag{2.6}$$

The reader can imagine that as particles become smaller, there will be a point

Table 2.1. Cunningham Slip Correction in Air for Various Particle Diameters[a]

$d\ (\mu m)$	C_c
0.01	22.7
0.05	5.06
0.1	2.91
0.5	1.337
1.0	1.168
5.0	1.034
10.0	1.017

[a]Pressure = 1 atm; temperature = 25°C.
Source: Flagan and Seinfeld, *Fundamentals of Air Pollution Engineering,* © 1988, p. 453. Reprinted by permission of Prentice Hall, Upper Saddle River, New Jersey.

at which noncontinuum effects become important (Chapter 1). Since the particle Reynolds numbers can be expected to be small at this limiting condition, the correction takes the form of an additional coefficient in the Stokes drag law:

$$F_D = \frac{3\pi\mu U d}{C_c} \qquad (2.7)$$

where C_c is known as the *Cunningham slip function*:

$$C_c = 1 + \mathrm{Kn}\left[\alpha + \beta \exp\left(-\frac{\gamma}{\mathrm{Kn}}\right)\right] \qquad (2.8)$$

In Eq. 2.8, $\alpha = 1.257$, $\beta = 0.40$, $\gamma = 1.10$, and Kn is the *Knudsen number* defined by Eq. 1.28. Flagan and Seinfeld (1988) provide values for the Cunningham slip correction for air, and these are reproduced in Table 2.1.

Slip correction factors are generally not needed in water, as noted in Section 1.5.

2.3. DRAG FORCE ON NONSPHERICAL PARTICLES

Many natural particles in the environment have irregular shapes (see Section 3.4). For example, certain fibers, clay minerals, and bacteria have cylindrical or cigar shapes; precipitates like calcium carbonate or particulate combustion products like soot generally have rugged surfaces that are difficult to characterize. In fact, many of the "particles" in the environment are actually agglomerates of smaller particles (see Section 3.4). For simplicity, we often assume that particles have a spherical shape; however, it is occasionally useful

Table 2.2. Shape Factors for Nonspherical Particles

Shape	χ
Cube	1.08
Cylinder ($L/D = 4$, axis normal to flow)	1.32
Aggregates of spheres	
Chain of 2	1.12
Chain of 3	1.27
Chain of 4	1.32
Compact aggregate of 3	1.15
Compact aggregate of 4	1.17
Bituminous coal	1.05–1.11
Quartz	1.36
Sand	1.57
Talc	2.04

Source: Hinds, *Aerosol Technology: Properties, Behavior, and Measurement of Airborne Particles,* © 1982, p. 48. Reprinted by permission of John Wiley & Sons, Inc.

to characterize the drag and sedimentation velocity of nonspherical particles. Drag coefficients for disks and long cylinders are presented graphically in Fig, 2.2, and can be incorporated into expressions for sedimentation velocity given in Section 2.4. However, another approach has proved useful for characterizing drag on nonspherical particles. Note that Eq. 2.2 can be rearranged as

$$\chi = \frac{F_D}{3\pi\mu U d_e} \tag{2.9}$$

In this equation, χ is called the *shape factor* or *dynamic shape factor*, and d_e is called the *equivalent-volume diameter*. For spherical particles χ is equal to unity. However, for particles other than spheres, the shape factor is generally greater than one. The equivalent-volume diameter is defined as the diameter of a spherical particle with the same volume as the irregular particle. Table 2.2 gives some values of χ determined through actual measurement of drag forces on nonspherical model particles. Given the values in Table 2.2, we can calculate drag force if we also know the equivalent-volume diameter (see Exercise 2.5).

2.4. LOW REYNOLDS NUMBER PARTICLE DYNAMICS AND STOKES LAW

The equation developed in this section require a fundamental application of *Newton's second law* to a particle:

$$\sum_i \mathbf{F}_i = M_p \mathbf{a} \tag{2.10}$$

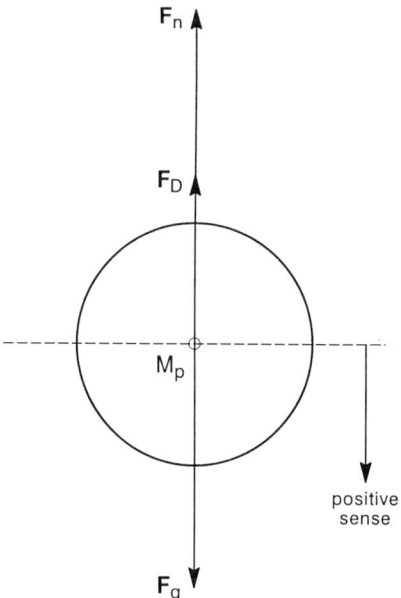

Figure 2.3. Forces acting on a suspended sphere.

Here the F_i are the forces acting on the particle, M_p is the mass of the particle, and **a** is the acceleration of the particle. Now imagine a smooth spherical particle falling in a still-fluid medium (Figure 2.3). In the figure, $\mathbf{F}_g$ is the net force of gravity acting on the particle, $\mathbf{F}_D$ is the steady drag force acting on the particle, and $\mathbf{F}_n$ is the nonsteady drag term. Summing forces in the vertical direction, the vector notation can be dropped, and Eq. 2.10 yields

$$F_g - F_D - F_n = M_p \frac{dv}{dt} \tag{2.11}$$

where the acceleration **a** has been replaced by dv/dt. From fluid statics, the net force of gravity on the particle is the difference between the gravitational force on the particle and the *buoyancy force* (see Section 5.4):

$$F_g = M_p g - M_f g \tag{2.12}$$

Here M_p is the particle mass, g is the acceleration of gravity, M_f is the mass of fluid displaced by the particle, and the product $M_f g$ is the buoyancy force.

For the steady drag force on the particle at low Reynolds number, we can use the results obtained in Section 2.2. Recall that we imagined the particle stuck in a flow with free-stream velocity U. We must now determine the drag force on a particle with velocity v in a quiescent fluid. Intuition serves us well in this case: if $v = U$, the drag force in the two cases is identical; therefore, the drag is given by Eqs. 2.2, 2.7, or 2.9, with v substituted for U.

The last force required to complete the analysis is F_n. It is related to two nonsteady contributions to the total particle drag (Maxey and Riley, 1983):

$$F_n = \frac{1}{2} M_f \frac{dv}{dt} + \frac{3}{2} d^2 \sqrt{\pi \rho_f \mu} \int_0^t \frac{dv(s)}{ds} \frac{ds}{\sqrt{t-s}} \qquad (2.13)$$

The first term in the nonsteady drag force is called the "added" or "virtual" mass term. When a particle is accelerating in a still fluid, fluid in the region of the particle is also being accelerated. This is not taken into account by the steady drag term, so it must be added in a nonsteady analysis. Note that this term is equal to one half the particle acceleration times M_f, the mass of fluid displaced by the particle. The second term is called the "history" or "Basset history" term. Note that the integral is over the time domain 0 to the present time t. In other words, it is an integral over the past history of the particle acceleration weighted by the inverse square root of $(t - s)$ [s is a dummy integration variable, so $t - s$ is the elapsed time since a particular acceleration $dv(s)/dt$.] This term arises because the velocity field in the region of the accelerating particle (and hence the drag) depends on the past history of the acceleration.

As shown by Cliff et al. (1978) and Näslund and Thaning (1991), the nonsteady drag terms are much less important in air than in water. For example, note that in air M_f in the added mass term is much less than in water. Also, the square root of the product of ρ_f and μ in the history term is much less in air than in water. Since evaluation of the nonsteady term (especially the history term) is a little more involved than necessary for our purposes, it is ignored in this chapter. Hence, the following analysis of the dynamics of a spherical particle released from rest is much more appropriate for particles in air. Knowing that the mass of the spherical particle (M_p) is $\pi d^3 \rho_p/6$ and M_f is $\pi d^3 \rho_f/6$, Eqs. 2.7, 2.11, and 2.12 yield, after rearrangement,

$$\frac{dv}{dt} + \left(\frac{18\mu}{C_c \rho_p d^2}\right) v = \left(\frac{\rho_p - \rho_f}{\rho_p}\right) g \qquad (2.14)$$

Equation 2.14 is a nonhomogeneous, ordinary differential equation with constant coefficients. Note that the group of parameters in parentheses in the second term in Eq. 2.14 has the units of inverse time (T^{-1}). Inverting this group of parameters yields the *particle characteristic time* τ:

$$\tau = \frac{C_c \rho_p d^2}{18\mu} \qquad (2.15)$$

We want to solve Eq. 2.14 for the situation in which we have just released the particle from rest, and it is allowed to rise or fall. For the initial condition $v = 0$

at $t = 0$, the solution to Eq. 2.14 is

$$v = \frac{\rho_p - \rho_f}{\rho_p} g\tau \left[1 - \exp\left(-\frac{t}{\tau}\right) \right]$$

(2.16)

Some interesting features of the solution can be seen without much effort. First, at $t = 0$, Eq. 2.16 predicts that $v = 0$; hence, our required initial condition is indeed satisfied. As $t \to \infty$, the term in square brackets approaches unity asymptotically; hence, the particle velocity approaches a constant value known as the *terminal velocity*:

$$v_t = \frac{\rho_p - \rho_f}{\rho_p} \tau g = \frac{C_c(\rho_p - \rho_f)d^2}{18\mu} g$$

(2.17)

Example 2.1. What relevance does the particle characteristic time τ have in Eq. 2.16? Calculate the particle characteristic time for a 1-μm sphere settling in air. The density of the sphere is 1.1×10^3 kg/m^3.

SOLUTION. Even though we already know approximately what the function looks like, it is interesting to plot Eq. 2.16. Using Eq. 2.17, Eq. 2.16 can be rearranged to

$$\frac{v}{\left(\dfrac{\rho_p - \rho_f}{\rho_p}\, g\tau\right)} = \frac{v}{v_t} = \left[1 - \exp\left(-\frac{t}{\tau}\right) \right]$$

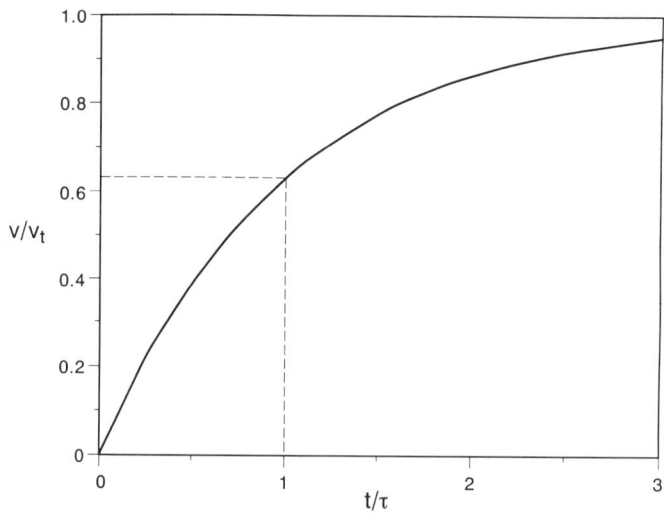

Example 2.1

Note that in this form, the equation is "nondimensional" — the dimension of the left- and the right-hand side is unity. It is often advantageous to develop functions in this way, since the functions become more "universal": If we plot v/v_t versus t/τ as shown above, the plot is the same for any parameter values we might choose.

Note that when $t = \tau$ (i.e., $t/\tau = 1$), the y-axis has the value $1 - \exp(-1) = 0.632$. In other words, when the characteristic time is reached, the particle reaches 63.2% of its terminal velocity. Since Eq. 2.16 implies that the particle takes an infinite amount of time to reach the final terminal velocity, it is convenient to characterize the process with a "characteristic time." Although the value 63.2% is not particularly special, the characteristic time derived above does seem to be a fundamental parameter grouping that derives from our physical and mathematical analysis. Throughout this book we will keep an eye out for these kinds of guideposts.

To calculate the characteristic time, we need to know the viscosity and density of air. Taking the ambient temperature to be 20°C, the viscosity of air is $1.8 \times 10^{-5}\,\text{kg-m}^{-1}\text{-s}^{-1}$. Also, recall from Table 2.1 that the Cunningham slip function for a 1-μm particle in air at 25°C is 1.168. The characteristic time in air is then

$$\tau = \frac{\rho_p d^2 C_c}{18\mu} = \frac{1.1 \times 10^3\,\text{kg/m}^3(1.0 \times 10^{-6}\,\text{m})^2 1.168}{18(1.8 \times 10^{-5}\,\text{kg-m}^{-1}\text{-s}^{-1})} = 4.0 \times 10^{-6}\,\text{s}$$

It is apparent that the characteristic time is very small. The practical significance of such a small characteristic time is that in many of our analyses, we can argue that particles are essentially always sedimenting at their terminal velocities.

The terminal velocity expression, Eq. 2.17, is what most people call *Stokes law*. Therefore, *Stokes law refers to the low Reynolds number (Re < 1) terminal settling velocity of a spherical particle in a quiescent fluid.*

Two other forms of Stokes law are sometimes used. For an aerosol particle, ρ_p is usually much greater than $\rho_f (\rho_p \gg \rho_f)$; therefore,

$$v_t = \frac{C_c \rho_p d^2}{18\mu}\, g = \tau g \tag{2.18}$$

In liquids, there is no slip correction ($C_c = 1$); therefore, Eq. 2.17 yields

$$v_t = \frac{(\rho_p - \rho_f)d^2}{18\mu}\, g \tag{2.19}$$

Finally, note that Stokes law can be derived in a slightly different manner. If we assume *steady-state* conditions from the outset, dv/dt in Eq. 2.14 is zero. In fact, since steady-state conditions are assumed, we do not have to worry about the nonsteady drag terms (Eq. 2.13). In Figure 2.3, at steady state, the gravity force is exactly balanced by the steady drag force. Stokes law is easily derived without solving a differential equation.

Recall that ignoring the nonsteady drag terms could make it more difficult to understand particle dynamics in liquids. However, the alternate derivation proves that Stokes law (Eq. 2.19) is also perfectly valid for low Reynolds number, steady-state, particle sedimentation in water (because we are only considering the steady state).

2.5. PARTICLE MOTIONS IN ELECTRICAL FIELDS

If a particle suspended in a still gas or liquid has a surface charge, the particle may be forced into motion by an electrical field. If the forces of gravity are an insignificant driving force for particle motion, Eq. 2.10 suggests that the equation of particle motion is

$$F_e - F_D - F_n = M_p \frac{dv}{dt} \qquad (2.20)$$

Here, F_e is the force on the particle caused by the electrical field:

$$F_e = qE \qquad (2.21)$$

where q is the particle charge (coulombs) and E is the electrical field strength (volts per meter).

Example 2.2. Determine the steady-state velocity of a 1.0-μm diameter aerosol particle in the electrical field shown in the device below at 20°C. The particle charge is 4×10^{-17} C.

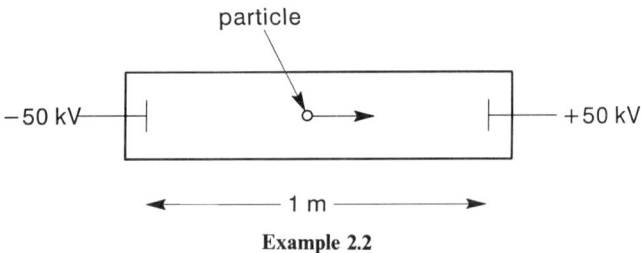

Example 2.2

SOLUTION. Assuming steady state (and $F_n = 0$), Eq. 2.20 reduces with some simplification to

$$v = \frac{C_c qE}{3\pi\mu d}$$

Assuming an air temperature of 20°C, we find

$$v = \frac{1.168 \times 4.0 \times 10^{-17}\,\text{C} \times 100 \times 10^3\,\text{V/m}}{3\pi \times 1.8 \times 10^{-5}\,\text{kg/m-s} \times 1 \times 10^{-6}\,\text{m}} = 0.028\,\frac{\text{C-V-s}}{\text{m-kg}}$$

The units in the answer were expected to be m/s. Therefore, a conversion is apparently necessary (see Appendix I). Recall that 1 volt is the potential difference between two points in an electric field such that 1 coulomb of charge is moved between the points by 1 joule of work; therefore, $1\,\text{V} = \text{J/C}$. Recall also that 1 joule = N-m. Finally, $1\,\text{N} = \text{kg-m-s}^{-2}$. Therefore,

$$v = 0.028\,\frac{\text{C-V-s}}{\text{m-kg}} \times \frac{\text{J/C}}{\text{V}} \times \frac{\text{N-m}}{\text{J}} \times \frac{\text{kg-m-s}^{-2}}{\text{N}} = 0.028\,\frac{\text{m}}{\text{s}}$$

We should also check to see that the requirement of $\text{Re}_p < 1$ has not been violated. (Recall that we assumed Stokes drag law to pertain.) For a dry-air density of $1.2\,\text{kg/m}^3$, the particle Reynolds number (Eq. 2.1) is given by

$$\text{Re}_p = \frac{\rho_f v d}{\mu} = \frac{1.2\,\text{kg/m}^3 \times 0.028\,\text{m/s} \times 1.0 \times 10^{-6}\,\text{m}}{1.8 \times 10^{-5}\,\text{kg/m-s}} = 1.9 \times 10^{-3}$$

So we are in good shape as far as the assumptions of the model.

Note that the steady-state velocity derived above (sometimes called a "drift" velocity) can be rearranged as

$$v = \frac{C_c}{3\pi\mu d}\,qE = \frac{C_c M_p}{3\pi\mu d}\,\frac{qE}{M_p} = \tau\,\frac{qE}{M_p}$$

where τ is the particle characteristic time discussed in Section 2.4. The term qE/M_p can be interpreted as the electrical force per unit mass of particle. Therefore, the steady-state velocity is the product of the characteristic time and the unit mass force on the particle. Note the similarity to the equation for aerosol sedimentation in Section 2.4—the term qE/M_p in the equation above plays a role analogous to gravity in Eq. 2.18.

2.6. QUIESCENT AND PERFECT-MIX BATCH SEDIMENTATION

Stokes law was developed by focusing on a single particle. However, in environmental studies, we are most often interested in systems of particles. For example, to prevent the eutrophication of a lake, we are interested in the average rate of sedimentation of particles (e.g., dead algae) on the bottom of the lake. To understand aerosol dry deposition on vegetation or cities, we are likewise interested more in average particle deposition or particle flux due to sedimentation. Several special cases are examined here and in Section 2.7,

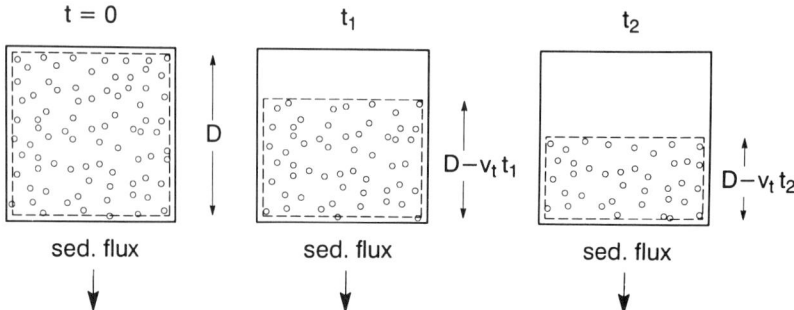

Figure 2.4. A box of initially homogeneously mixed particles experiencing batch sedimentation at three different times. The bottom of the box has area A. Note that the control volume shrinks over time.

where Stokes law can be used to predict straightforwardly the settling of low-concentration particle *monodispersions* (i.e., all particles the same size). The analysis can be extended to the sedimentation of *polydispersions*, and this is covered in Exercise 2.4.

First, imagine a square box that is initially filled with a monodispersion of small spherical particles (Figure 2.4). There are no fluid flows into or out of the box; hence we refer to this as "batch" sedimentation. Since the particles are all of the same size and density, they have the same terminal settling velocity, v_t.

Note also that since all particles have the same settling velocity, a clear region develops at the top of the box. The interface falls with the Stokes velocity v_t. But what is happening to particles settling on the bottom of the box? Here we assume that they are magically removed from the system, or that they are passing through the bottom of the box. The *sedimentation rate* can be considered to be this imaginary mass transfer through the bottom of the box.

The mass transfer due to sedimentation can be calculated using the control volume approach discussed in Chapter 1. However, rather than use a control volume of fixed volume, we imagine here that the control volume shrinks over time and corresponds exactly to the region beneath the interface in the box in Fig. 2.4 (note dotted control volumes). Then, from Eq. 1.14:

$$\begin{bmatrix} \text{rate of} \\ \text{accumulation} \\ \text{of particles} \end{bmatrix} = \frac{\partial}{\partial t} \iiint_{\text{control volume}} C\, dV = -\iint_{\text{control surface}} C\, \mathbf{n} \cdot \mathbf{v}_t\, dA \quad (2.22)$$

Note also in Eq. 2.22 that we are using the Stokes velocity v_t in place of the fluid velocity in Eq. 1.14, because the fluid is still and the particles are moving. (Don't forget, our mass balance here is on the particles, not the fluid.)

Note that since there is particle flux only through the bottom of the box, the surface integral must equal the sedimentation rate. And since the surface integral is equal to the negative of the volume integral, it is obvious that

$$\frac{\partial}{\partial t} \iiint_{\text{control volume}} C \, dV = - \begin{bmatrix} \text{sedimentation} \\ \text{rate} \end{bmatrix} \qquad (2.23)$$

Now note that particle concentration within the shrinking control volume is constant. Therefore, C can come out of the triple integral, and the partial derivative can be changed to a total derivative, yielding, with some simplification:

$$\frac{\partial}{\partial t} \iiint_{\text{control volume}} C \, dV = C \frac{d}{dt} \iiint_{\text{control volume}} dV = C \frac{dV}{dt} \qquad (2.24)$$

Now the volume of the control volume is easy to calculate at any point in time. It is simply the area of the bottom of the box, A, times the depth of the control volume, $D - v_t t$. Substituting for V,

$$\begin{bmatrix} \text{sedimentation} \\ \text{rate} \end{bmatrix} = - C \frac{dV}{dt} = - C \frac{d}{dt} [A(D - v_t t)] = C A v_t \qquad (2.25)$$

Thus, sedimentation rate in this ideal case is simply concentration times surface area times sedimentation velocity. Notice that the units of a concentration term (M/L^3) times a velocity term (L/T) are $M\text{-}L^{-2}\text{-}T^{-1}$. These are typical units of flux, which is the transport rate of mass to (or through) a surface (see Section 1.3). From Eq. 2.25 then, sedimentation rate is the product of flux and area, which is also considered a mass transfer rate in this book.

The motion of particles in the box in this derivation was only due to sedimentation. The flux of particles will last for an amount of time equal to D/v_t. We refer to this situation as *quiescent batch sedimentation*.

Now consider that the particles in the box are continuously and perfectly mixed (Figure 2.5). Particles can still sediment to the bottom of the box, and when they do, they are assumed to be magically lost or to have passed through the bottom of the box.

But as particle mass is lost due to sedimentation, we imagine that somehow the rest of the particles still in suspension are continuously rehomogenized or mixed. (We will encounter a similar concept at several junctures in this book.) We want to determine the concentration of particles remaining in the box at any time. Before attacking the problem with physics and mathematics, we can learn quite a bit about the solution by reflecting on Figure 2.5. At any time, the rate of sedimentation of particles to the floor of the box is given by Eq. 2.25; that is, sedimentation rate is proportional to the surface area of the bottom of the box (a constant) times the sedimentation velocity (a constant)

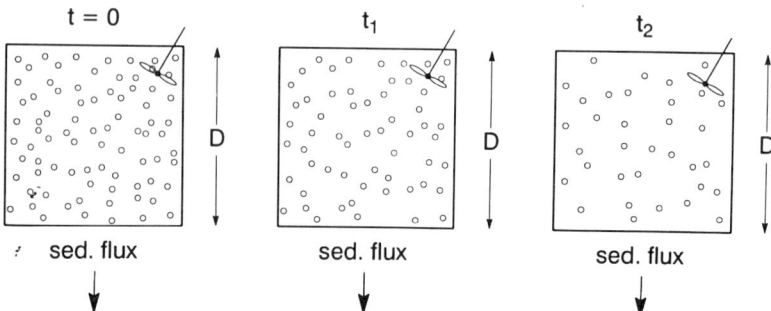

Figure 2.5. Perfect-mix batch sedimentation at three different times. The little mixing impeller indicates that the contents of the box are perfectly mixed or homogenized. The area of the bottom of the box is A.

times the particle concentration. A short, fixed-time step after the beginning of sedimentation, the homogeneous concentration of particles in the box will have decreased owing to the flux of particles to the floor of the box. Therefore, at this time (second picture in Figure 2.5), the sedimentation rate will also have decreased according to Eq. 2.25. During the next time interval, more particles will have been lost from the system because of sedimentation; therefore, the sedimentation rate decreases a little more (third picture in Figure 2.5), although not by as much as during the first time step (there were fewer particles available to sediment). We can continue this thought experiment and see that during each time interval, the rate of sedimentation is decreasing, and that the decrease lessens at each time step. We therefore predict that the sedimentation flux is highest in the beginning, decreases over time, and should eventually approach zero as we run out of particles. The reader should instantly recognize this as a classic decaying exponential flux rate.

We now write a differential equation for conservation of mass for this perfectly mixed case. Note that this system is a simplified box model (Section 1.6) with no flows into or out of the system, but with a single sink for mass which is the sedimentation rate. We can then consider the sedimentation rate (Eq. 2.25) to correspond to the sink term Vr in Eq. 1.30. So from Eqs. 1.30 and 2.25, we find for a constant volume system:

$$AD \frac{dC}{dt} = -CAv_t \qquad (2.26)$$

Simplifying this equation yields

$$\frac{dC}{C} = -\frac{1}{D} v_t \, dt \qquad (2.27)$$

Integrating Eq. 2.27 with the initial condition $C = C_0$ at $t = 0$, we find:

$$\frac{C}{C_0} = \exp\left(\frac{-v_t t}{D}\right) \tag{2.28}$$

which is the prediction for decay in particle concentration during *perfect-mix batch sedimentation*. The flux of particles to the bottom of the system box is the particle concentration times the sedimentation velocity:

$$\text{flux} = v_t C = v_t C_0 \exp\left(\frac{-v_t t}{D}\right) \tag{2.29}$$

Notice that the flux for quiescent sedimentation discussed after Eq. 2.25 does not require knowledge of the depth of the box, while the expression for particle flux in perfect-mix batch sedimentation (Eq. 2.29) does require the depth (D). Also note that the flux predicted by Eq. 2.25 is a constant (which lasts only a finite amount of time — see Exercise 2.14), whereas the flux predicted by Eq. 2.29 is an exponentially decreasing function.

2.7. CONTINUOUS SEDIMENTATION PROCESSES

In Section 2.6, we did not examine any situation in which there were fluid flows into or out of the system volume. Batch calculations could pertain, for example, to aerosol sedimentation in closed rooms, or to sedimentation in a lake during dry weather conditions of low or no flow into or out of the lake. When flows into or out of the system cannot be ignored, an expanded analysis is required. In this section we examine sedimentation in steady, continuous-flow systems, that is, systems in which the flow inputs and outputs are equal and do not change over time. Therefore, our definition of *continuous sedimentation* requires that the system volume is also constant.

We begin with an analysis of continuous sedimentation in an *ideal plug-flow* sedimentation system. Although the important concept of plug flow is developed more thoroughly in another chapter (especially in Section 10.5), for now we can imagine a plug-flow reactor or system as one in which the flow enters with a uniform velocity, continues through the system with uniform velocity, and exits the system with the same uniform velocity. We imagine that there is no turbulent motion or mixing within the system. A rectangular plug-flow sedimentation system is shown in Figure 2.6.

This is another good opportunity to use the box model approach described in Section 1.6, where (as in Section 2.6), the sedimentation rate is considered a sink for mass. Therefore,

$$\frac{d(VC_{\text{sys}})}{dt} = QC - Q\left(\frac{\tilde{D}}{D}\right)C - C(lw)v_t \tag{2.30}$$

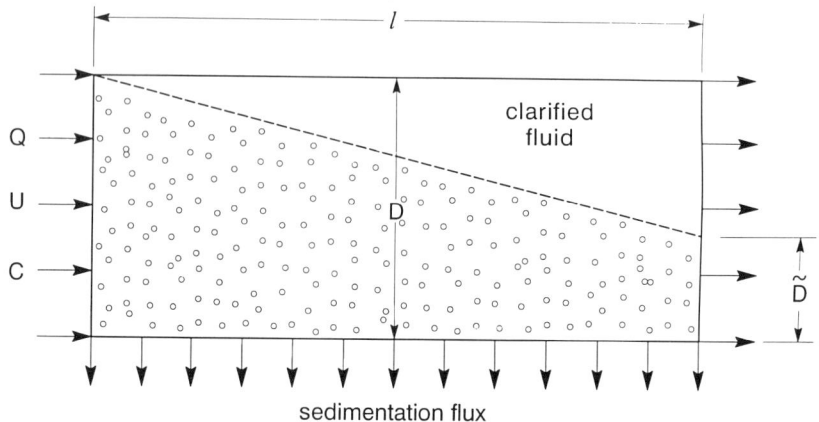

Figure 2.6. Schematic for continuous quiescent sedimentation.

In Eq. 2.30, C is the constant particle concentration in the unclarified fluid, C_{sys} is the fictitious concentration in the system,* $\tilde{D}$ is the depth of uncleared fluid leaving the system, w is the width of the system box (into the page), and lw is the area of the bottom of the system. The reader will recognize that the first term is the mass accumulation term, the second term is the rate of mass inflow, the third term is the rate of mass outflow, and the fourth term is the rate of removal of mass by sedimentation (the sink term). Note that the *fractional removal* of particles must be equal to the mass removed through sedimentation divided by the amount of mass entering the system. For steady-flow situations, the left-hand side of Eq. 2.30 is zero; hence,

$$\text{fractional removal} = R = \frac{v_t l w C}{QC} = \frac{QC - Q(\tilde{D}/D)C}{QC} = 1 - \frac{\tilde{D}}{D} \qquad (2.31)$$

which is fairly obvious from an examination of Figure 2.6. But we can also see that

$$\text{fractional removal} = R = \frac{v_t l w C}{QC} = \frac{v_t}{Q/lw} = \frac{v_t}{Q/A} = \frac{v_t}{V_0} \qquad \text{for } v_t \leqslant V_0 \qquad (2.32)$$

where A is the area of the bottom of the system. If $v_t > V_0$, the fractional

---

*Note that C_{sys} is called a fictitious concentration because part of the fluid in the box is totally clear and part has the fixed concentration C. An average concentration can be defined rigorously, but its characterization of either the clear or uncleared fluid may not be very good.

removal is, of course, always equal to unity, that is, R must be less than or equal to 1. The term V_0 is called the *overflow velocity* by chemical and environmental engineers. Note that the overflow velocity is equal to the system flow Q divided by the area of the bottom (or top) of the system (lw). Although we do not imagine here a flow out through the top of the box (the "overflow velocity"), V_0 would characterize that flow if fluid were forced to flow out over the top of the box. The overflow velocity is often used by engineers to characterize systems in which sedimentation is being examined. The simplicity of Eq. 2.32 shows why it is so useful. Note that as in the case of quiescent-batch sedimentation (see Section 2.6), removal efficiency in *quiescent continuous sedimentation* also does not appear to depend on the depth of the box. This point is examined more thoroughly in the exercises at the end of this chapter.

We now consider the continuous-flow analogy of perfect-mix batch sedimentation examined in Section 2.6 by referring to Figure 2.7. Again, using the box model approach, we have to modify Eq. 2.30 only slightly for the changed outlet condition:

$$\frac{d(VC_{sys})}{dt} = QC_{in} - QC_{out} - v_t(lw)C_{sys} \qquad (2.33)$$

For steady-state conditions, the left-hand side is zero. Noting that in perfect mix sedimentation $C_{sys} = C_{out}$ (Section 10.5), we rewrite Eq. 2.33 as

$$QC_{out} = QC_{in} - v_t lw C_{out} \qquad (2.34)$$

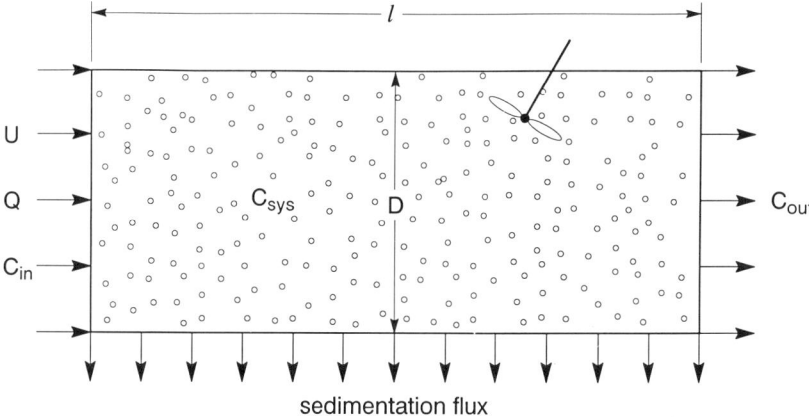

sedimentation flux

Figure 2.7. Schematic for development of continuous, complete-mix sedimentation relations.

Rearranging, simplifying, and using our previous definition of the overflow velocity, we find that

$$\frac{C_{out}}{C_{in}} = \frac{1}{1 + \dfrac{v_t}{V_0}} \tag{2.35}$$

The fractional removal for *continuous perfect-mix sedimentation* can be defined as

$$\begin{array}{c} \text{fractional} \\ \text{removal} \end{array} = R = 1 - \frac{C_{out}}{C_{in}} = 1 - \frac{1}{1 + \dfrac{v_t}{V_0}} \tag{2.36}$$

The fractional removal for quiescent and perfect-mix continuous sedimentation can be compared by plotting Eqs. 2.32 and 2.36 as a function of v_t/V_0 (Figure 2.8).

Sedimentation performance in continuous-flow systems with dilute mono-dispersions of particles can be expected to fall between the upper and lower curves in Figure 2.8. Clearly, the most efficient sedimentation occurs for the case of ideal (quiescent) plug flow. Considerably lower performance is found for perfect-mix sedimentation. This observation makes perfect sense: If we want efficient sedimentation, there should be as little mixing as possible.

We have been using the term "perfect mix" in describing the mixed case. The conventional interpretation is that our perfect-mix case pertains to sedimentation under turbulent-flow conditions. (Characteristics of turbulent flows are described in Chapter 5.) But what conditions pertain to sedimentation efficiencies falling between the two extremes shown in Figure 2.8? For now, we can imagine that in many systems, the mixing is intermediate between the perfect mix limit and the quiescent limit. For example, sedimentation in natural systems such as lakes or estuaries might be expected to fall in the intermediate region. Even in large tanks designed especially for quiescent plug-flow conditions, turbulence is generally unavoidable because of the scale of the flow, winds, imperfect inlet and outlet conditions, and thermal stratification/destratification. In these systems, performance generally falls between the quiescent sedimentation curve and the intermediate curve marked "plug-flow vertical mixing." The functional relationship for the intermediate curve is examined in Exercise 2.15.

The reader should not get the idea that sedimentation systems are sometimes designed with little mixing impellers (or fans for gaseous flows), as is perhaps suggested in Figures 2.5 and 2.7. The mixer represents a *theoretical extreme limit* of mixing in certain sedimentation systems. It is generally not a situation we would design; however, some real systems (often natural systems) with poor sedimentation performance may approach this theoretical limit.

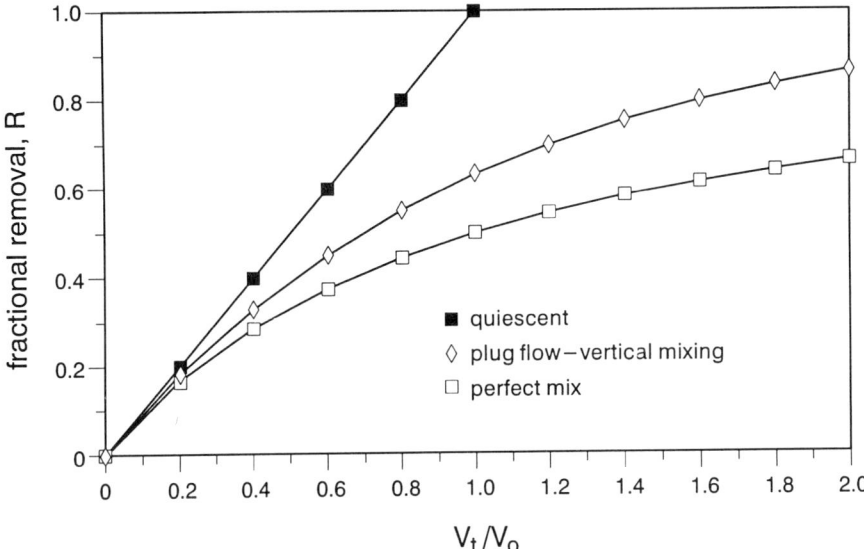

Figure 2.8. Dimensionless plot for rectangular systems comparing perfect-mix, plug flow with vertical mixing, and quiescent continuous sedimentation.

Students new to the concepts of this section often raise the following objection: If the mixing is so strong (i.e., "perfect"), why are particles settled on the bottom of the system not reentrained or resuspended into the overlying fluid? This is in fact a reasonable objection, and resuspension (and even rebound in gas systems) of particles must be considered in some liquid- and gas-sedimentation systems (see Exercise 2.16 and Chapter 5). However, as stressed above, students are asked to suspend their objections and simply consider the perfect-mixing models for batch and continuous sedimentation to be theoretical limits of estimates of poor sedimentation performance.

Example 2.3. Compare the theoretical fractional removal in a rectangular, continuous-sedimentation system with a same-size particles, for $v_t/V_0 = 0.6$. Also compare the size (bottom area A) of a continuous, rectangular, perfect-mix system and a continuous, rectangular, quiescent sedimentation system of the same depth or height when a fractional removal of 0.6 is desired.

SOLUTION. For the first part of the problem, simply draw a vertical line on Fig. 2.8 passing through $v_t/V_0 = 0.6$. The line pierces the quiescent sedimentation line at about $R = 0.6$ and the perfect mix curve at about $R = 0.38$. Therefore, the theoretical performance limits are $0.38 \leqslant R \leqslant 0.6$. This may not seem like a very satisfying answer; it does not seem like a very accurate answer.

Yet, in environmental and other types of modeling, a quick estimate of the anticipated range of performance is often helpful.

For the second part of the problem, simply draw a horizontal line on Figure 2.8 passing through $R = 0.6$. The line pierces the quiescent sedimentation curve at $v_t/V_0 = 0.6$ and the perfect mix curve at about $v_t/V_0 = 1.5$. For a fixed v_t,

$$\frac{v_t}{V_0} \text{ (quiescent)} \times V_0(\text{quiescent}) = \frac{v_t}{V_0} \text{ (perfect mix)} \times V_0(\text{perfect mix})$$

or

$$0.6 \times V_0 \text{ (quiescent)} = 1.5 \times V_0 \text{ (complete mix)}$$

or

$$\frac{V_0(\text{perfect mix})}{V_0(\text{quiescent})} = \frac{0.6}{1.5} = 0.4$$

If we assume that flow rate Q through the system is constant, and if we recall that $V_0 = Q/A$, the equation above implies that

$$\frac{A \text{ (quiescent)}}{A \text{ (perfect mix)}} = 0.4$$

or, for rectangular sedimentation systems of the same height, the perfect-mix sedimentation system must be 2.5 times as big as the quiescent sedimentation system to achieve a fractional removal of 0.6. This shows again why sedimentation tanks and chambers are designed with as little mixing as possible.

The superiority of plug flow in aiding certain environmental reactions is demonstrated at several junctures in this book; the difference in chemical "conversion" of various reactors is the special focus of Chapter 10.

2.8. INERTIAL FORCES ON PARTICLES AND THE STOPPING DISTANCE

Thus for, we have imagined particles attaining their terminal velocity after being released from rest. Now we imagine that the particle already has a velocity relative to the still fluid (and therefore inertia), and ask how long it will take for the velocity to reach zero because of fluid drag on the particle. This analysis is particularly relevant to aerosols, so we will ignore nonsteady drag terms (Eq. 2.13). Also, since we are primarily interested in "inertial forces," the gravity force is ignored. Referring to Fig. 2.3 and Eq. 2.10:

$$M_p \frac{dv}{dt} = -F_D \tag{2.37}$$

If we assume that Stokes drag law is applicable, Eq. 2.7 can be substituted into Eq. 2.37 (along with $M_p = \rho_p \pi d^3/6$) to yield

$$\frac{dv}{v} = -\frac{18\mu}{C_c \rho_p d^2} dt = -\frac{1}{\tau} dt \tag{2.38}$$

For this case, the initial condition is that at $t = 0$, $v = v_0$ (the initial velocity). Integrating Eq. 2.38 with these initial conditions,

$$\frac{v}{v_0} = \exp\left(-\frac{t}{\tau}\right) \tag{2.39}$$

Equation 2.39 simply states that if a particle is released with an initial velocity v_0, the velocity of the particle will decay exponentially over time. How far will the particle penetrate the fluid before its velocity goes to zero? Noting that velocity is simply dx/dt (x = distance of particle travel after $t = 0$),

$$dx = v_0 \exp\left(-\frac{t}{\tau}\right) dt \tag{2.40}$$

Integrating Eq. 2.40 from the limits $x = 0$ at $t = 0$, and $x = S$ at $t = \infty$, we find that

$$S = v_0 \tau \tag{2.41}$$

where S is called the *stopping distance*. Equation 2.41 again emphasizes the importance of the characteristic time τ. The stopping distance becomes an important consideration in whether flowing particles will impact a solid surface as the particles pass the surface.

2.9. INERTIAL FORCES IN PARTICLE FLOWS

We have thus far considered situations in which the particle was fixed and the fluid was moving uniformly past the particle, or the fluid was still and the particle was moving in a fixed direction relative to the still fluid. (The perfect-mix sedimentation example does not violate this statement, since we ignored all fluid motions and considered only particle motions due to gravity.) As the reader can imagine, for typical situations like particle transport in the atmosphere or air-pollution control devices, we actually have a particle (or particle-laden) flow. At a particular point, the fluid may have a velocity **u**, whereas a particle may have a different velocity **v**. If we ignore gravity and nonsteady drag forces, Eqs. 2.7 and 2.37 yield

$$M_p \frac{d\mathbf{v}}{dt} = \frac{3\pi\mu d}{C_c} (\mathbf{u} - \mathbf{v}) \tag{2.42}$$

if we consider the velocity term in Stokes drag law (Eq. 2.7) to be the *particle-fluid relative velocity*, **u** − **v**. Hence, when the particle and fluid are both in motion, there is only a drag force on the particle if there is relative motion between fluid and particle. Dividing Eq. 2.42 by $3\pi\mu d/C_c$, and rearranging a bit, we get

$$\tau \frac{d\mathbf{v}}{dt} + \mathbf{v} = \mathbf{u} \tag{2.43}$$

Equation 2.43 can be decomposed further to the following two-dimensional component equations:

$$\tau \frac{dv_x}{dt} + v_x = u_x \tag{2.44}$$

and

$$\tau \frac{dv_y}{dt} + v_y = u_y \tag{2.45}$$

Exact solutions of Eqs. 2.44 and 2.45 can be difficult when u_x and u_y are not constants. However, an important *dimensionless parameter* has emerged. To see this, nondimensionalize velocity and time as follows:

$$\tilde{v}_x = \frac{v_x}{U_0}; \qquad \tilde{u}_x = \frac{u_x}{U_0}; \qquad \tilde{t} = \frac{tU_0}{L} \tag{2.46}$$

Note that the three dimensionless parameters require the definition of a *characteristic fluid velocity scale*, U_0, and a *characteristic flow length scale*, L. Substituting the dimensionless parameters into Eq. 2.44 and rearranging, we find:

$$\frac{\tau U_0}{L} \frac{d\tilde{v}_x}{d\tilde{t}} + \tilde{v}_x = \tilde{u}_x \tag{2.47}$$

Note that there is now a single parameter grouping called the *Stokes number*:

$$\text{St} = \frac{\tau U_0}{L} = \frac{S}{L} \tag{2.48}$$

Equation 2.48 shows that the Stokes number is the ratio of the stopping distance (characteristic of the particle inertia) to the characteristic length scale

small free stream velocity, small St

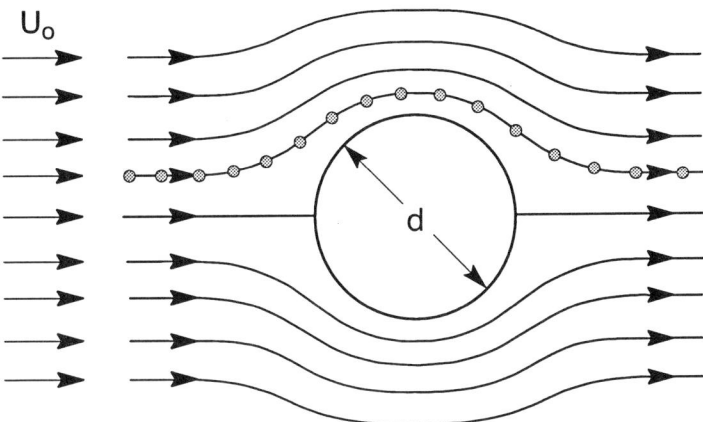

larger free stream velocity, larger St

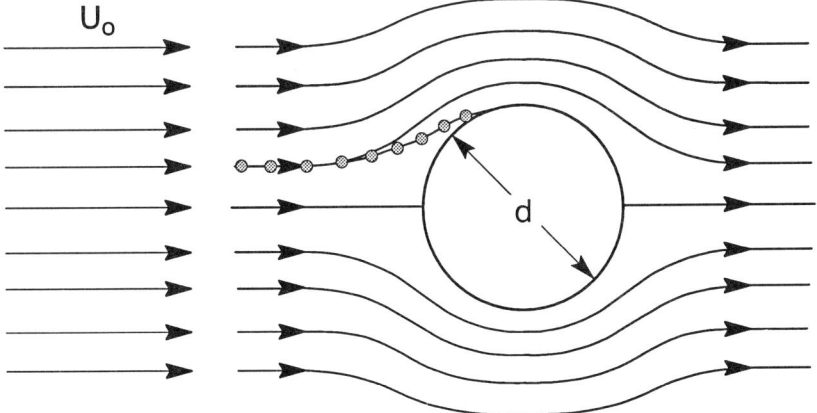

Figure 2.9. Particle transport past a stationary obstacle: small Stokes number (top) and larger Stokes number (bottom).

of the flow. In flows where gravity, thermal motion, and other external particle forces are unimportant, the Stokes number will determine particle motion and whether particles impact solid surfaces.

Consider, for example, the two flow situations shown in Figure 2.9. Here the characteristic flow velocity scale is the free stream velocity U_0, and the characteristic flow scale L is the obstacle diameter d. In the upper picture, the small particle follows the streamline around the cylindrical obstacle without impacting the obstacle. However, in the lower picture, the particle diverges from the flow streamline and impacts the obstacle. What is the difference between the two flow situations? The obvious answer is that the free-stream velocity U_0 is higher in the lower picture. Since, initially, the particle is assumed to be traveling at the free-stream velocity, it also has the velocity $v_x = U_0$. Therefore, the particle *inertia* is higher in the lower picture. In fact, the particle inertia is high enough in the lower picture that the particle is not able to make the turn around the obstacle; it diverges from the flow streamline and impacts the obstacle. Thus, as the Stokes number increases, from the upper to lower picture in Figure 2.9, the likelihood of impact increases.

The effect of inertial forces on particles is discussed in Chapter 8, where we develop an expression for the efficiency of collection of particles on fibers due to inertial impaction.

2.10. ROTATING FLOWS

For our present purpose, we need to imagine a flow determined by a fluid rotating steadily with a so-called "solid body rotation." This important concept from fluid mechanics is illustrated by Figure 2.10, in which a point is described by the cylindrical coordinates r and θ. From dynamics theory, the velocity at this point in the flow is given by

$$u_\theta = \frac{d\theta}{dt} r = \omega r \tag{2.49}$$

The *angular velocity*, ω, is defined by Eq. 2.49. Note that since the units of θ are radians (dimensionless), the units of ω are T^{-1}. Therefore, from Eq. 2.49, the units of u_θ are L/T, as required for a velocity term.

The *centrifugal acceleration* of the fluid at the point is given by

$$\left(\frac{d\theta}{dt}\right)^2 r = \omega^2 r \tag{2.50}$$

The reader should verify that the units of centrifugal acceleration are indeed the units of acceleration (L/T^2).

We will now apply Newton's second law, Eq. 2.10, to a small particle of diameter d and spatial coordinates r and θ immersed in a solid-body rotation

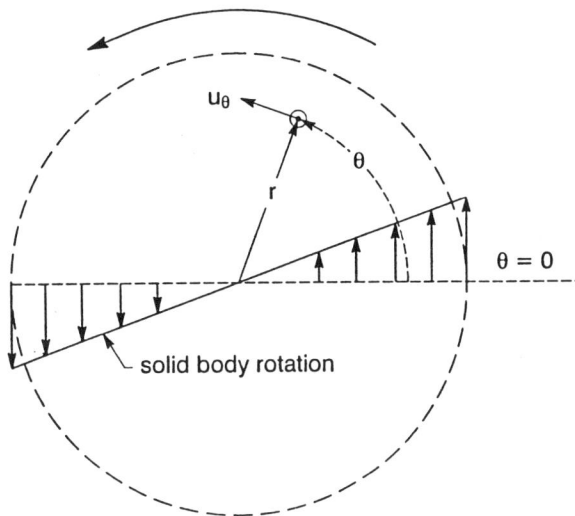

Figure 2.10. Solid-body rotation and cylindrical coordinates.

flow (Figure 2.11). The particle velocity and acceleration in the r direction are simply

$$v_r = \frac{dr}{dt}; \qquad a_r = \frac{d^2r}{dt^2} \tag{2.51}$$

If we ignore the nonsteady drag term (Eq. 2.13), the equivalent expression for the r-directed component of Newton's second law is

$$M_p \frac{d^2r}{dt^2} = M_p \frac{dv_r}{dt} = \sum_i F_i \tag{2.52}$$

We now assume that gravity forces on the particle are negligible so that only forces in the plane of rotation are considered. The neglect of gravitational acceleration is appropriate when another acceleration has a much larger magnitude than the acceleration of gravity.

We will consider three forces to be acting on the particle (Eqs. 2.53–2.55). Since we will assume that the particle is moving at its terminal velocity in the θ direction, the only particle acceleration term is the *particle centrifugal acceleration*. The centrifugal force is simply the particle mass times the centrifugal acceleration:

$$M_p \omega^2 r \tag{2.53}$$

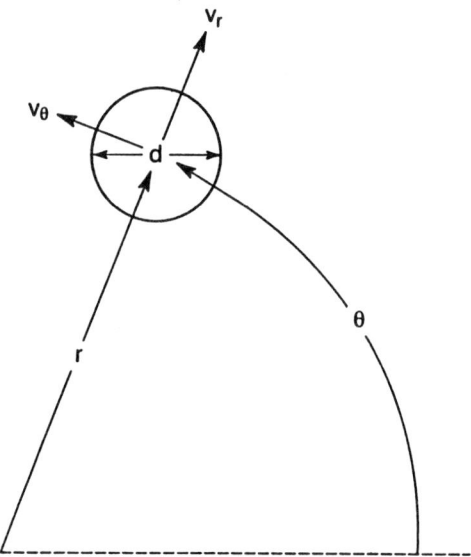

Figure 2.11. Nomenclature for particle dynamics in solid-body rotation.

Resisting the centrifugal migration of the particle is the fluid drag. We assume here that Stokes drag law (Eq. 2.2) is appropriate; therefore, the drag term in cylindrical coordinates is

$$3\pi\mu v_r d \tag{2.54}$$

The third force may be a surprise. Although we assumed the force of acceleration of gravity on the particle to be negligible, clearly there are now forces induced in the radial direction by the imposed centrifugal acceleration. Note that Eq. 2.53 seems to be analogous to the first term on the right-hand side of Eq. 2.12 (let $\omega^2 r$ take the place of g). However, is there a term analogous to the second term on the right-hand side of Eq. 2.12, the so-called buoyancy force? There is such a force, which is equal to the mass of fluid displaced by the particle (times the centrifugal acceleration):

$$M_f \omega^2 r = M_p \frac{\rho_f}{\rho_p} \omega^2 r \tag{2.55}$$

Collecting Eqs. 2.52–2.55 and rearranging slightly, we find that the radial component dynamical equation for a particle in a rotating flow is

$$M_p \frac{dv_r}{dt} = M_p \omega^2 r \left(1 - \frac{\rho_f}{\rho_p}\right) - 3\pi\mu v_r d \tag{2.56}$$

For an aerosol, the density of the particles is usually much greater than the density of the air (i.e., $\rho_p \gg \rho_f$); therefore, Eq. 2.56 can be simplified to

$$M_p\left(\frac{dv_r}{dt} - r\omega^2\right) = -\frac{3\pi\mu d_p}{C_c} v_r \tag{2.57}$$

where we have also modified Eq. 2.54 for noncontinuum effects (recall Eq. 2.7).

2.11. CENTRIFUGATION

Centrifugation is used to concentrate or separate particles and macromolecules suspended in liquids. Figure 2.12 provides an approximate idea of the design of a common centrifuge. The axis of rotation is connected to an electric motor and the sample is "spun up" to a specific value of angular velocity. We imagine that the sample container is composed of a suspending liquid which contains a layer of macromolecules or colloidal particles. Once the centrifuge is spun up to a constant angular velocity, we can imagine that the layer of material migrates slowly through the sample liquid in the radial direction. This is sometimes called a "drift velocity." If we assume that at any instant the migrating particles are moving at their terminal velocities, that is, if we assume steady state,* Eq. 2.56 reduces to

$$M_p r\omega^2\left(1 - \frac{\rho_f}{\rho_p}\right) = 3\pi\mu dv_r \tag{2.58}$$

Substituting $M_p = \pi d^3 \rho_p/6$ in Eq. 2.58 and rearranging, the drift velocity is found to be

$$v_r = \frac{(\rho_p - \rho_f)d^2}{18\mu} r\omega^2 \tag{2.59}$$

which is reminiscent of Stokes law in liquids (Eq. 2.19). Rearranging Eq. 2.59, we find that the (spherical) particle or molecule diameter is given by

$$d = \left[\frac{18\mu \dfrac{dr}{dt}}{r\omega^2(\rho_p - \rho_f)}\right]^{1/2} \tag{2.60}$$

---

*Note that the situation here is different than in sedimentation. The drift velocity in sedimentation was constant; the drift velocity in centrifugation is slowly changing because the acceleration term is equal to $\omega^2 r$. We assume that the rate of change in acceleration is so slow that the particle always has plenty of time to adjust to the local terminal velocity. This is our first encounter with the so-called *pseudosteady-state* assumption.

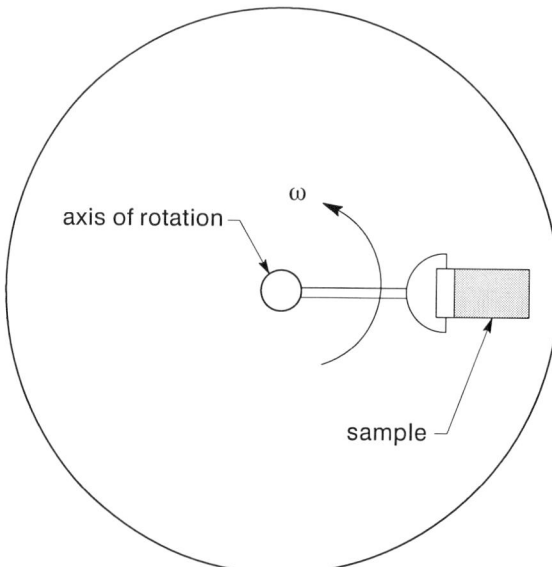

Figure 2.12. Schematic of common centrifuge.

where the substitution $v_r = dr/dt$ has been made. Referring to Figure 2.13, we can set up the limits of integration of Eq. 2.60. At the initial time (t_1), the radial position of the sample particles (or molecules) is r_1; at a later time t_2, the band of particles has migrated to radial position r_2 as a result of the centrifugal acceleration created in the centrifuge. Integration of Eq. 2.60 is then set up as follows:

$$\frac{d^2\omega^2(\rho_p - \rho_f)}{18\mu} \int_{t_1}^{t_2} dt = \int_{r_1}^{r_2} \frac{dr}{r} \tag{2.61}$$

Integration of Eq. 2.61 results in

$$d = \left\{ \left(\frac{18\mu}{\rho_p - \rho_f} \right) \left[\frac{\ln\left(\frac{r_2}{r_1}\right)}{\omega^2(t_2 - t_1)} \right] \right\} \tag{2.62}$$

Notice that the first group of terms in parentheses contains parameters related only to the physical characteristics of the fluid and particles, whereas the second group of parameters in parentheses contains only parameters related to the coordinates of the particle band. This group is called the *sedimentation*

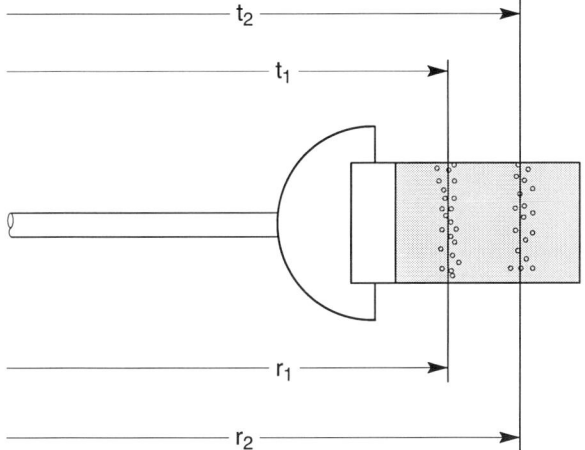

Figure 2.13. Radial and time coordinates for integration of Eq. 2.60.

coefficient by physical chemists:

$$S_d = \frac{\ln\left(\frac{r_2}{r_1}\right)}{\omega^2(t_2 - t_1)} \tag{2.63}$$

The sedimentation coefficient is used by chemists to characterize physically unknown macromolecule and particle suspensions. Although the units of S_d could be in seconds, it is traditional to express S_d in "Svedbergs," where 1 Svedberg $= 10^{-13}$ sec. Svedberg was the scientist who won the Nobel prize for the invention of the ultracentrifuge. Ultracentrifugation is discussed in Chapter 6, which considers the effects of diffusion.

2.12. SUMMARY

This chapter applied Newton's second law of motion to small particles in gases and liquids. From these analyses, the nonsteady and steady forms of the equation of motion for particles were developed. The important concept of particle characteristic time was developed. The characteristic time was shown to be a useful way of predicting how quickly a particle achieves its steady state of motion. The box model and control volume approaches of Chapter 1 were very useful in developing equations for the performance of several types of batch and continuous-flow sedimentation processes. Finally, the equation of motion for the particle was used in analyzing inertial forces on particles and the movement of particles in rotating flows.

An important assumption of this chapter was that the concentration of particles was small enough that any one particle was not affected by the motion of any other particle (thus the chapter title, Low-Concentration Particle Suspensions and Flows). In some real-world applications (especially concentrated sediments or sludge particles in water), this assumption is not a good one. Unfortunately, the analysis of particle transport in concentrated suspensions is much less developed than that for dilute suspensions. Therefore, in this chapter we have chosen to stick with the more elegant analysis of low-concentration particle suspensions.

Finally, even though the analyses of this chapter were aimed at quite small particles, aerosols, and macromolecules, we ignored the possibility of particle diffusion. Diffusion is important in many situations where particle diameters are smaller than about 10^{-6} m. Since this is the size range of many particulate systems of environmental importance, diffusion is a concern in subsequent chapters (e.g., Chapters 6, 7, and 10).

EXERCISES

2.1. Three rectangular sedimentation systems are shown below. Analyze fractional removal for the three cases assuming quiescent plug-flow sedimentation of a monodispersion. The systems have the same width into the page.

Some engineers say that this type of sedimentation does not depend on the system depth, but only on the system bottom surface area (or

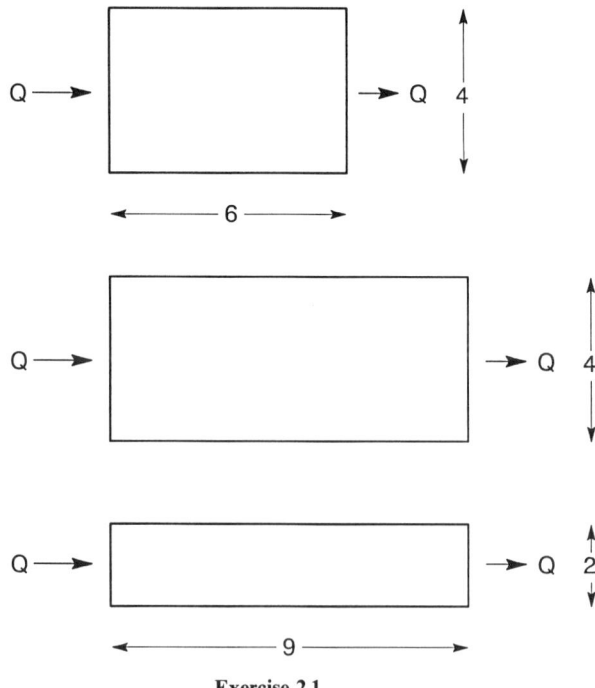

Exercise 2.1

equivalently the fluid surface area at the top of the tank). Why do they say this? Can you think of any practical limit(s) to this design truism?

2.2. Sometimes trays or plates are placed inside sedimentation chambers. Use the equations of this chapter to explain why trays or plates increase the efficiency of quiescent continuous sedimentation.

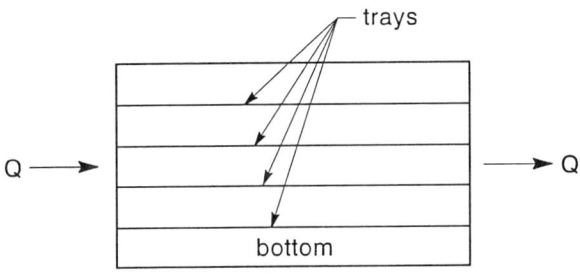

Exercise 2.2

2.3. Determine the terminal settling velocity in water of a spherical sand particle of 0.5 mm diameter. The density of the sand is 2.65×10^3 kg/m^3, and the viscosity of water at 25°C is 1.0×10^{-3} kg-m^{-1}-s^{-1}. Hint: Stokes law is not appropriate.

2.4. Throughout this chapter we have generally considered a single-particle size-class. But for dilute suspensions, there is no reason we cannot apply our analysis to polydispersions, which are suspensions with a distribution of particle sizes. A dilute (noninteracting) suspension of particles (in water) with an average density of 1.4×10^3 kg/m^3 is flowing through a plug-flow sedimentation tank. Assuming quiescent sedimentation, determine the outflow particle size distribution and plot it on the same figure with the inflow particle size distribution provided below. (Plot log concentration vs. average diameter.) The overflow velocity is 25 m-day^{-1}, the temperature is 25°C, and you can assume that Stokes law is appropriate. State in lay terms the effect of sedimentation on a particle-size distribution.

Average Diameter (mm)	Concentration (10^4/cm^3)
0.010	1.91
0.015	1.13
0.020	0.560
0.025	0.183
0.030	0.0354
0.035	0.0669
0.040	0.0149
0.045	0.0104

2.5. In natural water environments, some microorganisms survive by staying suspended in the water column. If the microorganism is not motile, it must be able to depend on its small size (and small settling velocity) and natural mixing phenomena to remain suspended. Compare spherical and cylindrical microorganisms of the same mass. Does a cylindrically shaped organism have any advantage over a spherically shaped organism in being able to stay suspended in the water column? Assume the spherical particle diameter is 25 μm, the organism density is 1.05×10^3 kg/m^3, the water density is 1.00×10^3 kg/m^3, and the length to diameter ratio of the cylindrical organism is $4:1$. Hint: assume the cylindrical organism settles with its main axis in a horizontal position.

2.6. A process is being designed to remove negatively charged aerosol particles from an air stream. The particle diameter is 1.0 μm, the air viscosity is 1.8×10^{-5} kg-m^{-1}-s^{-1}, and the particle charge is 4×10^{-17} C. Assuming that gravity is not a significant mass-transfer mechanism, consider the single particle shown in the figure below:

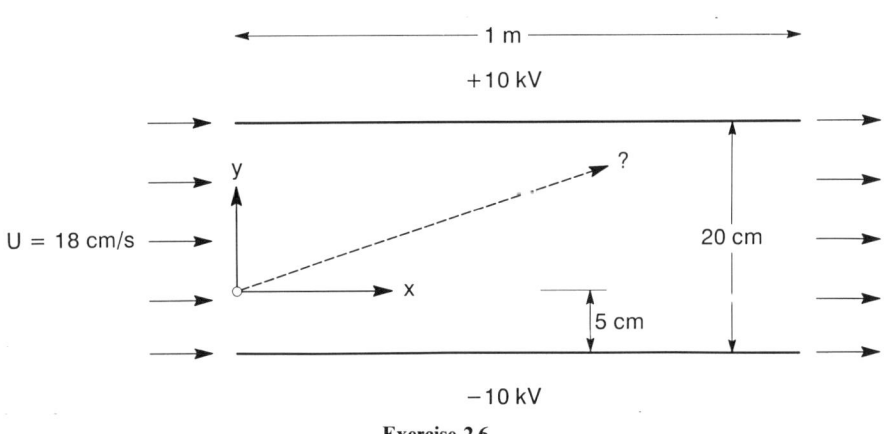

Exercise 2.6

As the uniform air flow passes through the rectangular duct, an electrical field is impressed across the duct. Assume that when the particle enters the duct, it begins to be affected by the electrical field. Just prior to entering the duct, the particle's y-direction velocity v_y is zero, and the x-direction velocity v_x is equal to the stream velocity U. Determine an equation for v_y in the duct, and determine the particle's coordinates x and y as it moves through the device. You can assume that v_x is always equal to the stream velocity U, and that the particle is drifting in the y direction at the steady-state velocity.

Are the predictions of this problem significantly in error from ignoring nonsteady effects (i.e., the particle acceleration in the y direction)? You may assume the aerosol particle has "unit density" — 1000 kg/m^3.

2.7. A process is being designed to remove charged aerosol particles from an air stream. As the air passes through a rectangular duct, an electrical field is impressed across the duct. The aerosol particles move because of the electrical field, and are considered removed if they impact either wall of the duct. The electrical field strength is given by

$$E = E_{max} \sin(\omega_0 t)$$

Assume that the particle characteristic time is small relative to the characteristic time of the electrical field oscillations. Hence, it can be assumed that the particle responds quickly enough to changes in the electrical field that the particle motion can be presumed to always be at steady state. Determine the following:

(a) Assuming a uniform mean fluid (air) velocity U in the duct, derive an equation describing the motion of the aerosol particles in the duct.

(b) Assuming fixed particle diameter, temperature, particle charge q, and ω_0, size the duct for 100% removal of entering particles. Sizing requires determining the distance separating the duct walls and the overall length of the duct.

(c) Using the parameters ω_0 and τ, how could one quantify the specification that the particle characteristic time must be much smaller than the characteristic time of changes in the electrical field?

2.8. For the case of batch quiescent sedimentation, analyze sedimentation rate in the odd-shaped system shown below. Consider only sedimentation rate during the period when the particle interface is above the sloped part of the system. Hints: The suspension is initially homogeneously mixed. The control volume approach is useful, although other approaches may be valid.

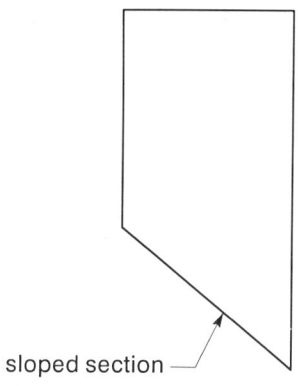

sloped section

Exercise 2.8

Can you draw any general conclusions about the impact of the bottom profile on sedimentation rate under these conditions? For example, if the sloped section was wavy, what effect would this have on the sedimentation rate?

2.9. For the case of batch quiescent sedimentation, analyze sedimentation rate in the odd-shaped system shown below. Hints: The suspension is initially homogeneously mixed. The control volume approach may be useful, although other approaches may be valid.

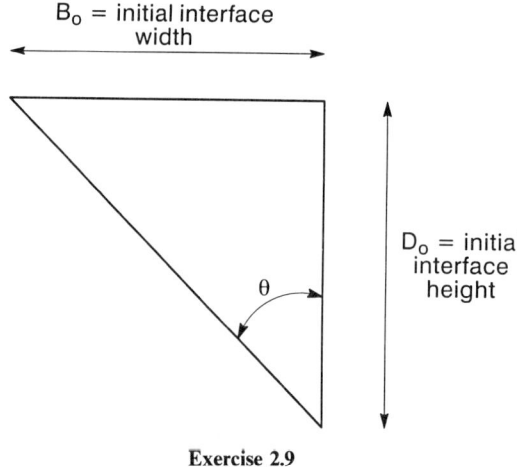

Exercise 2.9

2.10. Analyze perfect-mix batch sedimentation in the system from Exercise 2.8.

2.11. Analyze quiescent plug flow sedimentation in the triangular-section system shown below. Hint: the result of Exercise 2.9 may be useful in solving this problem.

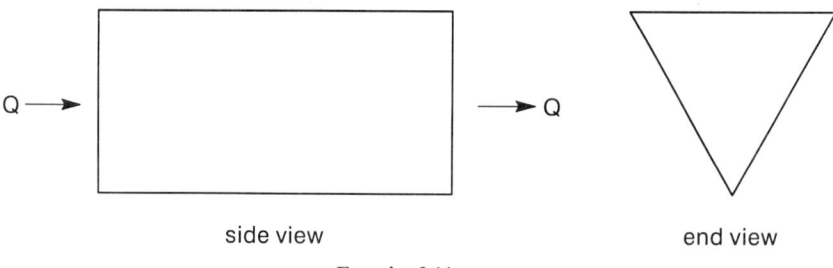

Exercise 2.11

2.12. Analyze perfect-mix continuous sedimentation in the three systems shown on p. 63. On the basis of these results, can you infer a general

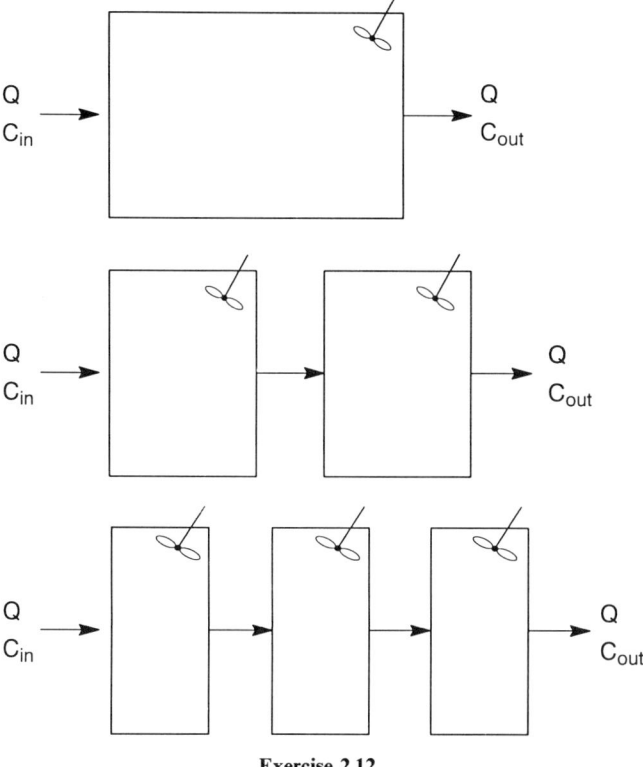

Exercise 2.12

formula for fractional removal in a system composed of *n* chambers? Note: the sum of the volume of the subsystems in both the second and third pictures is the same as the volume of the system in the first picture.

2.13. An inversion layer has formed over a city located in a mountainous area. Because of the local geography and the stable inversion layer, the air above the city becomes trapped.

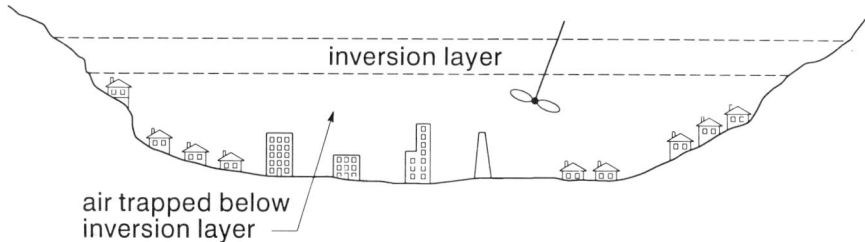

Exercise 2.13

Since air is trapped above the city, particulate pollutants (generated by industry and automobiles) accumulate and a pollution crisis begins. The particles can only leave the air through sedimentation. Assuming *perfect mixing* below the inversion layer, complete the following:

(a) Using a mass balance, write a differential equation governing the concentration of particulate pollutants in the city.

(b) If the inversion continues for an indefinite time period, what will be the steady-state particulate concentration in the city?

Data

1. Rate of generation of particulate pollutants $= 900\,\text{kg/day}$
2. Volume of air trapped above city $= 5 \times 10^{10}\,\text{m}^3$
3. Surface area of city $= 1.5 \times 10^8\,\text{m}^2$ (as measured on map)
4. Particle diameter $= 1.0\,\mu\text{m}$
5. Density of particles $= 1.5 \times 10^3\,\text{kg/m}^3$
6. Temperature $= 25°\text{C}$

2.14. What are the characteristic times of quiescent and perfect-mix *batch* sedimentation?

2.15. Consider fractional removal for the case of vertical-mix, plug-flow sedimentation in a rectangular chamber by using the following picture and differential element.

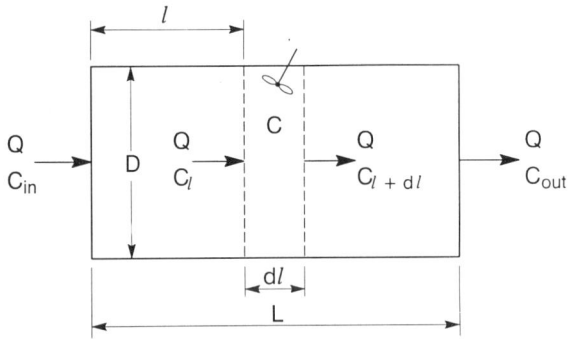

Exercise 2.15

(a) Show that the fractional removal in this system can be expressed as

$$R = 1 - \exp\left(-\frac{v_t}{V_0}\right)$$

Hint: First write a differential mass balance on the differential element, and then integrate it over the entire system. Take special note that this model imagines mixing *only* in the vertical direction.

(b) If Exercise 2.12 was completed, compare the fractional removal for very large n with the formula developed in this problem. What does this comparison show?

2.16. In the dry deposition of some aerosol particles on surfaces, some of the deposited particles rebound from the surface and/or are resuspended owing to the turbulence of the suspending air. This rate of resuspension or rebound has sometimes been modeled as a fixed fraction β of the mass transfer due to sedimentation. Therefore, the net sedimentation rate is $(1 - \beta)CAv_t$ (recall the nomenclature in Section 2.7).

(a) Develop an equation analogous to Eq. 2.36 for fractional removal in continuous perfect-mix sedimentation with correction for resuspension/rebound.

(b) Using your new equation, plot fractional removal vs. v_t/V_0 for $\beta = 0$, 0.2, 0.4, 0.6, 0.8, and 1. Use same axes as in Figure 2.8.

2.17. During warm weather, many lakes can thermally stratify. This means that the water warmed by the summer sunlight and winds lies on top of the colder and denser water left from the winter. In this condition, there can be limited mixing between the warm- and cold-water bodies. However, there is apparently nothing to prevent sedimentation from the warm- to cold-water bodies.

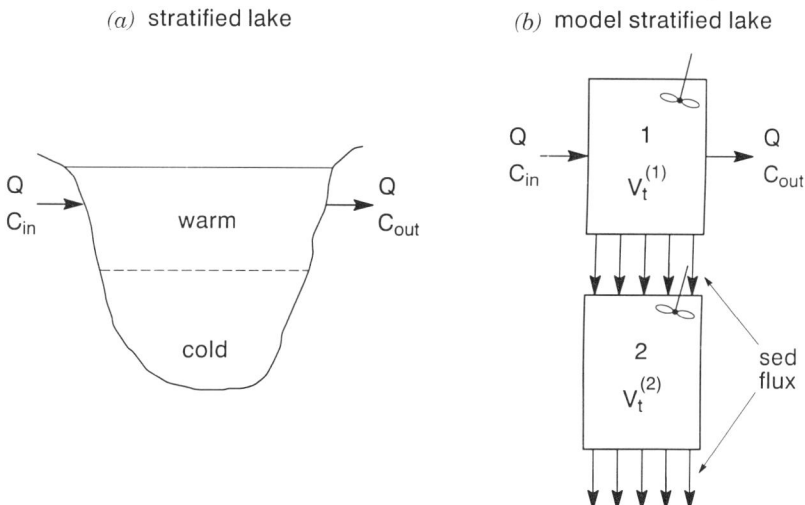

Exercise 2.17

In an effort to model the system, we might consider that the two water bodies are effectively separate, perfectly mixed systems, lying on top of one another. As shown in part b of the figure, the liquid flows are only into the overlying body (system 1). There is a flux of particulate matter from system 1 into system 2 due to sedimentation. Also, the liquid temperatures may be different; hence, we have allowed for different terminal settling velocities in the two systems, $v_t^{(1)}$ and $v_t^{(2)}$.

(a) Considering a monodisperse particle distribution, what is the flux of particulate matter from system 1 into system 2 at steady state? What must be the flux of particulate matter out of system 2 at steady state?

(b) Using a mass balance on system 2, show that the ratio of particle concentrations in the two systems at steady state is given by

$$\frac{C_{sys}^{(2)}}{C_{out}^{(1)}} = \frac{v_t^{(1)}}{v_t^{(2)}}$$

Does this prediction make sense? Explain in words using the equation.

2.18. In many environmental systems, we are often interested in the flux of material across certain boundaries. For example, the flux of dead algae to the bottom of a lake may be of interest in studies of eutrophication, and the dry deposition of solids could be of interest in air-pollution studies. In any event, consider Section 2.7 and prove that the normalized average sedimentation flux of particles in continuous sedimentation is equal to the normalized fractional removal, that is,

$$\frac{J}{C_{in} V_0} = R$$

where J is the average mass flux of particulate matter ($M\text{-}L^{-2}\text{-}T^{-1}$) and V_0 is the overflow velocity. Note: If the statements above are true, the fractional removal curves in Figure 2.8 are proportional to the average sedimentation flux out of the system.

2.19. Aerosol particles are flowing through a small two-dimensional duct elbow shown on p. 67. Some of the particles will strike the vertical wall because they will not be able to exactly follow the streamline pattern indicated. For the idealized streamline pattern shown below, estimate the average concentration of particles leaving the duct. The mean velocity is 20 m/s and the viscosity of air is 1.8×10^{-5} kg-m^{-1}-s^{-1}. The particles have a diameter of $5\,\mu$m and a density of 2.0×10^3 kg/m^3. Ignore gravitational effects, and assume that all particles that strike the vertical wall are removed.

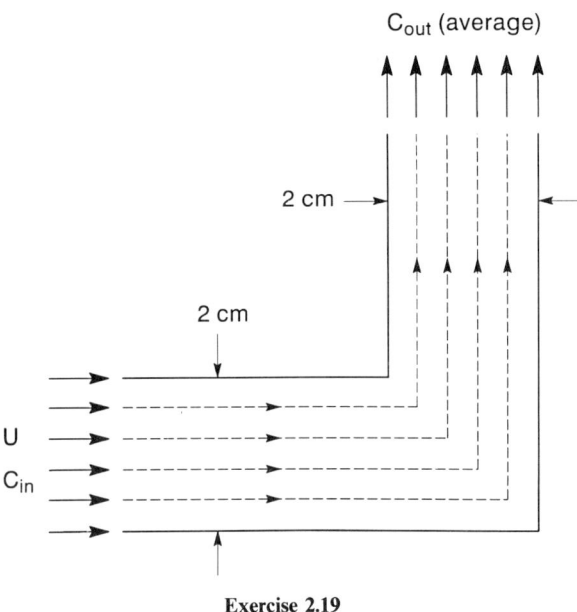

Exercise 2.19

2.20. Referring to Fig. 2.9, draw pictures of small and large cylindrical obstacles in flow fields with the *same* free-stream velocity. According to Eq. 2.48, the situation with the smaller obstacle would have a greater Stokes number. Should we expect that the probability of particle impaction would increase with larger Stokes numbers corresponding to the smaller obstacle? Why?

2.21. An organic compound is being studied using centrifugation. The particle boundary moves from $r = 6.314$ cm to $r = 6.367$ cm over a 10-min period. The density of the compound and suspending fluid are known to be 0.998 g/cm^3 and 0.728 g/cm^3, respectively. If the centrifuge is running at 4.47×10^3 rad/s, calculate the apparent molar mass of the compound, assuming a spherical molecule. The viscosity of the suspending medium is 0.0009 kg-m^{-1}-s^{-1}.

2.22. Cyclones are engineered to remove suspended particles from air or gas streams (as shown on p. 68). In the outer vortex of a "laminar flow" cyclone, the flow is approximated by a solid-body rotation; hence, the velocity u_θ at any point r is given by $r\omega$.

    (a) Assume that particles in the outer vortex have the same θ-component velocity as the fluid (i.e., $v_\theta = u_\theta$). Manipulate Eq. 2.58 and make necessary substitutions to show that the r-component velocity of the

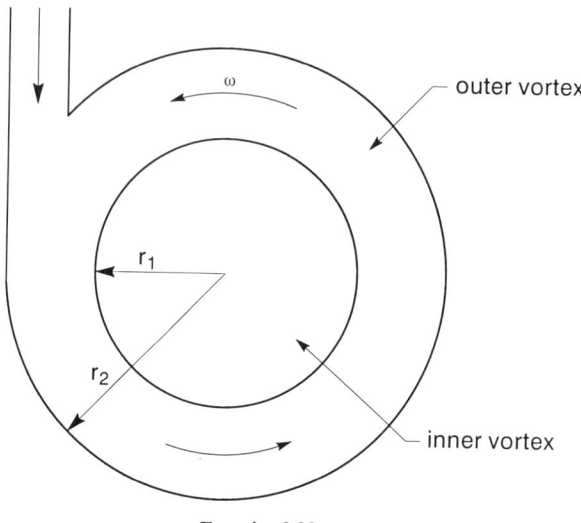

Exercise 2.22

particles is

$$v_r = \frac{dr}{dt} = \frac{C_c d^2 \rho_p}{18\mu} r\omega^2$$

(b) Flagan and Seinfeld (1988) demonstrated that for the outer laminar vortex, the following equation must be true:

$$\frac{d\theta}{dr} = \frac{1}{r} \frac{v_\theta}{v_r}$$

Show that the number of radians a particle travels going from $r = r_1$ to $r = r_2$ is

$$\theta = \frac{18\mu}{C_c d^2 \rho_p \omega} \ln\left(\frac{r_2}{r_1}\right)$$

REFERENCES

Cliff, R., Grace, J. R., and Weber, M. E., *Bubbles, Drops, and Particles*, Academic Press, New York, 1978.

Flagan, R. C. and Seinfeld, J. H., *Fundamentals of Air Pollution Engineering*, Prentice Hall, Englewood Cliffs, NJ, 1988.

Green, D. W., and Maloney, J. O. (Eds.), *Perry's Chemical Engineers' Handbook*, 6th ed., McGraw-Hill, New York, 1984.

Hinds, W. C., *Aerosol Technology: Properties, Behavior, and Measurement of Airborne Particles*, Wiley-Interscience, New York, 1982.

Maxey, M. R. and Riley, J. J., "Equation of Motion for a Small Rigid Sphere in a Nonuniform Flow," *Phys. Fluids*, **26**, 4:883–889 (1983).

Näslund, E. and Thaning, L., "On the Settling Velocity in a Nonstationary Atmosphere," *Aerosol Sci. and Technol*, **14**, 247–256 (1991).

BIBLIOGRAPHY

Batchelor, G. K., *An Introduction to Fluid Dynamics*, Cambridge University Press, 1967 (available in paperback).

Hazen, A., "On Sedimentation," *Trans. Am. Soc. of Civil Eng.*, **53**, 63 (1901).

Hiemenz, P. C., *Principles of Colloid and Surface Chemistry*, Marcel Dekker, New York, 1986.

Probstein, R. F., *Physicochemical Hydrodynamics: An Introduction*, Butterworths, Boston, 1989.

John, W., "Particle-Surface Interactions: Charge Transfer, Energy Loss, Resuspension, and Deagglomeration," *Aerosol Sci. and Technol.*, **23**, 1:2–24 (1995).

3

INTERACTION OF SMALL CHARGED PARTICLES

3.1. INTRODUCTION

In this chapter we investigate the interactions of small charged particles in water and air environments; hence, our focus is on interactions of particles in either *hydrosols* or *aerosols*. Generally, these particles fall in the size range of *colloids*, about 1×10^{-9} to 1×10^{-6} m; therefore, most of this chapter concerns classic topics from colloid and aerosol science, for example, acquisition of surface charge, hydrosol double layers, and coagulation. These are very important considerations in environmental science, pollution control engineering, and industry. Some of the traditional applications of colloid science are listed in Table 3.1.

Until about 150 years ago, most physicists and chemists believed that all systems could conveniently be classed as solids, liquids, or gases, and that all important physical and chemical phenomena could be understood by looking at the system as a pure material phase, a classical solution, or a mixture of gases. However, it became apparent that when a liquid or solid phase is dispersed in another phase as a very small particle dispersion, the surface of the particles and the *interfacial area* between the *dispersed* and *continuous* phases could dominate many observed phenomena. The relatively new area of study called *colloid and surface (or interface) science* has been successful in shedding light on many of the phenomena listed in Table 3.1. Several terms are commonly encountered in the colloid science literature for aqueous systems, and a brief definition and discussion of these terms are provided below:

Lyophobic/Lyophilic Colloids. Lyophobic colloids are "solvent-fearing" colloids, whereas *lyophilic* colloids are "solvent-loving" colloids. This chapter deals mostly with lyophobic colloids, particularly these suspended in water. These systems are commonly called *hydrophobic* colloids. One of the best ways to distinguish between lyophobic and lyophilic colloids is by their preparation. For example, *hydrophilic* colloids are usually large macromolecules that tend to disperse spontaneously. In fact, there are many similarities between a hydrophilic colloidal *dispersion* and a true solution. The thermodynamic or

Table 3.1. Environmental and Industrial Applications of Colloid Science

Field of Study or Industry	Problems or Application
Air-pollution control	Aerosol coagulation; deposition of contaminants in lungs; light scattering; scavenging of contaminants
Water-pollution control	Coagulation of water contaminants; membrane separation and fouling; flotation separation
Environmental science	Coagulation/deposition of sediments; structure of soils
Membrane industry	Structure of ceramic membranes; solid–liquid separation
Food industry	Fabrication of dairy products (ice cream, butter, cheese); membrane fouling
Paint and cosmetics industries	Dispersion of pigments and emulsions; manufacture of creams
Oil industry	Oil-in-water emulsions; structure of drilling muds
Biomedical and biotechnology industries	Blood structure and rheology
Mineral processing	Flotation separation

natural tendency in lyophilic systems seems to be toward dispersion or solution. Hydrophobic colloids, on the other hand, are composed of such materials that spontaneous solutions or dispersions are not likely. In fact, the thermodynamically favored change for hydrophobic colloidal dispersions is aggregation (called *flocculation* or *coagulation*) and eventual separation of phases.

Stability. Stability and instability refer to both the thermodynamically favored change and the rate of change or phase separation in lyophobic colloidal solutions. For example, some hydrophobic hydrosols can be placed in a jar on a shelf, and virtually no changes in the dispersion will occur over tens or even hundreds of years. Even though the dispersion is thermodynamically unstable (i.e., the natural change favors phase separation), the process may be so slow that the dispersion is *kinetically* stable. Other colloidal dispersions are much less stable kinetically; we have to control solution chemistry and temperature carefully, and limit agitation to maintain a disperse phase.

Indifferent Electrolytes. In Section 3.5, important models of the effect of salt solutions (electrolytes) on colloid stability are developed. These models all assume that the salt does not adsorb or otherwise specifically interact with the colloid surface. In effect, the salt ions behave as simple point charges. A salt such as NaCl can usually be assumed to form an indifferent electrolyte in solution. A counterexample would be hydrolyzing metal salts like aluminum and iron salts, commonly used in water and wastewater treatment. These molecules form charged hydrolysis species that are known to adsorb at colloid interfaces and alter the colloid surface potential and charge. Certain derivations also assume *symmetrical* electrolytes, which here are indifferent electrolytes in which the magnitude of the valence of positive and negative ions is the same. For example, NaCl solutions are called a 1:1 electrolytes and $CaSO_4$ solutions are called 2:2 electrolytes.

3.2. IMPORTANCE OF SURFACE

Before looking at particle interaction phenomena, we will develop some simple demonstrations of the importance of the interfacial region or area between two phases. Suppose a particular system is composed of n uniform spherical particles of volume $4/3\pi r^3$ and surface area $4\pi r^2$. The ratio of the total particle area to volume is

$$\frac{A_t}{V_t} = \frac{4n\pi r^2}{(4/3)n\pi r^3} = \frac{3}{r} \tag{3.1}$$

If we now imagine that the n particles are cut into smaller and smaller sizes (which maintain the spherical shape), n approaches infinity, r approaches zero, and by Eq. 3.1, the ratio of particle surface to volume in the system of particles (the dispersion) approaches infinity. As the disperse particles become smaller, we can imagine that phenomena active at the surface (e.g., surface charge, adsorption of solutes or gases, surface tension, or light scattering) can begin to dominate or control the behavior of the particle-containing system.

As a further example of the importance of surface, consider in detail the interfacial region between a liquid and gas phase (Figure 3.1). Molecules in the liquid phase have a tendency to want to stay in the liquid phase; *hydrogen bonds* and intermolecular forces called *van der Waals forces* cause a relatively strong attraction between the liquid molecules. However, although liquid-phase molecules far from the interface tend to have forces of attraction distributed uniformly around them, the liquid-phase molecules at the interface are only pulled away from the gas phase. In a sense then, we can imagine a tensile force in the plane of the interface — the *interfacial tension*. Just as a balloon requires energy to increase its volume (higher pressure), energy is also required to expand the gas–liquid interface. Now imagine that a liquid is dispersed in the gas phase as a small droplet dispersion (or aerosol): what is the most likely configuration of the dispersed phase droplets? The simple answer is that since a spherical shape allows most molecules to be liquid rather than interface molecules, the natural tendency is for the dispersed droplets to attain spherical shapes. We can then imagine that in order to increase the interfacial area of the droplet (equivalent to distorting the droplet), work will

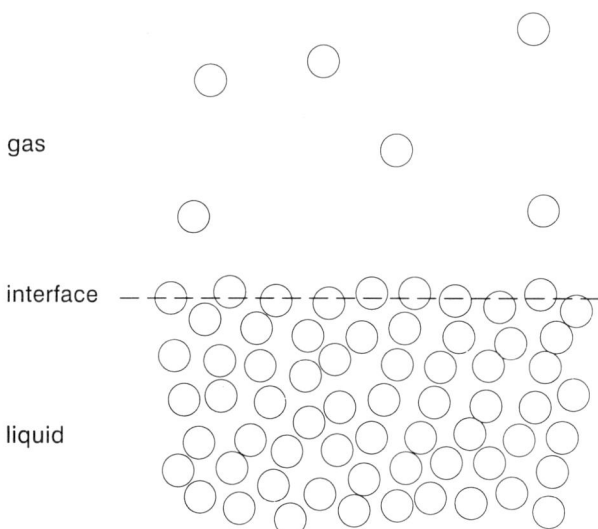

Figure 3.1. Interface between gas and liquid.

have to be done (energy will be required). Physical chemists have found that the differential work dw required to expand interfacial area by the differential amount dA is given by

$$dw = \sigma dA \tag{3.2}$$

where σ is the interfacial tension (or surface tension), which can be defined by Eq. 3.2. To increase interfacial area, we can imagine that work must be done against a hypothetical tension in the interface. Table 3.2 provides several values of interfacial tension.

Of course Equation 3.2 also has a more quantitative connection to classical thermodynamics. For a system containing an interface, the change in the surface Gibbs function (Section 4.2) is

$$dG^{(s)} = -S^{(s)}dT + \sum_i \mu_i^{(s)}dn_i^{(s)} + \sigma dA \tag{3.3}$$

where S is the entropy, T is the temperature, the μ_i are the chemical potentials of the i chemical species, the n_i are the amounts of the i species, and the lower-case s indicates the equation is only valid at the surface. Equation 3.3 shows that a decrease in surface area $(-dA)$ is the thermodynamically favored change, since systems try to minimize the Gibbs function. This can be interpreted as the tendency of a dispersed system to become less dispersed, and to minimize its surface area.

Table 3.2. Interfacial Tension (dyne/cm) in Several Systems

System	Temperature (°C)	Interfacial Tension
Air–water	25	72.0
Water–benzene	20	35.0
Water–carbon tetrachloride	20	45.0
Water–mercury	20	375.0

Source: Reprinted with permission from R. C. Weast, Ed., *Handbook of Chemistry and Physics*, 56th ed., CRC Press, Cleveland, OH, 1975. Copyright CRC Press, Boca Raton, Florida.

A class of compounds called surfactants (surface active agents) tends to accumulate at interfaces and to lower the interfacial tension. In oil-in-water emulsions, water-soluble surfactants (e.g., detergents) tend to accumulate at the oil–water interface, the polar part of the molecule sticking into the water phase, and the nonpolar part sticking into the oil phase (see Section 4.2.2). The accumulation of these molecules at the interface lowers the interfacial tension, decreases the Gibbs function, and stabilizes the emulsion (see Atkins, 1978). An emulsion is "broken" when the stability is decreased, the fine oil droplets coalesce into larger droplets, and eventually, the oil phase completely separates from the liquid phase.

Example 3.1. Compare the surface free energy of three pure water aerosols with droplet sizes of 10, 1, and 0.1 μm. The total amount of water is 1 l.

SOLUTION. If the water is pure, there is no change in the chemistry of the interface and the derivatives in the chemical potential terms in Eq. 3.3 are zero. If the temperature is assumed to be constant, Eq. 3.3 reduces to

$$dG^{(s)} = \sigma dA$$

Hence, we find that the surface free energy is given by

$$G^{(s)} = \sigma A$$

Note from Eq. 3.1 that

$$A_t = \frac{3V_t}{r}$$

Substituting for the 10-μm droplet aerosol (with $\sigma = 72$ dyne/cm from Table 3.2)

$$G^{(s)} = \sigma A_t = \left[\frac{72 \frac{\text{dyne}}{\text{cm}} (3)(1\,\text{L}) \left(\frac{100\,\text{cm}^3}{1\,\text{L}} \right)}{(5\,\mu\text{m}) \left(\frac{\text{cm}}{10^4\,\mu\text{m}} \right)} \right] = 4.32 \times 10^8 \text{ dyne-cm}$$

Similarly, for the other aerosols:

$$1\text{-}\mu\text{m droplets: } G^{(s)} = 4.32 \times 10^9 \text{ dyne-cm}$$

$$0.1\text{-}\mu\text{m droplets: } G^{(s)} = 4.32 \times 10^{10} \text{ dyne-cm}$$

Note that as the aerosol droplets decrease in diameter by factors of 10, the surface free energy increases by factors of 10.

The discussion above is but one demonstration of the importance of surface in environmental studies. In this book, we discuss many other phenomena active at interfaces.

3.3. ACQUISITION OF SURFACE CHARGE

In Section 3.2 we noted that surfactants can stabilize oil-in-water emulsions by lowering the interfacial tension. In a much larger class of solid-particle colloids in the environment, the *surface charge* or *potential* of the particle is responsible for the stability of the particle dispersion.* Such particle systems include all the nonliquid aerosols, as well as all the solid colloids dispersed in water (hydrosols).

3.3.1. Hydrosols

Five sources of hydrosol surface charge are discussed in this section (Everett, 1988; Van Olphen, 1977; Stumm and Morgan, 1981).

Ionization of Surface Groups. Many colloidal surfaces can react with water, and the surface charge depends on pH. Figure 3.2a shows the ionization of surface silanol groups on a silica oxide surface, and Figure 3.2b shows the progressive ionization of amino and carboxyl groups on an organic particle surface. Note that the solution pH determines the surface charge: at low pH, the surfaces have net positive charges; at some higher pH, the surface has a net negative charge. At another pH value, the surface charge is expected to have a (net) zero charge. This pH is called the *point of zero charge*, or pzc.

Isomorphous Substitution. In several of the clay minerals, imperfections in sheets of tetrahedrally or octahedrally coordinated silicon and aluminum allow the substitution of aluminum (for silicon) and magnesium (for aluminum). Because of the lower valence of the substitute ions, the tetrahedral or octahedral sheets can take on a net negative charge. An example of aluminum

---

*Steric stabilization, a type of stabilization caused by adsorbed polymers and natural organic matter, is not discussed in this book. See Stumm and Morgan, 1981.

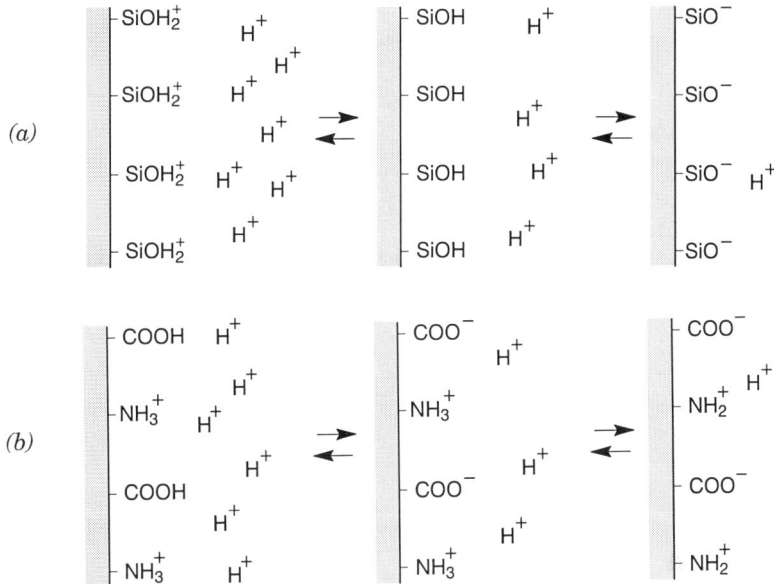

Figure 3.2. Ionization of (a) surface silanol and (b) amino and carboxyl groups.

(valence $+3$) substitution for silicon (valence $+4$) in a tetrahedrally coordinated silicon sheet is illustrated in Figure 3.3.

Fractured Crystal Surfaces. Another type of charge on the clay minerals occurs where tetrahedral silica and octahedral alumina sheets are fractured. This results in a totally different surface than that along the flat surface of the clay particles, which is more similar to the surface of silica (Figure 3.2) and alumina particles. In the natural environment, this edge surface normally carries a positive charge. The possibility of negative surface charges caused by isomorphic substitution (above) and positive edge charges can lead to interesting aggregation behavior of the clay minerals, including the so-called edge-to-face flocculation and resultant "house of cards" structures (Van Olphen, 1977).

Specific Ion Adsorption. Specific adsorption of cationic or anionic polymers, surfactants, and natural organic matter can lead to modification (and even reversal) of surface charge. In the clay mineral field, "peptizing" ions like polymetaphosphate are known to strongly adsorb to the edges of clay particles, thus neutralizing their positive charge, and in some cases, yielding even a net negative edge charge.

Variable Potential Surfaces. Colloidal dispersions made of AgI are known to have variable potential surfaces, which are determined by the relative amount of Ag^+ and I^- ions in solution. The AgI colloids behave like reversible electrodes; they have a positive charge for Ag concentrations greater than

(a)   no net charge

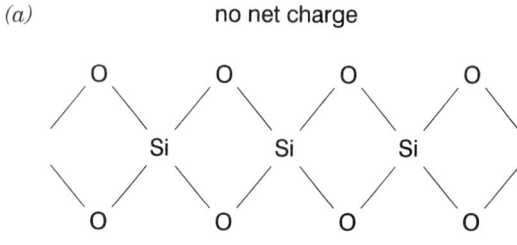

(b)   negatively charged

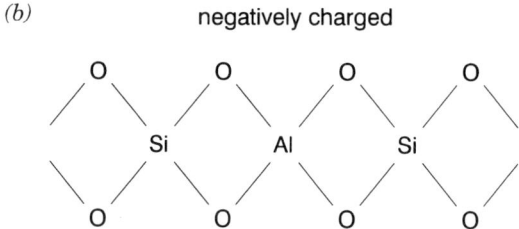

Figure 3.3. Illustration of isomorphic substitution in silicon tetrahedra of clay.

3.0×10^{-6} M, and a negative charge for concentrations less than 3.0×10^{-6} M (Hiemenz, 1977).

3.3.2. Aerosols

Aerosol particles acquire their charge in a fundamentally different way from hydrosols. In *diffusion* and *field* charging, the aerosol particle acquires surface charge by transfer from gas ions in the environment around the particle. The charged ions may result from natural processes like UV radiation in the atmosphere, or from engineered processes like corona discharge in an electro-static precipitator. In diffusion charging, charge is transferred to the particle as a result of the Brownian diffusion of gas ions to the surface of the particle. In field charging, the particle is charged as a result of ions being transported in the vicinity of the particle by an electrical field. The two mechanisms can occur simultaneously, as in an electrostatic precipitator. The relative magnitude of the two processes depends strongly on the particle size: Particles smaller than $1 \, \mu m$ are affected more by diffusion charging; particles bigger than $1 \, \mu m$ are affected more by field charging.

Static aerosol charging can occur during aerosol formation from liquids, for example, when an aerosol particle is popped out of a liquid that contains a surface charge. Static charging can also occur when there are high-velocity impacts between aerosol particles and solid surfaces in a relatively dry environment. Here a charge can be transferred from the solid to the aerosol by direct contact.

3.4. PARTICLE SIZE, SHAPE, AND POLYDISPERSITY

At this point the reader might be thinking that colloids are easily characterized, uniform, spherical particles. Although the size and shape of manufactured colloidal particles can be controlled rather precisely, the variety in size, shape, and surface properties of natural colloidal particles is astounding. Figure 3.4 shows electron micrographs of a few different colloidal particles. Polystyrene latex particles (Fig. 3.4a) can be manufactured to precise specifications of size and shape. Figures 3.4b and c show two naturally occurring colloids, chrysotile asbestos and kaolinite clay. Note that because of the extreme shapes of some of these natural particles, all the particle dimensions are not strictly in the

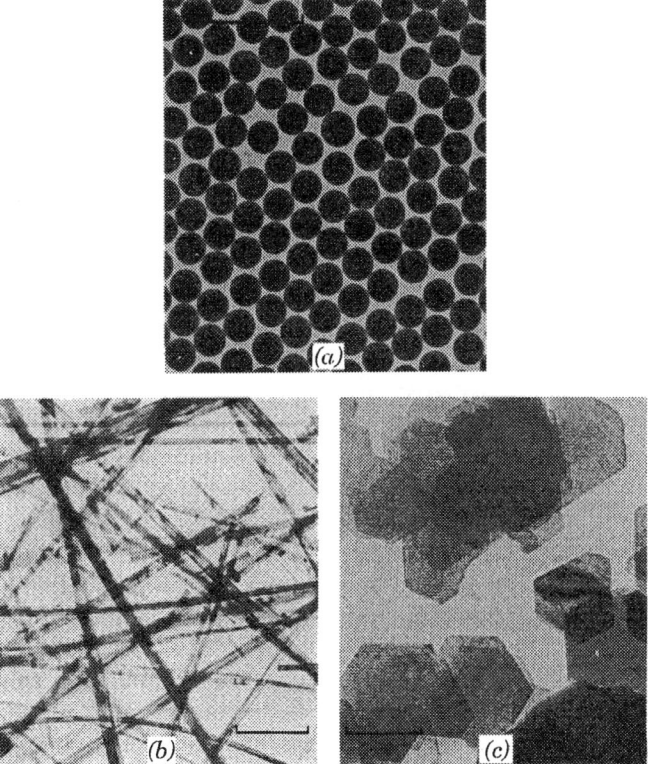

Figure 3.4. Electron micrographs of several colloids: (a) polystyrene latex spheres, (b) chrysotile asbestos fibers, and (c) kaolinite particles. The scale (1 μm) for all figures is shown in b. *Source*: Reproduced by permission from Everett, *Basic Principles of Colloid Science*, Royal Society of London, Cambridge, 1988.

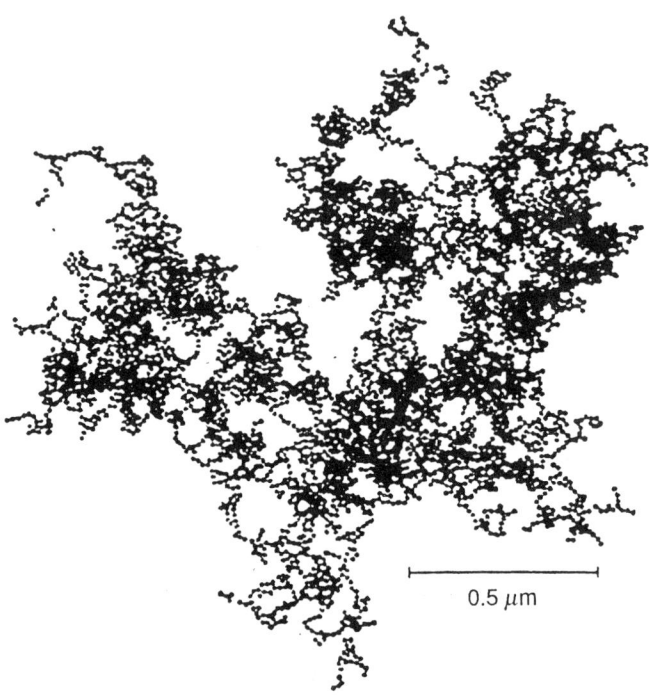

0.5 μm

Figure 3.5. A floc composed of gold particles that are 14.5×10^{-9} m in diameter. The floc contains 1739 particles. *Source*: Reproduced from Weitz and Oliverla, *Phys. Rev. Lett.*, **52**, 16:1434 (1984). Permission granted by Exxon Research and Development Company, Florham Park, New Jersey, 1996.

colloidal size range as defined above. However, at least one of the dimensions is in the colloidal range, thus leading to the kind of stability discussed above.

Natural colloids also exist in a range of sizes, and we speak of the *polydispersity* of colloidal dispersions. Because of the polydispersity and irregular shape of many natural colloids, the characterization of hydrosols and aerosols can be a formidable problem in detection and statistical representation. For example, how does one express the size of a cigar-shaped chrysotile fiber or kaolinite platelet? It seems clear that a "diameter" may be difficult to define. In fact, there has been no completely satisfactory resolution of the problem of simply characterizing extreme shapes. Several of the detectors used to measure particle size of aerosols and hydrosols register either the volume or mass of the particle, and it is sometimes possible to express the size of a particle as the diameter of an equivalent volume sphere (see Section 2.3). In aerosol science, the concept of an *aerodynamic diameter* is often used. This is a measurement-specific diameter that refers to the diameter of a sphere of unit density ($1000 \, \text{kg/m}^3$) which impacts on a collector or settles at the same rate as the actual aerosol particle.

   The problem of characterizing particle size becomes somewhat more complicated when we are faced with evaluating the size of aggregates of either natural or manufactured particles. For example, Figure 3.5 shows an aggregate (or floc) of gold particles. The aggregate has a tenuous and "random" shape, with many voids between individual particles. What is the size of this object? In fact, we might also ask, what is the aggregate and what is not the aggregate? Some scientists would consider the interstitial water to be part of the floc; therefore, the floc would be defined by an "enclosing volume" which wraps around the floc in the photograph. Others would say that the interstitial water is part of the floc only if it is immobile, essentially trapped because of the high fluid drag in the small interstitial caverns in the aggregate.

   Some scientists have even begun to wonder whether we should give up trying to characterize aggregates like that shown in Figure 3.5 as three-dimensional objects. This has been motivated in part by the realization the such disperse, random aggregates contain a much higher surface area than any equivalent sphere or other shape. In a sense, the floc shown in Figure 3.5 might be considered more of a tenuous surface (two-dimensional) rather than a homogenous three-dimensional object. The new theory of fractals is being used to characterize such random aggregates (e.g., Meakin, 1991).

   The scope of this book does not permit a thorough review of the fascinating science of particle characterization. We will assume that accurate measurements of the *particle size distribution* are available and that, generally, the particles can be treated individually as spheres. However, Table 3.3 indicates the variety of sizing methods available and their approximate ranges of size resolution.

Table 3.3. Summary of Some Available Methods for Sizing and Counting Aerosol and Hydrosol Particles

Method	Application	Resolution Range (μm)
Static and dynamic light-scattering methods	Aerosols/hydrosols	0.01–30
Differential mobility/condensation nuclei counter	Aerosols	0.01–1
Diffusion batteries	Aerosols	0.001–0.1
Cascade impactors	Aerosols	0.06–30
Microscopic methods	Aerosols/hydrosols	0.01–1000
Electrical sensing zone	Hydrosols	0.5–500
Light blockage	Hydrosols	0.5–500 μm
Sedimentation	Hydrosols	0.1–500
Microphotography	Hydrosols/aerosols	3–1000 μm

3.5. THE DOUBLE LAYER AND COLLOIDAL STABILITY

In this section classic theories on the distribution of electrical potential and charge in the region of colloidal surfaces in water, and the resulting stability that this charge brings to a colloidal dispersion are developed. Figure 3.6 shows the distribution of an indifferent electrolyte around a charged colloid.

In the figure, a negatively charged spherical particle (indicated by negative signs just inside the border of the particle) is surrounded by a diffuse ion atmosphere. Positive solution ions, called *counterions* [valence $(z) = +1$ in this case], are attracted to the particle surface, and their concentration is highest just adjacent to the surface. Negative solution ions, called *coions* ($z = -1$ in this case), are repelled from the surface, and their lowest concentration is just adjacent to the particle surface. As the distance from the particle surface increases, the concentration of both the counter- and coions approaches the same value, because the solution cannot have a net charge. This is known as the *law or principle of solution electroneutrality*. Also, the sum of the counterion excess (the area between the top curve and the horizontal line) and the coion defect (the area between the horizontal line and the lower curve) is the same

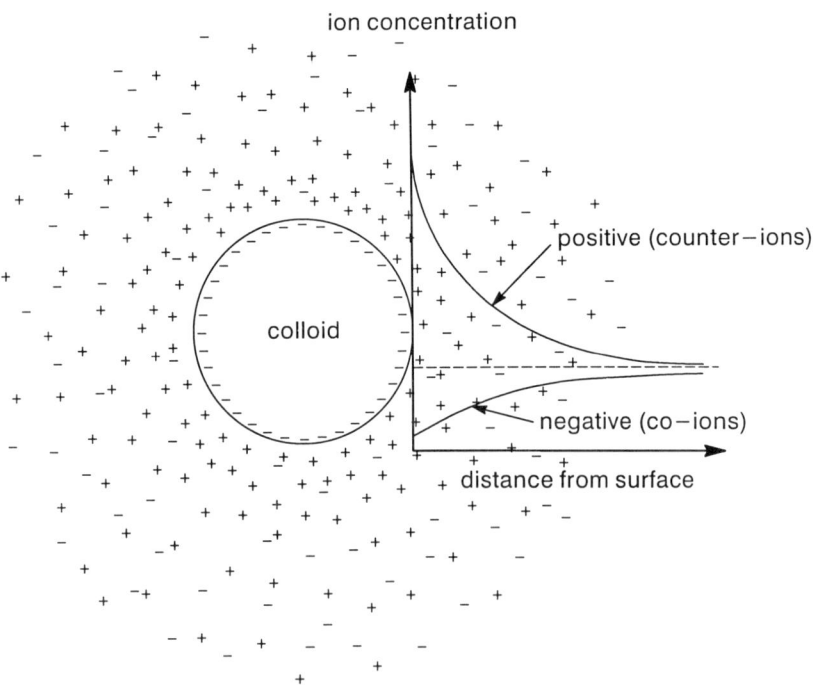

Figure 3.6. Negatively charged particle surrounded by diffuse layers of counterions and coions.

(but opposite in sign) as the *surface excess* of the particle negative charge — thus fulfilling the requirement that the whole solution (particle surface plus solution) be electrically neutral.

Looking a bit closer at the shape of the positive counterion concentration curve, we note that it has the form of a decaying exponential. This is sometimes called an atmospheric distribution because it resembles the distribution of gas molecules around the earth. Analogously to the electrostatic attraction of ions to the colloid surface, gas molecules in the earth's atmosphere are attracted to earth by gravity. In the absence of diffusion, the gas molecules would all collapse to the earth's surface. But diffusion acts to spread the gas molecules out from the surface of the earth; hence, gas molecules are acted upon by the opposing forces of gravity and diffusion. As we show in Exercise 5.1, an approximate relationship for the concentration (pressure) of gases in an isothermal atmosphere is

$$\frac{p}{p_0} = \exp\left(-\frac{gm_a z}{RT}\right) \tag{3.4}$$

where p_0 is the gas pressure at the surface, g is the acceleration of gravity, m_a is the molar mass of the gas, z is the height above the surface, R is the gas constant, and T is the absolute temperature.

3.5.1. Debye–Hückel Model of the Double Layer

The fundamental equation relating electrical potential to charge in the diffuse ion atmosphere is called the Poisson equation:

$$\nabla^2\Psi = \frac{\partial^2\Psi}{\partial x^2} + \frac{\partial^2\Psi}{\partial y^2} + \frac{\partial^2\Psi}{\partial z^2} = -\frac{\rho}{\epsilon\epsilon_0} \tag{3.5}$$

Here Ψ is the electric field potential (V), ρ is the charge density (C/cm^3), ϵ_0 is the permittivity in a vacuum, and ϵ is the relative permittivity of the medium (dimensionless). For a spherical particle, Eq. 3.5 should be solved in its spherical coordinate form. However, if we assume that the thickness of the counterion atmosphere is small relative to the diameter (or radius of curvature) of the colloid, we can argue that the one-dimensional form of Eq. 3.5 is a good approximation of reality:

$$\frac{d^2\Psi}{dx^2} = -\frac{\rho}{\epsilon\epsilon_0} \tag{3.6}$$

Here the x direction is normal to the particle surface. Unfortunately, the charge density is a function of the potential itself, so that we need further equations to relate Ψ and ρ. The classic Boltzmann distribution relates ion concentration to potential:

$$\frac{n_i}{n_{i0}} = \exp\left(\frac{-z_i e \Psi}{kT}\right) \tag{3.7}$$

where n_i is the number of ith-type ions per cubic centimeter, n_{i0} is the number of ith-type ions per cubic centimeter at an infinite distance from the surface, z_i is the valence of the ith-type ion, and e is the electron charge. A simple relationship exists between charge density and ion concentration:

$$\rho = \sum_i z_i e n_i \tag{3.8}$$

Combining Eqs. 3.6–3.8 yields

$$\frac{d^2\Psi}{dx^2} = -\frac{1}{\epsilon\epsilon_0} \sum_i z_i e n_{i0} \exp\left(\frac{-z_i e \Psi}{kT}\right) \tag{3.9}$$

As the reader will recall from calculus, exponential functions can be expressed as an infinite series of terms. For example,

$$e^x = 1 + x + \frac{x^2}{2!} + \frac{x^3}{3!} + \cdots \tag{3.10}$$

We recognize that if the argument (x) of the exponential is small, the infinite series can be represented by just the first few terms in the series (recall the Taylor series example, Section 1.4). In fact, if the potential is low, that is,

$$\left|\frac{z_i e \Psi}{kT}\right| \ll 1 \qquad \text{or} \qquad |z_i e \Psi| \ll kT \tag{3.11}$$

we retain only the first two terms in the series and Eq. 3.9 becomes

$$\frac{d^2\Psi}{dx^2} = -\frac{1}{\epsilon\epsilon_0} \sum_i \left(z_i e n_{i0} - \frac{z_i^2 e^2 n_{i0}\Psi}{kT}\right) \tag{3.12}$$

However, note that because of the condition of solution electroneutrality, the first term in the summation must equal zero. Hence, Eq. 3.12 simplifies to

$$\frac{d^2\Psi}{dx^2} = \frac{e^2\Psi}{\epsilon\epsilon_0 kT} \sum_i z_i^2 n_{i0} \tag{3.13}$$

With the boundary conditions that $\Psi = \Psi_0$ at $x = 0$ and $d\Psi/dx = 0$ as $x \to \infty$, the solution of Eq. 3.13 is

$$\Psi = \Psi_0 \exp(-\kappa x) \tag{3.14}$$

Table 3.4. Constants Used in Double-Layer Calculations

Description	Symbol	Value
Proton change	e	1.60219×10^{-19} C
Boltzmann constant	k	1.38066×10^{-23} JK^{-1}
Avogadro's number	L_A	6.02205×10^{23} mol^{-1}
Permittivity in vacuum	ϵ_0	8.854×10^{-14} C^2 cm^{-1} J^{-1}
Relative permittivity of H$_2$O at 25°C	ϵ_{H_2O}	80

where κ is given by

$$\kappa^2 = \frac{e^2 \sum_i z_i^2 n_{i0}}{\epsilon \epsilon_0 kT} \tag{3.15}$$

Note from Eq. 3.14 that κ must have units of inverse length. Also recall from our discussion of Example 2.1 in Chapter 2 that $1/\kappa$ is apparently a "characteristic" length of some kind; the distance from the colloid surface at which the potential has decayed to 37% of the surface potential. Traditionally, $1/\kappa$ is called the *double-layer thickness*. Table 3.4 provides values for several of the constants required in double-layer problems.

Recall that Eq. 3.14 is the solution of the Poisson equation only for the approximation expressed by Eq. 3.11 (i.e., small potentials). This is known as the *Debye–Hückel* approximation or theory. The assumption of low potentials can be relaxed in an analysis known as the *Gouy–Chapman* theory. The derivation of the Gouy–Chapman theory is not conceptually difficult—but it is long. Thus, only the final results are presented here (Hiemenz, 1977). The analogous result to Eq. 3.14 is

$$\frac{\exp\left(\frac{ze\Psi}{2kT}\right) - 1}{\exp\left(\frac{ze\Psi}{2kT}\right) + 1} = \frac{\exp\left(\frac{ze\Psi_0}{2kT}\right) - 1}{\exp\left(\frac{ze\Psi_0}{2kT}\right) + 1} \exp(-\kappa x) \tag{3.16}$$

This is written more compactly as

$$\gamma = \gamma_0 \exp(-\kappa x) \tag{3.17}$$

where γ and γ_0 are defined by Eqs. 3.16 and 3.17. Note also that these two equations hold only for symmetrical electrolytes (Section 3.1).

Example 3.2. Calculate the double-layer thickness around colloids in 1×10^{-3} M and 1×10^{-2} M solutions of sodium chloride at 25°C using the Debye–Hückel approximation.

SOLUTION. Substituting into Eq. 3.15 for the 10^{-3} M sodium chloride

$$\kappa^2 = \frac{(1.6 \times 10^{-19}\,\text{C})^2\{(-1)^2 \times 1 \times 10^{-3} + (1)^2 \times 1 \times 10^{-3}\}\dfrac{\text{mol}}{\text{L}} \times \dfrac{\text{L}}{1000\,\text{cm}^3} \times \dfrac{6.02 \times 10^{23}}{\text{mol}}}{80 \times 8.854 \times 10^{-14}\dfrac{\text{C}^2}{\text{J-cm}} \times 1.38 \times 10^{-23}\dfrac{\text{J}}{\text{K}} \times 298\,\text{K}}$$

which yields a double-layer thickness:

$$\frac{1}{\kappa} = 9.72 \times 10^{-7}\,\text{cm} = 97.2\,\text{Å}$$

In a similar manner, for the 0.01 M NaCl solution we find:

$$\frac{1}{\kappa} = 3.07 \times 10^{-7}\,\text{cm} = 30.7\,\text{Å}$$

By increasing the ionic strength, the double-layer thickness is decreased. This is sometimes referred to as "double-layer compression." Equation 3.14 is plotted below for the κ values calculated above.

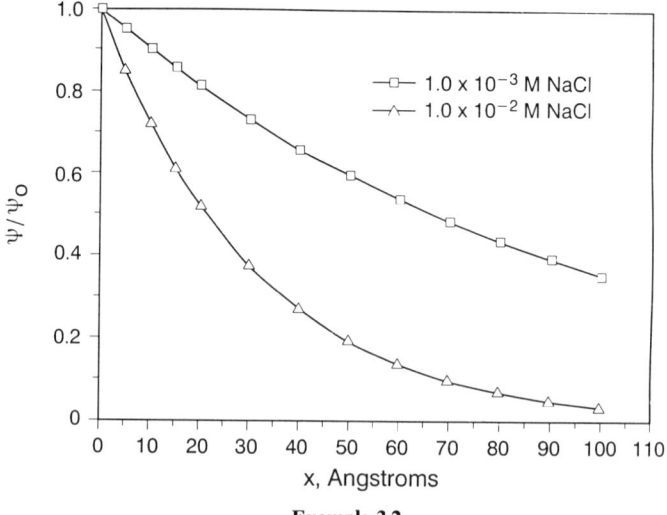

Example 3.2

Note also that these double-layer thicknesses are the same order of magnitude as the smaller colloid particles mentioned in Section 3.1. The exercises will demonstrate the importance of the valence of the salt ions in affecting κ and the double-layer profile.

Using the approach described above, we can also calculate the surface *charge density* or *surface excess*, σ:

$$\sigma = \int_0^\infty \rho dx \qquad (3.18)$$

Substituting Eq. 3.6,

$$\sigma = -\epsilon\epsilon_0 \int_0^\infty \frac{d^2\Psi}{dx^2} dx \qquad (3.19)$$

Integrating Eq. 3.19 we find:

$$\sigma = -\epsilon\epsilon_0 \frac{d\Psi}{dx}\bigg|_0^\infty = -\epsilon\epsilon_0\left(\frac{d\Psi}{dx}\bigg|_0\right) \qquad (3.20)$$

since $d\Psi/dx$ is zero at $x = \infty$. However, the term in parentheses is easily evaluated from Eq. 3.14:

$$\frac{d\Psi}{dx}\bigg|_0 = -\kappa\Psi_0 \qquad (3.21)$$

Substituting Eq. 3.21 into Eq. 3.20 we find:

$$\sigma = \epsilon\epsilon_0\kappa\Psi_0 \qquad (3.22)$$

which shows that the surface charge density is simply related to the surface potential and double-layer thickness.

3.5.2. Overlapping Double Layers and Interparticle Repulsion

Although the calculation of potential in the region of the particle is illuminating, we are ultimately interested here in the interaction of particles. Imagine that as two particles move toward each other, there is a point at which the individual particles begin to sense each other's double layers. Continuing with our previous simplification of the particle surface as a flat surface, we can draw a picture of the electrical potential between two particles if we assume that the overall potential is the sum of the potential of the two particles (Figure 3.7). Hiemenz (1977) shows that the force of repulsion between the two surfaces

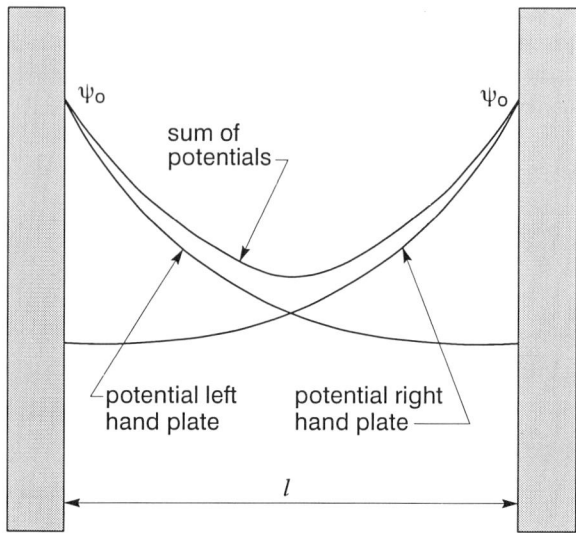

Figure 3.7. Overlap of two-plane double layers.

(symmetrical electrolyte) depends on separation distance l as follows:

$$F_R \simeq 64 n_0 k T \gamma_0^2 \exp(-l\kappa) \tag{3.23}$$

Customarily, the energy of interaction rather than the interparticle force is plotted versus the separation distance. This energy is given by

$$d\phi_R = -F_R dl \tag{3.24}$$

where the negative is added because the interaction energy must increase for decreasing separation distance l. Substituting Eq. 3.23 into Eq. 3.24 yields

$$d\phi_R = -64 n_0 k T \gamma_0^2 \exp(-l\kappa) dl \tag{3.25}$$

and integrating with the limit that $\Phi_R = 0$ at $l = \infty$ results in

$$\phi_R = \frac{64 n_0 k T \gamma_0^2}{\kappa} \exp(-l\kappa) \tag{3.26}$$

3.5.3. Van der Waals Forces and the DLVO Theory

Because of the very small size of colloids, they are influenced by forces that apply to molecules and atoms. One of these fundamental forces is called the *van der Waals* force, which is caused by the interaction of induced and permanent dipole moments in the constituent molecules of two approaching colloidal particles. This attractive force between colloidal particles can be expressed as an energy of interaction, as was done for the repulsive electrostatic forces studied in Section 3.5.2. The attractive van der Waals energy is given as (Hiemenz, 1977)

$$\phi_A = -\frac{A}{12\pi} l^{-2} \tag{3.27}$$

where l is the interparticle separation distance and A is the *Hamaker* constant,

$$A = \left(\frac{\rho L_A \pi}{m_a}\right)^2 \beta \tag{3.28}$$

Here ρ is the particle density, L_A is the Avogadro number, m_a is the molar mass of the particle material, and β is a composite van der Waals parameter related to the induced or permanent dipole moments.

The *DLVO Theory*, named for the Dutch and Russian scientists credited with its development (Derjaguin, Landau, Vervey, and Overbeek), incorporates the attractive and repulsive forces examined in this chapter into an overall picture of the factors involved in the stability of lyophobic (and hydrophobic) colloids. For example, Eqs. 3.26 and 3.27 can be combined to yield an estimate of the overall energy of interaction between two colloidal particles:

$$\phi_{net} = \phi_R + \phi_A = \frac{64 n_0 k T \gamma_0^2}{\kappa} \exp(-l\kappa) - \frac{A}{12\pi} l^{-2} \tag{3.29}$$

Equation 3.29 is plotted in Figure 3.8 for different values of κ, while holding constant the Hamaker constant A and the surface potential Ψ_0. Figure 3.8 demonstrates an important result because of its strong implications for the stability of colloids in the natural environment and pollution control processes — situations ranging from the deposition of sediments in estuaries to the treatment of drinking water. First note that positive values of ϕ_{net} represent energy required to overcome net repulsive forces between the particles; negative values of ϕ_{net} represent net attractive forces. The peaks in positive net interaction energy are typically interpreted as an energy barrier to particle contact, agglomeration, and flocculation. If two particles have enough kinetic energy to surmount the peak in positive net interaction energy, they may get close enough that their net interaction energy is negative. The particles should then come in contact and remain agglomerated. However, if two particles do not have enough kinetic energy, they will be at such separation distances where interaction energy curve will be positive and the collision will be unsuccessful. Therefore, the curves in Figure 3.8 describe the stability of a colloidal

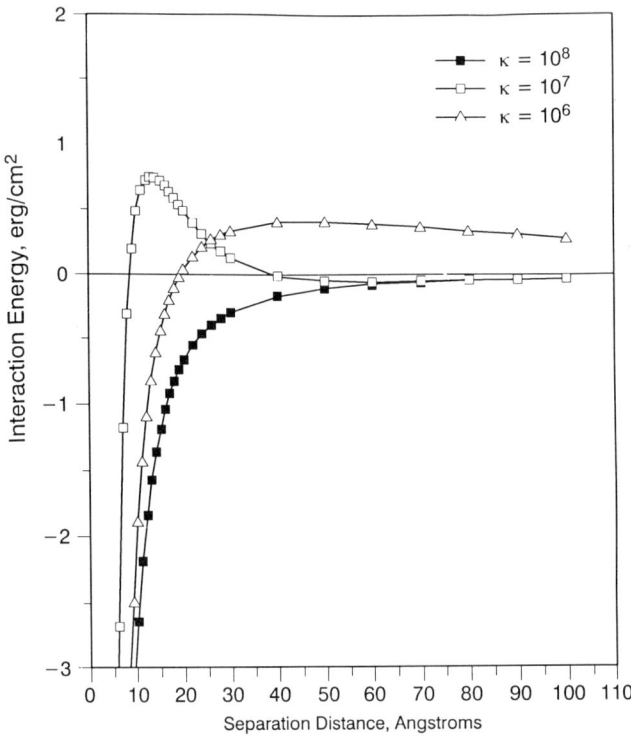

Figure 3.8. Net interaction energy between two flat plates for different values of κ. $A = 10^{-12}$ erg and $\psi_0 = 103$ mV.

dispersion. The other important observation about Figure 3.8 is the effect of the double-layer thickness $(1/\kappa)$ on the net energy barrier. Note that as κ increases and the double-layer thickness decreases, the net interaction energy curves decrease. In fact, for the largest κ value shown, it appears that the net interaction energy is negative (attractive) at all separation distances. In a dispersion of particles, we must imagine that there is a distribution of particle kinetic energies. The kinetic energies are a result of Brownian motion, fluid shear, and even sedimentation (Section 3.7). At any instant, only a certain fraction of all particle collisions will be energetic enough to result in agglomeration. However, Figure 3.8 suggests that if κ could be increased, the energy barrier would be lowered, and a higher proportion of collisions would be successful; therefore, as κ increases, the dispersion becomes less stable. It has long been known that where sediment-carrying rivers enter the sea, there is a zone of high sedimentation. We now understand that the increased ionic strength (salt concentration), resulting from the mixing of fresh and sea waters, compresses the colloid (sediment) double layers, decreases colloidal stability,

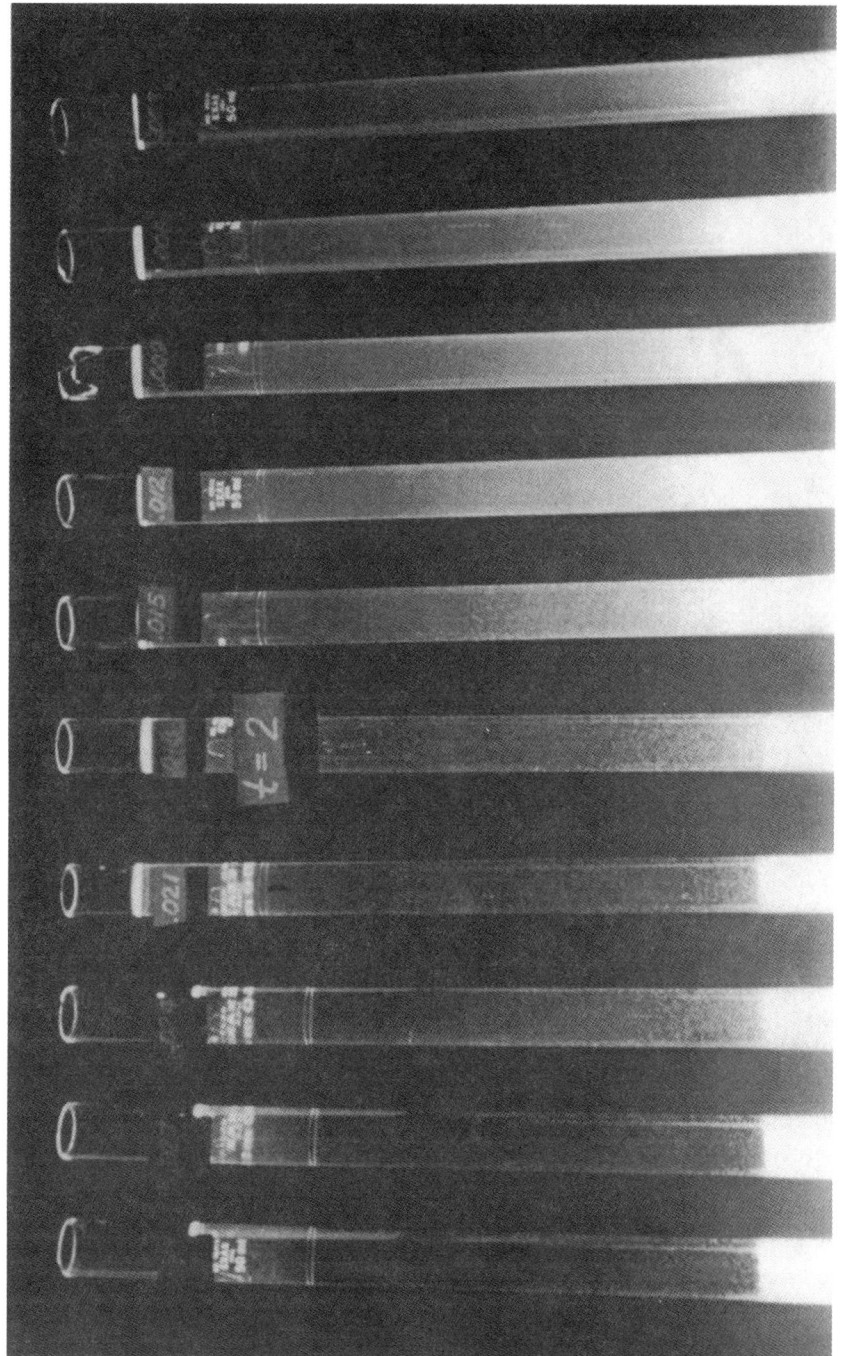

Figure 3.9. Flocculation series test. The tubes contain 50 mg/L Na-montmorillonite. The concentration of NaCl decreases from left to right in uniform increments from 0.03 M to 0.003 M. The reaction time is 2 h. *Source:* Clark (1978).

enhances flocculation, and results in sedimentation of the flocculated sediment material (see also Exercise 3.5).

This phenomenon can be studied in the laboratory and quantified to a certain extent with a *flocculation series* test. In this simple test, increasing concentrations of salt are added to a series of nestler tubes containing a stable colloidal solution. Over time, the effect of higher salt concentrations is obvious in the speed of flocculation and floc growth. For very long experiments, the effect of salt concentration on floc sedimentation and compaction in the bottom of the tube can be observed. An example of such a test is shown in Figure 3.9.

Note that we have maintained constant values of Ψ_0 and A in Figure 3.8. The exercises will show that varying these parameters also affects colloid stability.

3.6. THE SCHULZE–HARDY RULE

Section 3.5 provides background for understanding one of the most important qualitative rules of colloid science. Note in Eq. 3.15 that κ depends even more strongly on ion valence than on ionic concentration. This indicates that two particle-containing electrolyte solutions having the same molar salt concentration, but different salt ion valences, may have much different colloid stability. In 1900, Schulze formulated his observations of the importance of counterion valence on colloid stability and flocculation into what has become known as the *Schulze–Hardy Rule* (van Olphen, 1977):

> The ability of a particular salt to induce coagulation is determined by the valency of one of its ions. This ion is the one with charge opposite to that of the particle.

The Schulze–Hardy Rule is only strictly true for salt solutions of so-called indifferent electrolytes, that is, salts that do not form any specific interaction with the colloid surface.

3.7. PARTICLE COLLISION AND FAST COAGULATION

3.7.1. The General Dynamic Equation

The previous sections established a foundation for understanding why a particular dispersion of small particles will be stable or unstable. But how quickly will a certain instability condition make itself felt as an actual change in the composition of the dispersion? Imagine a particle dispersion which is initially a monodispersion. These equal-sized particles are often called *primary particles*. If particle contacts are frequent (unstable dispersion), each contact decreases the number of free primary particles, while simultaneously, larger particle aggregates (sometimes called flocs in the study of hydrosols) are formed. Furthermore, the aggregates can contact remaining primary particles and other aggregates. This process occurs in a variety of environmental processes: coagulation of aerosols, growth of water droplets in a cloud, coagulation of particles in lakes, flocculation of river sediments in estuaries, and coagulation of destabilized particles in water-treatment plants.

We will develop an equation describing the change in concentration of particles of all sizes because of coagulation. But first, we need a kinetic expression governing the rate of particle impact during *fast coagulation*. Fast coagulation occurs in very unstable dispersions, where there is essentially no energy barrier to interparticle contacts. Under these conditions, coagulation rate is governed by classic bimolecular kinetics (Section 9.3):

$$N_{ij} = \beta(r_i, r_j)n_i n_j \tag{3.30}$$

Here N_{ij} is the rate of collisions between spherical particles of size classes i and j ($\#\text{-}T^{-1}\text{-}L^{-3}$), β is the collision frequency function for class i and j particles, r_i and r_j are the radii of class i and j particles, and n_i and n_j are the concentrations of class i and j particles ($\#/L^3$). If we consider an interaction between two specific size classes, β is a constant and Eq. 3.30 simply says that the collision rate is proportional to the product of the concentration of the *two* interacting particle classes (*bimolecular kinetics*).

Now we need to describe the interaction of particles of all different sizes with each other. The equation is known as the *general dynamic equation* (GDE) of coagulation:

$$\frac{dn_k}{dt} = \frac{1}{2} \sum_{V_i+V_j=V_k} \beta(r_i, r_j)n_i n_j - n_k \sum_{i=1}^{\infty} \beta(r_i, r_k)n_i \tag{3.31}$$

Equation 3.31 is actually an infinite system of discrete, nonlinear differential equations. The left-hand side of the equation represents the rate of change of concentration of the k-class particles. The first term on the right-hand side of Eq. 3.31 sums the rates of formation of k-class particles from all collisions of particles smaller than the k-size particles; the summation is over all particle classes i and j such that the volume of the sum of the i and j class particles equals the volume of the k-class particle:*

$$V_k = V_i + V_j \tag{3.32}$$

In the particle conjunction represented by Eq. 3.32, volume is conserved. We can imagine two types of spherical particle conjunctions, depending upon

---

*The factor $\frac{1}{2}$ in front of the first term on the right-hand side of Eq. 3.31 is required because of our method of counting collisions. Since we let i and j take on all values such that the volume of the i- and j-class particles equal the volume of the k-class particle, we end up double counting collisions.

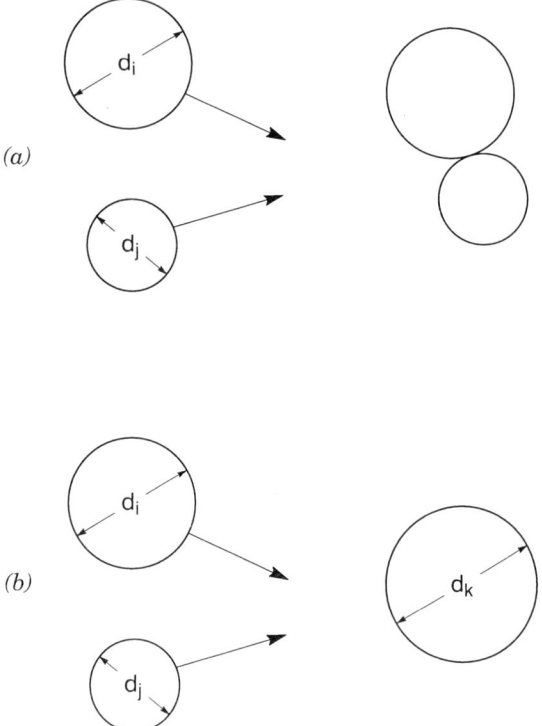

Figure 3.10. Two possible particle collision outcomes: (*a*) solid particles and (*b*) liquid particles.

whether the particles are solids or liquids. If the particles are solids, the only imaginable result of a conjunction between a class *i* and a class *j* particle is shown in Figure 3.10*a*. Note that while we would expect conservation of volume, the diameter of the aggregate is difficult to characterize. For liquid drop collision, however, we can imagine that the final result is a spherical liquid particle with diameter equal to a sphere having the same volume as the sum of class *i* and *j* particles (Figure 3.10*b*). As we will see shortly, the theories for the mechanisms of particle transport and collision presented here require spherical particles. Therefore, even for solid particles, we assume that a conjunction results in an equivalent spherical particle. This is known as the *coalescing drop assumption.*

We have said little about the particular particle size classes under consideration. In fact, the particle size classes can have any values. But a pleasing simplification comes about if we imagine that each class size (volume) is an integer multiple of the smallest, base, or primary particle size (volume). If this is true, we can simplify the limit in the first summation on the right-hand side of Eq. 3.31 as

$$\frac{1}{2} \sum_{V_i + V_j = V_k} \beta(r_i, r_j) n_i n_j = \frac{1}{2} \sum_{i+j=k} \beta(r_i, r_j) n_i n_j \tag{3.33}$$

The second term on the right-hand side of Eq. 3.31 is needed to account for all losses of class k-size particles because of their collisions with any other size particle.

3.7.2. Collision-Frequency Functions

In the environment, particle collisions generally have four causes: laminar shear, turbulent mixing, Brownian motion, and differential sedimentation. Laminar shear refers to particle collisions induced by a uniform laminar shearing motion. Although it does not occur frequently in the environment, the derivation is particularly enlightening and useful. Turbulent flows are most common in nature, and are the significant mechanism for particle collision when particles are larger than about $1 \, \mu m$. Brownian motion, the random colloid motion caused by the uneven bombardment of water molecules, is very significant for particles smaller than about $1 \, \mu m$. Finally, when large particles have large relative settling velocities, the particles can impact one another. This is called differential sedimentation.

To derive β for the case of laminar shear, we must develop the concept of a *collision sphere*. Note in Figure 3.11 that when two spherical particles of radii r_i and r_j make contact, it is possible to imagine a new sphere of radius $r_i + r_j$. This is called the collision sphere. In a collision sphere, one particle must be considered the central particle around which the collision sphere is centered. In Figure 3.11, the class i particle is considered the central sphere. Note that if any class j particle passes across the surface of the collision sphere, contact with the central class i particle is unavoidable. Since the collision sphere is a control volume, we can use the approach described in Section 1.3. If n_j is substituted for density in Eq. 1.11, the following equation results:

$$-\iint_{\text{control surface}} n_j \mathbf{n} \cdot \mathbf{u} \, dA = \frac{\partial}{\partial t} \iiint_{\text{control volume}} n_j \, dV \tag{3.34}$$

Equation 3.34 states that the accumulation of class j particles in the collision-sphere control volume is simply equal to the mass transfer of class j particles across the control surface.* To evaluate the left-hand side of Eq. 3.34, the notation in Figure 3.12 should be studied. Note that flux into the control volume occurs in the northwest and southeast quadrants of the control surface.

---

*Since the concentration units are number per volume, this is not strictly a "mass transfer." Nevertheless, we will continue to refer to it as mass transfer.

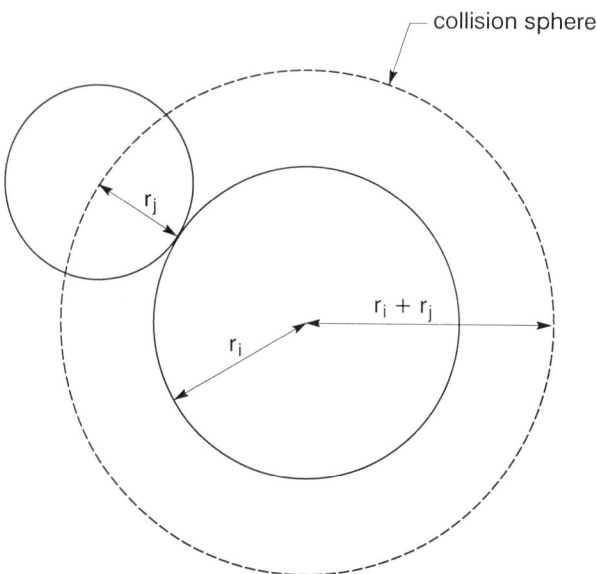

Figure 3.11. Collision sphere formed around a class i particle.

The reason for using a differential strip rather than a differential element is that the double integral in Eq. 3.34 can be reduced to a single integral. In the northwest quadrant detailed in Figure 3.11, the term $\mathbf{n} \cdot \mathbf{u}$ in Eq. 3.34 is given by

$$(\mathbf{n} \cdot \mathbf{u}) = -[(1) \sin \phi]\left[(r_i + r_j) \cos \phi \frac{du}{dx}\right] \tag{3.35}$$

In Eq. 3.35, $(r_i + r_j) \cos \phi(du/dx)$ is the local velocity at the differential element in Figure 3.12. The term $-(1) \sin \phi$ accounts for the dot product of the local velocity with the outward normal vector at the differential element. The term dA in Eq. 3.34 is given by

$$dA = [2(r_i + r_j) \sin \phi][(r_i + r_j)d\phi] \tag{3.36}$$

Combining Eqs. 3.34–3.36, and noting that the flux into the southeast quadrant will have the same magnitude as the northwest quadrant, we can write the total mass transfer ($\#/T$) into the control volume as

$$\begin{array}{c}\text{mass transfer into} \\ \text{control volume}\end{array} = 4n_j \frac{du}{dx}(r_i + r_j)^3 \int_0^{\pi/2} \sin^2\phi \cos \phi d\phi \tag{3.37}$$

Evaluating the integral using a table of definite integrals (see Spiegel, 1968)

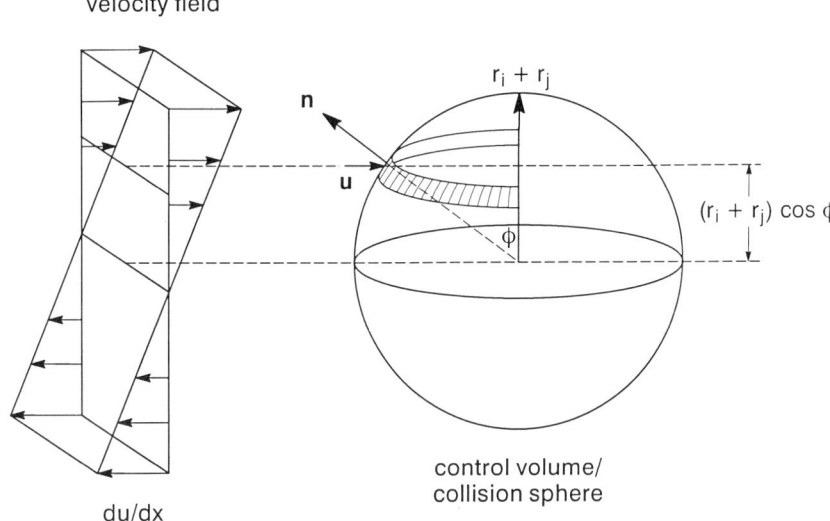

velocity field

$r_i + r_j$

n

u

ϕ

$(r_i + r_j)\cos\phi$

control volume/
collision sphere

du/dx

Figure 3.12. Three-dimensional collision sphere in two-dimensional velocity field.

results in

$$\frac{\text{mass transfer into}}{\text{control volume}} = \frac{4}{3}(r_i + r_j)^3 \frac{du}{dx} n_j \tag{3.38}$$

The final trick in this classical analysis is to multiply Eq. 3.38 by the concentration of class i particles, yielding the final result for the number of collisions due to laminar shear:

$$N_{ij} = \frac{4}{3}(r_i + r_j)^3 \frac{du}{dx} n_i n_j \tag{3.39}$$

Comparing Eqs. 3.30 and 3.39, the collision frequency function for *laminar shear* is found to be

$$\beta(r_i, r_j)_{ls} = \frac{4}{3}(r_i + r_j)^3 \frac{du}{dx} \tag{3.40}$$

The collision frequency function for *turbulent mixing* (isotropic turbulence) is derived in Chapter 5. It is given by

$$\beta(r_i, r_j)_t = 1.294(r_i + r_j)^3 \left(\frac{\varepsilon}{\nu}\right)^{1/2} \tag{3.41}$$

Here ε is the local average unit mass energy dissipation rate in the turbulence (L^2/T^3), and v is the kinematic viscosity (L^2/T). The β for *Brownian motion* is

$$\beta(r_i, r_j)_{bm} = \frac{2kT}{3\mu}\left(\frac{1}{r_i} + \frac{1}{r_j}\right)(r_i + r_j) \tag{3.42}$$

which is derived in Chapter 6. Finally, in the exercises at the end of this chapter, the β for *differential sedimentation* is derived:

$$\beta(r_i, r_j)_{ds} = \pi(r_i + r_j)^2 |(v_i - v_j)| \tag{3.43}$$

Here v_i and v_j are the settling velocities of the class i and j particles.

At this point, the reader may be wondering which β to use for modeling of coagulation. Equations 3.40–3.43 indicate different dependencies on the colli-

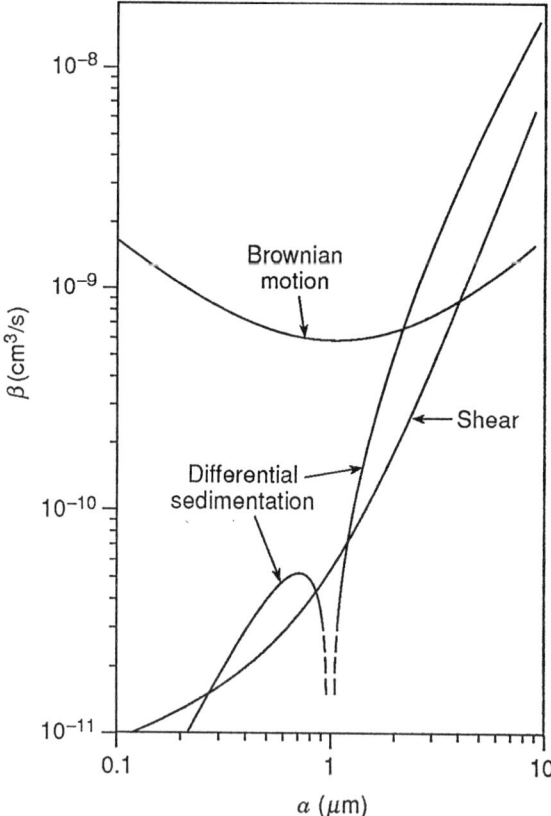

Figure 3.13. Comparison of collision frequency functions for a 1-μm-radius particle interacting with other particles in air. The x axis refers to the size of the other particle. *Source*: Reproduced from Friedlander, *Smoke, Dust and Haze*, John Wiley and Sons, New York, 1977. Used with permission of S.K. Friedlander, 1995.

sion-sphere dimension, temperature (viscosity), energy-dissipation rate, and even the relative sedimentation velocities of the particles. Figure 3.13 is a plot of the β functions for Brownian motion, differential sedimentation, and laminar shear in air.

From Figure 3.13 it becomes obvious that the dominant mechanism depends fairly strongly on relative particle size. For small-particle interactions, the dominant mechanism is Brownian motion; for larger particles, the dominant mechanisms are shear and differential sedimentation. Similar plots could be made for different velocity gradients (or energy-dissipation rates) and temperature; however, for normal ranges of these parameters, the aerosol curves have the general shape indicated in Figure 3.13. (Exercise 3.3 develops a similar plot for a hydrosol.)

There are of course regions shown in Figure 3.13 where two or more collision-frequency mechanisms have about the same magnitude. The conventional assumption is that the different mechanisms are independent of one another, and can therefore be added. In other words, in the GDE (Eq. 3.31), β can be considered the sum of collision-frequency mechanisms.

One final comment on Figure 3.13 concerns the apparent minimum collision frequency around 1 μm. One interpretation of the minimum is that particles in this range of sizes are the most difficult to coagulate. This observation is surprisingly universal: we will find that particles of this approximate size are the most difficult to remove by conventional mechanisms like coagulation and filtration.

3.7.3. Simplified Discrete Particle Dynamics

Although the discrete form of the GDE can be solved using numerical methods, an approximate solution can be seen for the case of Brownian motion (and laminar shear—see Exercise 3.4) if an assumption is made about the interacting particles. Note that in Eq. 3.42, if the i and j class particles are required to be the same size, $i = j$ and Eq. 3.42 reduces to

$$\beta = \frac{8kT}{3\mu} \tag{3.44}$$

Furthermore, the GDE reduces to

$$\frac{dn_k}{dt} = \frac{4kT}{3\mu} \sum_{i+j=k} n_i n_j - \frac{8kT}{3\mu} n_k \sum_{i=1}^{\infty} n_i \tag{3.45}$$

Equation 3.45 can now be summed over all values of k:

$$\sum_{k=1}^{\infty} \frac{dn_k}{dt} = \frac{d}{dt} \sum_{k=1}^{\infty} n_k = \frac{4kT}{3\mu} \sum_{k=1}^{\infty} \sum_{i+j=k} n_i n_j - \frac{8kT}{3\mu} \sum_{k=1}^{\infty} n_k \sum_{i=1}^{\infty} n_i \tag{3.46}$$

We define the total number of particles at any time as N_∞:

$$N_\infty = \sum_{i(\text{or } j \text{ or } k)=1}^{\infty} n_{i(\text{or } j \text{ or } k)} \tag{3.47}$$

It is easy to see that the product of summations in the last term on the right-hand side of Eq. 3.46 is equal to N_∞^2. However, by carefully expanding the double summation in the first term on the right-hand side of Eq. 3.46, the reader will find that this too is equal to N_∞^2. Hence, Eq. 3.46 reduces to

$$\frac{dN_\infty}{dt} = -\frac{4kT}{3\mu} N_\infty^2 \tag{3.48}$$

Equation 3.48 is easily integrated:

$$N_\infty = \frac{N_\infty(t = 0)}{1 + \{[4kTN_\infty(t = 0)]/[3\mu]\}t} \tag{3.49}$$

Note that the term in braces is a constant for a specific temperature and initial total particle concentration $N_\infty(t = 0)$. Note also that this group of parameters

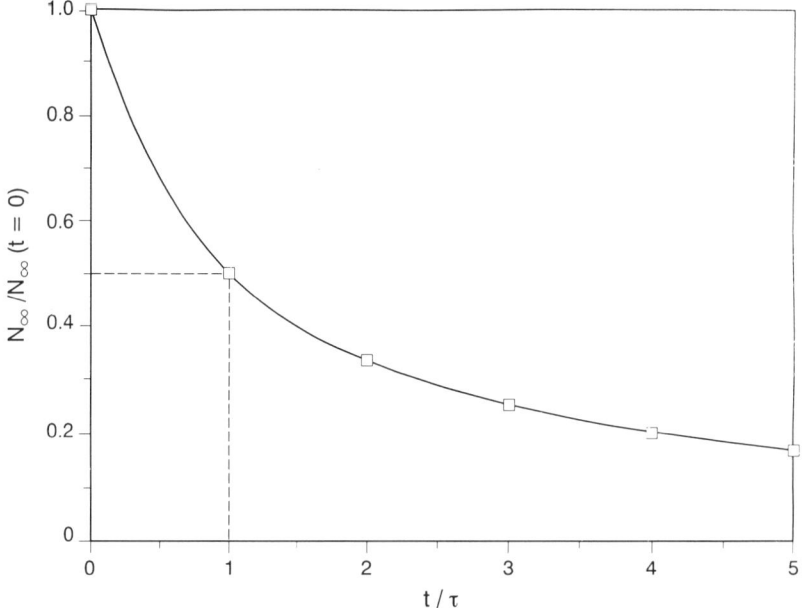

Figure 3.14. Decrease in total particle concentration for simplified Brownian kinetics. $\tau = 3\mu/[4kTN_\infty(t = 0)]$.

has the units of inverse time; it is the inverse of the characteristic time of Brownian coagulation. This characteristic time will be simply called τ. Figure 3.14 is a dimensionless plot of $N_\infty/N_\infty(t = 0)$ versus t/τ. At the characteristic time ($t = \tau$ or $t/\tau = 1$), the total concentration of particles has been reduced to one half of its initial value. The exercises at the end of this chapter includes an investigation of the dynamics of the concentration of individual particle size classes, which will reveal some additional interesting phenomena.

3.8. SLOW COAGULATION

The derivations of collision frequency performed in Section 3.7 were for the case of fast coagulation, where there is essentially no barrier to particle collision. In reality, there is almost always some barrier to coagulation. For example, in Section 3.5.3, we found that even though two particles may be on a collision course, they may not have enough kinetic energy to surmount the interaction energy barrier, such as that shown in Figure 3.8. In fact, the coagulation "reaction" can be thought to have an "activation energy" barrier analogous to the activation energy of chemical reactions. This situation has been analyzed for coagulation of hydrosols and aerosols by the Brownian motion mechanism. The ratio of the number of successful slow coagulation collisions to the number predicted by fast coagulation conditions is known as the *stability ratio*, $1/W$. For particles of the same radius R:

$$W = 2R \int_{2R}^{\infty} \exp\left(\frac{\phi_{net}}{kT}\right) r^{-2} dr \tag{3.50}$$

This expression could be evaluated with the aid of plots like Figure 3.8. Note that since $1/W$ is the fraction of successful collisions, we might also be able to estimate W from experimental determination of the kinetics of Brownian coagulation. The fraction of successful collisions is also called the collision efficiency in colloid science, and is not to be confused with the collision frequency defined earlier.

Note that in addition to the electrostatic energy barrier to coagulation just described, two particles can also be prevented from colliding by *hydrodynamic retardation*. When two particles approach each other in a fluid, one can imagine that as the particles become very close, fluid must be squeezed out from between them. This can be considered a viscous retardation and it is known as hydrodynamic retardation. Since the viscosity of air is much less than that of water, this type of retardation is generally of much less concern in modeling the coagulation of aerosols.

Progress has been made in understanding the so-called hydrodynamic interactions in hydrosols. For example, Adler (1981) studied the interaction of two spheres in a laminar shear flow. Based on Adler's work, Han and Lawler (1992) constructed Figure 3.15, which estimates the collision efficiency based

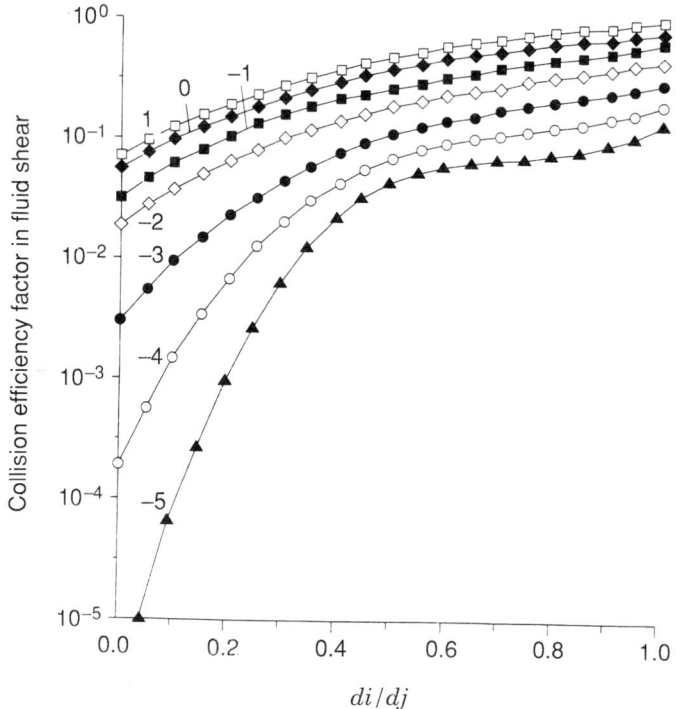

Figure 3.15. Effect of relative particle size and attraction group H_A on collision efficiency in laminar shear. Numbers on curves are $\log H_A$. *Source*: Reproduced from *Journal AWWA*, **84**, 10, by permission. Copyright © 1992, American Water Works Association.

on the effects of (attractive) van der Waals forces and hydrodynamic interactions. In the figure, the x axis is the diameter ratio of the interacting spherical particles. The curves represent different values of the log of the dimensionless *attraction or adhesion group*:

$$H_A = \frac{A}{36\mu r_i^3 \dfrac{du}{dx}} \qquad (3.51)$$

Here A is the Hamaker constant (Section 3.5.3), μ is the viscosity, du/dx is the velocity gradient (Section 3.7.2), and r_i can be interpreted as the radius of the central particle in the collision. The attraction group is essentially the ratio of the attractive and viscous forces; attractive forces enhance the collision efficiency, and viscous forces diminish the collision efficiency. Note that as the Hamaker constant increases, attractive forces dominate, and the collision efficiencies are greatest at any size ratio; as viscous forces become larger and $\log H_A$ decreases, collision efficiencies also decrease. Note also that for any given H_A, the effect of hydrodynamic retardation decreases as the interacting particles approach the same size.

The sum of the effects of electrostatic repulsion, van der Waals attraction, and hydrodynamic retardation determine the kinetic stability of a colloidal dispersion. In reality, there is often not enough data available to accurately determine the influence of each on collision efficiency. One approach has been to lump the energy barrier and retardation effects into a single parameter, $\alpha(r_i, r_j)$, which is called the *collision efficiency* or *alpha* function. The collision efficiency function has a value between one and zero; it describes the probability of success of a collision between class i and class j particles. The GDE can then be rewritten as

$$\frac{dn_k}{dt} = \frac{1}{2} \sum_{i+j=k} \alpha(r_i, r_j)\beta(r_i, r_j)n_i n_j - \sum_{i=1}^{\infty} \alpha(r_i, r_k)\beta(r_i, r_k)n_i n_k \qquad (3.52)$$

Note that the alpha value is equal to $1/W$ in the absence of hydrodynamic retardation.

The exercises at the end of this chapter include some approaches to help simplify finding the value of alpha. As suggested above, we find that in some studies, information on the kinetics of coagulation can be used to back-calculate the collision efficiency function (and perhaps the stability ratio). This is one approach to characterizing colloidal stability.

3.9. SUMMARY

This chapter reviewed aspects of colloid and aerosol science relevant to the environmental stability of small charged particles. For hydrosols, we showed that the chemistry (ionic strength, salt composition) of the solution phase has a great deal to do with the stability of colloids. Stability in turn determines whether particles will flocculate. Stability was characterized in two senses: thermodynamic and kinetic stability. Although some colloidal dispersions and aerosols are thermodynamically unstable, the kinetics of coagulation and phase separation can be slow. The rate of coagulation depends on the energy barrier to particle collision, hydrodynamic retardation (for hydrosols), and the dominant mechanism of relative particle motion. Because large particles normally settle faster than small particles, flocculation and coagulation can be expected to affect sedimentation and the ultimate disposition of the colloidal material and aerosols.

Colloids are transported in air, groundwater, and surface water. Colloids are major components of many wastewaters and industrial gases. The environmental significance of colloid transport may not depend on the particle itself. Clay particles may be relatively inert; however, important contaminants are often associated with colloids and are thus transported by the colloids. For example, synthetic organic contaminants (like pesticides) adsorb to the surface of clay particles. To a certain extent, the movement of these contaminants in the environment can be related to the movement of the clay colloids in the environment. The same is true for aerosol particles, since important contaminants may associate with or partition into the aerosol particles or droplets. The next chapter focuses on equilibrium adsorption and partitioning phenomena.

EXERCISES

3.1. Calculate the double-layer thickness around colloidal particles in (a) 1×10^{-3} M NaCl and (b) 1×10^{-3} M $CaSO_4$ at 25°C using the Debye–Hückel theory. Discuss whether your results are consistent with the Schulze–Hardy rule.

3.2. Calculate the net interaction energy ϕ_{net} versus separation distance l for colloids using the flat-plate version of the DLVO theory; Ψ_0 is 103 mV, κ is 10^7 cm^{-1}, and the temperature is 25°C. Plot three curves: for $A = 5 \times 10^{-13}$ erg; for $A = 1 \times 10^{-12}$ erg; and for $A = 2 \times 10^{-12}$ erg. Assume a 1:1 electrolyte. You can use the same x axis as Figure 3.8 if you like.

3.3. Create a figure like Figure 3.13 for water. Use the same x-axis range and a velocity gradient of 40 s^{-1}. The temperature is 20°C. Assume a collision efficiency factor of unity and a particle density of 1.5 g/cm^3. Assume Stokes law is okay for modeling sedimentation rates.

3.4. (a) Show how to compute the simplified differential equation for laminar shear coagulation, which is analogous to Eq. 3.48 for Brownian motion. You can assume that all particles are about the same size, $r_i \approx r$. The answer is

$$\frac{dN_\infty}{dt} = -\frac{16}{3}\frac{du}{dx}r^3 N_\infty^2$$

(b) Now note that the total volume of particulate matter (per unit volume of solution) is a constant which can be written as

$$V = \frac{4}{3}\pi r^3 N_\infty = \text{constant}$$

Substitute this expression for V into the differential equation you derived for laminar shear coagulation and show the following:

$$\frac{dN_\infty}{dt} = -\frac{4V}{\pi}\frac{du}{dx}N_\infty$$

What is the integrated form of this equation for the initial condition $N_\infty = N_\infty(0)$ at $t = 0$?

3.5. In their study of coagulation in estuaries, Edzwald et al. (1974) performed experiments on the coagulation of kaolinite solutions under conditions of various ionic strength. This was thought to simulate the coagulation of natural river suspensions in saline estuaries. An electrical sensing zone instrument (the Coulter Counter) was used to determine total particle concentrations in stirred flocculation vessels over time. Assuming "slow coagulation," the authors attempted to use a modified version of the first-order differential equation presented in Exercise 3.4 which contained the collision efficiency factor α (recall Section 3.8):

$$\frac{dN_\infty}{dt} = -\alpha \frac{4V}{\pi} \frac{du}{dx} N_\infty$$

(a) What is the integrated form of this equation for the initial condition $N_\infty = N_\infty(0)$ at $t = 0$?

(b) Some of the coagulation data collected in the experiments by Edzwald et al. for synthetic esturine solutions of three different ionic strengths (0.036, 0.087, and 0.343 M) is shown in the figure below. Using the

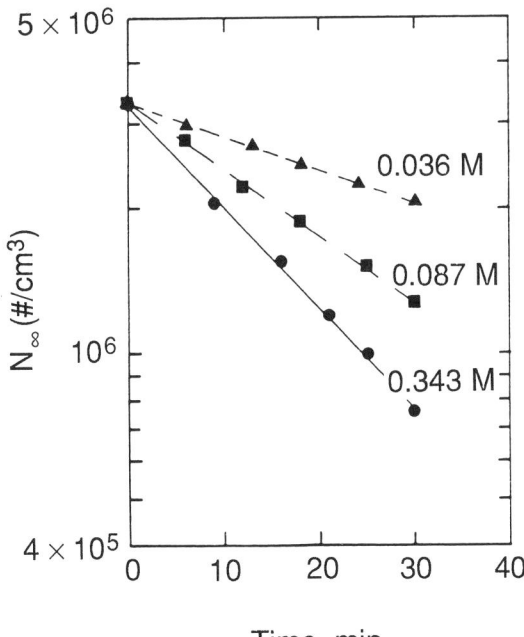

Exercise 3.5 (b). *Source*: Reprinted with permission from Edzwald et al., *Environ. Sci. Technol.*, **8**, 1:60 (1974). Copyright 1974, American Chemical Society.

integrated equation and the experimental results above, compute the apparent collision efficiency values for the three experiments. The characteristic velocity gradient was 52.3 s^{-1} and V was 8.67×10^{-5}. [This is the first opportunity in this book to observe what is generically referred to as "data fitting." In this problem we are in a sense overlaying our model (the straight lines) on the experimental data, and determining the model parameter (α) which results in the best overlap of the model and data.]

(c) Based on your results, what general comments can you make about the effect of ionic strength on coagulation kinetics? Is this consistent with our discussion of colloidal stability?

3.6. Develop the collision frequency function, $B_{ds}(r_i, r_j)$, for the case of differential sedimentation of spheres. It may be helpful to consider the general solution procedure for laminar shear, as well as the following picture of the control volume for differential sedimentation. The sedimentation velocities of the spheres are also shown in the figure.

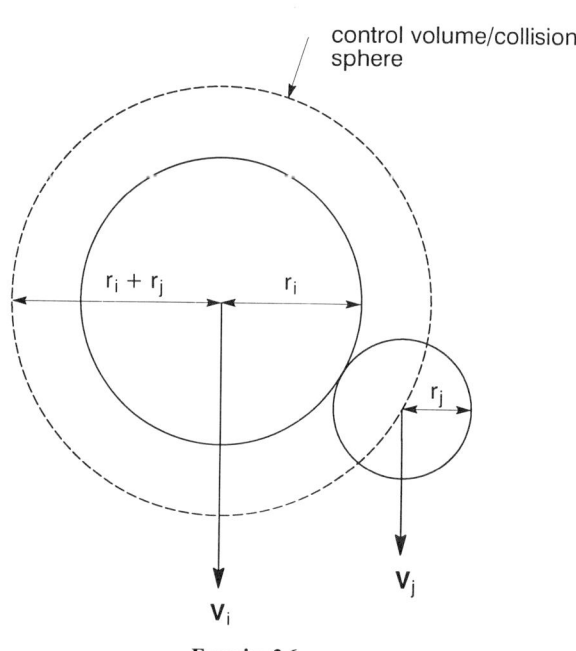

Exercise 3.6

Hint: The correct final answer is provided in Section 3.7.2.

3.7. In Section 3.7.3, we noted that simple expressions for n_1, n_2, and so on could be developed. Imagine that at the beginning of coagulation, all particle are of size class 1 (sometimes called the "primary particles"), that is, $n_1(t = 0) = N_\infty(t = 0)$. Therefore, it follows that at $t = 0$, the concentration of all larger-size-class particles is zero.

(a) Examine Eq. 3.45 and show that the differential equation for n_1 is

$$\frac{dn_1}{dt} = -Kn_1 N_\infty$$

where

$$K = \frac{8kT}{3\mu}$$

Next show that this differential equation integrates to

$$n_1 = \frac{N_\infty(0)}{(1 + t/\tau)^2}$$

where τ is given by

$$\tau = \frac{2}{KN_\infty(0)}$$

Hint: You may need the following relationship from calculus, $a^x = \exp(x \ln a)$.

(b) Likewise show that the differential equation for n_2 is

$$\frac{dn_2}{dt} = \frac{K}{2} n_1 n_1 - Kn_2 N_\infty$$

and that the solution is

$$n_2 = \frac{N_\infty(0)t/\tau}{(1 + t/\tau)^3}$$

Hint: The integrating-factor method of solving differential equations is useful in this problem.

(c) Plot $N_\infty/N_\infty(0)$, $n_1/N_\infty(0)$, and $n_2/N_\infty(0)$ versus t/τ on the same figure. Do the plots make sense? Explain.

REFERENCES

Adler, P. M., "Heterocoagulation in Shear Flow," *J. Colloid Interface Sci.*, **83**, 1: 106–115 (1981).

Atkins, P. W., *Physical Chemistry*, W. H. Freeman and Company, San Francisco, CA, 1978.

Clark, M. M., "Brownian Motion Coagulation of Clay Hydrosols," Master's thesis, Department of Civil Engineering, University of Missouri, Columbia, MO, 1978.

Edzwald, J. K., Upchurch, J. B., and O'Melia, C. R., "Coagulation in Estuaries," *Environ. Sci. Technol.* **8**, 1:58–62 (1974).

Everett, D. H., *Basic Principles of Colloid Science*, Royal Society of Chemistry Paperbacks, London, 1988.

Friedlander, S. K., *Smoke, Dust, and Haze*, Wiley, New York, 1977.

Han, M. and Lawler, D. F., "The Relative Insignificance of G in Flocculation," *J. Am. Water Works Assoc.*, **84**, 10:79–91 (1992).

Hiemenz, P. C., *Principles of Colloid and Surface Chemistry*, Marcel Dekker, Inc., New York, 1977.

Meakin, P., "Fractal Aggregates in Geophysics," *Rev. Geophys.*, **29**, 3:317–354 (1991).

Spiegel, M. R., *Mathematical Handbook of Formulas and Tables* (*Schaum's Outline Series*), McGraw-Hill, New York, 1968.

Stumm, W. and Morgan, J. J., *Aquatic Chemistry: An Introduction Emphasizing Chemical Equilibria in Natural Waters*, Wiley, New York, 1981.

Van Olphen, H., *Clay Colloid Chemistry for Clay Technologists, Geologists, and Soil Scientists*, Wiley, New York, 1977.

Weitz, D. A. and Oliveria, M., "Fractal Structures Formed by Kinetic Aggregation of Aqueous Gold Colloids," *Phys. Rev. Lett*, **52**, 16:1433–1436 (1984).

BIBLIOGRAPHY

Flagan, R. C. and Seinfeld, J. H., *Fundamentals of Air Pollution Engineering*, Prentice-Hall, Englewood Cliffs, NJ, 1988.

Hahn, H. H. and Stumm, W., "The Role of Coagulation in Natural Waters," *Am. J. Sci.*, **268**, 354–368 (1970).

Hinds, W. C., *Aerosol Technology: Properties, Behavior, and Measurement of Airborne Particles*, Wiley-Interscience, New York, 1982.

Probstein, R. F., *Physicochemical Hydrodynamics: An Introduction*, Butterworths, Boston, MA, 1989.

Russel, W. B., Saville, D. A., and Schowalter, W. R., *Colloidal Dispersions*, Cambridge University Press, Cambridge, 1989.

Seinfeld, J. H., *Atmospheric Chemistry and Physics of Air Pollution*, Wiley, New York, 1986.

Swift, D. L. and Friedlander, S. K., "The Coagulation of Hydrosols by Brownian Motion and Laminar Shear Flow," *J. Colloid Interface Sci.*, **19**, 7 (1964).

Verwey, E. J. W. and Overbeek, J. T. G., *Theory of the Stability Lyophobic Colloids*, Elsevier, New York, 1948.

4

ADSORPTION, PARTITIONING, AND INTERFACES

4.1. INTRODUCTION

Chapters 2 and 3 dealt primarily with the dynamics of systems of particles and large molecules. In this chapter we shift to an equilibrium viewpoint and examine how certain molecules preferentially accumulate at interfaces or in one or more phases of multiphase systems. In the equilibrium condition, there is no net mass transfer between two phases, or between one phase and an interface. In effect, we ignore all the system dynamics and focus only on the equilibrium condition. Although ignoring dynamics can sometimes be hazardous, the methods utilized in this chapter have proved extremely useful in analyzing environmental systems. For example, in biology, we have learned that when ether is inhaled, it enters the bloodstream and flows to the brain. There the ether strongly *partitions* into the brain, because of the large lipid and fat content of the brain matter. In effect, the nonpolar ether molecules prefer to be in the lipid phase of the brain rather than the more watery blood phase. The dynamics of this mass transport from the blood into the brain occur rather quickly, practically speaking; for this reason, as the reader may be aware, ether was once used as an anesthetic. As a final example, when it is necessary to remove volatile organic contaminants from air and water waste streams, the stream is often passed through a bed or filter containing activated carbon grains or particles. The carbon is engineered to have an extremely porous structure with thousands of square meters of surface area per gram of carbon. The contaminants are removed from the waste stream through *adsorption* (not absorption) on the internal surface of the carbon grains or particles.* In effect, the contaminants prefer to concentrate at the *interfacial* region between the surface and the gas or liquid phase, rather than stay in either the gas or liquid phase. Although we said that the techniques of this chapter are limited mostly

---

*The term *absorption* is also used in environmental engineering and science. In this book, absorption refers to the transfer of gas into a liquid (e.g., absorption of O_2 by water). Absorption is covered in Section 7.9.

109

to equilibrium configurations, they almost always tell us the natural direction of change in systems not at equilibrium. This is also important information for environmental scientists and engineers.

When equilibrium is studied in physical chemistry and environmental science, the laws of thermodynamics are usually invoked. We do not have the space to develop the main results of thermodynamics, and the reader is referred to some of the references at the end of the chapter for a development of the laws of thermodynamics. Rather, we begin with one of the most important results of thermodynamics, the Gibbs function:

$$dG = V \, dp - S \, dT + \sum_i \mu_i \, dn_i \tag{4.1}$$

where G is the Gibbs function, V is volume, p is pressure, S is entropy, T is temperature, μ_i is the chemical potential of species i, and n_i is the number of moles of species i. One of the most enlightening results of classical thermodynamics is that the Gibbs function for any system tends to minimize itself. In other words, nature dictates that dG tends to negative values. At equilibrium, dG is zero. These two lessons of the Gibbs equation—prediction of the direction of change and the equilibrium state—are extremely powerful tools in physical chemistry and environmental science.

4.2. ACCUMULATION OF SOLUTES AT INTERFACES

4.2.1. Gibbs Adsorption Isotherm

It is well known that certain chemical components of liquid and gaseous phases will accumulate at interfaces. For example, surfactants in detergents accumulate at the interface of soils and water, thereby allowing dirt to be removed from soiled clothing. Some organic molecules with hydrophobic ("water fearing") characteristics can accumulate at the air–water interface to such an extent that very few solvent molecules are present at the interface. These surface films have been studied extensively by physical chemists, and in some cases, the mechanics of the films has yielded information on molecular dimensions. Nitrogen gas is also known to accumulate at solid–gas interfaces, and this property allows us to use nitrogen to determine the surface area of very fine or porous surfaces or adsorbents. Likewise, some molecules or ions may be *depleted* (negatively adsorbed) at interfaces. In either case, something about the interface is either liked or disliked by the molecules in question. Accumulation of molecules at the interface at these three examples suggests that this configuration somehow minimizes the Gibbs function for these systems.

The thermodynamics of accumulation of molecules at interfaces can be appreciated using the following hypothetical example of the interface between

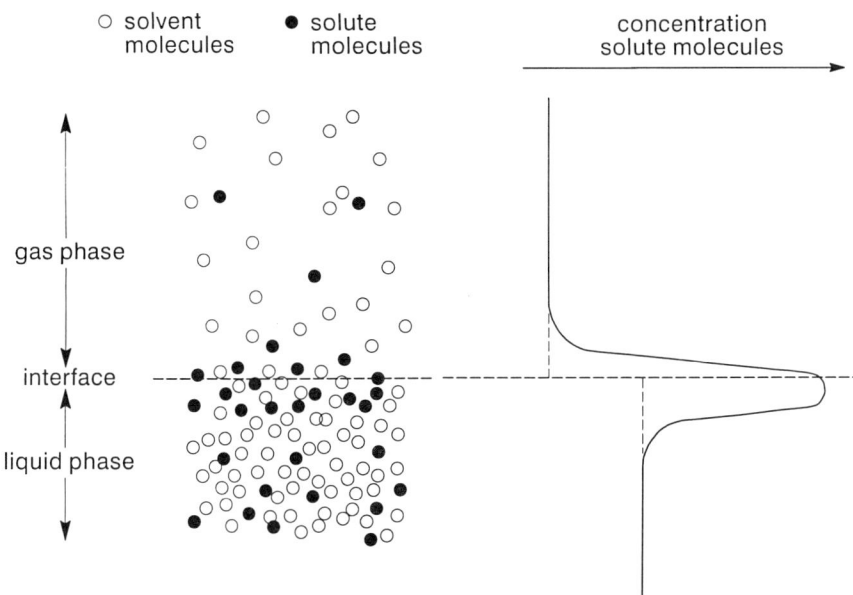

Figure 4.1. The interface between a liquid and its vapor, showing an excess of solute molecules accumulated at the interface.

a liquid (solvent) and its vapor (Figure 4.1). Also shown in the figure are some solute molecules that are primarily soluble in the liquid phase, and that for some reason have preferentially accumulated near the interface. It is useful now to provide a more thorough definition of a molecular excess at the interface. Considering only the solute molecules, note that the solute molecules have uniform concentrations in the vapor and liquid phases, just up to the interface region of finite thickness. Now imagine that there is no excess concentration of solute molecules at the interface; then the concentration would be uniform through the two phases, with a sharp discontinuity in concentration at an infinitesimal interface (note dotted extensions to concentrations in both phases). For this imaginary case, we define $n_i^{(v)}$ and $n_i^{(l)}$ as the amounts of the ith-type molecules in the vapor and liquid phase, respectively. (The index i has a single value at this point in the discussion, since we are considering a single solute species.) Now maintaining these definitions, but considering the actual interfacial concentration of solute molecules in Figure 4.1, the excess of molecules in the interface region is defined by

$$n_i^{(s)} = n_i^{(T)} - (n_i^{(v)} + n_i^{(l)}) \qquad (4.2)$$

Here $n_i^{(T)}$ is the total amount of ith-type molecule in the actual system (i.e., the liquid, vapor, and interfacial regions), and s stands for surface or interface. So Eq. 4.2 says that the interfacial excess of solute molecules is the difference between the actual total amount of solute molecules and the total amount that would be present if the concentrations in the liquid and vapor phases were uniform all the way up an infinitesimally thin interface. As we stated in Section 3.2, the Gibbs function for the interface can be written as

$$dG^{(s)} = -S^{(s)}dT + \sum_i \mu_i^{(s)}dn_i^{(s)} + \sigma dA \tag{4.3}$$

where σ is the interfacial tension, A is the interfacial area, and s indicates the equation only applies to the surface or interface. At constant temperature, Eq. 4.3 can be integrated to

$$G^{(s)} = \sum_i \mu_i^{(s)}n_i^{(s)} + \sigma A \tag{4.4}$$

Taking the total derivative of Eq. 4.4, we find that

$$dG^{(s)} = \sum_i \mu_i^{(s)}dn_i^{(s)} + \sum_i n_i^{(s)}d\mu_i^{(s)} + \sigma dA + A d\sigma \tag{4.5}$$

Comparing Eqs. 4.3 and 4.5 at constant temperature, we discover that

$$\sum_i n_i^{(s)}d\mu_i^{(s)} + A d\sigma = 0 \tag{4.6}$$

Dividing through by A, we can state

$$d\sigma = -\sum_i \Gamma_i d\mu_i^{(s)} \tag{4.7}$$

where Γ_i is the so-called *surface excess*:

$$\Gamma_i = \frac{n_i^{(s)}}{A} \tag{4.8}$$

Equation 4.7 has an elegant interpretation for a simple system like that in Figure 4.1. There are only two types of molecules, the solvent (open circles) and solute (dark circles). Hence the subscript in the equations above takes on only two values, referred to here as 1 for solvent molecules and 2 for solute molecules. It turns out that the interface position in Figure 4.1 can be adjusted slightly so that the surface excess of the *solvent* molecules is zero. To see this, consider Figure 4.2. Note that the position of the interface is such that the surplus of solvent molecules on the vapor side of the interface is balanced by the deficiency of solvent molecules on the liquid side; hence, the surface excess is exactly zero (note that the two shaded areas are the same). However, the surface excess of solute molecules at the interface is still clearly positive. In this

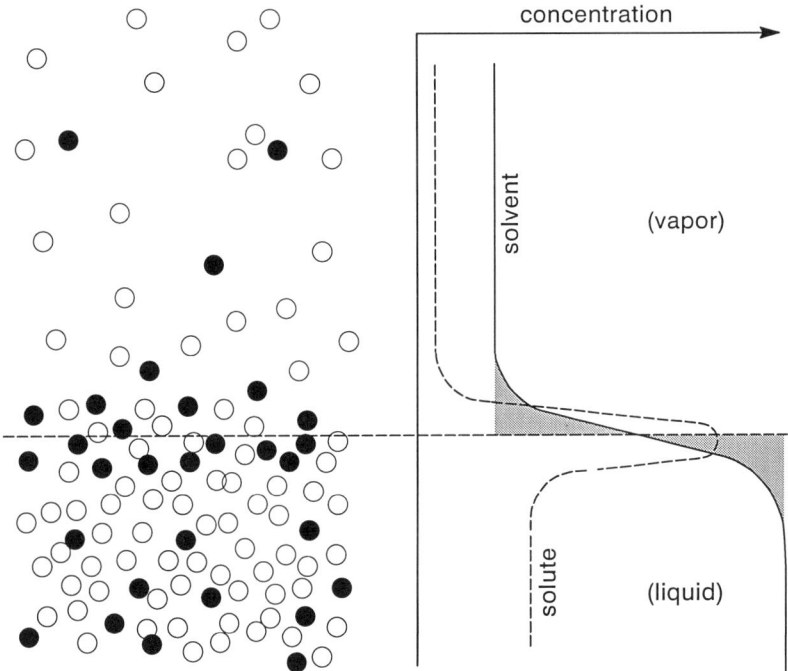

Figure 4.2. Concentration of solute and solvent molecules in the region of the interface.

case, since the surface excess of solvent molecules is zero ($\Gamma_1 = 0$), we conclude the following from Eq. 4.7:

$$d\sigma = -\Gamma_2 d\mu_2^{(s)} \tag{4.9}$$

or rearranging

$$\Gamma_2 = -\frac{d\sigma}{d\mu_2^{(s)}} \tag{4.10}$$

We can assume that the solute concentration is sufficiently low that the chemical potential can be expressed in terms of concentration (e.g., Atkins, 1978):

$$\mu_2^{(s)} = RT \ln C_2^{(s)} \tag{4.11}$$

where R is the universal gas constant. One of the important lessons of thermodynamics is that at equilibrium, the chemical potential of any species is the same in adjacent phases like the liquid, vapor, and interface shown in our illustration. Therefore, the surface concentration and chemical potential in Eq.

4.11 can be considered the chemical potential and concentration in the liquid phase, and using the superscript l for liquid, we find, upon substituting Eq. 4.11 into Eq. 4.10:

$$\Gamma_2 = \frac{-d\sigma}{RTd(\ln C_2^{(l)})} = -\frac{C_2^{(l)}}{RT}\frac{d\sigma}{dC_2^{(l)}} \tag{4.12}$$

Equation 4.12 is a classic result of physical chemistry, called the *Gibbs adsorption isotherm*. The equation implies that in order for there to be a surface excess of solute molecules, the change in interfacial tension with a positive change in solute concentration must be negative. Or, in other words, accumulation of solutes at an interface lowers the interfacial tension. We can also see that the reverse is true — a surface depletion of solute molecules will raise the interfacial tension.

At this point, it is instructive to look at some data. Although quantitative measurement of surface excess has been a challenge in physical chemistry, there have been extensive measurements of the effect of solutes on interfacial tension. In Figure 4.3a, we see the effect of increased concentration of sodium lauryl sulfate on the interfacial tension between air and water, which is 72 dyne/cm in pure water (Table 3.2). The interfacial tension steadily decreases up to about a solute concentration of 10^{-2} M. Equation 4.12 then suggests an increasing surface excess. Sodium lauryl sulfate is in fact a well-known surface-active agent or *surfactant*, so the theoretical prediction is verified. On the other hand, Figure 4.3b shows that increasing the NaCl concentration increases the air–water interfacial tension, suggesting that Na^+ and Cl^- are *negatively* concentrated or depleted at the air–water interface. A simple experiment has

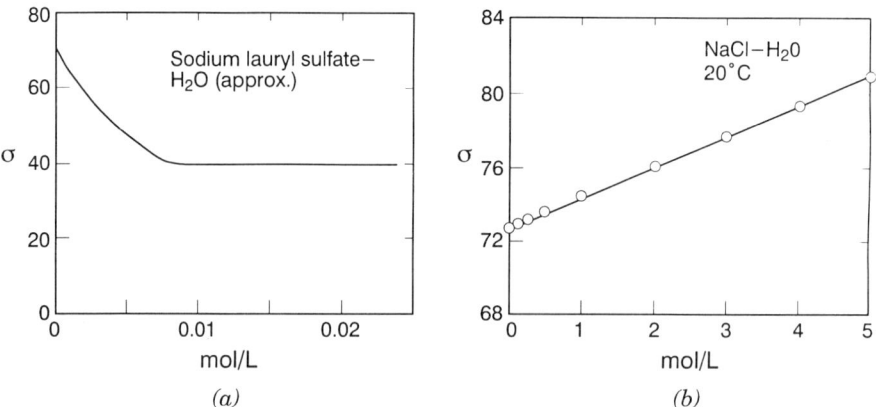

Figure 4.3. Effect of (a) sodium lauryl sulfate and (b) NaCl on air–water interfacial tension σ. Units of interfacial tension are dyne per centimeter. *Source*: Reproduced from Adamson, *Physical Chemistry of Surfaces*, p. 67. © 1976. Reprinted by permission of John Wiley & Sons, Inc.

shown that the prediction is correct: When air is bubbled through water containing NaCl, if the water popped out of the surface from the bursting bubbles is collected, the NaCl concentration in the collected water is lower than in the bulk solution, demonstrating a negative surface excess and a possibly interesting way to desalinate water.

One last comment will be made about Eq. 4.12. Note that as solute concentration increases, the surface excess either increases or decreases, depending on whether the solute is surface active. The change in interfacial concentration with increasing solute concentration is known as an *isotherm*. Note also that if the term $d\sigma/dC_2^{(l)}$ is assumed to be constant, the surface excess either increases or decreases *linearly* with solute concentration. This is our first exposure to a so-called *linear isotherm*.

4.2.2. Orientation of Large Molecules at Interfaces

In the previous section, we looked at some of the macroscopic thermodynamics of interfaces, and were able to predict the effect of a surface excess (or depletion) of solute molecules on interfacial tension. Altough we have suggested that these phenomena are related to minimization of the Gibbs function for the system containing an interface, the approach did not yield much insight into *why* a particular molecule would concentrate at an interface. (In fact, we found that some molecules are depleted at interfaces.)

The accumulation of molecules like fatty acids, proteins, and polyelectrolytes at interfaces is now understood in relation to a pheomenon known as the *hydrophobic* force. Although we should imagine that individual water molecules are in thermal motion (hence the Brownian motion of colloidal particles suspended in a water), there is a degree of order in water because it is a highly *polar* solvent and because of strong *hydrogen bonding* between water molecules. Both the O and H atoms are capable of hydrogen bonding with other like atoms. When a compound like glucose (a sugar) is dissolved in water, the glucose molecule interacts strongly with water because of hydrogen bonding between water and the numerous -OH groups on the glucose molecule. Hence, sugars like glucose are very soluble in water (good for sweetening our foods and drinks). Contrast this with a hydrocarbon like benzene, which is hardly soluble at all in water. When we try to dissolve an aromatic compound like the very *nonpolar* benzene molecule in water, it displaces water molecules and forms a hole in the water structure. In fact, it is believed that water molecules form an even more ordered structure around the *hydrophobic* (solvent fearing) benzene molecule. Since this increased ordering would result in an entropy *decrease*, it is not favored thermodynamically. As a result, benzene molecules associate with each other to form a separate benzene phase where there are strong van der Waals forces on interaction between the benzene molecules. Therefore, the hydrophobic "force" is not a force in the conventional sense; rather, it is the tendency of the polar water molecules to reject nonpolar molecules from their environment.

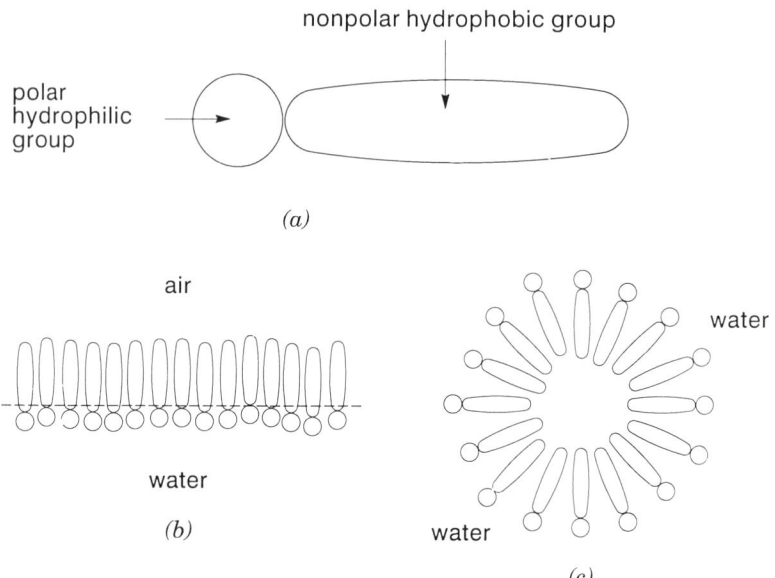

Figure 4.4. (*a*) A long-chain fatty acid molecule showing polar and nonpolar groups, (*b*) a surface film of the molecules, and (*c*) a micelle formed from the molecules.

Now consider a long-chain fatty acid like stearic acid or the surfactant sodium lauryl sulfate mentioned in the last section. These molecules have both hydrophobic nonpolar regions and *hydrophilic* polar groups like carboxyl groups. When these types of molecules are dissolved in water, the hydrophilic groups can hydrogen bond with water (thus aiding solubility), whereas the hydrophobic groups want to be removed from solution. Nature yields two extremely interesting solutions to the problem faced by a molecule with both hydrophilic and hydrophobic functional groups (Figure 4.4). For example, at an air-water interface, these molecules form a surface layer or film with their hydrophilic groups in the water phase, and their hydrophobic groups sticking out of the solution (Figure 4.4*b*).

The tendency to form surface films has been useful in determining molecular sizes and in engineering thin films of pure compounds. Figure 4.4*c* shows another possibility that often occurs with surfactants, the formation of *micelles*. When the concentration of some surfactants reaches a high enough concentration in aqueous solutions (the *critical micelle concentration*), the molecules can aggretate with the polar groups extending into the solution, and the nonpolar groups buried on the inside of the molecule. Hence, micellization solves the hydrophobic problem by essentially creating a new interface—a new phase of colloidal micelles.

In this section, we touched only lightly on the importance of hydrophobic interactions. Chemists and biologists have found that hydrophobic interactions are important in explaining a wide range of macromolecular phenomena, from the structure and chemistry of membranes to the structure of amino acids and the coiling of DNA.

4.3. ADSORPTION AT SOLID–LIQUID AND SOLID–GAS INTERFACES

Adsorption at solid–liquid interfaces is ubiquitous in the environment and pollution-control engineering. As mentioned above, activated carbon is used to adsorb a variety of liquid-phase organic contaminants. Table 4.1 lists several volatile organic compounds (VOCs) now controlled through U.S. drinking-water regulations; activated carbon is thought to be effective in adsorption of these compounds. In the treatment of some toxic leachates from hazardous waste facilities, activated carbon is used to adsorb contaminants. In a different type of adsorption process, large-molecular-weight synthetic organic polymers are added to dirty water from lakes and rivers to neutralize colloid surface charge, decrease stability, and promote flocculation through interparticle bridging by the polymers. Natural organic compounds like humic substances can adsorb to the surface of natural colloidal material to such an extent as to stabilize the colloids and prevent them from getting close enough to aggregate, which may be a type of *steric* stabilization. Finally, hydrophobic organic contaminants (like DDT and PCBs) are known to adsorb to the outside of natural colloidal material. The contaminant is then essentially carried by the particle as it moves down rivers and through lakes or groundwater.

Adsorption at gas–solid interfaces is also very important to the environment. For example, activated carbon can be used to remove odors from stack gases in various industrial processes. To alleviate indoor air-pollution problems, activated carbon can be used to adsorb various VOCs and other gases like benzene and formaldehyde. Finally, it has been verified that particulate matter in the atmosphere can adsorb a variety of gaseous pollutants like PCBs and polyaromatic hydrocarbons.

Adsorption at solid–liquid and solid–gas interfaces is often categorized as being by *physisorption* or *chemisorption*. Physisorption is generally thought to arise from three types of forces between the *adsorbent* (e.g., a surface) and the *adsorbate* (the adsorbing ion or molecule):

1. *Van der Waals forces*. These are exactly the same as the forces between closely approaching particles discussed in Chapter 3. Imagine that dipoles in surfaces can interact with dipoles in adsorbates if they are close enough. Equation 3.27 predicts that these forces depend on the inverse square of the separation distance.
2. *Electrostatic forces*. If an adsorbate and adsorbent have opposite charge, there can be an attractive force.

3. *Solvent ordering and the hydrophobic "force."* In aqueous systems, certain organic molecules with hydrophobic functional groups cause a high degree of ordering in the surrounding water molecules when the organic material is dissolved in water. As we discussed in Section 4.2.2, this results in an entropy (*decrease*) in the system. If possible, these organic molecules will cluster with each other or accumulate at hydrophobic interfaces since this decreases the surface area in contact with the water and thereby (*increases*) the entropy of the system (recall that nature favors entropy increases); therefore, the compound becomes less soluble in water and more likely to adsorb at a hydrophobic interface.

Adamson (1976) summarized the hydrophobic phenomenon as follows: A nonpolar adsorbent will preferentially adsorb the more nonpolar component of a polar solution. An extension of this truism is known as *Traube's rule*: "The adsorption of organic substances from aqueous solutions increases strongly and regularly as we ascend the homologous series" (Freundlich, 1926).* Traube's rule assumes that as we consider compounds of greater molecular weight in the homologous series, a nonpolar compound becomes increasingly nonpolar and less able to form hydrogen bonds with a polar solvent (i.e., water).

After considering the three general forces described above, the reader might wonder if two types of force can come into conflict, for example, whether an unfavorable electrostatic interaction (e.g., negatively charged surface/negatively charged adsorbate) could dominate a favorable hydrophobic interaction. This type of question describes a lively area of study in surface and environmental science. Shirahama et al. (1990) studied the adsorption of proteins on hydrophilic silica and hydrophobic (polystyrene-coated silica) surfaces. The proteins have both hydrophobic and hydrophilic functional groups; the proteins also carry a net charge, which can be varied from positive to negative by changing the pH. Since both surfaces had a net negative charge over the pH range studied, Shirahama et al. could study favorable and unfavorable electrostatic interactions on both hydrophilic and hydrophobic surfaces. The investigators found that at the hydrophobic surface, an unfavorable electrostatic interaction did not really affect the mass adsorbed in single-protein systems. However, at the hydrophilic surface, the amount of protein adsorbed increased as the electrostatic interaction became more favorable.

Finally, physisorption is thought to permit the formation of multilayers on the adsorbent surface. In fact, there is no reason why the formation of second layers cannot start before the first layer is complete. Also, desorption can occur in physisorption, that is, the adsorption is reversible. Chemisorption is thought to result from the actual formation of a chemical bond, such as a covalent

---

*A homologous series refers to a class of organic compounds composed of the same functional groups, but with steadily increasing molecular weight and smoothly changing physical properties (e.g., melting point and boiling point).

bond, between the surface and the adsorbate. Since a chemical bond is required, chemisorption should only result in a monolayer of adsorbate molecules on the surface. Chemisorption may also be slower than physisorption if the chemical reaction is slow. Chemisorption may involve an activation energy, and may not be readily reversible. Despite our attempt to classify adsorption as either physisorption or chemisorption, the reader should be aware that both types of adsorption can occur simultaneously. Thus, experimental results may not indicate clearly what type of adsorption is occurring.

4.4. ADSORPTION ISOTHERMS

As we saw in Section 4.2.1, the Gibbs isotherm was developed to help explain the accumulation of solutes at the air–water interface, and the effect this acumulation had on the air–water interfacial tension. In gas–solid and liquid–solid systems, it is often impossible to directly measure the accumulation of adsorbate at the interface. Therefore, with these systems, we have relied on solution or gas–phase measurements of adsorbate concentration to make inferences about the amount of material adsorbed at the interface. Three other types of isotherms have been useful in environmental modeling, in explaining experimental adsorption data, characterizing surfaces, and designing pollution-control equipment: the Langmuir, BET, and Freundlich isotherms.

The *Langmuir* isotherm is thought to govern physisorption of gases or liquid solutes on solids when only a monolayer of adsorbate is formed. The derivation of the Langmuir isotherm starts with the simple physical picture of Figure 4.5. The section of surface shown has S adsorption sites or positions, indicated by the little sticks coming out of the surface. S_1 of the S sites are occupied by an adsorbate molecule, leaving $S - S_1$ sites free. For a gas–solid system, we assume that the rate of departure of adsorbed molecules from the surface is the evaporation rate and is proportional to the number of molecules occupying surface sites:

$$\text{rate of evaporation} = k_1 S_1 \tag{4.13}$$

Likewise, we assume that the rate of arrival of molecules at the adsorption sites is the condensation rate and is proportional to the number of free sites:

$$\text{rate of condensation} = k_2 p(S - S_1) \tag{4.14}$$

In a gas, p can be thought of as the pressure or as some function of the pressure. A differential equation can therefore be written for the change in number of adsorbed molecules over time:

$$\frac{dS_1}{dt} = k_2 p(S - S_1) - k_1 S_1 \tag{4.15}$$

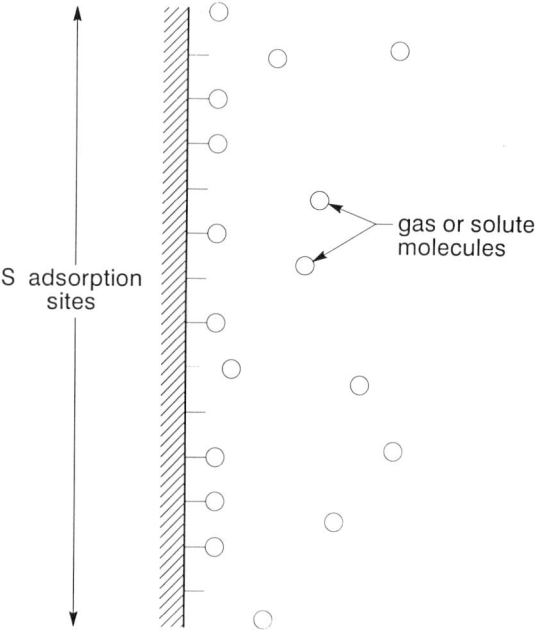

Figure 4.5. Adsorption of molecules on an ideal surface.

At steady state (or in this case equilibrium), the left-hand side of Eq. 4.15 is zero, and with a little rearrangement we find:

$$\theta = \frac{S_1}{S} = \frac{\dfrac{k_2}{k_1}p}{1 + \dfrac{k_2}{k_1}p} = \frac{bp}{1 + bp} \qquad (4.16)$$

From Eq. 4.16, the fractional surface coverage S_1/S is given the symbol θ, and the ratio k_2/k_1 is given the symbol b. (Note that θ is analogous to Γ in the Gibbs isotherm, Eq. 4.12.) In gas adsorption studies, S_1/S is often replaced by an equivalent expression, V/V_m, were V is the total volume of adsorbed molecules, and V_m is the volume of adsorbed molecules required to form a monolayer.

Often in this book we examine the limiting behavior of equations, since these equations are often more simple and often approximate real situations. For example, as $p \to \infty$, the numerator and denominator in Eq. 4.16 converge to the same value; hence, we expect that

$$\theta = \frac{S_1}{S} \to 1 \qquad (4.17)$$

This is the limit of *complete coverage*. This limiting condition is expected since the Langmuir isotherm model only allows for a single monolayer of adsorbate; the adsorbent is effectively saturated at this point. Another interesting limit occurs as $p \rightarrow 0$. Then,

$$\theta = \frac{bp}{1} = bp \tag{4.18}$$

This simple result says that if p is low enough, the surface coverage increases linearly with p (see also Exercise 4.3). This is our second example of a linear isotherm. Contaminants and adsorbates in the environment are often present at low concentrations (or equivalently, low partial pressures in gases). Because of the simplicity of Eq. 4.18, the assumption of a linear isotherm can come in very handy in environmental modeling and pollution control.

In liquid–solid systems, the equation analogous to Eq. 4.16 is

$$\theta = \frac{S_1}{S} = \frac{bC_e}{1 + bC_e} \tag{4.19}$$

where C_e is the (*equilibrium*) solute concentration. Note that both sides of Eq. 4.19 can be divided by the mass of adsorbent m and Avogadro's number L_A, yielding

$$\frac{S_1}{L_A m} = \frac{S}{L_A m} \frac{bC_e}{1 + bC_e} \tag{4.20}$$

The equation is finally written as

$$q_e = Q^0 \frac{bC_e}{1 + bC_e} \tag{4.21}$$

Note by comparison of Eqs. 4.20 and 4.21 that q_e is *the number of moles (or sometimes mass) of adsorbate adsorbed per unit mass of adsorbent,* and Q^0 is *the number of moles (or sometimes mass) of adsorbate adsorbed per unit mass of adsorbent at complete surface coverage.* The reader should confirm that at very high and very low solute concentrations, equations analogous to Eqs. 4.17 and 4.18 will result. The general shape of the Langmuir isotherm is shown in Figure 4.6a.

Equations 4.16 and 4.21 have been useful for describing some experimentally determined adsorption results. However, the parameters in these questions (i.e., b in Eq. 4.16 and Q^0 and b in Eq. 4.21) vary with temperature and the specific chemistry of the solution or gas and the surface. For example, in liquid–solid systems Q^0 and b determined for a certain adsorbent and adsorbate will generally be a function of temperature, pH, ionic strength, and so forth.

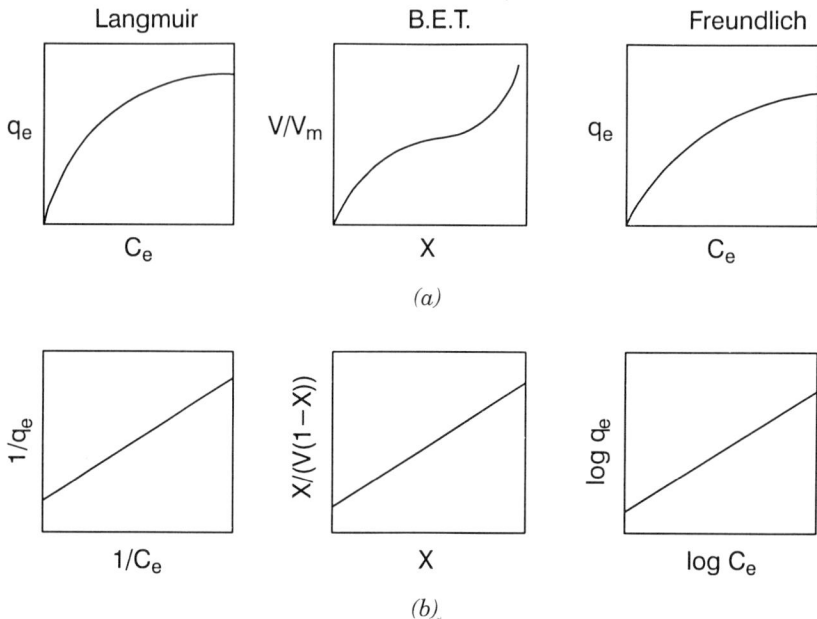

Figure 4.6. Schematic diagrams of Langmuir, BET, and Freundlich isotherms: (*a*) plotted isotherms, and (*b*) linearized forms of isotherms used in parameter fitting.

Often, experimental adsorption data are recorded in the form of a series of measured q_e values for increasing C_e values. Then, to determine the isotherm parameters, it is helpful to rearrange Eqs. 4.16 and 4.21 in a *linearized* form (*not* a linear isotherm). For liquid–solid systems, Eq. 4.21 can be rearranged to

$$\frac{1}{q_e} = \frac{1 + bC_e}{Q^0 b C_e} \tag{4.22}$$

which can be rearranged further to

$$\frac{1}{q_e} = \frac{1}{Q^0 b C_e} + \frac{bC_e}{Q^0 b C_e} \tag{4.23}$$

or

$$\frac{1}{q_e} = \left(\frac{1}{Q^0 b}\right)\left(\frac{1}{C_e}\right) + \frac{1}{Q^0} \tag{4.24}$$

Note that Eq. 4.24 is the equation of a straight line, $y = $ (slope) $x + $ intercept, where y is $1/q_e$, x is $1/C_e$, the slope is $1/(Q^0 b)$, and the intercept is $1/Q^0$. The linearized Langmuir isotherm is shown in Figure 4.6*b*.

Some subtle assumptions of the Langmuir model deserve consideration:

1. The Langmuir model assumes physisorption of a monolayer (no specific chemical reactions, but we have already stated that physisorption can lead to multiple layers of adsorbate. So this could be a fundamental problem with the Langmuir isotherm model.

2. In the liquid or gas phase, the adsorbate behaves as an ideal gas or solute. This should be an accurate assumption for gases at low pressure. The accuracy of the assumption is less clear in liquids because of the complexity and large size of some organic molecules. In any event, the lower the solute concentration, the more accurate this assumption will be in liquids.

3. The adsorbate molecules remain at specific adsorption sites. Although this is an accurate assumption for chemisorption, physisorbed molecules are thought to diffuse along the surface of the adsorbate.

4. Adsorption sites are uniformly distributed on the adsorbent surface, and the energy of interaction of each adsorption site is the same. Many adsorbing surfaces (especially in the environment!) have heterogeneous surfaces with different interaction energies, so this assumption can also be violated. These heterogeneities can result in isotherms which are not as smooth as suggested by Eqs. 4.16 and 4.21.

5. There are no interactions between adsorbed molecules. Molecules adsorbed on a surface are subject to electrostatic, van der Waals, and other forces from surrounding molecules. Newer adsorption simulation models are taking these types of lateral interactions into consideration.

Another well-known adsorption isotherm is the *BET* (Brunauer, Emmet, and Teller) isotherm. It does not suffer from problem 1 noted above, since the BET isotherm model envisages the formation of multilayers of adsorbate. The BET model further assumes that individual layers need not be complete and that the Langmuir model applies to adsorption in each layer, although the interaction of the first layer with the adsorbent surface can have a different heat of adsorption (adsorption enthalpy) than that between successive layers (whose heat of adsorption is assumed to be constant). The BET model takes the following form in gas adsorption studies (temperature = constant):

$$\theta = \frac{V}{V_m} = \frac{cX}{(1 - X)[1 + (c - 1)X]} \tag{4.25}$$

Here V_m is the volume of a monolayer, $X = p/p_0$ is the ratio of the adsorbate gas pressure p to the vapor pressure of the pure liquid adsorbate p_0, and c has the following definition:

$$c = \exp\left(\frac{\Delta H_{ads} - \Delta H_{vap}}{RT}\right) \tag{4.26}$$

Here ΔH_{ads} is the heat (enthalpy) of adsorption and ΔH_{vap} is the heat (enthalpy) of vaporization of the pure liquid adsorbate. The general shape of the BET isotherm is shown in Figure 4.6a. To determine the BET parameters X and c from experimental data, it is convenient to linearize Eq. 4.25 as

$$\frac{X}{V(1-X)} = \frac{1}{cV_m} + \frac{(c-1)X}{cV_m} \tag{4.27}$$

Note that Eq. 4.27 is the equation of a straight line if $X/(V(1-X))$ is taken as y, X is taken as x, the slope is taken as $(c-1)/(cV_m)$, and the intercept is taken as $1/(cV_m)$. The linearized form of the BET isotherm is illustrated in Figure 4.6b. One of the exercises at the end of this chapter examines how the BET isotherm is used to determine the surface area of porous adsorbents. Another exercise illustrate that the BET isotherm can mimic the Langmuir isotherm under certain conditions.

The final adsorption isotherm is called the *Freundlich* isotherm. It is typically used as an *empirical* adsorption model for solid–liquid systems. In other words, the model is used to fit data rather than to verify an adsorption mechanism.* The model takes the following form:

$$q_e = K_f C_e^{1/n} \tag{4.28}$$

where K_f is an empirical constant related to the capacity of the adsorbent material to adsorb the adsorbate (the higher the K_f value, the more material potentially adsorbed) and n is a constant related to the affinity of the adsorbate for the surface (the smaller the n value, the greater the affinity of the adsorbate for the surface). Table 4.1 gives some typical K_f and $1/n$ values for the adsorption of six regulated VOCs on granular-activated carbon. The general shape of the Freundlich is shown in Figure 4.6a.

To determine the empirical parameters K_f and $1/n$, it is convenient to linearize Eq. 4.28 by taking the logarithm of both sides of the equation:

$$\log q_e = \log K_f + \frac{1}{n}\log C_e \tag{4.29}$$

The linearized form of the Freundlich isotherm is shown in Figure 4.6b.

---

*Empirical equations are typically considered to be inferior in physics and chemistry since they often do not yield an understanding of basic phenomena. However, in engineering applications, empirical equations are often useful in correlating (organizing) data, problem solving, and design.

Table 4.1. Freundlich Carbon Adsorption Isotherm Parameters of Several Volatile Organic Compounds Regulated in the United States[a]

Compound	K_f Value $(mg/g)(L/mg)^{1/n}$	$1/n$ Value
Trichloroethylene	28	0.62
Carbon tetrachloride	11	0.83
1,2-Dichloroethane	3.6	0.83
Benzene	1.0	1.6
1,1-*trans*-Dichloroethylene	14	0.45
1,1,2-Trichloroethane	5.8	0.60

[a]Compounds were in distilled water. Values provided depend on type of carbon and exact water chemistry.
Source: Pontius, *Water Quality and Treatment*, McGraw-Hill Companies, © 1990. Reprinted with permission of The McGraw-Hill Companies.

Example 4.1. Use the Freundlich isotherm to determine the concentration of activated carbon (C_c) required to reduce the concentration of trichloroethylene (TCE) in a 55-gal container to 1% of its original concentration of $C_0 = 500\,\mu g/L$.

SOLUTION. If the initial concentration of $500\,\mu g/L$ must be reduced by 99%, then the equilibrium concentration C_e is $5\,\mu g/L$. Examining Eq. 4.28,

$$q_e = \frac{\text{mass adsorbate (mg)}}{\text{mass adsorbent (g)}} = \frac{(C_0 - C_e)V}{C_c V} = K_f \left(\frac{mg}{g}\right)\left(\frac{L}{mg}\right)^{1/n}\left(C_e\frac{mg}{L}\right)^{1/n}$$

Rearranging and substituting for K_f from Table 4.1:

$$C_c(g/L) = \frac{(500 - 5)\dfrac{\mu g}{L}\left(\dfrac{mg}{1000\,\mu g}\right)}{28\left(\dfrac{mg}{g}\right)\left(\dfrac{L}{mg}\right)^{0.62}\left(5\dfrac{\mu g}{L}\dfrac{mg}{1000\,\mu g}\right)^{0.62}} = 0.47\,g/L$$

Adsorption isotherm parameters depend on temperature and solution chemistry. Another important effect occurs in some gas and aqueous systems with multiple adsorbates. Sometimes, the adsorption isotherm parameters for one adsorbate depend on the concentration of the other adsorbate. This effect arises from *competitive adsorption* effects (Snoeyink, 1990). For example, Qi et al. (1994) studied adsorption of the pesticide Atrazine on powdered activated carbon. The Atrazine was dissolved in a natural water source (i.e., a water source containing naturally occurring organic compounds like humic and fulvic acid). Since the natural organic compounds also adsorb strongly to the carbon, these researchers suspected that there would be competition between the Atrazine (present in microgram per liter concentrations) and the natural

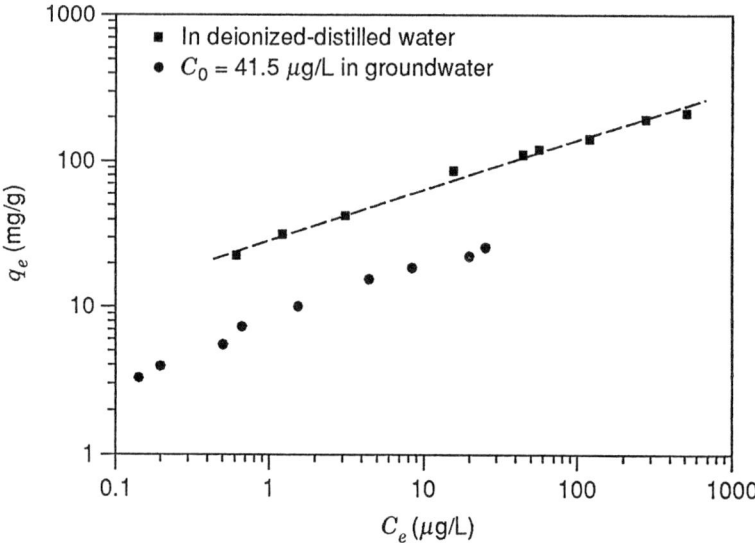

Figure 4.7. Adsorption of Atrazine on powdered activated carbon. The initial Atrazine concentration was 41.5 µg/L in the groundwater experiments. *Source*: Qi et al., *J. Environ. Eng*, **120**, 1:210 (1994). Copyright 1994 American Society of Civil Engineers.

organic matter (present in milligram per liter concentrations). Figure 4 7 shows some of their experimental results (Qi et al., 1994). The data are assumed to be described by the Freundlich isotherm; hence, the plot is log q_e vs. log C_e. Note first that the isotherm for Atrazine in distilled water (no competing background organic matter) predicts the highest solid-phase concentration for a given equilibrium concentration. The Freundlich isotherm parameters for the distilled water adsorption isotherm were $K_f = 797$ $(\mu mol/g)(L/\mu mol)^{1/n}$ and $n = 3.0$. However, when the isotherm for Atrazine in the presence of the natural water organics was measured, the capacity of the adsorbent for the Atrazine decreased (filled circles). This is interpreted as a competition effect between the Atrazine and the natural organic matter for adsorption sites on the powdered activated carbon adsorbent. In Exercise 4.7, you will determine the Freundlich isotherm parameters for the adsorption of Atrazine in natural water.

4.5. LINEAR EQUILIBRIUM PARTITIONING BETWEEN TWO PHASES

4.5.1. Dalton's and Raoult's Laws

From physical chemistry, *Dalton's law* states that for an ideal mixture of gases, the total pressure p of the mixture is the sum of the pressures exerted by the individual gases occupying the same total volume by themselves. This can be written as

$$p = p_A + p_B + \cdots \tag{4.30}$$

The *partial pressure* of gas A can be defined by the ideal gas law:

$$p_A = \frac{n_A RT}{V} \tag{4.31}$$

where n_A is the number of moles of component A in the gas phase, R is the universal gas constant, T is the absolute temperature, and V is the volume of the gas phase. Combining Eqs. 4.30 and 4.31, we see that

$$p = p_A + p_B + \cdots = \frac{(n_A + n_B + \cdots)RT}{V} \tag{4.32}$$

The *mole fraction of a component A in the gas phase* is defined simply as

$$y_A = \frac{p_A}{p} \tag{4.33}$$

A second important law, called *Raoult's law*, concerns the equilibrium partial pressure of a solvent A above a liquid phase containing the solvent and other possible components:

$$p_A = x_A p_{0,A} \tag{4.34}$$

Here $p_{0,A}$ is the pure vapor pressure of A, and x_A is the *mole fraction of A in the liquid phase*:

$$x_A = \frac{c_A}{c} \tag{4.35}$$

where c_A is the molar concentration of the solvent A and c is the total molar concentration of the solution. Thus, Raoult's law states that the partial pressure of a solvent above a solution is equal to the mole fraction of the solvent in the solution. The range of p_A is from 0 ($x_A = 0$) to $p_{0,A}$ ($x = 1$). Raoult's law is only generally accurate when the different compounds in the solution are chemically similar (Atkins, 1978). If this condition does not hold, Raoult's law can only be guaranteed to work when x_A approaches unity. In other words, in a solution of a solvent (e.g., water) and other dissolved chemicals (unlike the solvent), the partial pressure of the solvent above the solution will only obey Raoult's law in the limit where the mole fraction of the other dissolved components approaches zero. This may seem like a severe restriction, but many environmentally important dissolved components are at low concentrations; hence, the solvent may well obey Raoult's law.

Example 4.2. Use Raoult's law to estimate (a) the equilibrium partial pressure in a closed container containing $\frac{1}{2}$ mole fraction benzene and $\frac{1}{2}$ mole fraction toluene, and (b) the partial pressure of benzene directly above the surface of a benzene spill.

SOLUTION. For part *a*, consider the following closed container:

```
+-----------------------------+
|                             |
|         gas phase           |
|     (benzene + toluene)      |
|                             |
+-----------------------------+
|                             |
|        liquid phase          |
|     (benzene + toluene)      |
|                             |
+-----------------------------+
```

Example 4.2

The pure vapor pressures of toluene and benzene at 25°C are (Table 4.2) $p_{0,\text{tol}} = 0.0375$ atm and $p_{0,\text{ben}} = 0.125$ atm. (1 standard atmosphere of pressure is 1.013×10^5 Pa — Appendix I.) If we assume that mixtures of toluene and benzene follow Raoult's law, the partial pressure of toluene above the mixture is calculated as

$$p_{\text{tol}} = x_{\text{tol}}p_{0,\text{tol}} = 0.5 \times 0.0375\,\text{atm} = 0.0188\,\text{atm}$$

and the partial pressure of benzene is calculated as

$$p_{\text{ben}} = x_{\text{ben}}p_{0,\text{ben}} = 0.5 \times 0.125\,\text{atm} = 0.0627\,\text{atm}$$

The total pressure in the vapor phase of a system containing only benzene and toluene is the sum of the partial pressures:

$$p = p_{\text{tol}} + p_{\text{ben}} = 0.0188 + 0.0627 = 0.0815\,\text{atm}$$

For part *b*, we assume that the benzene is lying on a flat surface and that it is present as a pure component. Therefore, $x_{\text{ben}} = 1$. From Raoult's law, the

Table 4.2. Properties of Selected Organic Chemicals at 25°C

Chemical	Molar Mass (g/mol)	Solubility (g/m^3)	Vapor Pressure, (Pa)	log K_{ow}
n-Hexane	86.2	9.5	20,200	4.11
Cyclohexane	84.2	55.	12,700.	3.44
n-Octane	114.2	0.66	1,880	5.18
Benzene	78.1	1780	12,700	2.13
Toluene	92.1	515	3,800	2.69
p-Xylene	106.2	185	1,170	3.15
Naphthalene	128.2	31.7	10.4	3.35
Biphenyl	154.2	7.48	1.2	4.03
Anthracene	178.2	0.041	0.0008	4.63
Phenanthrene	178.2	1.29	0.0161	4.57
Pyrene	202.3	0.135	0.0006	5.22
Benzo(a)pyrene	252.3	0.0038	0.0000007	6.04
Chloroform	119.4	8200	23,080	1.97
Trichloroethylene	131.4	1100	9,870	2.29
1,1,1-Trichloroethane	133.4	730	12,800	2.47
Trichlorofluoromethane	137.4	1100	91,600	2.53
Chlorobenzene	112.6	472	1,580	2.84
1,4-Dichlorobenzene	147.0	83.1	90.2	3.40
1,2,4-Trichlorobenzene	181.5	34.6	60.6	4.00
Hexachlorobenzene	284.8	0.005	0.0023	5.50
2-Chlorobiphenyl	188.7	1.3	2.04	4.54
2,2',4,4'-Tetrachlorobiphenyl	291.9	0.068	0.02	5.90
2,2',4,4',6,6'-Hexachlorobiphenyl	360.9	0.0007	0.0016	7.00
2,3,7,8-TCDD	322.0	0.0193	0.0000001	6.80
2,3,7,8-TCDF	306.0	0.419	0.000002	6.10
DDT	354.5	0.0031	0.00002	6.19
Mirex	545.6	0.00007	0.0001	6.89
Phenol	94.1	82000	70.6	1.46
Pentachlorophenol	266.4	14	0.0147	5.01
p-Cresol	108.1	16800	14.67	1.95
Quinoline	129.2	60000	133	2.03
Diethylhexylphthalate	390.6	0.4	0.0000267	5.30
Chlorpyrifos	350.6	0.4	0.0015	5.11
2,4-D	221.0	890.	0.000056	2.81

Source: Mackay, *Multimedia Environmental Models*, Lewis Publishers, Chelsea, Michigan, pp. 45–46, © 1991. Copyright Lewis Publishers, an imprint of CRC Press, Boca Raton, Florida.

partial pressure of benzene right above the surface of the spill is

$$p_{ben} = x_{ben} p_{0,ben} = 1 \times 0.125\,atm = 0.125\,atm$$

The mole fraction of benzene in the vapor at the surface of the spill is, therefore,

$$y_{ben} = \frac{p_{ben}}{p} = \frac{0.125\,atm}{1\,atm} = 0.125$$

Note that when the spill is open to the atmosphere, we assume that the total pressure at the benzene surface is still atmospheric pressure.

4.5.2. Henry's Law and Partition Coefficients

Raoult's law was concerned with the vapor pressure of a solvent above a solution, but how do we estimate the vapor pressure of a dissolved component above a solution? *Henry's law* states that at equilibrium, the ratio of the concentration c_A of a volatile solute in a solvent to the solute's partial pressure in the vapor phase above the solvent-solute mixture is a constant. Although the rule is obeyed for low solute concentrations only, it will apply to a wide range of gases and VOCs of environmental interest. Mathematically, Henry's law can be expressed as

$$p_A = Hx_A \tag{4.36}$$

where H is the *Henry's law constant*. Many different units are used for H in this chapter the units of H are atmosphere per mole fraction. We see that solute concentration in the gas phase is a *linear* function of the solute concentration in the liquid phase if Henry's law is obeyed. Henry's law can be called a linear (equilibrium) *partition coefficient*. In other words, the volatile component partitions linearly between the liquid and vapor phases.

Example 4.3. Use Henry's law to estimate the solubility of oxygen in water at 25°C.

SOLUTION. At 20°C, the Henry's law constant for O_2 in water is $H = 4.38 \times 10^4$ atm/mol fraction (Thibodeaux, 1979). Now dry air contains 21 mole % O_2, so $p_{O_2} = 0.21$ atm. From Henry's law, the mole fraction of O_2 in the water at equilibrium is

$$x_A = \frac{p_{O_2}}{H} = \frac{0.21\,atm}{4.38 \times 10^4\,(atm/mole\ fraction)} = 4.79 \times 10^{-6}$$

Clearly, not much O_2 dissolves in water. The number of moles of pure water in a liter of pure water is the mass of 1 L of water at 25°C (1000 g) divided by the molar mass of H_2O (i.e., 18 g/mol), or 1000 g/(18 g/mol) = 55.6 mol. Hence, the molar concentration of O_2 in water is $4.79 \times 10^{-6} \times 55.6$ mol/L = 2.66×10^{-4} mol/L. The saturation concentration of O_2 in mass units is 2.66×10^{-4} mol/L $\times$ 32 g/mol = 8.52×10^{-3} g/L or 8.52 g/L.

In environmental science, where water is often the liquid phase and air is the gaseous phase, the *air–water partition coefficient* K_{AW} is frequently used instead of Henry's law. The two concepts are actually quite similar, with K_{AW} given by

$$c_A^{air} = K_{AW} c_A \qquad (4.37)$$

Here c_A^{air} is the solute concentration in moles per liter of air. Often, to estimate values of K_{AW}, it is necessary to measure various combinations of c_A^{air} and c_A, to plot these values on a graph of c_A^{air} versus c_A, and to take K_{AW} as the slope of the resulting line. However, if the solubility of the solute in the solvent is small, Henry's law should be obeyed up to the range of the (*solubility*) limit. Then K_{AW} can be estimated by taking the ratio of the maximum solute concentrations in air and water. The maximum solute concentration in water is the solubility; the maximum solute concentration in air is related to the partial pressure of the pure solute compound. The relevant data can often be found in reference books, and some selected data are given here in Table 4.2.

Example 4.4. From Table 4.2, the solubility of DDT in water is found to be 0.003 g/m^3 and the vapor pressure of pure DDT is 2×10^{-5} Pa (both at 25°C). What is the air–water partition coefficient K_{AW} at 25°C?

SOLUTION. The ideal gas law is given by

$$pV = nRT$$

where p is the pressure (*Pa*), V is the volume (m^3), n is the number of moles (dimensionless), R is the universal gas constant (8.314 J-K^{-1}-mol^{-1}), and T is the absolute temperature (K). The gas concentration c_{DDT}^{air} is simply n/V, which from the equation above is equal to p/RT (units of mol/m^3). Thus, K_{AW} can be calculated as

$$K_{AW} = \frac{c_{DDT}^{air}(@ \text{ saturation})}{c_{DDT}(@ \text{ saturation})} = \frac{p_{DDT}/RT}{c_{DDT}(@ \text{ saturation})}$$

The molar mass of DDT is 354.5 g/mol, and substituting other numerical values yields

$$K_{AW} = \frac{2 \times 10^{-5}\,\mathrm{Pa}\left(\dfrac{\mathrm{kg\text{-}m^{-1}\text{-}s^{-2}}}{\mathrm{Pa}}\right)}{8.314\dfrac{\mathrm{J}}{\mathrm{Kmol}}\left(\dfrac{\mathrm{kg\text{-}m^2\text{-}s^{-2}}}{\mathrm{J}}\right)(298\,\mathrm{K}) \times 0.003\dfrac{\mathrm{g}}{\mathrm{m}^3} \times \dfrac{1\,\mathrm{mol}}{354.5\,\mathrm{g}}} = 9.2 \times 10^{-4}$$

Another important partition coefficient is the *octanol–water partition coefficient*, which is given by

$$K_{OW} = \frac{c_A^{\mathrm{oct}}}{c_A} \tag{4.38}$$

where c_A and c_A^{oct} are the solute concentrations in water and octanol respectively (mol/m^3 or mol/L). In this case, octanol and water are mixed together until the solute (which could initially be present in either or both liquid phases) partitions between the octanol and water phases. Octanol was selected for one phase because it has an atomic carbon to oxygen ratio similar to the lipid material in animal fats. Therefore, we believe that the partitioning between octanol and water indicates the tendency of a solute to accumulate in the body. The reader will recall that we have been stressing the importance of hydrophobic organic chemicals in this chapter, that is, chemicals that prefer to accumulate at solid–water interfaces. Octanol is a nonpolar solvent, while water is a polar solvent. So when a compound partitions strongly into the octanol phase, we can assume that the molecule has a significant concentration of nonpolar (or hydrophobic) functional groups. These types of chemicals are common pollutants, and the octanol–water partition coefficient is useful in predicting the tendency of a chemical to concentrate in animal fat or accumulate at solid–water interfaces. Table 4.2 contains a listing of selected octanol–water partition coefficients. Note that, generally speaking, a lowering of water solubility results in an increase in K_{ow}. Lyman (1982) states that generally chemicals with $\log K_{ow}$ between 4 and 7 are very hydrophobic, chemicals with $\log K_{ow}$ between 1 and 4 are relatively hydrophobic, and chemicals with $\log K_{ow}$ between -3 and 1 are relatively hydrophilic.

Many other partition coefficients have been proposed (Mackay, 1991). These include the organic carbon–water, aerosol–air, mineral–water, and even the activated carbon–water partition coefficients. A critical question in all these proposed coefficients is whether they represent physical reality. For example, consider the general partition coefficient

$$K_{12} = \frac{c^{(1)}}{c^{(2)}} \tag{4.39}$$

where $c^{(1)}$ is the concentration in phase 1, $c^{(2)}$ is the concentration in phase 2, and K_{12} is the phase 1–phase 2 partition coefficient. Rearranging the equation, we find that the concentration in phase 1 is predicted to be linearly related to the concentration in phase 2:

$$c^{(1)} = K_{12}c^{(2)} \tag{4.40}$$

Is this prediction correct? Extreme care must be taken here. For example, if one inserts for $c^{(2)}$ a concentration that exceeds the phase solubility, the equation can give an erroneous prediction. Also, for partition coefficients like the activated carbon–water, aerosol–air, and mineral–water, it is likely that the "partitioning" here is essentially interfacial adsorption (as we studied earlier in the chapter). As we have seen above, the adsorption isotherm is generally only linear at low solute concentrations. So the linear partitioning may only hold for chemicals of low concentration.

4.5.3. Mass Balances and the Partition Coefficient

While the partition coefficient compactly expresses the equilibrium ratio of the concentration of a gas or solute in two phases, we have to incorporate a mass balance in order to predict multiphase concentrations when we are only told the total mass of solute or gas present and the volumes of the two phases. Suppose that two liquid phases under consideration are called phases A and B. Then the total mass of partitioned solute (or gas) in the two phases is the sum of the mass in phases A and B:

$$M_T = M^{(A)} + M^{(B)} = V_A c^{(A)} + V_B c^{(B)} \tag{4.41}$$

where $c^{(A)}$ is the concentration in phase A, $c^{(B)}$ is the concentration in phase B, V_A is the volume of phase A, and V_B is the volume of phase B. However, if there is a partition coefficient

$$K_{AB} = \frac{c^{(A)}}{c^{(B)}} \tag{4.42}$$

the mass balance can be rewritten as

$$M_T = V_A c^{(B)} K_{AB} + V_B c^{(B)} = c^{(B)}(V_A K_{AB} + V_B) \tag{4.43}$$

Or solving for $c^{(B)}$:

$$c^{(B)} = \frac{M_T}{V_A K_{AB} + V_B} \tag{4.44}$$

By a similar approach, we find that

$$c^{(A)} = \frac{M_T}{V_A + V_B K_{AB}^{-1}} \tag{4.45}$$

Example 4.5. A tube containing 60 mL of octanol, 40 mL of water, and 0.5 g of trichloroethylene (TCE) is shaken until the TCE reaches its equilibrium concentration in the octanol and water phases. What is the concentration of TCE(mg/L) in the water and octanol phases if the octanol–water partition coefficient is $10^{2.29}$?

SOLUTION. Associate A with octanol and B with water in the equations above; then the concentration of TCE in the water is from Eq. 4.44,

$$c_{TCE}^{(w)} = \frac{M_T}{V_o K_{ow} + V_w} = \frac{0.5\,g}{60\,mL \times 10^{2.29} + 40\,mL} = 4.26 \times 10^{-5}\,\frac{g}{mL} = 42.6\,\frac{mg}{L}$$

and the concentration of TCE in the octanol is, from Eq. 4.45,

$$c_{TCE}^{(o)} = \frac{M_T}{V_o + \dfrac{V_w}{K_{ow}}} = \frac{0.5\,g}{60\,ml + \dfrac{40\,mL}{10^{2.29}}} = 8.30 \times 10^{-3}\,\frac{g}{mL} = 8300\,\frac{mg}{L}$$

We can use Eq. 4.41 to perform a mass balance and see if these numbers are reasonable:

$$M_T = 0.5\,g = V_o c_{TCE}^{(o)} + V_w c_{TCE}^{(w)} = 8300\,\frac{mg}{L} \times 0.06\,L$$

$$+ 42.6\,\frac{mg}{L} \times 0.04\,L = 500\,mg \quad (OK)$$

We can also see that we get back the correct K_{ow}:

$$K_{ow} = \frac{c_{TCE}^{(o)}}{c_{TCE}^{(w)}} = \frac{8300\,\dfrac{mg}{L}}{42.6\,\dfrac{mg}{L}} = 1.95 \times 10^2 = 10^{2.29} \quad (OK)$$

4.6. PARTITIONING AND SEPARATION IN FLOW SYSTEMS

Up to this point we have only thought of adsorption and partitioning in batch (nonflow) systems containing two phases. But what if one of the phases is stationary and the other is flowing? We now consider two such systems, as shown in Figure 4.8.

In these figures, there is flow through the system in one phase only. Although the second phase is stationary, it is in intimate contact with the first phase, and dissolved chemicals or other components are assumed to be

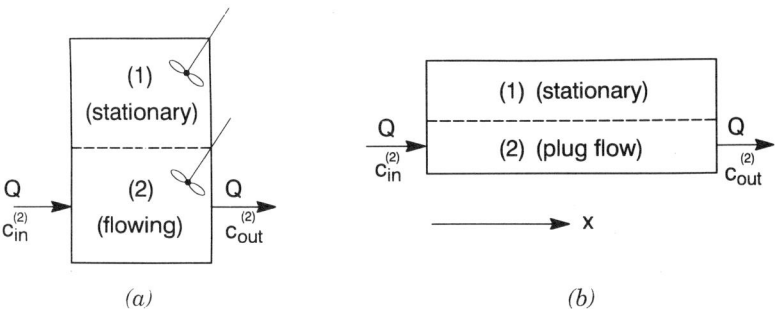

Figure 4.8. Examples of partitioning in flow systems: (a) a perfectly mixed system, and (b) a plug-flow system.

partitioned between the two phases. System a in Figure 4.8 is easily analyzed with the tools developed in Chapters 1 and 2. For example, using the mass balance approach of Section 1.2, we can write the following:

$$V_2 \frac{dc_{out}^{(2)}}{dt} + V_1 \frac{dc_{sys}^{(1)}}{dt} = Qc_{in}^{(2)} - Qc_{out}^{(2)} \tag{4.46}$$

The left-hand side of Eq. 4.46 contains two mass accumulation terms, one for each of the two phases. As usual, the right-hand side represents the difference between mass input and output to the overall system. However, we have made the bold assumption that any dissolved component is at its equilibrium concentration in the two phases (partitioned). From Eq. 4.39 we note that

$$dc_{sys}^{(1)} = K_{12} dc_{out}^{(2)} \tag{4.47}$$

Substituting into Eq. 4.46 yields

$$V_2 \frac{dc_{out}^{(2)}}{dt} + V_1 K_{12} \frac{dc_{out}^{(2)}}{dt} = Qc_{in}^{(2)} - Qc_{out}^{(2)} \tag{4.48}$$

Rearranging, we quickly find the compact first-order differential equation:

$$\frac{dc_{out}^{(2)}}{dt} = \frac{1}{\bar{t}R} [c_{in}^{(2)} - c_{out}^{(2)}] \tag{4.49}$$

Here, $\bar{t}$ is defined as the *mean residence time* (see also Chapters 7 and 10):

$$\bar{t} = \frac{V_2}{Q} \tag{4.50}$$

and R is a type of *retardation coefficient* (see also Chapters 6 and 8):

$$R = \left(1 + K_{12} \frac{V_1}{V_2}\right) \tag{4.51}$$

To solve Eq. 4.49, imagine that at $t = 0$, the inflow concentration switches from 0 to some fixed concentration C_0. This is called a positive step change (Section 10.3). By separating Eq. 4.49 and integrating, we find the solution for the specified initial condition:

$$\frac{c_{out}^{(2)}}{C_0} = 1 - \exp\left(-\frac{t}{\bar{t}R}\right) \tag{4.52}$$

Figure 4.9 is a plot of Eq. 4.52 for different values of R and $t/\bar{t}$. Note that for a constant $t/\bar{t}$, as the retardation coefficient increases, the exit concentration profile is lowered. This means that as the chemical or component of interest partitions more and more strongly into phase 1 (the stationary phase), the passage of the chemical through the system is more and more retarded. Although the system in Figure 4.8a is very idealized, it can serve as a model for understanding more complicated natural systems. For example, compounds like pesticides enter lakes and partition into sediments at the bottom of the lake. The "washout" of the pesticide from the lake can then be delayed or retarded for months, years, or longer. In homes, volatile organic carbon compounds can adsorb or partition onto internal surfaces, thus delaying the ventilation of the compound from the home (Chang and Guo, 1992).

Another important facet of Eq. 4.52 becomes evident when we ask what would happen to two different dissolved chemicals entering the system at the same time, but which have two very different partition coefficients? From the analysis described above, the chemical with the smallest partition coefficient would exit the system earlier than the chemical with the higher partition coefficient. This is a classical aspect of chemical analysis called separation. Chemicals with different adsorptive or partitioning characteristics move through separatory systems at different speeds, and this characteristic can be used to quantitatively differentiate the chemicals (Section 8.6).

For the system in Figure 4.8b, we find in Exercise 4.11 that the differential equation describing the transport of some partitioning chemical through the system is

$$\frac{\partial c^{(2)}}{\partial t} + \left(\frac{U}{R}\right)\frac{\partial c^{(2)}}{\partial x} = 0 \tag{4.53}$$

Here U is the velocity in the flowing liquid (phase 2). We find an almost identical equation in Section 8.5 where we study nondispersive transport of an adsorbing compound in a porous medium (Eq. 8.27). The reader will also note

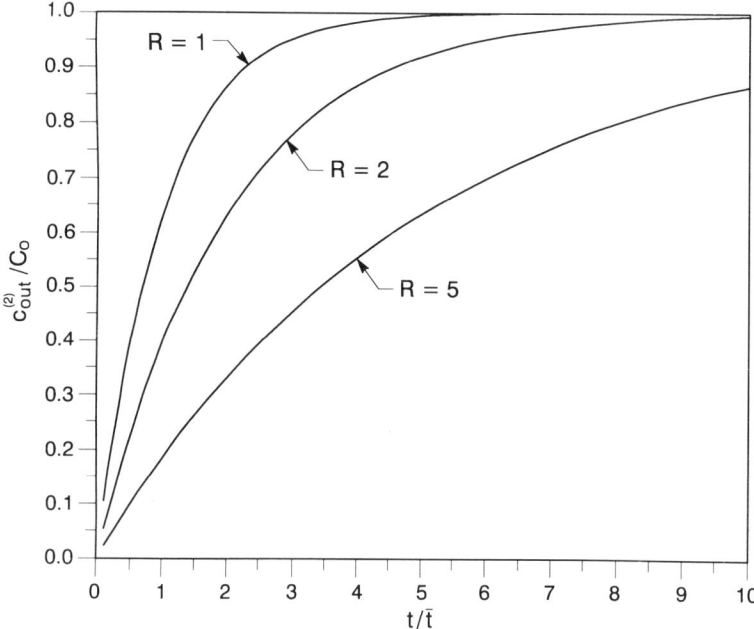

Figure 4.9. Normalized exit concentration of system in Fig. 4.8a, showing the effect of the retardation coefficient R.

that Eq. 4.53 is the one-dimensional version of Eq. 1.26, which is the differential expression for conservation of mass. Note that any retardation coefficient greater than unity effectively lowers the velocity in the system (from the standpoint of the dissolved or transported component). The solution for Eq. 4.53 is also studied in Section 8.5. The main result is very similar to the analysis completed for the system in Figure 4.8a: Partitioning of a chemical slows the transport of the chemical through a two-phase system. Just as in the previous example, the retardation can have environmental significance, and can sometimes be used to advantage in chemical analysis (Chapter 8).

4.7 SUMMARY

In this chapter we considered the equilibrium accumulation of material between two phases. We developed the Gibbs, Langmuir, Freundlich, and BET isotherms to quantify accumulation of material at the gas–liquid, gas–solid, and liquid–solid interfaces. We also pointed out that the adsorption of many environmentally important chemicals can be described by linear isotherms and linear partitioning, and we developed the concept of the equilibrium partition coefficient to describe these situations. Finally, for flowing systems, we found that partitioning could result in the slowing down or retardation of the chemical's transport through multiphase systems.

The equilibrium approach used in this chapter has proved very powerful in solving several types of environmental transport and pollution-control problems. The exercises provide further examples of the usefulness of the equilibrium assumption. However, assumptions in science can be both our salvation and our curse. By accepting the equilibrium assumption in this chapter, we also implicitly assumed that the time required to reach the equilibrium state was of little concern. Or, as we begin to say in later chapters, the time scale of the adsorption or partitioning process is assumed to be small relative to some other time scale of interest. For instance, in Example 4.1, it was implicitly assumed that there was plenty of time for the adsorption reaction between the carbon and TCE to go to completion: the time scale of adsorption was short relative to the time allowed for contact of activated carbon and TCE. But what if we were trying to decontaminate millions of gallons of contaminated solution or hundreds of thousands of cubic meters of contaminated air? We probably would not want to build a million-gallon processing tank or a thousand-cubic-meter chamber because of the capital cost of such a system. In fact, we would prefer a smaller continuous-flow treatment system, or a smaller batch-treatment system that would be used to process many small batches of the contaminated system. We also probably do not want to wait an indefinite time for each batch to come to equilibrium, since we have to pay people to wait around while the treatment system is being operated.

This leads us to new questions: What are the kinetics of the adsorption process? (How long do I have to wait?) If the initial adsorption rate is fast and later adsorption is slow, might we want to stop the process early, discard the adsorbent, and start with new adsorbent? However, if we use lots of adsorbent material, our operational costs are higher (adsorbents like activated carbon are not free!). Understanding the system dynamics (or kinetics) will help us to sort our these questions.

We return to these dynamics questions at several points in this book. The material on diffusive mass transport in Chapter 6 is particularly relevant to the adsorption and partitioning kinetics questions we raise here.

EXERCISES

4.1. Referring to Figure 4.3, determine which compound — sodium lauryl sulfate or sodium chloride — would be expected to yield a "linear" Gibbs adsorption isotherm?

4.2. The following data were measured in a study of the interfacial tension at the water–decane surface in the presence of various concentrations of the surfactant sodium dodecyl sulfate (NaDS) in the water phase (Cockbain, E. G., *Trans. Faraday Soc.*, **50**, 874, 1954):

Concentration NaDS (mol/L)	Interfacial Tension (dyne/cm)
0.0079	8.5
0.00694	10.8
0.00521	15.3
0.00347	20.8
0.001735	28.3

(a) As in Figure 4.3, plot the interfacial tension vs. the concentration of NaDS. Is the plot linear?

(b) Making any necessary unit conversions, determine the surface excess (mol/cm^2) for each of the five measurements listed above. Assume $T = 25°C$.

4.3. At low surface coverage (i.e., $S_1 \approx 0$), the rate of condensation should be approximated sufficiently by

$$\text{rate of condensation} = k_2 P(S - S_1) \approx k_2 PS$$

Show that with this assumption, we also end up with Eq. 4.18. (Comment: this shows that when the rates of evaporation and condensation are simple linear functions, a linear isotherm will always result.)

4.4. The following data were gathered in a laboratory study of the adsorption of trichlorophenol (TCP) on a commercial powdered, activated carbon:

Solvent: Central Illinois groundwater
Solute: 2,4,6-trichlorophenol (molar mass = 197.5 g/mol)
Initial concentration of solute: 470 μg/L
Volume of solvent: 500 mL

Carbon Dose (mg)	C_e (μg/L)
13.4	4.1
9.3	6.2
5.6	27.9
3.3	91.1
2.0	168.0

Fit these data to the Langmuir and Freundlich isotherms, and determine the constants (parameters) of both models. On the basis of the model fit to the data, does one of these isotherm models seem more appropriate than the other?

4.5. It is necessary to determine the surface area of a sample of activated carbon. This is often done to compare different types of activated carbon (e.g., from different manufacturers), because carbons with greater surface area generally adsorb more aqueous or gaseous pollutants. A classic experiment for determining the surface area of activated carbons and other surfaces is the N_2-BET surface area determination. In this test, nitrogen is adsorbed on the sample, and the resulting data are fit to the BET isotherm model. From the parameter fit to the BET isotherm, the surface area of the carbon can be estimated. The following data were determined in such an experiment for a granular-activated carbon sample:

Relative Pressure (P/P_0)	Volume N_2 Adsorbed per Gram Adsorbent (cm^3/g)
0.0054	154.69
0.0085	161.52
0.0101	164.10
0.0312	184.77
0.0505	195.48
0.0804	206.85
0.0993	212.14
0.1253	218.17

Note that both sides of Eq. 4.27 can be divided by the mass of adsorbent (mass of carbon in this case) to yield

$$\frac{X}{V'(1-X)} = \frac{1}{cV'_m} + \frac{(c-1)X}{cV'_m} \tag{4.27a}$$

Here $V' = V/\text{mass adsorbent}$ and $V'_m = V_m/\text{mass adsorbent}$. Note that the second column of data given above are actually V'.

(a) Use the equation above and experimental data to determine V'_m.

(b) Determine the specific surface area (m^2/g) of the activated carbon. To do this, note that the specific surface area should equal the volume of the monolayer (per gram of carbon), V'_m, times the number of moles of gas per unit volume times the surface area occupied by an N_2 molecule on the surface. If the area of an N_2 molecule is $16.2 \times 10^{-20}\ m^2/\text{molecule}$, the required calculation is

$$\frac{\text{specific surface}}{\text{area of solid}}\left(\frac{m^2}{g}\right) = V'_m\left(\frac{cm^3}{g}\right) \times \frac{L}{1000\ cm^3} \times \frac{6.02 \times 10^{23}\ \text{molecule}}{22.4\ L}$$

$$\times\ 16.2 \times 10^{-20}\ \frac{m^2}{\text{molecule}}$$

The answer to this problem should demonstrate why activated carbon is such a useful adsorbent.

4.6. Can you describe any limiting conditions under which the BET isotherm equation (Eq. 4.25) has a mathematical form similar to the Langmuir isotherm equation (Eq. 4.16)?

4.7. Fit a straight line to the data in Figure 4.7 for adsorption of Atrazine in groundwater. The molar mass of Atrazine is 215.7 g/mol. From the straight-line fit, determine the Freundlich adsorption parameter $1/n$. Then determine the Freundlich K_f using a point on the straight line. How does the Freundlich capacity parameter K_f compare with the K_f value determined in distilled water. Why are the two capacity values different?

4.8. To a 1-L container holding 0.01 mg/L carbon tetrachloride, 25 mg/L of activated carbon was added. Determine the equilibrium concentration of tetrachloride in the water if the Freundlich isotherm parameters for this carbon are those given in Table 4.1. Hint: perform a mass balance on carbon tetrachloride.

4.9. A tube containing 50 mL of octanol, 50 mL of water, and 0.5 g of DDT is shaken until the DDT reaches its equilibrium concentration in the octanol and water phases. What is the concentration of DDT (mg/L) in the water phase? The octanol–water partition coefficient for DDT is given in Table 4.2.

4.10. A 1000-m³ lake contains 1000 fish and 1 g of DDT. The average volume of the fish is 100 cm³. If the fish are 5% lipids, and if we can assume that the DDT partitions into the fish exactly as it does into octanol, what is the equilibrium DDT concentration in the fish? Assume that there is no dilution (or concentration) of the lake water nor any interactions of DDT with the lake sediments.

4.11. To derive Eq. 4.53, draw a differential element across the phases 1 and 2, make a mass balance around the differential element, and substitute for the concentration in phase 1 as was done in the development of Eq. 4.49. Finally, take the limit of the equation as the volume of the differential element shrinks to zero. Hint: you will need to express the concentration at the right side of the differential element in terms of the concentration at the left side using a truncated Taylor expansion:

$$c_{x+\Delta x}^{(2)} = c_x^{(2)} + \frac{\partial c^{(2)}}{\partial x} \Delta x$$

Also assume a uniform velocity u in the plug-flow region 2.

4.12. For the system examined in Figure 4.8a plot the normalized concentration in phase 1, $c_{\text{sys}}^{(1)}/C_0$, vs. $t/\bar{t}$ for $V_1 = V_2$ and $R = 1, 2,$ and 5. What values of K_{12} are used in the plots?

REFERENCES

Adamson, A. W., *Physical Chemistry of Surfaces*, 3rd ed., Wiley, New York, 1976.

Atkins, P. W., *Physical Chemistry*, W. H. Freeman, San Francisco, CA, 1978.

Chang, J. C. S. and Guo, Z., "Modeling of the Fast Organic Emissions from a Wood-Finishing Product — Floor Wax," *Atmos. Environ.*, **26A**, 13:2365–2370 (1992).

Freundlich, H., *Colloid and Capillary Chemistry*, Meuthen, London, 1926.

Lyman, W. J., "Octanol/Water Partition Coefficient," in *Handbook of Chemical Property Estimation Methods. Environmental Behavior of Organic Compounds*, Lyman, W. J., Reehl, W. F., and Rosenblatt, D. H., Eds., McGraw-Hill, New York, 1982.

Mackay, D., *Multimedia Environmental Models: The Fugacity Approach*, Lewis Publishers, Chelsea, MI, 1991.

Qi, S., Adham, S., Snoeyink, V. L., and Lykins, B. W., "Prediction and Verification of Atrazine Adsorption by PAC," *J. Environ. Eng.*, **120**, 1:202–218 (1994).

Shirahama, H., Lyklema, J., and Norde, W., "Comparative Protein Adsorption in Model Systems," *J. Colloid Interface Sci.*, **139**, 1:177–187 (1990).

Snoeyink, V. L., "Adsorption of Organic Compounds," in *Water Quality and Treatment*, 4th ed. Pontius, F. W., McGraw-Hill, New York, 1990, Chapter 13.

Thibodeaux, L. J., *Chemodynamics: Environmental Movement of Chemicals in Air, Water, and Soil*, Wiley, New York, 1979.

BIBLIOGRAPHY

Adamczyk, B., "Localized Adsorption of Particles on Spherical and Cylindrical Interfaces," *J. Colloid Interface Sci.*, **146**, 1:123–136 (1991).

Baum, S. J. and Hill, J. W., *Introduction to Organic and Biological Chemistry*, Macmillan, New York, 1993.

Cohen Stuart, M. A., Fleer, G. J., Lyklema, J., Norde, W., and Scheutjens, J. M. H. M., "Adsorption of Ions, Polyelectrolytes and Proteins," *Adv. Colloid Interface Sci.*, **34**, 477–535 (1991).

Freifelder, D., *Principles of Physical Chemistry with Applications to the Biological Sciences*, 2nd ed., Jones and Bartlett, Boston, MA, 1982.

Hiemenz, P. C., *Principles of Colloid and Surface Chemistry*, Marcel Dekker, New York, 1977.

Roberts, P. V., Goltz, M. N., and Mackay, D. M., "A Natural Gradient Experiment on Solute Transport in a Sand Aquifer: 3. Retardation Estimates and Mass Balances for Organic Solutes," *Water Resour. Res.*, **22**, 13:2047–2058 (1986).

Stumm, W. and Morgan, J. J., *Aquatic Chemistry: An Introduction Emphasizing Chemical Equilibria in Natural Waters*, 2nd ed. Wiley, New York, 1981.

Weber, W. J., "Adsorption," in *Physicochemical Processes for Water Quality Control*, Weber, W. J., Ed., Wiley-Interscience, New York, 1972, Chapter 5.

Welty, J. R., Wicks, C. E., and Wilson, R. E., *Fundamentals of Momentum, Heat, and Mass Transport*, Wiley, New York, 1976.

Wu, S. and Gschwend, P., "Sorption Kinetics of Hydrophobic Organic Compounds to Natural Sediments and Soils," *Environ. Sci. Technol.*, **20**, 7:717–725 (1986).

5

BASIC FLUID MECHANICS
OF ENVIRONMENTAL
TRANSPORT

5.1. INTRODUCTION

Fluid mechanics is one of the oldest areas of study in applied mathematics and engineering. Ancient civilizations were much preoccupied with the movement of water in rivers and the transport of water through aqueducts, channels, and pipes. In the early twentieth century, humans became very interested in aerodynamics, or the fluid mechanics of air flows.* Advancements in this type of fluid mechanics sped the advent of powered flight. In environmental science and engineering, the implications and importance of fluid mechanics are immense and span scales from the submicron to the continental. For example, in Chapter 2, we examined the impact of fluid drag (a classic area of fluid mechanics) on the transport of small particles under the influence of gravitational, electrical, and centrifugal forces. However, fluid mechanics is involved in a far larger group of transport problems. When we can predict the movement of environmental fluids (the liquids and gases that make up our environment), we can often predict what will happen to particles, chemicals, and even energy that tend to follow (or be transported by) the flow. The goal of this chapter is to sharpen our understanding of how fluids flow. In later chapters we use this information to developed mass-transfer models.

In this relatively small chapter we cannot do justice to the topic of classical fluid mechanics or to the full range of possible scales of fluid mechanics and turbulence. Thus, our focus is on selected problems in fluid mechanics at very small scales, like those pertaining to colloidal phenomena, and at more intermediate scales, such as those pertaining to pollution-control equipment, ducts, pipes, and channels. This leaves out the fluid mechanics of large-scale phenomena, like flows in oceans, estuaries, and the atmosphere.

---

*Sometimes students have difficulty getting used to thinking of air and other gases as fluids. Air at normal temperatures and pressures is a fundamentally different type of medium than water. The intermolecular forces are different in air, and the molecular spacing in air is vastly greater than in water (recall the discussion of continua in Chapter 1). Nevertheless, air generally qualifies in all regards as a fluid, and is considered such in this book.

5.2. THE JOY OF FLUID MECHANICS

We live in and experience fluids throughout our lives. We begin life in a sac of fluid, spend our lives breathing the fluid air, feel the warm airs of summer against our skin, and marvel at humankind's ability to engineer craft that lift us into the air. In fact, we are largely composed of the fluid water (although we generally do not behave like fluids). The pleasure of fluid mechanics study comes when we realize that our everyday observations are so neatly predicted by theory.

Fluid mechanics is an area of *continuum mechanics.* Hence, there are many similarities between fluid mechanics and solid mechanics. We define a *fluid* as *a deformable material continuum that cannot withstand shear forces without experiencing continuous motion.* We see immediately the main practical difference between a solid and a fluid: A solid (although deformable) can withstand shear forces without experiencing continuous motion. The word "continuum" in the definition is important. In general, we want simple equations to describe the movement of fluids. We do not really care about the movement of individual fluid molecules; rather, we assume that fluid properties vary smoothly and that shear forces are smoothly transferred.

The equations of fluid mechanics are used to predict the movement of fluids. Their development relies on the continuum assumption (Section 1.5), knowledge of the forces acting on a differential element of fluid, and an assumption about the transfer of stresses between adjacent elements of fluid. In fact, the equation of motion of a viscous fluid (the Navier–Stokes equations), developed in Section 5.3, will remind the reader of the equation of motion of a particle

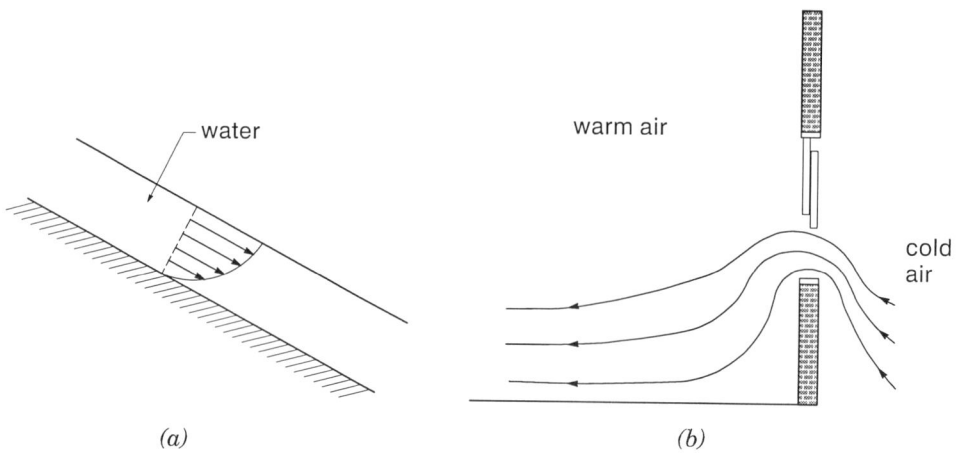

(a)   *(b)*

Figure 5.1. Examples of (*a*) water running downhill and (*b*) cold air entering a room through a window at night.

(Section 2.4). By and large, the theoretical predictions of the Navier–Stokes equations strongly support our everyday observations.

For example, we will find that water does indeed run downhill because of gravity (Figure 5.1a and Exercise 5.8). We also see in Figure 5.1b that cool night air will enter an open window (even without a breeze) because it is more dense than the warm air inside the house or building. Since the cooler night air has a greater density, the *hydrostatic pressure* (or better yet *aerostatic pressure*) is greater outside the window than inside. This phenomenon is closely related to water running downhill, because it would not occur in the absence of gravity (Section 5.4).

We also experience *dynamic pressures* in our everyday lives. Pumps or fans are very useful in circulating air in our homes and fluids in our bodies. A pump or fan uses mechanical energy to increase the pressure of a flow (Figure 5.2a). A spherical particle settling in a quiet atmosphere or still lake also experiences differential pressures around its surface. As expected, pressure is highest in front of the sphere and reaches a minimum at the back of the sphere (Figure 5.2b). We study the pressure distribution on the spherical particle in Section 5.9.2.

Finally, most of the fluid mechanics covered in this chapter requires modeling of viscous forces. Readers will already understand that viscosity can affect fluid motion. For example, we well know that it takes energy to force viscous honey through the small orifice in the container shown in Figure 5.3a. Another interesting experiment to demonstrate the effect of viscosity can be done with a coffee cup and a saucer (a clear glass cup is best). Add some heavy cream to a cup of coffee and slowly spin the cup in its dish. The careful observer will see during the first few seconds that only the fluid next to the walls of the

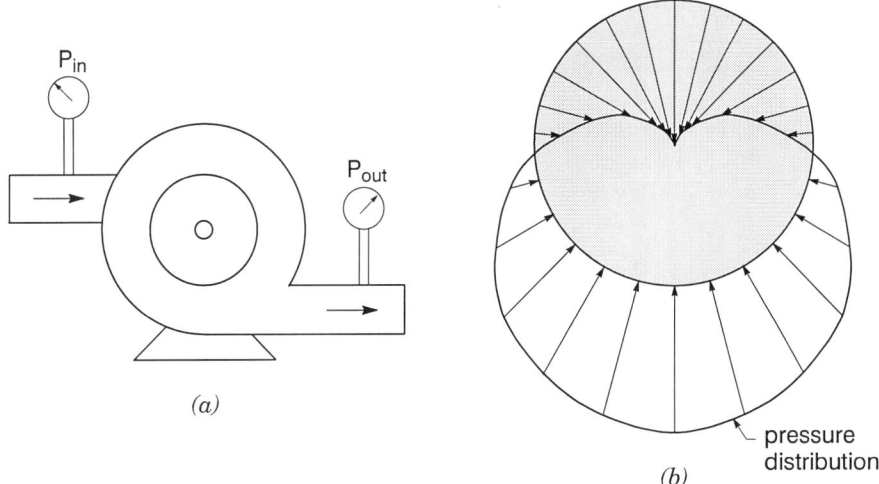

(a)

(b)

pressure distribution

Figure 5.2. Examples of (a) pressure generated by a pump or fan and (b) pressure distribution around a falling sphere.

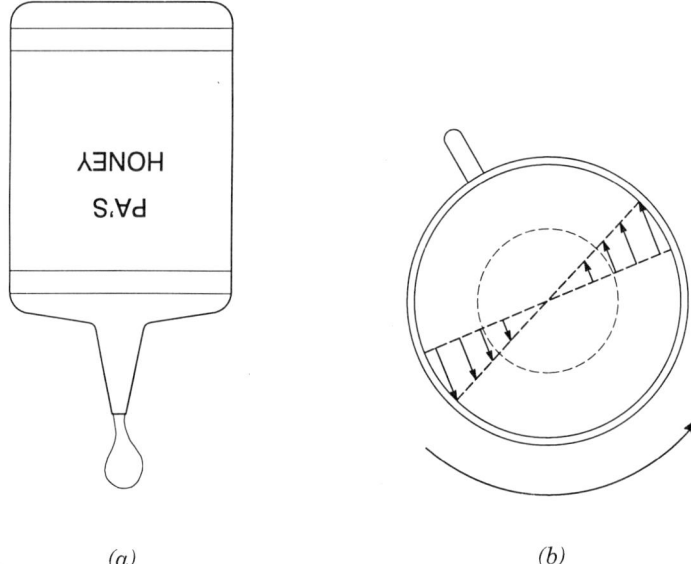

(a)   (b)

Figure 5.3. Examples of (a) the flow of viscous fluids through a small orifice and (b) viscous diffusion of vorticity in a spinning coffee cup.

cup will move — it seems to be dragged around by the walls of the coffee cup. After a while, careful observation will show that more of the fluid near the wall seems to be dragged into motion. And if the spinning goes on long enough, all the fluid in the cup will spin with the solid body rotation discussed in Section 2.10 (Figure 5.3b). It seems that the velocity or momentum of the outer fluid is slowly diffusing into the center of the cup, thereby adding momentum to the inner fluid. In fact, this is exactly what is happening, as shown in Section 5.8. We will also see that this diffusion of momentum increases with increasing fluid viscosity. (To achieve solid-body rotation, we would need to spin the cup fewer times if it were filled with coffee mixed with a lot of honey.)

5.3. THE NAVIER–STOKES EQUATIONS

We dive into our study of fluid mechanics with the most important equation describing incompressible viscous flows, the *Navier–Stokes* equation:

$$\rho \frac{\partial \mathbf{u}}{\partial t} + \rho \mathbf{u} \cdot \nabla \mathbf{u} = -\nabla P + \mu \nabla^2 \mathbf{u} + \rho \mathbf{g} \tag{5.1}$$

Here ρ is the fluid density (assumed constant throughout the fluid*), $\mathbf{u}$ is the fluid velocity, t is time, P is the pressure, μ is the absolute viscosity (assumed constant throughout the fluid), and $\mathbf{g}$ is the body force caused by gravity. This is a rather complicated partial differential equation; it is also a *nonlinear* differential equation (note the term involving a dot product). Because of the nonlinearities, the equation historically has been difficult to solve in the most general cases. Fortunately, it can be often simplified.

In many books, Eq. 5.1 is presented in an equivalent form:

$$\rho \frac{D\mathbf{u}}{Dt} = -\nabla P + \mu \nabla^2 \mathbf{u} + \rho \mathbf{g} \tag{5.2}$$

Here D/Dt is called the *material* or *substantial derivative* operator, which in rectangular coordinates is

$$\frac{D}{Dt} = \frac{\partial}{\partial t} + u_1 \frac{\partial}{\partial x_1} + u_2 \frac{\partial}{\partial x_2} + u_3 \frac{\partial}{\partial x_3} \tag{5.3}$$

Note that the left-hand sides of Eqs. 5.1 and 5.2 involve the products of density (or "unit mass") and acceleration-like terms. Since acceleration is the derivative of velocity, the Navier–Stokes equations are known as the "momentum equations" of a differential fluid element (or more properly the change of momentum equation). The density and acceleration terms are analogous to the products of mass and acceleration in the equation of motion of a particle (Section 2.4). Equations 5.1 and 5.2 are therefore also known as the "equations of motion" of a differential fluid element (recall Figure 1.4). The time derivative on the left-hand side is called the *local acceleration*; the nonlinear dot-product terms are called the *convective acceleration* terms.

The terms on the right-hand sides of the equations are the forces acting on the fluid element. The first is the pressure-gradient term: a fluid element will respond to a local pressure gradient (recall Figure 5.2a). The second term involves viscosity and the Laplacian of velocity. These are called the viscous terms and they arise because each face of the differential fluid element can be acted upon by viscous stresses from adjacent fluid (recall Figure 5.3b). The final term on the right-hand side is the gravity term, which as we know from everyday experience can lead to fluid motion. The gravity force is known as a *body* force, because gravity works equally on all molecules in the fluid element. The pressure and viscous terms are known as *surface* forces, because they interact with the surface of the fluid element.

---

*Air and other gases are much more compressible than liquids like water. Nevertheless, *in fluid-flow problems*, as long as the air velocity is less than about 50 m/s, the maximum change in density is less than about 1% (Lighthill, 1986). So none of the flow analyses in this book will suffer because of the assumption of incompressibility. We will need to consider compressibility when considering the vertical variations of pressure in a static atmosphere.

Equations 5.1 and 5.2 are vector equations; in reality, they represent three individual equations which are also called the Navier–Stokes equations. This is best seen with the aid of tensor notation:

$$\rho \frac{\partial u_i}{\partial t} + \rho u_k \frac{\partial u_i}{\partial x_k} = -\frac{\partial P}{\partial x_i} + \mu \frac{\partial^2 u_i}{\partial x_j \partial x_j} + \rho g_i \tag{5.4}$$

Tensor notation helps compactly express three equations as one. It is based on the *Einstein convention*, which means simply that repeated indices imply summation over that index. Thus, Eq. 5.4 is actually three equations. The possible index values are 1, 2, and 3, which stand for the three dimensions of rectangular cartesian space.* For example, let $i = 1$; the Einstein convention says that a summation occurs in the second term on the left-hand side (repeated k index) and the second term on the right-hand side of Eq. 5.4 (repeated j index):

$$\rho \frac{\partial u_1}{\partial t} + \rho \left(u_1 \frac{\partial u_1}{\partial x_1} + u_2 \frac{\partial u_1}{\partial x_2} + u_3 \frac{\partial u_1}{\partial x_3} \right)$$
$$= -\frac{\partial P}{\partial x_1} + \mu \left(\frac{\partial^2 u_1}{\partial x_1 \partial x_1} + \frac{\partial^2 u_1}{\partial x_2 \partial x_2} + \frac{\partial^2 u_1}{\partial x_3 \partial x_3} \right) + \rho g_1 \tag{5.5}$$

With $i = 2$, we find the x_2 component of the Navier–Stokes equation:

$$\rho \frac{\partial u_2}{\partial t} + \rho \left(u_1 \frac{\partial u_2}{\partial x_1} + u_2 \frac{\partial u_2}{\partial x_2} + u_3 \frac{\partial u_2}{\partial x_3} \right)$$
$$= -\frac{\partial P}{\partial x_2} + \mu \left(\frac{\partial^2 u_2}{\partial x_1 \partial x_1} + \frac{\partial^2 u_2}{\partial x_2 \partial x_2} + \frac{\partial^2 u_2}{\partial x_3 \partial x_3} \right) + \rho g_2 \tag{5.6}$$

Finally, for $i = 3$, we find the x_3 component of the Navier–Stokes equation:

$$\rho \frac{\partial u_3}{\partial t} + \rho \left(u_1 \frac{\partial u_3}{\partial x_1} + u_2 \frac{\partial u_3}{\partial x_2} + u_3 \frac{\partial u_3}{\partial x_3} \right)$$
$$= -\frac{\partial P}{\partial x_3} + \mu \left(\frac{\partial^2 u_3}{\partial x_1 \partial x_1} + \frac{\partial^2 u_3}{\partial x_2 \partial x_2} + \frac{\partial^2 u_3}{\partial x_3 \partial x_3} \right) + \rho g_3 \tag{5.7}$$

Note that by writing out the three equations implied by the tensor notation (Eqs. 5.5–5.7), we arrive at three *scalar* equations representing four unknowns

---

*Note that we are now using the subscripts 1, 2, and 3 to represent the three rectangular cartesian space dimensions, whereas in earlier chapters we used x, y, and z, respectively. We do this without excuse, because both forms are commonly used in the literature. The student must get used to both forms of presentation.

(u_1, u_2, u_3, and P). Since there are just three equations, the set of equations cannot be solved uniquely for the four unknowns. However, by adding the differential continuity equation from Chapter 1 (Eq. 1.24) and sufficient boundary and initial conditions, we can in principle solve the Navier–Stokes equations uniquely. In reality, we find that most of the solutions to the Navier–Stokes equations in this book do not require the continuity equation, because the equations greatly simplify in the examples and problems covered here.

Finally, the equations in this section have been presented largely for a rectangular coordinate system. The equivalent equations for cylindrical and spherical coordinates are presented in Appendix III.

5.4 FLUID STATICS AND THE BUOYANCY FORCE

The easiest solution to the Navier–Stokes equations occurs when there is no fluid motion at all, that is, $\mathbf{u} = 0$. Then Eq. 5.1 or 5.2 reduces to

$$\nabla P = \rho \mathbf{g} \tag{5.8}$$

If we consider the container of fluid shown in Figure 5.4, we find a simple demonstration of the importance of this equation.

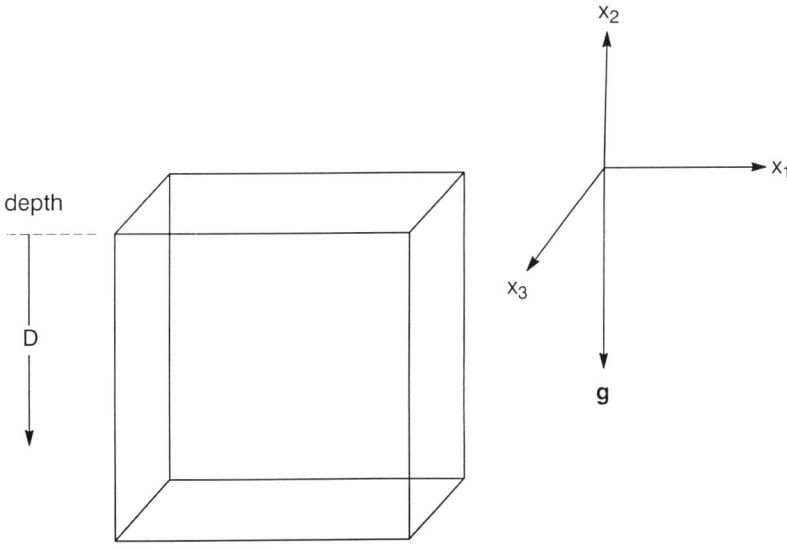

Figure 5.4. A box of quiescent fluid.

With the coordinate system shown, the gravity force has only one component in the negative x_2 direction. Therefore, $\partial P/\partial x_1 = 0$ and $\partial P/\partial x_3 = 0$, and Eq. 5.8 requires that

$$\frac{\partial P}{\partial x_2} = -\rho g_2 = -\rho g \tag{5.9}$$

where for simplicity we drop the subscript 2 in the gravity term. Separating and integrating Eq. 5.9 yields

$$P = -\rho g x_2 + \text{constant} \tag{5.10}$$

If we let $P = 0$ at the fluid surface ($x_2 = 0$), we find that the constant is equal to zero. Replacing x_2 with the negative of depth ($-D$), we find that

$$P = \rho g D \tag{5.11}$$

This is the basic equation of fluid statics, which says quite elegantly that the *hydrostatic* (or *aerostatic* for air) pressure increases linearly with depth for a fluid of constant density and constant body force (**g**).

Example 5.1. Calculate the total force acting on any of the side walls of the box shown in Figure 5.4 if the box contains water. The side walls have a width (into the page) of 10 m. Ignore atmospheric pressure.

SOLUTION. Examining the left-hand side of the box shown in Figure 5.4, we can make the sketch shown below.

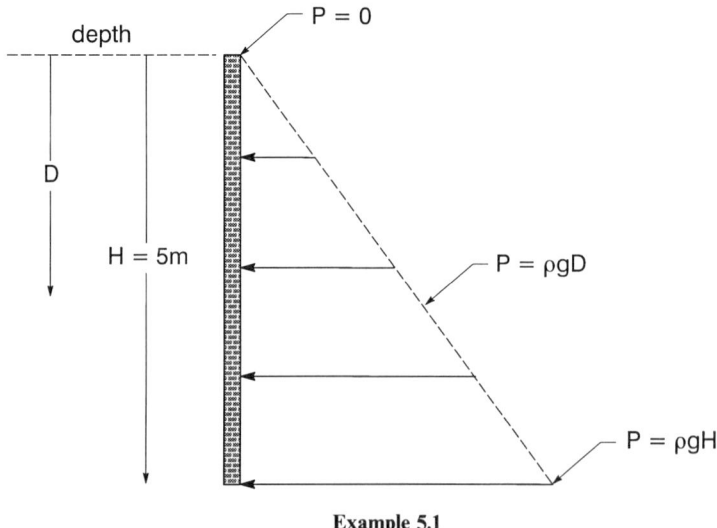

Example 5.1

Since force is equal to pressure integrated over an area, we quickly surmise that

$$\text{force} = \iint P dA = \int_0^H PwdD$$

where the surface integral is replaced by differentiation over a differential strip of width w and height dD. Substituting Eq. 5.11 for the pressure force term:

$$\text{force} = \rho g w \int_0^H DdD = \rho g w \frac{H^2}{2} = \left(\frac{\rho g H}{2}\right)(wH)$$

The final rearrangement shows the classic result that hydrostatic force on a rectangular surface is equal to the average pressure times the surface area.

The actual force on the side wall can be gotten by substituting in numerical values. For $\rho = 1000 \text{ kg/m}^3$, $g = 9.80 \text{ m}^2/\text{s}$, $w = 10 \text{ m}$, and $H = 5 \text{ m}$:

$$\text{force} = \rho g w \frac{H^2}{2} = 1000 \frac{\text{kg}}{\text{m}^3} \; 9.8 \frac{\text{m}}{\text{s}^2} \; 10 \text{ m} \frac{(5 \text{ m})^2}{2} = 1.23 \times 10^6 \frac{\text{kg-m}}{\text{s}^2} = 1.23 \times 10^6 \text{ N}$$

Using Newton's second law (force = mass × acceleration), we find that this force is equivalent to a mass of $(1.23 \times 10^6 \text{ kg-m-s}^{-2})/(9.80 \text{ m/s}^2) = 125,000 \text{ kg}$ ($= 276,000 \text{ lb}$).

Equation 5.8 can be derived in another way, and doing so will help us to understand the so-called buoyancy forces. Consider any closed region in a fluid. (This could be a sphere, a box, or any closed shape.) For there to be no translation of this fluid element, it must be true that the sum of the force of gravity (body force) on the element and the pressure forces acting normal to the surface of the element is zero. The gravity force is given by

$$\iiint \rho \mathbf{g} dV \tag{5.12}$$

and the surface force is given by*

$$-\iint \mathbf{n} P dA \tag{5.13}$$

where $\mathbf{n}$ is the outward unit normal vector. To achieve equilibrium, the surface

---

*The outward unit normal vector points away from the surface, but we know that the pressure force is actually directed toward the surface. Therefore, we need a negative sign before the surface integral to make the force have the correct direction.

force must equal the negative of the body force, or

$$\iiint \rho \mathbf{g} dV = \iint \mathbf{n} P dA \tag{5.14}$$

However, there is an integral vector theorem that is analogous to the divergence theorem (Batchelor, 1967), which says that for scalars like pressure

$$\iint P \mathbf{n} dA = \iiint \nabla P dV \tag{5.15}$$

Hence, substituting Eq. 5.15 into Eq. 5.14 we find that

$$\iiint_V (\rho \mathbf{g} - \nabla P) dV = 0 \tag{5.16}$$

However, since the region of volume V is arbitrarily chosen, the relation shown above can only be true in general if the term in parentheses is zero; hence, as desired, we recover Eq. 5.8:

$$\nabla P = \rho \mathbf{g} \tag{5.8}$$

Now assume that we somehow replace the fluid in region V with another fluid of density ρ', or even a solid of density ρ'. The surface force on the volumetric region must be the same as before, which in turn is equal to the negative of the body force on the old fluid of density ρ, that is,

$$\begin{array}{c}\text{surface force on}\\\text{element with new fluid}\end{array} = \begin{array}{c}-\text{body force on}\\\text{element with old fluid}\end{array} = -\iiint \rho \mathbf{g} dV \tag{5.17}$$

This is called the *buoyancy force*. As we noted in Section 2.4, Eq. 5.17 shows that the buoyancy force works in a direction opposite to the gravity force. In the case of particles, the buoyancy force can be described as follows: the magnitude of the buoyancy force is equal to the mass of fluid displaced by the particle ("region" or "element" in discussion above) times the acceleration of gravity, and the direction of the buoyancy force is in a sense opposite to the acceleration of gravity.

In Section 2.4 we required the *net force due to gravity* acting on a spherical particle of constant density ρ_p immersed in some medium with constant density ρ_f. The downward-acting force is simply the particle mass M_p times the acceleration of gravity. As stated above, from Eq. 5.17, the buoyancy force is simply the mass of fluid displaced by the particle, M_f, times the acceleration of gravity. The net force of gravity acting on the particle is then the difference of these two forces:

$$\mathbf{F}_g = M_p \mathbf{g} - M_f \mathbf{g} \tag{5.18}$$

This is called the net force of gravity here because it completely disappears in the absence of gravity.

5.5. THE MODIFIED PRESSURE AND FREE-SURFACE FLOWS

At this point, the reader might wonder what becomes of the hydrostatic pressure term uncovered above when there is movement of the fluid. The effect of this term is still felt, but it does not tell the whole pressure story. In fact, Batchelor (1967) shows that total pressure P in the Navier–Stokes equations can always be thought of as the sum of three terms:

$$P = p_0 + \rho \mathbf{g} \cdot \mathbf{x} + p \tag{5.19}$$

The term p_0 is a constant pressure which, for example, might be due to standard atmospheric pressure or an external pressurization of the system. This pressure would be constant throughout the fluid.* The second term involving the dot product of $\mathbf{g}$ and $\mathbf{x}$ is equivalent to the term from fluid statics (Eq. 5.11). The last term then is the part of the pressure that arises because of fluid motion. Batchelor calls this the *modified* pressure. Note that if we apply the gradient operation to Eq. 5.19, we find the following (see Exercise 5.2):

$$\nabla P = \rho \mathbf{g} + \nabla p \tag{5.20}$$

Substituting Eq. 5.20 into Eq. 5.2 yields a special form of the Navier–Stokes equations containing the modified pressure:

$$\rho \frac{D\mathbf{u}}{Dt} = -\nabla p + \mu \nabla^2 \mathbf{u} \tag{5.21}$$

This shows that when pressure is decomposed as suggested by Batchelor, gravity somehow magically drops out of the equations of motion. In fact, this is only true in viscous flows with no *free surface*. Even though it appears that we can always dispose of the gravity force, gravity will reenter the problem through boundary conditions if gravity is in fact a significant driving force for fluid motion. In other words, if gravity is important, we cannot get rid of it.

---

*The term p_0 was of course acting in Example 5.1. We simply were not interested in it. Our solution was for the so-called "gauge pressure." In Example 5.1, p_0 corresponded to the atmospheric pressure.

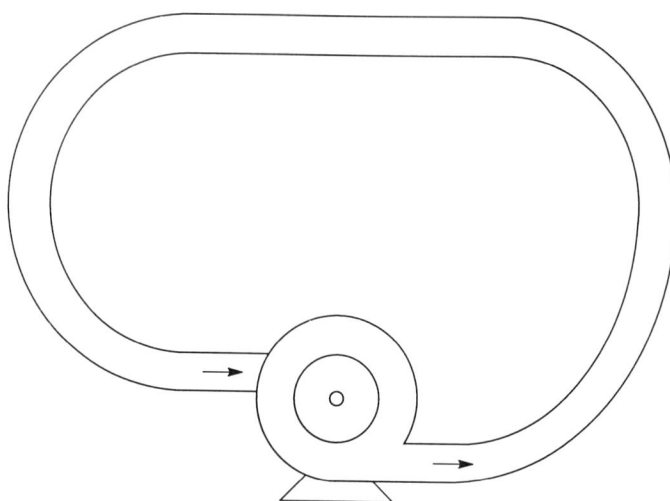

Figure 5.5. Pump and pipe loop completely filled with fluid.

Gravity is obviously an important factor in the two flows shown in Figure 5.1. Other examples can be imagined, such as water running out of a pipe from an open reservoir. Flows where gravity is important are often referred to as *free-surface flows*. It is also easy to find flows in which gravity has no effect. Figure 5.5 shows a pump connected to a closed-loop pipe. The fluid could be air, water, or most any other liquid. Since the pipe is completely filled with fluid, there is obviously no fluid motion in the pipe when the pump is turned off. When the pump is turned on, the fluid is set into motion. But does gravity affect fluid motion and velocity inside the pump–pipe system? The answer is no. In fact, we could spin the whole system at a steady rate, or turn it upside down; as long as the system geometry remained rigid, fluid motion inside the pipe would not be affected. This is a perfect example of a flow *without* a free surface. Gravity is irrelevant in such flows.

In general, we should examine a problem to see if in fact gravity is a mechanism of fluid motion, or if it is negligible compared with other mechanisms of fluid motion (e.g., pressure gradient or viscous stresses). If the gravity force is not applicable (as in flows without a free surface), or negligible compared with other forces, we can begin our analysis with Eq. 5.21.

5.6 STEADY UNIDIRECTIONAL FLOWS AND STEADY CIRCULAR STREAMLINE FLOWS

As stated earlier, the primary difficulty in solving the Navier–Stokes equations has been related to the nonlinearity of the convective acceleration terms. However, in *unidirectional* flows, all the flow is constrained to a single direction

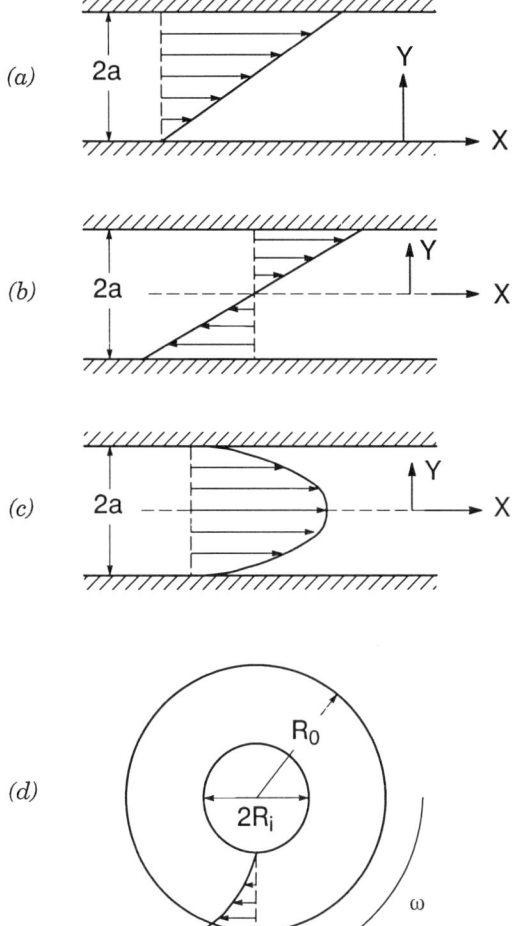

Figure 5.6. (*a*) Plane couette flow with one wall moving; (*b*) plane couette flow with walls moving at the same speed in opposite directions; (*c*) plane couette flow due to pressure gradient; and (*d*) couette flow in circular annulus with one wall moving.

and flow velocities do not change along streamlines. Then the gradient of velocity must be perpendicular to the velocity vectors, because the greatest change in velocity is perpendicular to the velocity vectors. Therefore the dot product in the nonlinear convective acceleration terms is identically zero (i.e., dot product of perpendicular vectors is zero) and we do not have to worry about the nonlinear terms! Figure 5.6 shows three straightforward examples of steady (i.e., steady-state) unidirectional flows. An additional example is the free-surface incline flow illustrated in Figure 5.1a. The flows demonstrated in Figure 5.6 are also called *couette* flows, which is the French word for bearing.

Figure 5.6a shows the steady flow pattern that results from the steady motion of the upper boundary. Figure 5.6b shows the steady flow pattern that results from the steady motion of both boundaries at the same speed but in opposite directions. Figure 5.6c shows the steady flow pattern that arises from a uniform pressure gradient in the streamwise (x) direction with both upper and lower boundaries stationary. In Figure 5.6a–c, we can imagine that the flow extends to infinity in the x direction. Figure 5.6a–c provides an excellent example of *two-dimensional* flows: all fluid motion is sufficiently described by the x and y dimension. In such two-dimensional flows, we can imagine that the flow also extends to infinity in the z direction (out of the page).

The solutions of the Navier–Stokes equations for the first two examples in Figure 5.6 are fairly straightforward and are left as Exercises 5.3 and 5.4. We study here in detail the third flow in Figure 5.6c. First note that since flow is only in the x (or x_1) direction, we probably only need to solve the x_1 component of the Navier–Stokes equations. As noted above, the convective acceleration terms are zero. Further, based on the assumed flow pattern in the figure, there are only velocity derivatives in the y direction. Finally, since we can assume that this is a *no*-free-surface flow, the x_1 component of the Navier–Stokes equations becomes

$$\frac{dp}{dx} = \mu \frac{d^2 u_x}{dy^2} \tag{5.22}$$

We will assume that the pressure gradient term is constant in this flow, because the flow is unidirectional and invariant in the x direction. Integrating Eq. 5.22 once, we find:

$$\frac{dp}{dx} y = \mu \frac{du_x}{dy} + A \tag{5.23}$$

where A is a constant of integration. This is a good point at which to apply the *first boundary condition*. Because of the symmetry of the flow about the y axis, it must be true that at $y = 0$, $du/dy = 0$. Hence we find that the constant of integration A is zero. Integrating Eq. 5.23 (with $A = 0$) yields

$$\frac{dp}{dx} \frac{y^2}{2} = \mu u_x + B \tag{5.24}$$

To evaluate the new integration constant B, we need the second boundary condition. At solid boundaries, the fluid just next to the boundary has the same velocity as the boundary; this is called the *no-slip condition*.* Hence, the *second boundary condition*: at $y = a$ (or $y = -a$), $u = 0$. Substituting into Eq. 5.24, we find that $B = (dp/dx)(a^2/2)$. Substituting this back into Eq. 5.23, we find, with

---

*Actually, for the flow of gases at very low pressures, the no-slip assumption may be erroneous. Fortunately, we rarely worry about such flows in environmental science. Also recall that the no-slip condition may be violated when the flow boundary is very small compared with gas molecule spacing, such as in the sedimentation of very small aerosol particles (Chapter 2). But we conquered that problem with the Cunningham slip correction.

a little rearrangement:

$$u_x = -\frac{1}{2\mu}\frac{dp}{dx}(a^2 - y^2) \tag{5.25}$$

At first glance, the reader might feel that there is an error in Eq. 5.25: it seems to predict negative velocities because $y \leqslant a$. However, we know that for the flow to exist, the pressure gradient term must be negative. In other words, pressure decreases uniformly in the streamwise direction. So the equation does in fact predict velocities in the direction assumed in Figure 5.6c.

Equation 5.25 is an elegant answer to a complicated problem. It says that velocity in the annular space is inversely proportional to viscosity, proportional to the pressure gradient, and proportional to the width of the annulus squared when $y = 0$.

A closely related problem is the steady viscous flow in a circular-section tube of radius a and constant streamwise pressure gradient — essentially the same problem solved above, but in a cylindrical coordinate system. The solution is also similar to that given above, and is left to the student as Exercise 5.5. The answer to the problem is

$$u_z = -\frac{1}{4\mu}\frac{dp}{dz}(a^2 - r^2) \tag{5.26}$$

where r is the radial distance from the center of the tube, a is the tube radius, and z is in the streamwise direction. Equation 5.26 is the famous *Poiseuille* or *Hagen–Poiseuille* equation. The derivation of the Poiseuille equation assumes laminar flow in the tube or capillary. For pipes with smooth inside surfaces, this equation should apply up to Reynolds numbers around 2000. A particularly useful form of the Poiseuille equation is obtained by deriving the average velocity across the section (Exercise 5.5):

$$u_{z,\text{ave}} = -\frac{1}{8\mu}\frac{dp}{dz}a^2 \tag{5.27}$$

Example 5.2. A liquid is flowing through a long narrow capillary tube of 0.5 cm diameter with an average velocity of 5 cm/s (see figure on p. 158). The static pressure drop between the measuring stations is 204 Pa. What is the viscosity of the fluid? The density of the fluid is known to be $1.05 \times 10^3 \text{ kg/m}^3$.

SOLUTION. We can rearrange Eq. 5.27, yielding

$$\mu = -\frac{1}{8}\frac{dp}{dz}\frac{a^2}{u_{z,\text{ave}}}$$

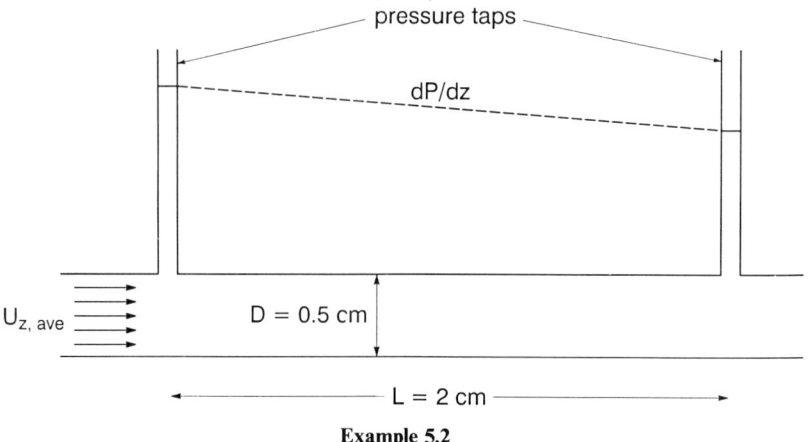

Example 5.2

The pressure gradient term dp/dz is simply the static pressure drop divided by the length separating the pressure taps, $-204 \, \text{Pa}/2 \, \text{m} = -102 \, \text{Pa/m}$. Substituting into the equation above, we find

$$\mu = -\frac{1}{8} \left(-102 \, \frac{\text{Pa}}{\text{m}} \right) \frac{(0.0025 \, \text{m})^2}{0.05 \, (\text{m/s})} = 1.59 \times 10^{-3} \, \text{Pa-s} = 1.59 \times 10^{-3} \, \frac{\text{kg}}{\text{m-s}}$$

We want to check the Reynolds number to make sure the flow is laminar:

$$\text{Re} = \frac{U_{\text{ave}} D \rho_f}{\mu} = \frac{0.05 \, (\text{m/s})(0.005 \, \text{m}) 1.05 \times 10^3 \, (\text{kg/m}^3)}{1.59 \times 10^{-3} \, (\text{kg/m-s})} = 165 \quad (\text{OK})$$

The flow shown in Figure 5.6d is also included in this section because circular streamline flows have many characteristics of unidirectional flows. The most important characteristic from our point of view is that in circular streamline flows like that shown in Figure 5.6d, the nonlinear terms in the Navier–Stokes equations drop out like they did in the problems considered above. We explore this characteristic further in Exercise 5.7.

5.7. FLUID SHEAR STRESSES AND THE VISCOSITY OF NEWTONIAN FLUIDS

We avoided a detailed development of the Navier–Stokes equations (e.g., see Batchelor, 1967; Currie, 1974; or Welty et al., 1976), preferring to focus on applications of the equations of motion to fluid-flow problems. Because we did not derive the Navier–Stokes equations, a thorough discussion of fluid stress was not required. Now that we have solved some basic fluid-flow problems, it is useful to think a little more about stress. Besides their important role in the equations of motion, fluid stresses are an important consideration in a number of environmental problems.

When we bend a twig in our hands, it is obvious that the stick is subjected to stress: The bending or twisting forces in our hands are transferred to the wood. In an analogous fashion, *viscous stresses* are exerted in fluids between adjacent molecules of fluid (and between fluid molecules and solid boundaries) when there is relative motion in the fluid. Molecular diffusion is thought to be responsible for momentum transport between the adjacent layers of fluid. A more basic equation of motion for a differential fluid element that considers local and convective acceleration, pressure forces, gravity forces, and fluid stresses is

$$\rho \frac{\partial u_i}{\partial t} + \rho u_k \frac{\partial u_i}{\partial x_k} = -\frac{\partial P}{\partial x_i} + \frac{\partial \tau_{ij}}{\partial x_j} + \rho g_i \tag{5.28}$$

Comparing with Eq. 5.4, the reader will note that the only difference is in the fourth term, which involves the gradient of the *stress tensor*, τ_{ij}. The stress tensor for a Newtonian fluid is presented in various coordinate systems in Appendix III. In three-dimensional, rectangular coordinates, the shear stress acting normal to the y axis and in planes parallel to the x axis is given by,

$$\tau_{yx} = \mu \left(\frac{\partial u_y}{\partial x} + \frac{\partial u_x}{\partial y} \right) \tag{5.29}$$

Applying this equation to the flow in Figure 5.6a, we find that since there is no y-direction component of velocity, the stress acting between layers of adjacent fluid in planes parallel to the x axis is

$$\tau_{yx} = \mu \frac{du_x}{dy} \tag{5.30}$$

With the aid of Figure 5.6a, Eq. 5.30 can be interpreted to mean that the viscous shear stress τ in planes parallel to the x axis is proportional to the fluid viscosity and the derivative of u_x in the y direction. Equation 5.30 is a famous equation in fluid mechanics known as *Newton's viscosity relationship*; Equation 5.30 thus defines a *Newtonian fluid*. The velocity derivative has a special name in fluid mechanics — the *strain rate*. The relationship between stress, viscosity, and strain rate exemplified in Eq. 5.30 (or between the stress and *strain rate* tensors; see Batchelor, 1967) is one of the basic underpinnings of the Navier–Stokes equations.

It is interesting to examine stress profiles in the slightly more complicated Poiseuille flow discussed in Section 5.6. In cylindrical coordinates appropriate to this problem, the stress in the z direction along cylindrical surfaces in the flow centered at the pipe axis is given by (Appendix III)

$$\tau_{rz} = \mu \left(\frac{\partial u_r}{\partial z} + \frac{\partial u_z}{\partial r} \right) \tag{5.31}$$

Now for the Poiseuille flow problem, since there is no velocity in the r direction, the equation above is simply $\tau_{rz} = \mu \partial u_z / \partial r$. Substituting the expression for u_z from Eq. 5.26, we find:

$$\tau_{rz} = \mu \frac{d}{dr} \left[-\frac{1}{4\mu} \frac{dp}{dz} (a^2 - r^2) \right] = \frac{1}{2} \frac{dp}{dz} r \tag{5.32}$$

This interesting relationship shows that in laminar-tube flow, the shear stress between adjacent layers of fluid increases linearly moving away from $r = 0$. The maximum stress occurs at the maximum r, that is, at $r = a$ (the inside surface of the pipe). This is called the *wall stress*:

$$\tau_{rz}^{(max)} = \tau_w = \frac{1}{2} \frac{dp}{dz} a \tag{5.33}$$

Using a force balance on a differential element of fluid, it can be shown that Eq. 5.33 gives the correct wall stress for the flow of any Newtonian fluid, laminar *or turbulent* (Streeter, 1971). Combining Eqs. 5.32 and 5.33, we can express the stress at any distance r as a function of the maximum stress:

$$\tau_{rz} = \frac{r}{a} \tau_{rz}^{(max)} \tag{5.34}$$

As stated above, fluid stress is important in a number of environmental problems. For example, since it is known that fluid stresses are transferred to solid boundaries, we predict that increasing strain rates in overlying fluid will cause increased stress at solid boundaries. If the solid boundary is covered with an erodible material (like gravel on a stream bed or aerosol solids deposited on vegetation), or if the surface is actually a deformable object (like a floc, microbe, or rain drop), models like that given above can be used to predict the surface stresses and the tendency of erosion and resuspension (for the deposited solids) or deformation (for the floc, biological cell, or rain drop). We touch on this problem again in our discussion of viscous boundary layers (Section 5.12). The related problem of mass transfer at interfaces between solids and moving fluids, which is critically dependent on the boundary stress, is treated in detail in Chapter 7.

5.8. NONSTEADY UNIDIRECTIONAL FLOWS: STOKES' FIRST PROBLEM

Thus far, we have largely ignored the nonsteady or time-varying nature of many flows of environmental interest. A particularly enlightening nonsteady flow problem is portrayed in Figure 5.7. Here we see the case of a fluid initially

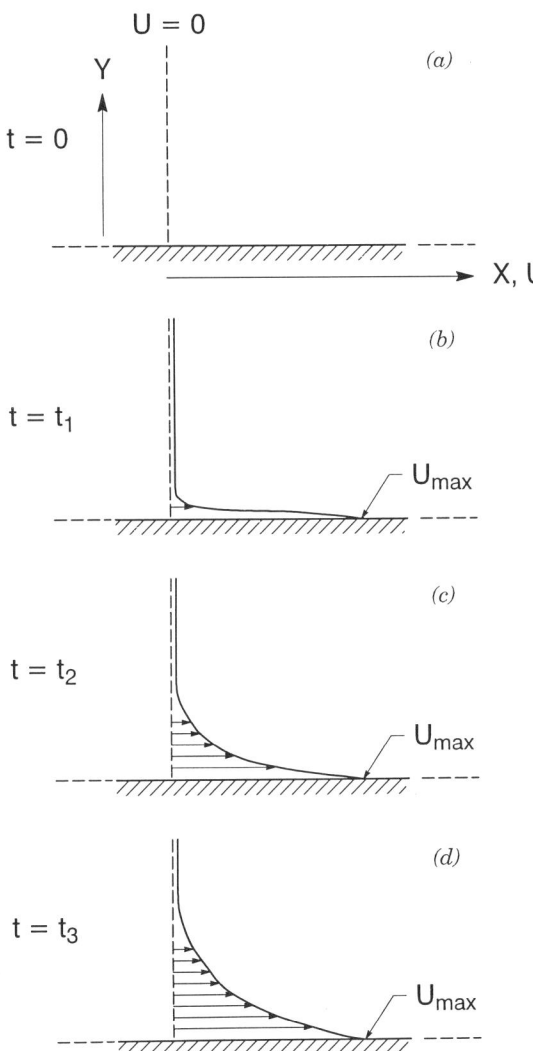

Figure 5.7. Velocity profiles observed at fixed observation point and at four different times near a flat infinite boundary moving to the right with constant velocity U_{max}.

at rest (Figure 5.7a) and bordered only by the single solid boundary shown at $y = 0$. After $t = 0$, the solid boundary is jerked into motion at constant velocity U_{max}.

It helps if you imagine that the solid boundary extends to infinity in both directions; therefore, the boundary can keep rolling by our observation point like the tape passing a cassette recording head, indefinitely, and at the same speed $U = U_{max}$. What are the governing equations for this flow? If we assume that the flow is unidirectional in the x direction and if we ignore gravity, the only relevant component of the Navier–Stokes equations is

$$\rho \frac{\partial u}{\partial t} = -\frac{\partial p}{\partial x} + \mu \frac{\partial^2 u}{\partial y^2} \tag{5.35}$$

However, can there be a pressure gradient in the x direction? Apparently not — we have not impressed a pressure gradient like in the *confined flow* shown in Figure 5.6c, and because u is not a function of x, pressure cannot be a function of x. Hence, the pressure gradient term is zero and we are left with

$$\frac{\partial u}{\partial t} = v \frac{\partial^2 u}{\partial y^2} \tag{5.36}$$

where v is the *kinematic viscosity*:

$$v = \frac{\mu}{\rho} \tag{5.37}$$

Equation 5.36 is a well-known partial differential equation in physics and engineering. The boundary conditions for solution are obvious from Figure 5.7 and our discussion:

$$u(0, t) = \begin{cases} 0 & \text{for: } t \leqslant 0 \\ U_{max} & t > 0 \end{cases} \tag{5.38}$$

There are a number of approaches to solving Eq. 5.36, including Laplace transforms and dimensional analysis; these techniques are pretty straightforward and within the grasp of the typical reader (see the reference section). We skip the details here and just provide the answer:

$$u(y, t) = U_{max} \left(1 - \frac{2}{\sqrt{\pi}} \int_0^{y/(2\sqrt{vt})} e^{-\zeta^2} d\zeta \right) \tag{5.39}$$

The integral expression is well known in mathematics as the *error function*, and is available in spreadsheets and mathematics references. Hence, Eq. 5.39 can be expressed more compactly as

$$u(y, t) = U_{\max} \left\{ 1 - \text{erf} \left(\frac{y}{2\sqrt{vt}} \right) \right\} \qquad (5.40)$$

Figure 5.7b–d provides sketches of the velocity profile at our fixed observation point (in the neighborhood of $x = 0$ and $y = 0$) in the fluid above the moving boundary at three successive times, t_1, t_2, and t_3. Note how, over time, the *viscous disturbance* caused by the boundary seems to diffuse out into the fluid: as time passes, more of the fluid above the moving boundary seems to be dragged into motion. This is one of the most important demonstrations of fluid mechanics — because of viscosity, momentum can be transferred from faster-moving fluid to slower-moving fluid. As time approaches infinity, the error function approaches zero and we predict that all the fluid above the moving boundary will approach the same velocity U_b.*

Another illuminating demonstration, and it is related to our later discussion of *boundary layers*, can be given using this problem. Scanning a table of the error function, one can quickly find that when the argument of the error function $[y/2(vt)^{0.5}]$ is around $3/2$, the value of the error function is about 0.96. For this value of the error function argument, $u(y, t) = 0.04\, U_b$. We might then find the following question of some interest: At what distance y above the solid boundary is the velocity just equal to $0.04\, U_b$? If we call this special distance the *boundary layer thickness* δ (i.e., $y = \delta$ when $u = 0.04\, U_b$), the following must be true:

$$\delta = 3\sqrt{vt} \qquad (5.41)$$

If we say that above $y = \delta$ the fluid is almost motionless, Eq. 5.41 gives a powerful and simple way of describing the progress of the viscous disturbance into the overlying fluid. It says that the thickness of the zone of fluid affected by the viscous disturbance increases as the square root of viscosity and as the square root of time.

We could also ask a somewhat equivalent question: At what time t_v has the viscous disturbance progressed a distance δ above the plate or boundary? Rearrangement of Eq. 5.41 yields the answer:

$$t_v = \frac{1}{9} \frac{\delta^2}{v} \qquad (5.42)$$

---

*There is an exact analogy between the spread (diffusion) of momentum from faster- to slower-moving fluid in a direction normal to the flow discussed here, and diffusion of mass from more- to less-concentrated regions. In Chapter 6, we find a differential diffusion equation that is mathematically identical to Eq. 5.36. This is the main reason fluid mechanicians talk about the "diffusion of momentum."

This is also a simple and elegant answer: The time for a viscous disturbance to progress a certain distance is proportional to the distance squared and inversely proportional to the kinematic viscosity.

5.9. LOW REYNOLDS NUMBER FLOWS

Most readers probably have some feel for the meaning of the Reynolds number and that it has something to do with fluid velocity, viscosity, and density. In this section, we define the Reynolds number and examine a number of flows where we can assume that the Reynolds number approaches zero.

5.9.1. The Reynolds Number and the Stokes Equation

Let's take another look at the Navier–Stokes equations for the case of no free surface:

$$\frac{\partial \mathbf{u}}{\partial t} + \mathbf{u} \cdot \nabla \mathbf{u} = -\frac{1}{\rho} \nabla p + \nu \nabla^2 \mathbf{u} \tag{5.43}$$

We now attempt to *nondimensionalize* the Navier–Stokes equations by considering *characteristic velocity* and *length scales*. Assume that the flow of interest has a characteristic length scale of L and a characteristic velocity scale of U. For example, in pipe flow, L could correspond to the pipe diameter and U to the average velocity (recall Section 1.5). Then we nondimensionalize the four dependent and independent variables in the Navier–Stokes equations in the following way. Let

$$\tilde{\mathbf{u}} = \frac{\mathbf{u}}{U}; \quad \tilde{\mathbf{x}} = \frac{\mathbf{x}}{L}; \quad \tilde{t} = \frac{\nu}{L^2} t; \quad \tilde{p} = \frac{L}{\rho \nu U} p \tag{5.44}$$

Substituting these quantities into Eq. 5.43 (see Exercise 5.10) yields a nondimensional form of the Navier–Stokes equations:

$$\frac{\partial \tilde{\mathbf{u}}}{\partial \tilde{t}} + \mathrm{Re}\,(\tilde{\mathbf{u}} \cdot \tilde{\nabla} \tilde{\mathbf{u}}) = -\tilde{\nabla} \tilde{p} + \tilde{\nabla}^2 \tilde{\mathbf{u}} \tag{5.45}$$

where $\tilde{\nabla} = \nabla L$ and Re is the Reynolds number:

$$\mathrm{Re} = \frac{\rho U L}{\mu} = \frac{U L}{\nu} \tag{5.46}$$

A number of important insights are gained from the result shown above. First, the Navier–Stokes equation that we started with had two parameters— density and kinematic viscosity. In the nondimensional form, the equation has only one parameter—the Reynolds number. This means that all flows with the same Reynolds number and the same nondimensional initial and boundary conditions have the same nondimensional solution to the Navier–Stokes equation: These flows are called *dynamically similar* flows because the ratio of forces at specific (nondimensional) points in the flows should be the same. This is a very powerful result in the history of fluid mechanics and aerodynamics. For example, consider the problem of trying to experimentally determine the drag on a small particle settling in air. It is very difficult to directly measure the drag force because of the particle's small size. However, the result given above suggests that as long as the Reynolds number is kept constant (perhaps by varying viscosity), a larger model of the particle can be studied.

The second insight results from a slight rearrangement of the Reynolds number. Note that

$$\text{Re} = \frac{UL}{v} = \frac{L^2/v}{L/U} = \frac{\tau_v}{\tau_c} \tag{5.47}$$

The term τ_c is called the *convective time scale*; it is the basic measure of the characteristic velocity and length scales. For example, consider a flow of characteristic velocity U passing a stationary sphere of characteristic dimension (diameter) L. Then τ_c is the characteristic time for the flow to pass the sphere. The term τ_v is called the *viscous diffusion time scale*. It is the characteristic time for a viscous disturbance to progress a certain distance. A good example of such a time scale was provided in the discussion of Eq. 5.42 above. The Reynolds number can then be considered a fundamental comparison of the viscous and convective time scales in a fluid-flow problem. Since the convective terms in the Navier–Stokes equations are related to the fluid inertia, the Reynolds number can also be considered a measure of the relative importance of inertial and viscous forces in a particular problem. Note also that in most types of modeling, the smaller the time scale of the phenomenon, the more important the phenomenon. For example, if the viscous time scale is much smaller than the convective time scale (hence a small Reynolds number), the viscous effects are more quickly felt. However, if the Reynolds number is large, the convective time scale is much smaller than the viscous time scale and inertial forces are much more important than viscous forces.

The third insight to be gained from the exercise above comes when we let the Reynolds number in Eq. 5.45 become very small (Re → 0). Then the nonlinear convective acceleration terms become very small relative to the other terms and we make the following estimate of the equation of motion:*

$$\frac{\partial \mathbf{u}}{\partial t} = -\frac{1}{\rho} \nabla p + v \nabla^2 \mathbf{u} \tag{5.48}$$

---

*We have jumped back to the dimensional form of the Navier–Stokes equations. If a term falls out of the nondimensional form of an equation, it also falls out of the dimensional form.

This is the so-called *Stokes* or *creeping-flow* equation. It is a much nicer equation than the original Navier–Stokes equation because it is no longer nonlinear. If we use the Stokes equation, we are in effect saying that viscous forces in the problem are much more important than inertial forces.

Finally, an interesting transformation comes about by taking the divergence of Eq. 5.48 for the case of steady flows:

$$\nabla \cdot \nabla p = \mu \nabla \cdot \nabla^2 u \tag{5.49}$$

Using the identities of Appendix II, we find:

$$\nabla^2 p = \mu \nabla^2 (\nabla \cdot \mathbf{u}) \tag{5.50}$$

However, the term in parentheses — the divergence of the velocity — is zero for an incompressible fluid (recall Eq. 1.24); hence, we find the following pleasing result:

$$\nabla^2 p = 0 \tag{5.51}$$

Equation 5.51 says that for very low Reynolds number (incompressible) flows, the Laplacian of the pressure field must be zero. In fact, when the Laplacian of a scalar is zero, the equation is called the *Laplace equation*. Solutions of the Laplace equation have been well studied in physics and mathematics. So we may have some valuable help in solving for the pressure field in certain low Reynolds number flows. With the pressure field determined, Eq. 5.48 should be much easier to solve.

5.9.2.   Sphere in Axisymmetric Low Re Flow: Stokes Drag Law

Although exact solutions to the Navier–Stokes equations and the Stokes equation for three-dimensional flows are somewhat rare, the Stokes equation can be solved fairly easily for certain *axisymmetric flows*. Consider a sphere of radius a immersed in the flow shown in Figure 5.8, where U is the *free-stream velocity*. Note that in axisymmetric flows, the flow depends only on r and θ; the flow does not depend on the angle ϕ originating in the plane of the paper (not shown). The flow is symmetric about the z axis; hence the name "axisymmetric." For the boundary conditions of no-slip at the surface on the sphere (i.e., at $r = a$) and for $u_r \to -U \cos \theta$ and $u_\theta \to U \sin \theta$ as $r \to \infty$, the solution for the flow field is (Panton, 1984)

$$u_r = -\frac{1}{2} U \cos \theta \left[\left(\frac{a}{r} \right)^3 - 3 \left(\frac{a}{r} \right) + 2 \right] \tag{5.52}$$

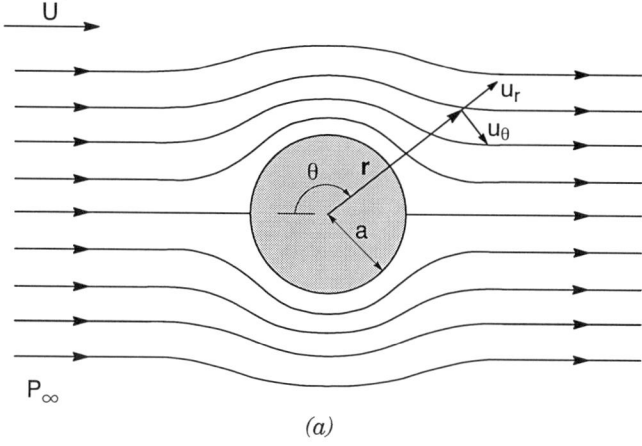

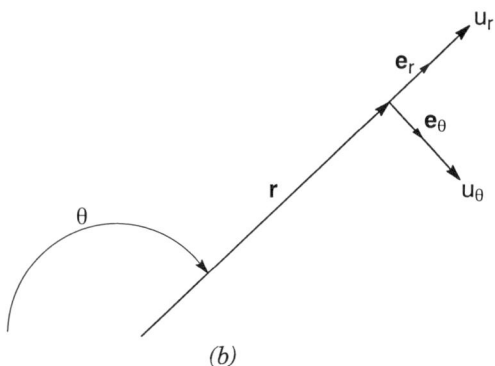

Figure 5.8. Sphere immersed in low Reynolds number axisymmetric flow. The unit vectors $\mathbf{e}_r$ and $\mathbf{e}_\theta$ define the positive sense for u_r and u_θ. The z axis is parallel to U and passes through the center of the sphere.

and

$$u_\theta = -\frac{1}{4} U \sin \theta \left[\left(\frac{a}{r}\right)^3 + 3\left(\frac{a}{r}\right) - 4 \right] \tag{5.53}$$

The reader should verify that the boundary conditions stated above are indeed met by the solution.

The pressure distribution at the surface of the sphere is known to be (Panton, 1984)

$$p - p_\infty = \frac{3\mu U}{2a} \left(\frac{a}{r}\right)^2 \cos \theta \tag{5.54}$$

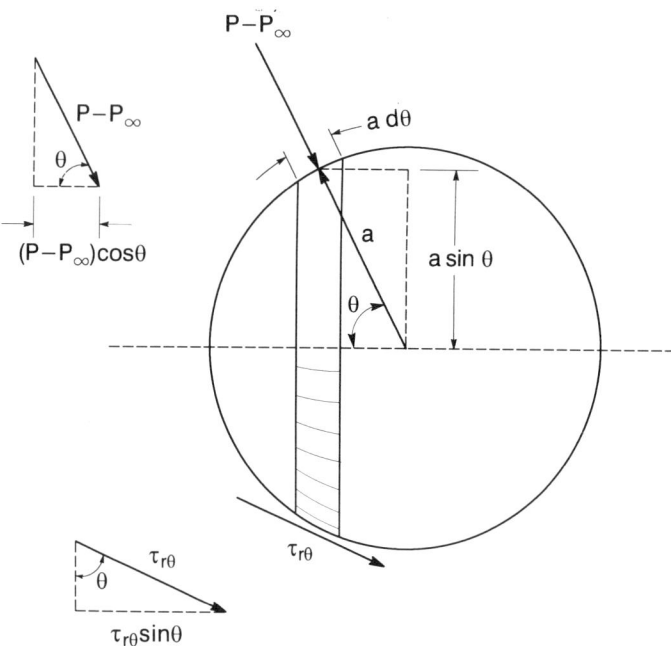

Figure 5.9. Setup for integration of forces along the surface of a sphere in low Reynolds number flow.

Here p_∞ is the static pressure far away from the sphere where the flow is unaffected by the presence of the sphere.

The low Reynolds number flow past the stationary sphere in Figure 5.8 causes a drag force on the sphere in the same sense as the direction of the free-stream velocity (to the right in Figure 5.8). This drag is considered to arise from two sources. First, as we saw in Section 5.7, velocity gradients near surfaces can cause shear stress to be transferred to the surface. If this stress can be integrated around the sphere, we can calculate the component of this type of drag in the z direction. This is called the *skin friction* component of the drag. The calculation of this component is easily set up as a surface integral of the z-directed shear force along the surface of the sphere:

$$F_D^s = \iint \tau_{r,\theta}|_{r=a} \sin \theta \, dS \qquad (5.55)$$

In this equation, the integral of τ over the surface gives the total force, while weighing with $\sin \theta$ assures that we only count the component in the z direction. This integral can be set up in spherical coordinates, but in the axisymmetric flow of Figure 5.8 it is simpler to use the differential strip shown in Figure 5.9. We can now develop further the different terms in Eq. 5.55. From the indicated geometry in Figure 5.9, the differential surface area represented by the ring is the product of the perimeter of the strip times the width of the strip:

$$dS = (2\pi a \sin \theta)(a d\theta) = 2\pi a^2 \sin \theta \, d\theta \qquad (5.56)$$

For the shear stress acting at the particle we find, from Appendix III:

$$\tau_{r\theta} = \mu \left[r \frac{\partial}{\partial r} \left(\frac{u_\theta}{r} \right) + \frac{1}{r} \left(\frac{\partial u_r}{\partial \theta} \right) \right] \qquad (5.57)$$

Now when we consider Eq. 5.57 at the surface of the sphere, the r component of velocity is zero all around the sphere (in the θ direction); hence, the derivative of u_r with respect to θ must be zero at $r = a$. Incorporating Eq. 5.53, Eq. 5.57 evaluated at $r = a$ reduces to:

$$\tau_{a\theta} = \mu r \frac{\partial}{\partial r} \left(\frac{u_\theta}{r} \right) \bigg|_{r=a} = -\frac{\mu U \sin \theta}{4} \left\{ r \frac{\partial}{\partial r} \left[\frac{a^3}{r^4} + 3 \frac{a}{r^2} - \frac{4}{r} \right] \right\} \bigg|_{r=a} = \frac{3\mu U \sin \theta}{2a}$$

$$(5.58)$$

Substituting Eqs. 5.56 and 5.58 into Eq. 5.55 we find:

$$F_D^s = \frac{3\mu U}{2a} \int_0^\pi \sin^2\theta(2\pi a^2) \sin \theta \, d\theta = 3\pi\mu Ua \int_0^\pi \sin^3\theta \, d\theta = 4\pi\mu Ua \quad (5.59)$$

where the final integral can be evaluated with the aid of almost any definite integrals table. The second component of drag is caused by differences in the pressure distribution around the sphere. The z component of the net pressure force is called the *form drag*. It is set up as

$$F_D^f = \iint (p - p_\infty) \cos \theta \, dS \qquad (5.60)$$

In this equation, the integral of the pressure $p - p_\infty$ over the sphere gives the total pressure force, while Figure 5.9 shows that weighing with $\cos \theta$ gives the net z-directed component of this force. Substituting Eqs. 5.54 and 5.56 into Eq. 5.60 yields:

$$F_D^f = \frac{3\mu U}{2a} \int_0^\pi \cos^2\theta(2\pi a^2 \sin \theta)d\theta = 3\pi a\mu U \int_0^\pi \cos^2\theta \sin \theta \, d\theta = 2\pi\mu Ua \quad (5.61)$$

The net drag force exerted by the fluid on the sphere is then the sum of the skin friction and form drag components:

$$F_D = F_D^s + F_D^f = 6\pi\mu U a \tag{5.62}$$

This is known as *Stokes drag law*. It is one of the most important results of classical fluid mechanics, and as we have seen in Chapter 2, it is of great importance in the transport of small particles in the environment.

5.9.3. Flow Through Porous Media

Porous media are ubiquitous in the environment. By far the most extensive porous medium is the one you are standing on if you are reading this book outside — the ground. The filters used in treating drinking water and some dense air filters can also be considered examples of porous media. As the reader can imagine after examining a small sample of earth or sand, the flow path through the medium must be tortuous and random. Figure 5.10 presents results of recent visualizations of liquid flow in a porous medium composed of irregularly shaped grains. These fascinating measurements show the tortuous and curvilinear flow around the different grains.

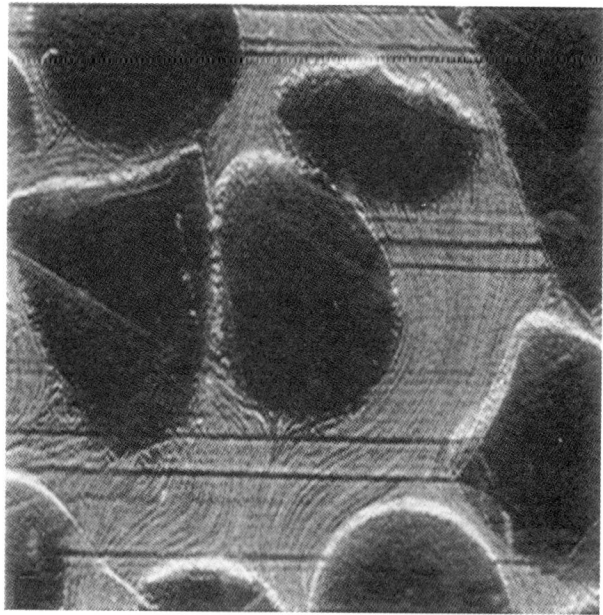

Figure 5.10. Streamlines in a porous medium were determined by a technique called particle image displacement velocimetry, which is based on a time-lapse exposure of illuminated particles following the flow. *Source*: Reproduced from Saleh, Thovert, and Adler, *Exp. Fluids*, **12**, 211 (1992). Reprinted with permission granted by Springer-Verlag.

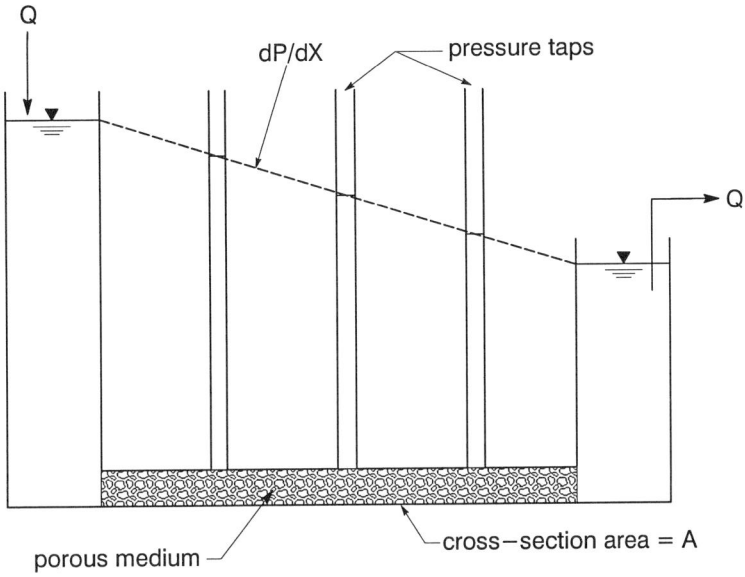

Figure 5.11. Experiment to measure flow through a porous plug.

During the last century, Darcy performed experiments on the flow through porous plugs of sand. His experiments can be represented by the experimental setup shown in Figure 5.11. The two fluid reservoirs are connected by the closed conduit filled with sand. We replenish water in the higher reservoir at exactly the same rate Q as the flow through the porous media (and as removed from the lower reservoir). Similar to pipe flow, we observe a linear decrease in hydrostatic pressure along the length of the sand-filled conduit, dp/dx. Darcy found that the average velocity through the porous conduit was described by the following relation, which is known as *Darcy's law*:

$$u = -\frac{k}{\mu}\frac{dP}{dx} \qquad (5.63)$$

Here k is the *permeability* (L^2) of the porous medium. Note that Darcy's law predicts the same dependence on pressure gradient and viscosity as the integrated Poiseuille law (Eq. 5.27). Based on this observation, scientists long ago wondered if in some sense flow through a porous medium could be considered as analogous to flow through a number of more or less parallel (but twisted) tubes of very small diameter. If this were true, the similarity in the predictions of the Poiseuille and Darcy equations would be understandable.

A famous, and somewhat empirical, hydraulic analysis makes the connection between the Poiseuille and Darcy equations. For a tube of any cross-sectional shape, Eq. 5.27 can be generalized to

$$v = \frac{m^2}{k_0 \mu} \frac{\Delta P}{L_e}$$
(5.64)

Here k_0 is a constant, L_e is the length of the tube, and m is known as the *hydraulic radius*:

$$m = \frac{\text{cross-sectional area of tube}}{\text{wetted perimeter of tube}}$$
(5.65)

Note that for a cylindrical tube of radius a, $m = \pi a^2/2\pi a = a/2$; hence, Eq. 5.64 is identical to Eq. 5.27 for $k_0 = 2$ and where $\Delta P/L_e$ is taken as the pressure gradient, dp/dz. In real porous media, the flow path is random and tortuous, so the hydraulic radius is estimated as

$$m = \frac{\text{void volume}}{\text{wetted surface of porous medium}} = \frac{\epsilon V}{V(1 - \epsilon)S_0}$$
(5.66)

Here V is the volume of the porous medium sample, ϵ is the porosity (fraction of porous medium composed of voids — see also Section 8.2), and S_0 is the *specific surface* of the porous medium:

$$S_0 = \frac{\text{porous medium surface area}}{\text{unit volume of porous medium solids}}$$
(5.67)

Now v is an average velocity in the pores of the medium, and this must be higher than the value u in Darcy's law; in fact, the two velocities are related simply by

$$v = \frac{u}{\epsilon}$$
(5.68)

Substituting Eqs. 5.66 and 5.68 into Eq. 5.64 yields

$$u = \frac{1}{k_0 \mu} \frac{\epsilon^3}{S_0^2(1 - \epsilon)^2} \frac{\Delta P}{L_e}$$
(5.69)

Now recall that L_e is the length of one of the average, twisted flow tubes in the porous medium. We would prefer to express the pressure drop in terms of the overall thickness of the section of porous medium under consideration (e.g., the length of the porous plug in Figure 5.11). Therefore, we invent a new constant:

$$K = \frac{L_e}{L} k_0 \qquad (5.70)$$

where L is the length of the porous medium section, and K is called the *Kozeny constant*. Substitution of Eq. 5.70 into Eq. 5.69 yields

$$u = \frac{\epsilon^3}{K\mu S_0^2 (1 - \epsilon)^2} \frac{\Delta P}{L} \qquad (5.71)$$

This equation is called the *Carman–Kozeny equation*; K has historically been taken to be equal to 5, although this value is not exact and could vary somewhat with different types of porous media. Finally, comparing Eqs. 5.71 and 5.63 yields an estimate of the permeability:

$$k = \frac{\epsilon^3}{K S_0^2 (1 - \epsilon)^2} \qquad (5.72)$$

Hence, the Poiseuille law and Darcy's law are shown to have a specific connection through the definition of the permeability in Eq. 5.72.

5.10.   IDEAL FLUIDS, POTENTIAL FLOWS, AND STREAM FUNCTIONS

5.10.1.   Inviscid Flows and the Euler Equation

In the last four sections, we examined flows in which the viscosity of the fluid played a large role in determining the velocity field. Especially in Section 5.9, we found that viscosity took on increasing importance as the Reynolds number became small (about $Re = 1$ and lower). Hence, these solutions had most relevance to flows with small characteristic length scales, flow at very low speeds, and/or flows of fluids of very high viscosity. But we might ask, are there flows in which the effect of viscosity is irrelevant? When we are studying in front of an open window on a breezy night, it is our experience that, with little warning, a gust of air can quickly sweep by and blow papers from our desk. Somehow, intuition tells us that the strong gust of air had little to do with viscosity. In fact our intuition is correct: such a gust of air is caused by pressure differences across different windows of the house, quickly setting up a flow between various windows.

   A large area of fluid mechanics has concerned itself with flows in which the effect of viscosity is negligible. In these *ideal fluid* flows or *inviscid* flows, we in fact assume that $\mu = 0$. Hence, in addition to the continuity equation (Eq. 1.24), the equations of motion are simply the old Navier–Stokes equations minus the viscous terms:

$$\rho \frac{\partial \mathbf{u}}{\partial t} + \rho \mathbf{u} \cdot \nabla \mathbf{u} = -\nabla P + \rho \mathbf{g} \qquad (5.73)$$

In classical fluid mechanics, this equation is called the *Euler equation*. Although the order of the differential equation has been reduced from second to first, it is still nonlinear. Hence, we do not seem to have improved the situation greatly from the pure standpoint of equation solving. Currie (1974) and Batchelor (1967) point out another concern with the Euler equation: because of the lower order of this differential equation, the no-slip condition so important in viscous fluid flows can no longer be mathematically satisfied. Therefore, one distinguishing characteristic of ideal fluid flows is that solid boundaries are also streamlines of the flow (i.e., $\mathbf{u}$ is not zero along solid boundaries).

5.10.2. Irrotational Flows and the Velocity-Potential Function

To see the advantage of the inviscid flow equations, we have to introduce a new characteristic of fluid flows: *vorticity*. The vorticity is defined as

$$\boldsymbol{\omega} = \nabla \times \mathbf{u} \qquad (5.74)$$

In other words, vorticity is the cross product of the gradient operator and velocity (see Appendix II); therefore, the vorticity vector is perpendicular to the local velocity vector. One can think of vorticity as being related to the turning of fluid elements in a flow. Application of Eq. 5.74 to the two-dimensional flows studied in the previous sections shows that these viscous flows are jam-packed with vorticity.

In contrast to the vorticity-filled viscous flows, an important subset of inviscid flows are flows with no vorticity, or *irrotational* ideal flows:

$$\nabla \times \mathbf{u} = 0 \qquad (5.75)$$

Kelvin's theorem states that for inviscid flows that originate in irrotational motions, the flow remains irrotational regardless of the flow path (Currie, 1974). For example, imagine that the sphere shown in Figure 5.8a is immersed in an inviscid instead of a viscous fluid. Since the flow field far upstream of the sphere is irrotational, Kelvin's theorem assures us that even the flow around the sphere is irrotational. This is in contrast to the viscous flow solution outlined in Section 5.9.2, which has a lot of vorticity near the sphere.

An expression for the velocity vector which satisfies the irrotationality condition (Eq. 5.75) involves the *velocity-potential function* ϕ:

$$\mathbf{u}(= u\mathbf{i} + v\mathbf{j} + w\mathbf{k}) = \nabla\phi \left(= \frac{\partial\phi}{\partial x}\mathbf{i} + \frac{\partial\phi}{\partial y}\mathbf{j} + \frac{\partial\phi}{\partial z}\mathbf{k} \right) \tag{5.76}$$

where ϕ is a scalar function. Equation 5.76 automatically satisfies Eq. 5.75, since the curl of the gradient of a scalar function is identically zero (Appendix II). The real advantage of the velocity-potential function is seen by substituting Eq. 5.76 into the continuity equation for incompressible fluids (Eq. 1.24):

$$\nabla \cdot \mathbf{u} = \nabla \cdot \nabla\phi = \nabla^2\phi = 0 \tag{5.77}$$

The Laplacian of the potential function is therefore zero by continuity.

The velocity-potential function is a very exciting development. We have already said that physicists and scientists have a lot of experience solving the Laplace equation. Hence, if Eq. 5.77 can be solved, then Eq. 5.76 can be used to simply retrieve the velocity components. This is generally simpler than solving the Euler equations directly. After the velocity components are found, the Euler equation can be used to solve for the pressure field (if that is of interest). The exercises provide some practice in solving equations of irrotational inviscid fluid flow.

5.10.3. The Stream Function

Stream functions do not necessarily belong in a section on inviscid flows, because they do not really require the assumption of a perfect fluid. Nevertheless, they are often presented with the two-dimensional potential function, because both the potential and stream function have some interesting relationships when the flow is inviscid.

Consider the total derivative of some function ψ:

$$d\psi = \frac{\partial\psi}{\partial x}\,dx + \frac{\partial\psi}{\partial y}\,dy \tag{5.78}$$

The total derivative above is called an exact differential only if $d\psi = 0$. An important theorem (Boyce and Diprima, 1969) of exact differential equations states that the general equation

$$M(x, y)dx + N(x, y)dy = 0 \tag{5.79}$$

is exact if and only if

$$\frac{\partial M}{\partial y} = \frac{\partial N}{\partial x} \tag{5.80}$$

That is to say, there exists a function ψ satisfying

$$\frac{\partial \psi}{\partial x} = M(x, y) \quad \text{and} \quad \frac{\partial \psi}{\partial y} = N(y, x) \tag{5.81}$$

if and only if the relationship in Eq. 5.80 holds true. How does this apply to fluid mechanics? In two dimensions, the equation of a streamline is

$$\frac{dx}{u} = \frac{dy}{v} \tag{5.82}$$

This is easy to derive, and means simply that if you are on a streamline, your change in position in the x direction is proportional to u and your change in position in the y direction is proportional to v. Rearranging Eq. 5.82 yields

$$-v\,dx + u\,dy = 0 \tag{5.83}$$

This is an exact differential (and hence there is some ψ such that $\partial \psi / \partial x = M$ and $\partial \psi / \partial y = N$) if and only if

$$\frac{\partial M}{\partial y} = -\frac{\partial v}{\partial y} = \frac{\partial N}{\partial x} = \frac{\partial u}{\partial x} \tag{5.84}$$

In fact, this relation is true; to see it, all we have to do is look at the two-dimensional continuity equation:

$$\frac{\partial u}{\partial x} + \frac{\partial v}{\partial y} = 0 \Rightarrow \frac{\partial u}{\partial x} = -\frac{\partial v}{\partial y} \tag{5.85}$$

Therefore, for two-dimensional, incompressible flows, there exists a stream function ψ where

$$M = -v = \frac{\partial \psi}{\partial x} \quad \text{and} \quad N = u = \frac{\partial \psi}{\partial y} \tag{5.86}$$

Note that all this effort has yielded a significant improvement: Where before we had two dependent variables (u and v), we now have one (ψ).

Example 5.3. The stream function for Stokes flow past a stationary sphere is given by

$$\psi = -\frac{1}{2} U r^2 \sin^2\theta \left[\frac{1}{2} \left(\frac{a}{r}\right)^3 - \frac{3}{2} \left(\frac{a}{r}\right) + 1 \right]$$

In spherical coordinates, the expressions for calculating u_r and u_θ (see Figure 5.8) are

$$u_r = \frac{1}{r^2 \sin \theta} \frac{\partial \psi}{\partial \theta}$$

and

$$u_\theta = -\frac{1}{r \sin \theta} \frac{\partial \psi}{\partial r}$$

Using these expressions, verify Eqs. 5.52 and 5.53.

SOLUTION. To calculate u_r, substitute for ψ in the equation for u_r:

$$u_r = \frac{1}{r^2 \sin \theta} \frac{\partial}{\partial \theta} \left\{ -\frac{1}{2} U r^2 \sin^2\theta \left[\frac{1}{2} \left(\frac{a}{r}\right)^3 - \frac{3}{2} \left(\frac{a}{r}\right) + 1 \right] \right\}$$

$$u_r = -\frac{U}{2 \sin \theta} \left[\frac{1}{2} \left(\frac{a}{r}\right)^3 - \frac{3}{2} \left(\frac{a}{r}\right) + 1 \right] \frac{\partial}{\partial \theta} (\sin^2\theta)$$

or

$$u_r = -\frac{2U \sin \theta \cos \theta}{2 \sin \theta} \left[\frac{1}{2} \left(\frac{a}{r}\right)^3 - \frac{3}{2} \left(\frac{a}{r}\right) + 1 \right]$$

$$= -\frac{1}{2} U \cos \theta \left[\left(\frac{a}{r}\right)^3 - 3 \left(\frac{a}{r}\right) + 2 \right] \qquad \text{(OK)}$$

To calculate u_θ, substitute u_θ for ψ in the equation for u_θ:

$$u_\theta = -\frac{1}{r \sin \theta} \frac{\partial}{\partial r} \left\{ -\frac{1}{2} U r^2 \sin^2\theta \left[\frac{1}{2} \left(\frac{a}{r}\right)^3 - \frac{3}{2} \left(\frac{a}{r}\right) + 1 \right] \right\}$$

$$u_\theta = \frac{U \sin \theta}{2r} \frac{\partial}{\partial r} \left(\frac{1}{2} \frac{a^3}{r} - \frac{3}{2} ar + r^2 \right)$$

or

$$u_\theta = \frac{U \sin \theta}{2r} \left(-\frac{1}{2} \frac{a^3}{r^2} - \frac{3}{2} a + 2r \right) = -\frac{1}{4} U \sin \theta \left(\frac{a^3}{r^3} + 3 \frac{a}{r} - 4 \right) \qquad \text{(OK)}$$

The stream function has some very special characteristics when the flow is two dimensional and irrotational. Note that in two dimensions, the only vorticity component is in the z direction (out of the paper):

$$\omega_z = \left(\frac{\partial v}{\partial x} - \frac{\partial u}{\partial y}\right) \tag{5.87}$$

Substituting the relationships for u and v from Eq. 5.86 and assuming an irrotational flow ($\omega_z = 0$), we find:

$$\frac{\partial^2 \psi}{\partial x^2} + \frac{\partial^2 \psi}{\partial y^2} = \nabla^2 \psi = 0 \tag{5.88}$$

Once again we find the Laplace equation in fluid mechanics.

As noted above, for two-dimensional, inviscid, irrotational flows, there are some interesting properties of stream and potential functions (see Batchelor, 1967, for details and proofs):

1. Lines of constant ψ are in fact streamlines of the flow.
2. The unit flow between two streamlines ψ_1 and ψ_2 is simply $\psi_2 - \psi_1$. Therefore, when streamlines converge, the local velocity is increasing, and when streamlines diverge, velocity is decreasing.
3. Lines of constant ϕ and constant ψ are orthogonal (i.e., perpendicular to each other).
4. Since solid boundaries are also streamlines in inviscid flows, point 3 implies that constant potential lines are normal to solid surfaces.

The last two characteristics are important in one of the older graphical methods of solving the inviscid flow equations, which is called the *flow net*. In this method, we simply sketch stream and potential lines so that neither point 3 nor point 4 is violated. Figure 5.12 is an example showing streamlines and potential lines for the inviscid flow past a cylinder.

The next section introduces the famous *Bernoulli* equation of fluid mechanics. It is the final and perhaps most useful result of inviscid flows to be studied here.

5.11. THE BERNOULLI EQUATION

Mathematics textbooks detail the many contributions to mathematics and physics made by eight members of a single Italian family, the Bernoullis, during the seventeenth and eighteenth centuries. One of the most important contribu-

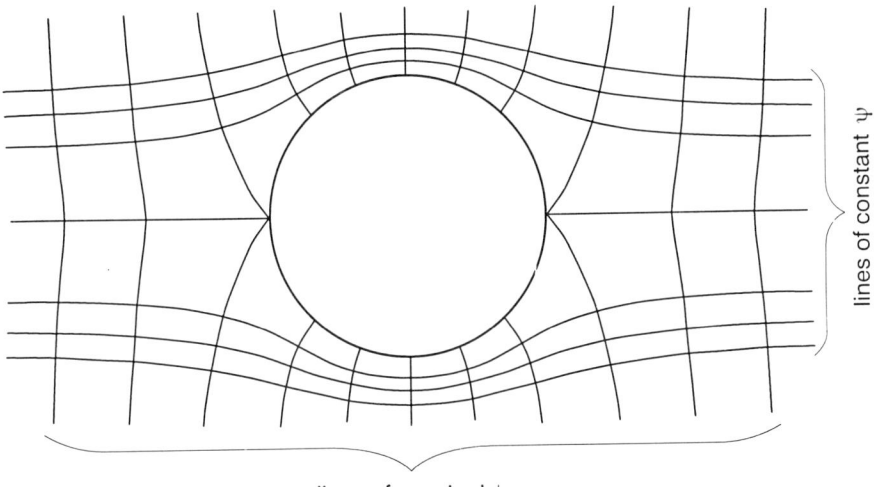

lines of constant ψ

lines of constant ϕ

Figure 5.12. Streamlines and lines of constant potential for inviscid, irrotational flow around a cylinder. Flow is from left to right.

tions with respect to fluid mechanics is the *Bernoulli* equation. We begin by writing the steady form of the Euler equation:

$$\mathbf{u} \cdot \nabla \mathbf{u} = -\frac{1}{\rho} \nabla P + \mathbf{g} \tag{5.89}$$

The next step involves a well-known fact from physics — that for *conservative body forces* like gravity, $\mathbf{g}$ can be expressed as the gradient of a scalar field, which we call G:

$$\mathbf{g} = \nabla G \tag{5.90}$$

(We will later note that G is the potential energy of a fluid particle, $G = -gz$.) Hence, the steady Euler equation can now be expressed as

$$\mathbf{u} \cdot \nabla \mathbf{u} = -\frac{1}{\rho} \nabla P + \nabla G \tag{5.91}$$

However, there is a vector algebra identity which says that

$$\mathbf{u} \cdot \nabla \mathbf{u} = \nabla \left(\frac{1}{2} \mathbf{u} \cdot \mathbf{u} \right) - \mathbf{u} \times (\nabla \times \mathbf{u}) \tag{5.92}$$

Hence, recalling the definition of vorticity, Eq. 5.74, the steady Euler equation can be expressed as

$$\nabla \left(\frac{1}{2} \mathbf{u} \cdot \mathbf{u} \right) - \mathbf{u} \times \omega = -\frac{1}{\rho} \nabla P + \nabla G \tag{5.93}$$

Using vector algebra, Currie (1974) showed that the pressure gradient term in the equation above can be expressed as follows when evaluated along a streamline in the flow:

$$\frac{1}{\rho} \nabla p = \nabla \left(\int \frac{dp}{\rho} \right) \tag{5.94}$$

Hence, Eq. 5.93 can be written as

$$\nabla \left(\int \frac{dp}{\rho} + \frac{1}{2} \mathbf{u} \cdot \mathbf{u} - G \right) = \mathbf{u} \times \omega \tag{5.95}$$

The last vector transformation required is to form the dot product of Eq. 5.95 with $\mathbf{u}$:

$$\mathbf{u} \cdot \nabla \left(\int \frac{dp}{\rho} + \frac{1}{2} \mathbf{u} \cdot \mathbf{u} - G \right) = \mathbf{u} \cdot (\mathbf{u} \times \omega) \tag{5.96}$$

However, the dot product on the right-hand side of Eq. 5.96 must be zero (do you see why?); hence, Eq. 5.96 simplifies to

$$\mathbf{u} \cdot \nabla \left(\int \frac{dp}{\rho} + \frac{1}{2} \mathbf{u} \cdot \mathbf{u} - G \right) = 0 \tag{5.97}$$

Now recall the definition of the substantial derivative from Section 5.3:

$$\frac{D}{Dt} = \frac{\partial}{\partial t} + u_1 \frac{\partial}{\partial x_1} + u_2 \frac{\partial}{\partial x_2} + u_3 \frac{\partial}{\partial x_3} \tag{5.3}$$

Note that Eq. 5.97 is just like the space derivative part of the substantial derivative. And because the flow is steady, we conclude that when the material derivative is zero, the quantity inside the parentheses in Eq. 5.71 must be a constant, or

$$\int \frac{dp}{\rho} + \frac{1}{2} \mathbf{u} \cdot \mathbf{u} - G = \begin{bmatrix} \text{constant along} \\ \text{a streamline} \end{bmatrix} \qquad (5.98)$$

If we note that the potential energy term is equal to $-gz$, where z is the elevation above some fixed datum, and if we assume an incompressible fluid (i.e., $\rho = $ constant), then Eq. 5.98 simplifies to a form familiar from hydraulics:

$$\frac{p}{\rho} + \frac{u^2}{2} + gz = \begin{bmatrix} \text{constant along} \\ \text{a streamline} \end{bmatrix} \qquad (5.99)$$

where the constant is sometimes known as the *Bernoulli constant*. It is important to summarize all the assumptions required in the development of this form of the Bernoulli equation: an inviscid steady flow, and an incompressible fluid. Currie (1974) and Batchelor (1967) show that an even stronger form of the Bernoulli equation can be derived if we also assume an irrotational flow. In this case, the Bernoulli constant is just not constant along a streamline, it is constant *throughout* the flow.

Example 5.4. Use the Bernoulli equation to derive a formula for the velocity of flow from the orifice of the tank shown below.

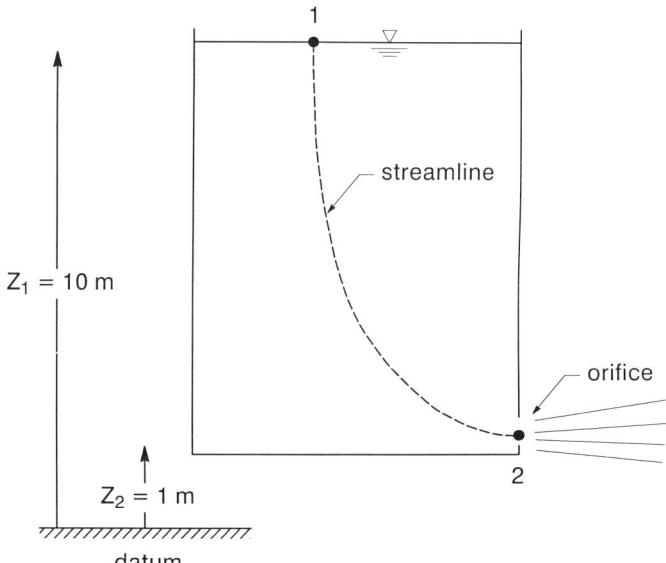

Example 5.4

SOLUTION. Since the Bernoulli constant is constant along a streamline, all we need to do is draw a streamline in the flow and apply the Bernoulli equation between two points on the streamline. We pick the streamline shown in the sketch. The reader might wonder how we knew that the line we picked was actually a streamline. We can cheat here. There must be a streamline connecting the free surface and the orifice, because there could not be a flow otherwise. Any streamline connecting the free surface and the orifice gives the same answer. So we arbitrarily chose the streamline shown.

Since the Bernoulli constant is constant along a streamline, we can write the following:

$$\frac{p_1}{\rho} + \frac{u_1^2}{2} + gz_1 = \frac{p_2}{\rho} + \frac{u_2^2}{2} + gz_2$$

However, if the flow is fairly small (as it would be for a small orifice), u_1 is zero. Also, p_1 is obviously atmospheric pressure; but since the orifice is open to the atmosphere, p_2 is also atmospheric pressure. Hence, we find the following simple and classic formula:

$$u_2 = \sqrt{2g(z_1 - z_2)} = \left(2 \times 9.80 \frac{m}{s^2} \times 9\,m\right)^{1/2} = 13.3 \frac{m}{s}$$

Example 5.5. The two devices shown below and on p. 183 have long been used to measure the velocity of fluids. The first device is a simple pitot or total head tube; the second device combines a pitot tube with a static pressure measurement and a differential manometer.

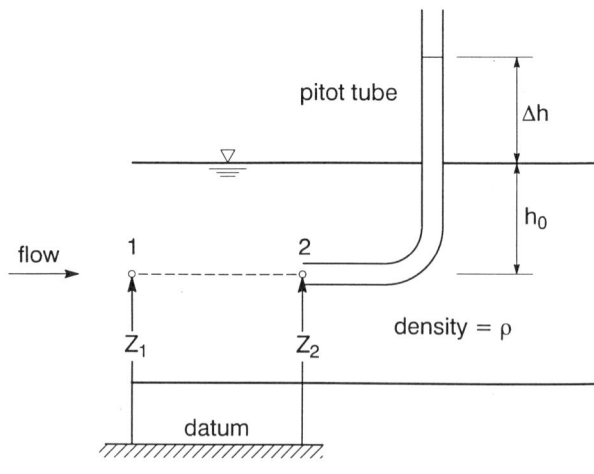

Example 5.5(a)

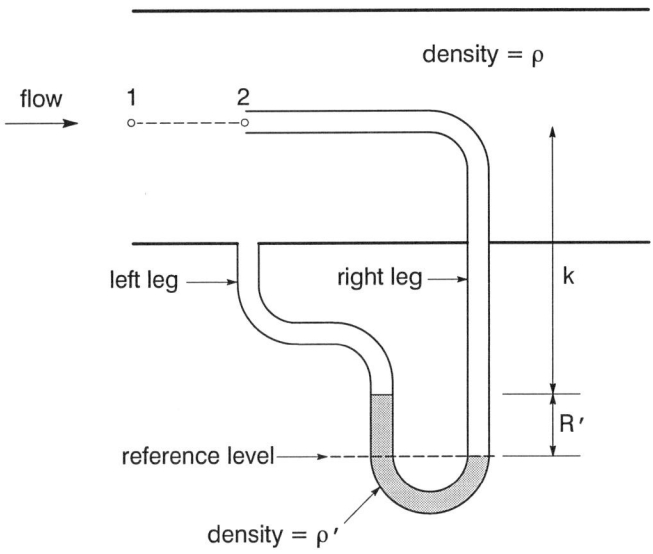

Example 5.5(b)

Develop equations to predict the fluid velocity with these two devices.

SOLUTION. For the first device, we again write the Bernoulli equation between points 1 and 2:

$$\frac{p_1}{\rho} + \frac{u_1^2}{2} + gz_1 = \frac{p_2}{\rho} + \frac{u_2^2}{2} + gz_2$$

However, $u_2 = 0$ at the entrance to the pitot tube (i.e., the *stagnation point*); and since $z_1 = z_2$, the equation simplifies to

$$\frac{u_1^2}{2} + \frac{p_1}{\rho} = \frac{p_2}{\rho}$$

In this free-surface flow, the pressure at point 1 is simply the hydrostatic pressure $\rho g h_0$ (Section 5.4). As indicated by the height of fluid in the pitot tube, the pressure at point 2 (the *stagnation pressure*) must be $\rho g(h_0 + \Delta h)$; hence, rearranging, we arrive at the final equation:

$$u_1 = \left(2\,\frac{p_2 - p_1}{\rho}\right)^{1/2} = (2g\Delta h)^{1/2}$$

(Note: $\rho g \Delta h$ is traditionally called the *dynamic pressure*; hence, here the stagnation pressure is the sum of the dynamic pressure and the hydrostatic pressure.)

As shown, the second device measures velocity in a closed pipe or duct. We can again write the Bernoulli equation between points 1 and 2, and simplify as above to

$$\frac{u_1^2}{2} = \frac{p_2 - p_1}{\rho}$$

We would like to express u_1 in terms of the easily measured parameter R'. We will need to develop an equation for pressure in the manometer section. This is most easily accomplished by considering the reference point shown in the manometer tube. Note that at the point shown, the pressure in both left and right legs of the manometer tube must be the same (otherwise we would predict different pressures at the very bottom of the manometer tube). Thus, considering pressure in the right leg above the reference point,

$$\left\{ \begin{array}{l} \text{pressure at reference} \\ \quad \text{point (right leg)} \end{array} \right\} = p_2 + k\rho g + R'\rho g$$

Similarly, considering the left leg,

$$\left\{ \begin{array}{l} \text{pressure at reference} \\ \quad \text{point (left leg)} \end{array} \right\} = p_1 + k\rho g + R'\rho' g$$

Equating the pressure expressions,

$$p_1 + k\rho g + R'\rho' g = p_2 + k\rho g + R'\rho g$$

or

$$p_2 - p_1 = R'g(\rho' - \rho)$$

Substituting into the Bernoulli equation from above yields the final equation:

$$u_1 = \left[2gR' \left(\frac{\rho'}{\rho} - 1 \right) \right]^{1/2}$$

Hence, by simply measuring R', we have a convenient method for measuring velocity in closed conduits.

Exercise 5.14 will show that the assumptions of the Bernoulli equation are not always justified, particularly the neglect of viscosity. Nevertheless, it is often possible to make corrections for viscous and other effects in engineering work and environmental modeling, and this approach can be pursued in other references (e.g., Welty et al., 1976).

5.12. STEADY VISCOUS MOMENTUM BOUNDARY LAYERS

The reader may be a bit surprised to see that we are returning to viscous flows, after spending considerable effort understanding inviscid flows. Recall that in many flows, a large portion of the flow behaves essentially as if it were inviscid. Again imagine the night air blowing across your desk, or water moving through a large settling tank under roughly plug-flow conditions (recall Section 2.7). Nevertheless, fluid mechanicians know that even if a large portion of the flow is essentially inviscid, there are always *viscous momentum boundary layers* at solid surfaces. In other words, real fluids cannot avoid the no-slip condition. In many environmental mass-transport problems dominated by convection of the main flow, these boundary layers may be almost irrelevant. However, in other problems where there is mass transport at solid boundaries, the details of the momentum boundary layers (and concentration boundary layers, to be developed in Chapter 7) may be critical.

We have already seen an excellent example of boundary layer development in Section 5.8, which concerned the viscous boundary layer along an infinite flat plate in uniform motion in an otherwise still fluid. Other examples include the so-called entrance region of flow development in pipe or Poiseuille flow (see Figure 5.14a). A particularly enlightening and important example of boundary layer development for environmental scientists is viscous flow over a flat plate, as illustrated in Figure 5.13. The boundary layer is generally considered the region of flow in which $u \leqslant 0.99\,U$. The boundary layer shown in Figure 5.13

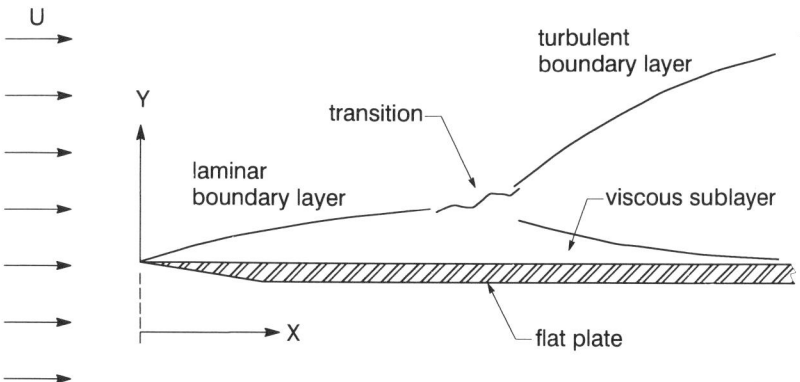

Figure 5.13. Laminar boundary layer along a flat plate immersed in a uniform flow of velocity U. Note that there might also be a boundary layer below the plate, but we are not interested in it here.

is *fully developed*. In other words, enough time has elapsed for the boundary layer to have reached its steady-state configuration. At first glance, this problem seems closely related to the problem considered in Section 5.8. However, there are some important differences. The "flow" above the plate in Section 5.8 was invariant in the x direction, because the plate was of infinite length and the boundary layer grew evenly in the y direction at any point along the plate surface. Contrast this with Figure 5.13. Now the plate is motionless and the fluid is moving, uniformly, until it encounters the "leading edge" of the plate. There can be no boundary layer at $x = 0$. However, as we move down the plate in the x direction, more fluid above the plate begins to feel the effect of the no-slip condition. Viscosity causes a transfer of negative momentum from the plate upward into the overlying fluid. (Note that the plate in Section 5.8 was transferring positive momentum to the overlying fluid.) In any event, we now find that our problem is not strictly unidirectional as in Section 5.8: The "growth" of the boundary layer causes the flow to be slightly nonuniform in the x direction. Hence, although small, there is a finite velocity in the positive y direction.

Therefore, our analysis begins with the x component of the Navier–Stokes equations, which for steady two-dimensional flow with no free surface is

$$u \frac{\partial u}{\partial x} + v \frac{\partial u}{\partial y} = -\frac{\partial p}{\partial x} + v \left(\frac{\partial^2 u}{\partial x^2} + \frac{\partial^2 u}{\partial y^2} \right) \tag{5.100}$$

The classic approach to boundary-layer calculations is to simplify the governing equations as much as possible. We begin by scrutinizing the pressure-gradient term. Note that two points labeled 1 and 2 are indicated in Figure 5.13 along the outside edge of the boundary layer. If we apply the Bernoulli equation between these two points, we find:

$$p_1 - p_2 = \rho g(y_2 - y_1) \tag{5.101}$$

where we used the fact that $u_1 = u_2 = U$ at the edge of the boundary layer. Now we anticipate that boundary layers are actually quite thin. (We exaggerated the boundary layer thickness δ is Figure 5.13 for clarity.) If the boundary layer is very thin, the difference $y_2 - y_1$ is negligible; hence, Eq. 5.101 says that there is essentially no pressure gradient in the x direction; therefore, we dispense of the pressure gradient term in Eq. 5.100.

Next we estimate the relative magnitude of the two viscous terms. We begin by trying to estimate u. We say that u is the "order of U." This is written as

$$u = O(U) \tag{5.102}$$

where the omicron stands for "order of." For the student who is just learning the order-of-magnitude estimation business, it may seem strange to say that any x component of velocity in the boundary layer can be approximated by U. Those students are encouraged to remember that we are trying to get rough estimates here to help us simplify the problem.

Now on to the gradient of u in the y direction. Here we estimate that

$$\frac{\partial u}{\partial y} = O\left(\frac{U}{\delta}\right) \tag{5.103}$$

This is also clearly a rough estimate, but it makes considerable sense: If u changes from zero at the plate surface to U at the edge of the boundary layer, the gradient should be roughly the value given in Eq. 5.103. Finally, having acquired the spirit of estimation, we estimate the second derivative in the y direction as

$$\frac{\partial^2 u}{\partial y^2} = O\left(\frac{U}{\delta^2}\right) \tag{5.104}$$

In a similar spirit, we estimate the derivatives in the x direction as

$$\frac{\partial u}{\partial x} = O\left(\frac{U}{x}\right) \tag{5.105}$$

and

$$\frac{\partial^2 u}{\partial x^2} = O\left(\frac{U}{x^2}\right) \tag{5.106}$$

Now, finally, we can compare the relative magnitude of the two second-derivative terms, Eqs. 5.104 and 5.106:

$$\frac{\dfrac{\partial^2 u}{\partial y^2}}{\dfrac{\partial^2 u}{\partial x^2}} = O\left(\frac{\dfrac{U}{\delta^2}}{\dfrac{U}{x^2}}\right) = \left(\frac{x}{\delta}\right)^2 \tag{5.107}$$

Clearly, since $x \gg \delta$ for the vast majority of the boundary layer, Eq. 5.107 says that the second derivative of u in the y direction will be much greater than the same derivative in the x direction. Hence, from our order-of-magnitude analysis, we approximate the Navier–Stokes equation in the boundary layer as

$$u\frac{\partial u}{\partial x} + v\frac{\partial u}{\partial y} = v\frac{\partial^2 u}{\partial y^2} \tag{5.108}$$

This equation, along with the two-dimensional continuity equation

$$\frac{\partial u}{\partial x} + \frac{\partial v}{\partial y} = 0 \tag{5.109}$$

are called the *boundary-layer equations* for this problem. The boundary conditions are:

$$\begin{aligned} u = v = 0 &\quad \text{at } y = 0 \\ u = U &\quad \text{at } y = \infty \end{aligned} \tag{5.110}$$

In Section 5.10.3, we have already seen that the stream function satisfies the two-dimensional continuity equation above. Substituting the relationships from Eq. 5.86, we find that Eq. 5.108 transforms into

$$\frac{\partial \psi}{\partial y} \frac{\partial^2 \psi}{\partial x \partial y} - \frac{\partial \psi}{\partial x} \frac{\partial^2 \psi}{\partial y^2} = v \frac{\partial^3 \psi}{\partial y^3} \tag{5.111}$$

A similarity solution to this differential equation exists for the following transformation of variables:

$$\psi(x, y) = \sqrt{vUx} f(\eta) \tag{5.112}$$

where the dimensionless independent variable η is given by

$$\eta = \frac{y}{2\sqrt{\dfrac{vx}{U}}} \tag{5.113}$$

With a little effort (and algebra), the following relationships can be calculated:

$$u = \frac{\partial \psi}{\partial y} = \frac{U}{2} f'(\eta) \tag{5.114}$$

$$-v = \frac{\partial \psi}{\partial x} = \frac{1}{2} \left(\frac{vU}{x} \right)^{1/2} (f - \eta f') \tag{5.115}$$

$$\frac{\partial^2 \psi}{\partial x \partial y} = -\frac{U\eta}{4x} f'' \tag{5.116}$$

$$\frac{\partial^2 \psi}{\partial y^2} = \frac{U}{4} \left(\frac{U}{vx} \right)^{1/2} f'' \tag{5.117}$$

and

$$\frac{\partial \psi^3}{\partial y^3} = \frac{U^2}{8vx} f'''$$ (5.118)

Substitution of these relationships into Eq. 5.111 yields the following ordinary differential equation:

$$f''' + ff'' = 0$$ (5.119)

with the following dimensionless boundary conditions:

$$\begin{array}{ll} f = f' = 0 & \text{at } \eta = 0 \\ f' = 2 & \text{at } \eta = \infty \end{array}$$ (5.120)

Equation 5.119 has been solved numerically, and the reults are presented in Table 5.1. These data allow us to determine the velocity anywhere in a viscous boundary layer, or to determine the boundary layer thickness at any point x.

Table 5.1. Solution of Laminar Boundary Layer Problem

$\eta = \dfrac{y}{2}\sqrt{\dfrac{U}{vx}}$	f'	$\dfrac{u}{U}$	f''
0	0	0	1.32824
0.2	0.2655	0.1328	1.3260
0.4	0.5294	0.2647	1.3096
0.6	0.7876	0.3938	1.2664
0.8	1.0336	0.5168	1.1867
1.0	1.2596	0.6298	1.9670
1.2	1.4580	0.7290	0.9124
1.4	1.6230	0.8115	0.7360
1.6	1.7522	0.8761	0.5565
1.8	1.8466	0.9233	0.3924
2.0	1.9110	0.9555	0.2570
2.2	1.9518	0.9759	0.1558
2.4	1.9756	0.9878	0.0875
2.6	1.9885	0.9943	0.0454
2.8	1.9950	0.9962	0.0217
3.0	1.9980	0.9990	0.0096
3.2	1.9992	0.9996	0.0039
3.4	1.9998	0.9999	0.0015
3.6	1.9999	1.0000	0.0005
3.8	2.0000	1.0000	0.0002
4.0	2.0000	1.0000	0.0000
5.0	2.0000	1.0000	0.0000

Source: Reproduced from Welty, Wicks, and Wilson, *Fundamentals of Momentum, Heat, and Mass Transport*, p. 173, © 1976. Reprinted by permission of John Wiley & Sons, Inc.

For example, let's say we define the limit of the boundary layer by $u = 0.99\,U$. Scanning Table 5.1, we estimate that u/U is approximately equal to 0.99 when η is equal to 2.5. Then setting $y = \delta$ in Eq. 5.113, we find:

$$\frac{\delta}{x} = 5\left(\frac{\nu}{Ux}\right)^{1/2} = \frac{5}{\sqrt{\text{Re}_x}} \tag{5.121}$$

where Re_x is a special *boundary-layer Reynolds number*:

$$\text{Re}_x = \frac{Ux}{\nu} \tag{5.122}$$

Again, from a complicated analysis, we find a simple and useful result. Consider a fixed point x along the plate. Equation 5.121 then says that as the free-stream velocity U is increased, the boundary layer Reynolds number *increases* and the boundary layer thickness *decreases*. This makes a lot of sense physically, when we consider the competing forces in the boundary layer. As we have said before, the free-stream flow has a lot of momentum. In the boundary layer, viscosity is trying to make sure the no-slip condition is met by smoothly matching the zero velocity at the plate surface with the free-stream velocity U. However, when the free-stream velocity is increased, the increased momentum of the free stream overwhelms the local viscous forces and "sweeps" the boundary layer further downstream. A new equilibrium is reached with a thinner boundary layer. Alternatively, if the free-stream velocity is reduced, the Reynolds number decreases and the local boundary layer thickness increases.

Finally, note that with the expression for stress given in Section 5.7 and the relation for the velocity gradient at the plate surface given above, we can calculate the stress at the plate surface. In Exercise 5.16, we show that the derivative of velocity at the plate surface is

$$\frac{\partial u}{\partial y} = \frac{U}{4}\left(\frac{U}{\nu x}\right)^{1/2} f''(0) \tag{5.123}$$

From Table 5.1, $f''(0) = 1.32824$; therefore, the shear stress at the surface is from Eqs. 5.30 and 5.123:

$$\tau_{yx}\big|_{y=0} = \mu\left(\frac{\partial u}{\partial y}\right)\bigg|_{y=0} = 0.332\mu U\left(\frac{U}{\nu x}\right)^{1/2} \tag{5.124}$$

Since there is no pressure gradient-related drag term, as there is for viscous flow past a sphere (Section 5.9.2), the drag along the flat-plate surface is only caused by the effect of τ_{yx} along the plate surface. As was the case for the sphere, this type of drag is called the *skin-friction* drag. Based on Newton's drag law (Section 2.2), it is traditional to define a skin-friction drag coefficient for the flat plate as

$$C_{fx} = \frac{\tau_{yx}|_{y=0}}{\rho(U^2/2)} = \frac{(0.664)\mu U}{\rho U^2} \left(\frac{U}{\nu x}\right)^{1/2} = 0.664 \left(\frac{\nu}{Ux}\right)^{1/2} = 0.664 \, \text{Re}_x^{-1/2} \quad (5.125)$$

Equation 5.125 is the local skin-friction coefficient along the flat plate, that is, at any specific point x. Often we want to know the average skin-friction coefficient over a plate of finite length L. This is easily set up as

$$C_{fL} = \frac{1}{L} \int_0^L C_{fx} dx = \frac{1}{L}(0.664) \left(\frac{\nu}{U}\right)^{1/2} \int_0^L x^{-1/2} dx = 1.32 \left(\frac{\nu}{UL}\right)^{1/2} = 1.32 \, \text{Re}_L^{-1/2}$$

$$(5.126)$$

5.13. TURBULENT FLOWS

Until this point, we have generally avoided the issue of turbulence. In this chapter we have so far considered flow to be *laminar*: smooth streamlines and very predictable flow velocities. Nevertheless, we must acknowledge that many (if not the majority) of fluid flows in the environment and in pollution-control equipment are *turbulent*. The Reynolds number is a guidepost for predicting whether flows will be turbulent. For example, Figure 5.14a shows the development of laminar flow in a duct when the Reynolds number is less than about 2000. Note how the boundary layer smoothly "invades" the core "potential flow" from the sides of the pipe. At a certain distance, the flow is fully developed. However, if we examine the same flow at a higher Reynolds number (e.g., Re = 3000), we find that just inside the entrance region, the flow develops instabilities and "turbulent puffs" begin to form—little regions of irregular or chaotic flow velocities (Figure 5.14b). A short distance down the duct, the turbulence has totally invaded the flow section. The resulting *average* velocity profile looks generally like that of Poiseuille flow, only flatter at the center. The onset of turbulence is somehow related to the greater instability of fluid masses at higher velocities. According to Stanley Corrsin, "Generally speaking, turbulence can be expected in a fluid whenever there is a shearing flow and the inertial effects are much larger than viscous effects" (Corrsin, 1961).

After reflecting on the turbulent flow in the pipe of Figure 5.14 (or in rivers or the atmosphere) the reader may appreciate that it will probably be very difficult to predict the random-like velocities at any specific time or place in turbulent flows. In fact, turbulence has been one of the main unsolved mysteries of physics. Fortunately, enough experiments and analysis have been done to aid in solving practical engineering problems and in the important area of turbulent transport of mass and energy. We will concern ourselves primarily with average qualities of the turbulent flow, like the average velocity profile in Figure 5.14b, or with average stresses at boundaries.

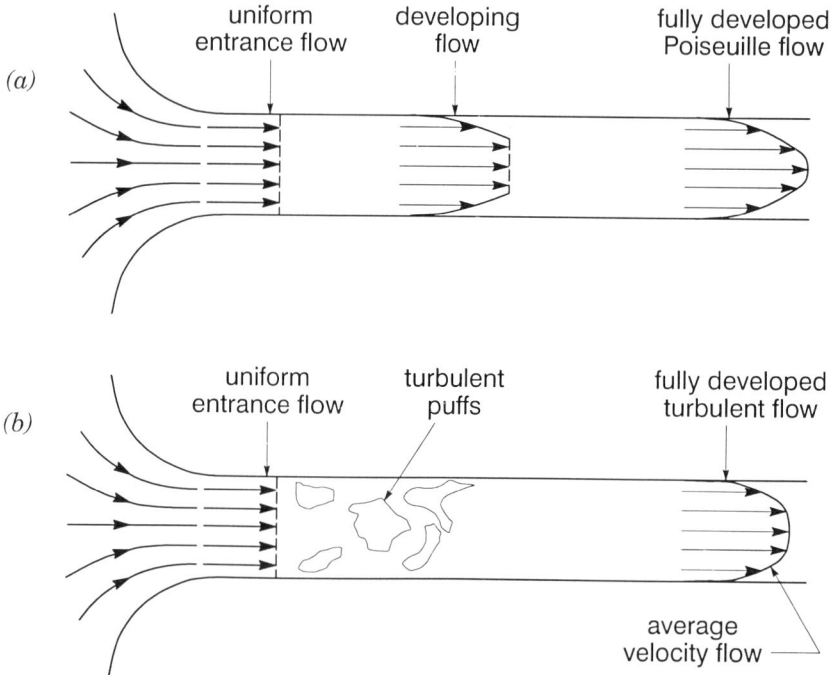

Figure 5.14. Development of (*a*) laminar and (*b*) turbulent flows in a duct. The duct surface is smooth.

5.13.1   Characteristics of Turbulence

Before demonstrating elements of the classical statistical approach to turbulence, it is instructive to describe some of the commonly observed characteristics of turbulence.

General Randomness. If a velocity probe were placed into the turbulent pipe flow shown in Figure 5.14*b*, we might get a signal something like that shown in Figure 5.15. The response of the velocity probe looks essentially random. The signal is random in the sense that the value at some time in the future cannot be predicted by the value at any previous point. Yet turbulence is not as random as white noise. Energy is spread over a range of frequencies; yet statistically speaking, there are correlations between velocities at two points in space, or at two points in time at a single point in space. This is assured by the existence of viscosity and eddy-like motions in the turbulence. Viscosity also assures us that the frequency of these motions does not extend to infinity.

Mean values can be calculated for the *statistically stationary* (*steady*) turbulence shown in Figure 5.15. For example, the mean value $\bar{U}_1$ of the x_1 (streamwise) component of velocity computed from a longer record of velocity

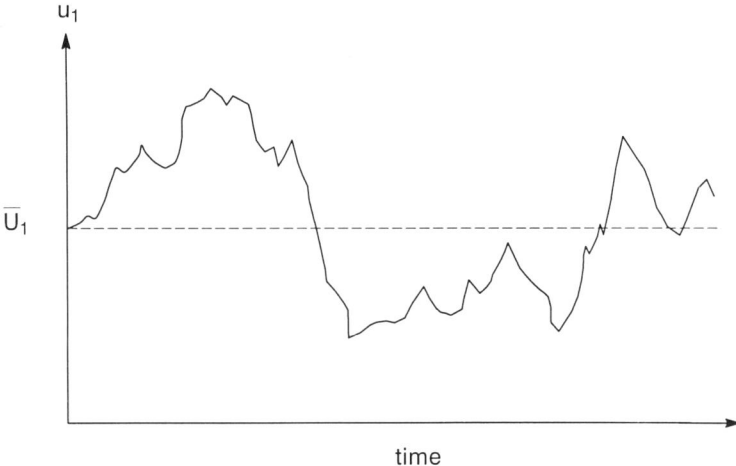

Figure 5.15. Possible response to measurement of the streamwise (x_1) component of velocity in a turbulent flow over a short period of time.

measurements is superimposed on the short velocity record shown in Figure 5.15. As suggested earlier, mean values of velocity are very important in many environmental transport problems.

In addition to the velocity field, we would find that pressure, strain rate, and vorticity also have similar random-looking values.

Three-Dimensionality. On reflection it seems almost obvious that two-dimensional turbulent flows do not exist. In turbulence, velocity fluctuations cause momentum to spill out in all directions. Hence, turbulence is always three dimensional.

Hierarchy of Eddy Sizes and Scales of Motion. One of the most exciting early discoveries about turbulence was that there were correlations between velocities at different points in turbulent flows at a given instant. This suggested that one might think of turbulence as being composed of a wide range of eddies, with smaller eddies contained within larger eddies. In fact, this characteristic is easy to observe in many natural phenomena. For example, look at the smoke rising from a smoke stack on a windy day. The trail of smoke seems to be composed of billows of different size. The largest billows or eddies lead to the most transport over large-length scales. But careful observation reveals smaller-scale motions or eddies that are twisting and turning about, seemingly on their own. These smaller eddies transport mass, momentum, and energy (e.g., heat) over smaller scales of motion.

In recent and exciting measurements, the trajectories of small particles suspended in turbulent flow have been measured, and from these traces, the

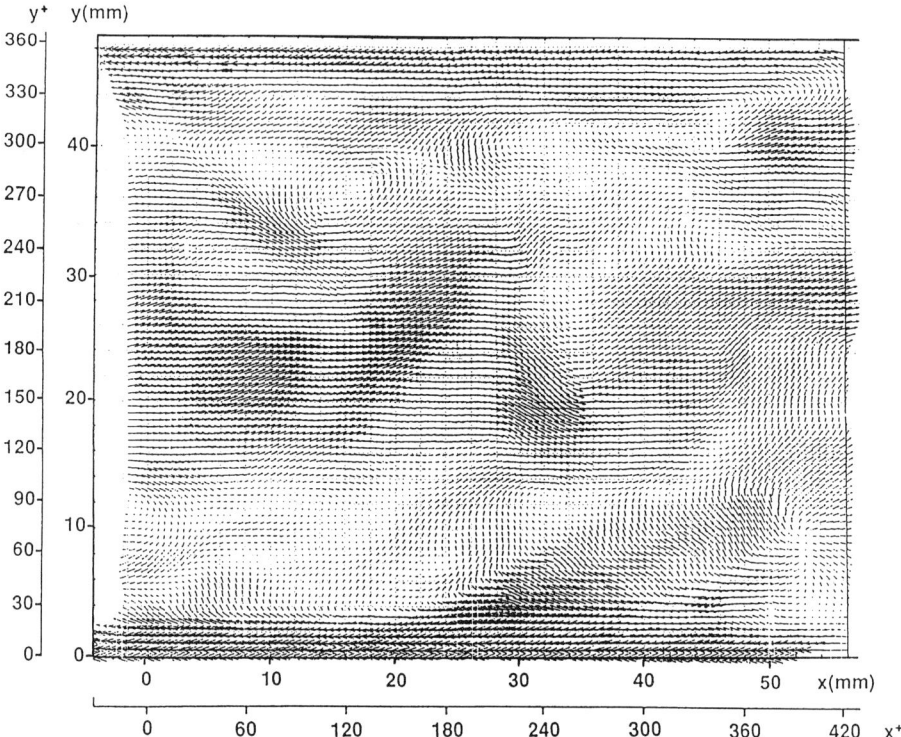

Figure 5.16. Determination of the velocity field in a rectangular region in a turbulent-channel flow using particle image velocimetry. *Source*: Reproduced from Liu, Landreth, Adrian, and Hanratty, *Exp. Fluids*, **10**, 305 (1991). Reprinted with permission granted by Springer-Verlag.

instantaneous velocity vector field at different instants in time and at different places in the flow has been reconstructed. Figure 5.16 shows the results of one of these measurements in a turbulent-channel flow in the laboratory. The reader must examine the figure closely to see that the instantaneous velocity vectors have been reconstructed over a certain two-dimensional region of the channel, and at a certain instant in time. The plot shows several things. First, it reveals the unpredictable way in which velocity changes in space in a turbulent flow. But it also suggests that turbulence is not totally random. Note that there seem to be organized regions in the flow, like swirling and shearing motions. These recent measurements show some of the inherent self-organization of turbulence.

Net Transfer of Kinetic Energy from Large- to Small-Scale Motions. Everyone has seen ice skaters performing during the Olympics or other events. Often, near the end of the performance, the skater initiates a spin that accelerates over time. The skater accomplishes this by pulling his or her arms in closer and closer. As the skater's body becomes more compact, the frequency of spinning increases. This is a result of the conservation of momentum principle of physics. In turbulence, we can imagine that eddies get twisted, stretched, and strained by even larger-scale motions or eddies. Owing to conservation of momentum, the frequency of the motions grows during this stretching and straining process. We say that kinetic energy is transferred from larger-scale motions to smaller-scale motions. Hence, there is a net flow of kinetic energy from large to small-scale motions. Turbulence researchers therefore talk about the *energy cascade*, or the feeding of energy from larger-scale, lower-frequency motions, to smaller-scale, higher-frequency motions.

Dispersion of Heat, Mass, and Momentum. Let's look again at the laminar and turbulent pipe flows discussed earlier. But now, at a short distance into the pipe and near the centerline of the pipe, we inject a tracer material like a dye. We would observe something like that shown in Figure 5.17.

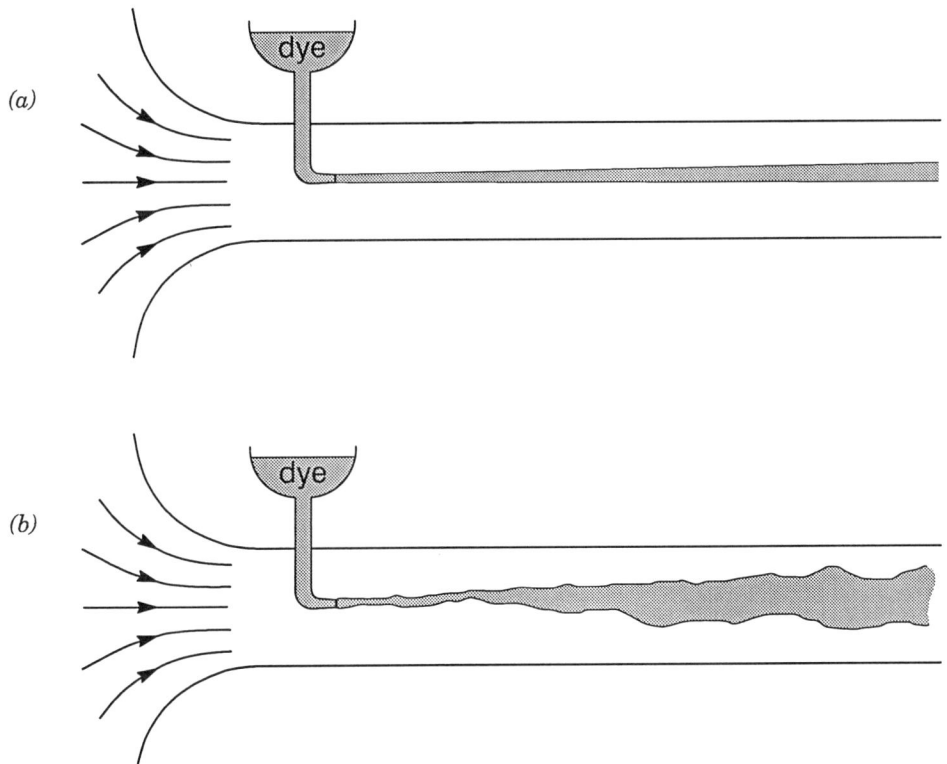

Figure 5.17. Comparison of dye mixing in (*a*) laminar and (*b*) turbulent flows.

Note that in laminar flow, the width of the tracer stream downstream from the injection point changes very little. In fact, the small growth in dye stream width is caused by molecular diffusion of the dye. Contrast this with the turbulent dispersion of the dye in Figure 5.17b. Here the dye is transported into a much larger region of the flow. Since the dye is being added at the same rate in both cases, the concentration of the dye with turbulent mixing is lower than with laminar flow. For environmental engineers, this dispersive characteristic of turbulent flows is one of the most important. Turbulent flows are always better than laminar flows at transporting heat, mass, and momentum.

In a more philosophical vein, we can think about this characteristic of turbulence in several ways. For example, some older environmental scientists may be familiar with the saying, "Dilution is the solution to pollution!" This means that by mixing a pollutant over a larger range of space (perhaps by turbulent mixing), the concentration of pollutant is decreased. But we recognize the cynicism of the comment when we acknowledge that the pollutant has simply been diluted in a larger region (volume) of the environment. It has not gone away, it only appears to have gone away because concentrations are lower. Of course, there are elements of truth in the saying. For example, when organic wastes from sewage-treatment plants are discharged into rivers (which are always turbulent), it could be argued that by dispersing the contaminant over larger regions, the concentration is diminished to levels at which the organic matter can be more effectively oxidized through natural microbial action.

In Chapter 7, we dive further into the problem of predicting the turbulent transport of contaminants. We find that, because of the random nature of turbulence, special approaches have to be developed to predict the turbulent dispersion of contaminants in the environment.

5.13.2.   Reynolds Averaging

We have already stated that all turbulent flows are three dimensional. Therefore, we have to examine the fluctuating velocity components in the three dimensions of cartesian space. For example, if we stick a velocity probe into a stationary turbulent flow, we might measure the three measurements shown in Figure 5.18. The three *instantaneous components* of velocity can be expressed as follows:

$$u_1 = \bar{U}_1 + u_1' \tag{5.127}$$

$$u_2 = \bar{U}_2 + u_2' \tag{5.128}$$

$$u_3 = \bar{U}_3 + u_3' \tag{5.129}$$

The *average velocities* in each direction are $\bar{U}_1$, $\bar{U}_2$, and $\bar{U}_3$. The components u_1', u_2', and u_3' are called the *fluctuating components* of velocity. Note that the

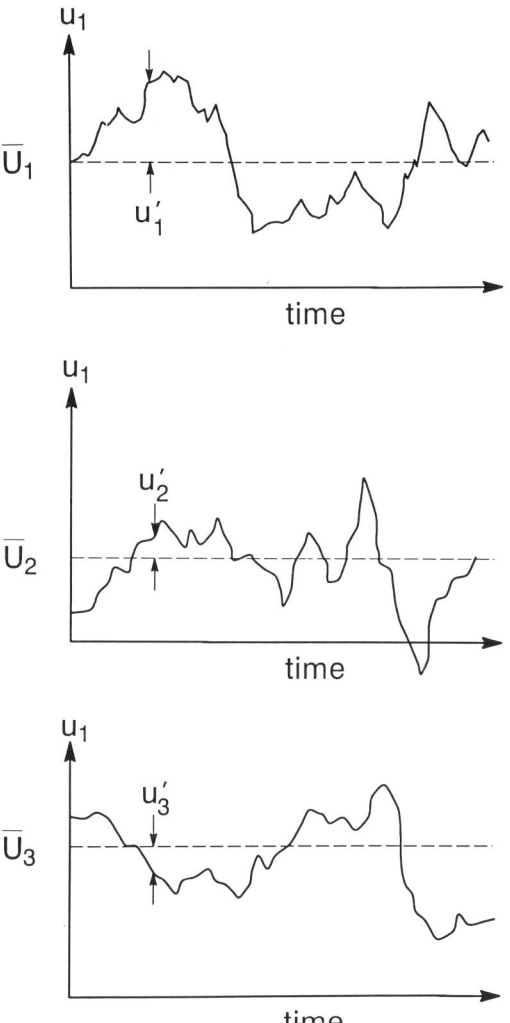

Figure 5.18. Records of three velocity components at a single point in a turbulent flow.

average velocity components are generally different, and could be negative. The fluctuating components have both positive and negative values.

Now let's take the average over time of one of the three equations above, for example, Eq. 5.127:

$$\frac{1}{T} \int_{t_0}^{t_0+T} u_1 dt = \frac{1}{T} \int_{t_0}^{t_0+T} \bar{U}_1 dt + \frac{1}{T} \int_{t_0}^{t_0+T} u_1' dt \qquad (5.130)$$

Here T is the time interval of averaging, which is long enough to get average quantities of the desired accuracy. The value t_0 is the starting time. In this analysis, we require that the average quantities do not vary with different starting times. This ensures a statistical property called *stationarity* (i.e., the average value is statistically steady — it does not vary with time). The term on the left-hand side is the definition of the average velocity, $\bar{U}_1$. The first term on the right-hand side of Eq. 5.130 is the average of the average velocity, which is of course still the average velocity. Hence, the operation performed in Eq. 5.130 implies the following:

$$\bar{U}_1 = \bar{U}_1 + \frac{1}{T} \int_{t_0}^{t_0 + T} u_1' dt \tag{5.131}$$

Note that Eq. 5.131 can only be true if the term involving the integral is exactly zero. Hence, the average of the fluctuating term must be zero. Using our averaging notation, we write this as

$$\overline{u_1'} = \frac{1}{T} \int_{t_0}^{t_0 + T} u_1' dt = 0 \tag{5.132}$$

In a similar way, we would find that the averages of the other two fluctuating components are zero, or

$$\overline{u_1'} = \overline{u_2'} = \overline{u_3'} = 0 \tag{5.133}$$

This exercise of *Reynolds averaging* demonstrates that averaging may reduce the complexity of our equations.

Let's next apply Reynolds averaging to the continuity equation for incompressible fluids (Eq. 1.24):

$$\frac{\partial u_1}{\partial x_1} + \frac{\partial u_2}{\partial x_2} + \frac{\partial u_3}{\partial x_3} = 0 \tag{5.134}$$

Substituting in Eqs. 5.127–5.129 and averaging, we find

$$\frac{1}{T} \int_{t_0}^{t_0 + T} \frac{\partial}{\partial x_1} (\bar{U}_1 + u_1') dt + \frac{1}{T} \int_{t_0}^{t_0 + T} \frac{\partial}{\partial x_2} (\bar{U}_2 + u_2') dt$$
$$+ \frac{1}{T} \int_{t_0}^{t_0 + T} \frac{\partial}{\partial x_3} (\bar{U}_3 + u_3') dt = 0 \tag{5.135}$$

This can be written in special averaging notation as

$$\frac{\partial(\bar{U}_1 + u_1')}{\partial x_1} + \frac{\partial(\bar{U}_2 + u_2')}{\partial x_2} + \frac{\partial(\bar{U}_3 + u_3')}{\partial x_3} = 0 \qquad (5.136)$$

However, the averaging procedure is distributive over addition, so the equation above can be also expressed as

$$\frac{\partial \bar{U}_1}{\partial x_1} + \frac{\partial u_1'}{\partial x_1} + \frac{\partial \bar{U}_2}{\partial x_2} + \frac{\partial u_2'}{\partial x_2} + \frac{\partial \bar{U}_3}{\partial x_3} + \frac{\partial u_3'}{\partial x_3} = 0 \qquad (5.137)$$

Next, because the averaging procedure is carried out over a long time, the space differentiation and time-averaging procedures can be interchanged. Therefore, it is eay to see that the terms involving the average velocities can be reduced, resulting in

$$\frac{\partial \bar{U}_1}{\partial x_1} + \frac{\overline{\partial u_1'}}{\partial x_1} + \frac{\partial \bar{U}_2}{\partial x_2} + \frac{\overline{\partial u_2'}}{\partial x_2} + \frac{\partial \bar{U}_3}{\partial x_3} + \frac{\overline{\partial u_3'}}{\partial x_3} = 0 \qquad (5.138)$$

However, if differentiation and the averaging procedure can be interchanged, by examining one of the fluctuating velocity terms, we see that

$$\frac{\overline{\partial u_1'}}{\partial x_1} = \frac{1}{T} \int_{t_o}^{t_o + T} \frac{\partial u_1'}{\partial x_1} dt = \frac{\partial}{\partial x_1} \left(\frac{1}{T} \int_{t_o}^{t_o + T} u_1' dt \right) \qquad (5.139)$$

But we already know that the term in parentheses is zero (Eq. 5.132); therefore, considering the two other fluctuating terms in a similar manner, we find the following pleasing result:

$$\frac{\partial \bar{U}_1}{\partial x_1} + \frac{\partial \bar{U}_2}{\partial x_2} + \frac{\partial \bar{U}_3}{\partial x_3} = 0 \qquad (5.140)$$

In other words, the mean velocity must also satisfy the famous continuity equation.

5.13.3. Reynolds Stresses and the Closure Problem

When we apply Reynolds averaging to the tensor form of the Navier–Stokes equation (Eq. 5.4), we find the following (assuming no free surface, i.e., no gravity term — Exercise 5.17):

$$\rho \frac{\partial \bar{U}_i}{\partial t} + \rho \bar{U}_k \frac{\partial \bar{U}_i}{\partial x_k} = -\frac{\partial \bar{p}}{\partial x_i} + \mu \frac{\partial^2 \bar{U}_i}{\partial x_j \partial x_j} - \overline{\rho u_j' \frac{\partial u_i'}{\partial x_j}} \qquad (5.141)$$

where $\bar{p}$ is the mean pressure.

Note that this equation is almost the same as the old Navier–Stokes equations if we substitute average velocities and pressure, that is, the first four terms involve the mean acceleration terms, the mean pressure terms, and the mean velocity-viscous stress terms. But the last term in Eq. 5.141 is totally new to our study of fluid mechanics. Traditionally, one further transformation of this equation is made, involving the last term. Note that the following identity must be true:

$$\frac{\partial}{\partial x_j}(\overline{u_i' u_j'}) = \overline{u_j' \frac{\partial u_i'}{\partial x_j}} + \overline{u_i' \frac{\partial u_j'}{\partial x_j}} \tag{5.142}$$

However, the repeated indices in the derivative in the last term imply summation, or

$$\overline{u_i' \frac{\partial u_j'}{\partial x_j}} = \overline{u_i' \left(\frac{\partial u_1'}{\partial x_1} + \frac{\partial u_2'}{\partial x_2} + \frac{\partial u_3'}{\partial x_3}\right)} \tag{5.143}$$

From continuity, the average of the terms in parentheses is zero (Exercise 5.18); using this fact and the remainder of the relation Eq. 5.142, Eq. 5.141 can be expressed as

$$\rho \frac{\partial \bar{U}_i}{\partial t} + \rho \bar{U}_k \frac{\partial \bar{U}_i}{\partial x_k} = -\frac{\partial \bar{p}}{\partial x_i} + \mu \frac{\partial^2 \bar{U}_i}{\partial x_i \partial x_i} - \rho \frac{\partial}{\partial x_j}(\overline{u_i' u_j'}) \tag{5.144}$$

These are the equations of motion for an incompressible turbulent flow. The last term in Eq. 5.144 involves derivatives of the *Reynolds stress*. The Reynolds stresses then are correlations between fluctuating velocity components:[*]

$$\tau_{ij}^t = -\rho \overline{u_i' u_j'} = -\rho \frac{1}{T} \int_{t_0}^{t_0 + T} u_i' u_j' dt \tag{5.145}$$

The time averaging in Eq. 5.145 is essentially the statistical correlation between two fluctuating velocity components. As we pointed out in the general discussion of turbulence, researchers have found that such correlations exist, even between velocity fluctuations along different axes, and for velocity fluctuations separated in time and/or space. Since the correlations only exist over finite spans of space and time, the correlation functions can be used to define various time and space scales for the turbulent flow (Tennekes and Lumley, 1972; Hinze, 1975).

---

[*]By convention, τ_{ij} is the stress that acts on a plane normal to the i axis (the ji plane) and acts in the j direction. When the subscripts are the same, the stresses are called normal stresses; when subscripts are different, they are called shear stresses.

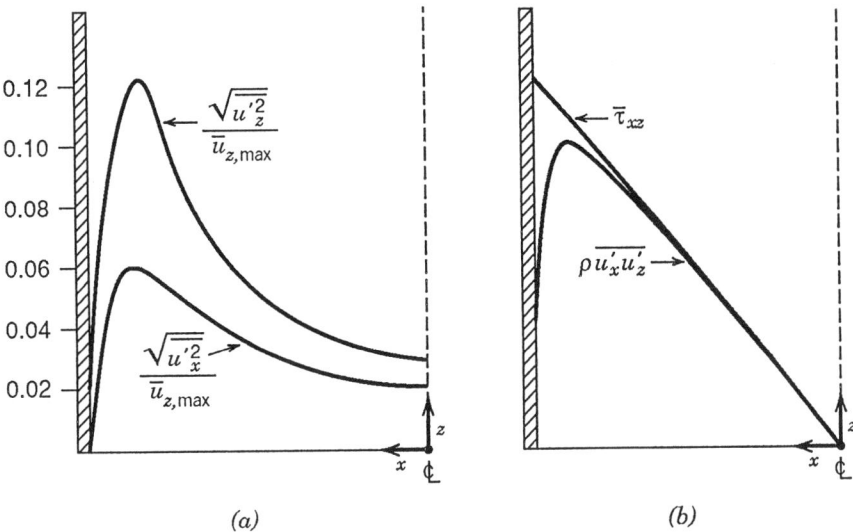

(a)   (b)

Figure 5.19. Turbulence properties in a rectangular channel: (a) mean-square turbulent velocity fluctuations, and (b) comparison of Reynolds stresses and total stress ($\bar{\tau}_{xy}$). Coordinate z is in the streamwise direction, and x is measured from the center of the channel toward the wall. *Source:* Reproduced from Bird, Stewart, and Lightfoot, *Transport Phenomena*, p. 158, © 1960. Reprinted by permission of John Wiley & Sons, Inc.

The Reynolds stresses appear to be rather original terms in the equation of motion. Recall that the terms on the right-hand side of the Navier–Stokes equations are considered forces on the fluid element. Hence Eq. 5.144 suggests that the Reynolds stresses are a new mechanism of momentum transport, which exist only in turbulent flows. We have already mentioned that turbulent mean-velocity profiles in pipes and ducts are flatter in the center than in viscous flows. This is directly related to the Reynolds stresses and their greater efficacy in momentum transport. Figure 5.19b compares Reynolds shear stress with the total shear stress in a rectangular channel. (The total shear stress is a sum of the Reynolds stresses and the viscous stresses due to the mean-velocity gradients in the channel.) Note how over almost all of the flow, the Reynolds stress greatly dominates the viscous stress. Recall the quote of Corrsin, cited in Section 5.13, regarding the importance of shearing flow in creating turbulence. Note also in Figure 5.19a and b how the Reynolds stresses increase from low values near the center of the channel (flat mean-velocity profile, low shear flow) to their maximum values near the channel wall (steep velocity profile, high shear flow). Hence, the Reynolds stresses seem to increase when the mean shear stresses increase. Turbulence researchers speak of the "production of turbulence" by shear flows; the measurements in Figure 5.19 are a perfect demonstration of turbulence production. Note in Figure 5.19b, however, that at positions close to the wall, the viscous stress due to the mean-velocity gradients does eventually dominate the Reynolds stress, because the closer we get to the wall, the smaller are the eddies contributing to the Reynolds stress. At distances very close to the wall, viscosity dampens out these small eddies and the turbulence production, and viscous stresses eventually dominate Reynolds stresses.

The Reynolds stresses τ_{ij}^t are a tensor quantity:

$$\tau_{ij}^t = -\overline{\rho u_i' u_j'} = -\begin{pmatrix} \overline{\rho u_1' u_1'} & \overline{\rho u_1' u_2'} & \overline{\rho u_1' u_3'} \\ \overline{\rho u_2' u_1'} & \overline{\rho u_2' u_2'} & \overline{\rho u_2' u_3'} \\ \overline{\rho u_3' u_1'} & \overline{\rho u_3' u_2'} & \overline{\rho u_3' u_3'} \end{pmatrix} \tag{5.146}$$

Since only six of the Reynolds stresses are unique (the tensor is symmetrical), the new form of the Navier–Stokes equation contains 10 unknowns (3 average velocities, the average pressure, and the 6 Reynolds stresses). Since there are still only four equations (three Navier–Stokes equations and continuity), we have a situation now in which the time-averaged Navier–Stokes equations cannot be solved in principle. This is referred to as the *closure problem* of turbulence. Scientists have occupied themselves during the better part of this century with attempts to circumvent the closure problem. The next section discusses one of those approaches.

5.13.4. Eddy Viscosity and the Mixing-Length Model

In our original discussion of the Navier–Stokes equations we mentioned that the viscous-stress model imagined that stresses were transferred between adjacent sheets of fluid by molecular transport. The greater the rate of strain (velocity gradient), the greater the viscous stress (recall Eq. 5.30). Early turbulence researchers attempted to estimate the Reynolds stresses by drawing an analogy between the Reynolds stresses and the viscous stresses. In this model, the Reynolds stresses are imagined to be driven by the mean-velocity gradients in the fluid and are estimated as

$$\tau_{ij}^t = \rho v_t \frac{\partial \bar{U}_i}{\partial x_j} \tag{5.147}$$

where v_t is called the *eddy diffusivity of momentum*. The eddy diffusivity of momentum is simply related to another common turbulence parameter, the *eddy viscosity*, μ_t: $v_t = \mu_t/\rho$. Substituting Eq. 5.147 into Eq. 5.144 and rearranging a bit, we find:

$$\frac{\partial \bar{U}_i}{\partial t} + \bar{U}_k \frac{\partial \bar{U}_i}{\partial x_k} = -\frac{1}{\rho} \frac{\partial \bar{p}}{\partial x_i} + (v + v_t) \frac{\partial^2 \bar{U}_i}{\partial x_j \partial x_j} \tag{5.148}$$

Note that the assumption of an eddy viscosity has apparently closed the equations of motion, since the correlations of fluctuating velocity in the Reynolds stress tensor have been done away with. However, there is a major problem here. Unlike the real fluid viscosity, *the eddy viscosity is not a characteristic of the fluid*. It should be a function of the flow. We should imagine that the eddy viscosity is not a constant as implied by Eq. 5.147, but rather varies within a flow and between different flows.

Hence, we have a problem with what to use for the eddy viscosity. Since it is not a characteristic of the fluid, it cannot be found in a reference book. The *mixing length* model imagines that turbulent momentum is transferred in an analogous way to momentum transfer in gases undergoing shear. The term corresponding to the mean free path in the gas is called the mixing length. For example, with this model τ_{12}^t is estimated as (Tennekes and Lumley, 1972)

$$\tau_{12}^t = C\rho\sqrt{\overline{u_2'^2}}\, l\, \frac{\partial \bar{U}_1}{\partial x_2} \tag{5.149}$$

where C is an unknown constant and l is the mixing length. This means that for this component of the turbulent shear stress, the eddy viscosity can be imagined to be

$$\mu_t = \rho C\sqrt{\overline{u_2'^2}}\, l \tag{5.150}$$

Unfortunately, the root-mean-square velocity fluctuation and the mixing length are still functions of the flow, and hence are not known a priori. Technically speaking, we have not yet closed the turbulent equations of motion. Nonetheless, researchers have found that in some flows (or some portions of some flows), the mean-square fluctuating quantities are approximately constant. If it is argued that the largest eddies transport most of the turbulent momentum, the mixing length could also be assumed to equal the size of the largest eddies. For example, in wall-bounded shear flows, the mixing length is often set equal to the distance from the wall or boundary. If we also realize that certain components of the stress are more important than others, there may be enough daylight to estimate the eddy viscosity and at least partially close the Navier–Stokes equations. For example, Grant and Madsen (1979) were interested in modeling the mean vertical-velocity profile on the ocean shelf (Figure 5.20).

For unidirectional flow, the relevant Navier–Stokes equation was simplified to

$$\frac{\partial \bar{U}}{\partial t} = -\frac{1}{\rho}\nabla\bar{p} + v_t\frac{\partial}{\partial z}\frac{\partial \bar{U}}{\partial z} \tag{5.151}$$

where the eddy diffusivity of momentum is estimated as

$$v_t = \kappa u_* z \tag{5.152}$$

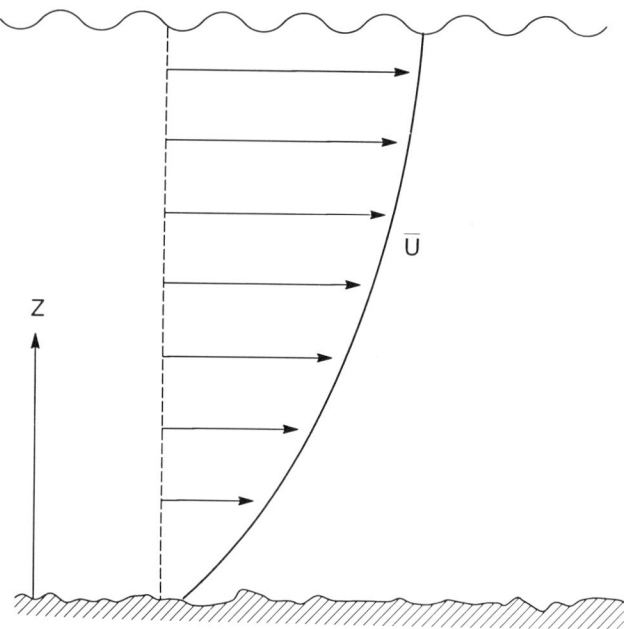

Figure 5.20. Mean velocity profile on the ocean shelf. *Source*: Grant and Madsen, *J. Geophys. Res.*, **84**, 1979.

Here κ is the von Kármán constant and u_* is the so-called *shear velocity*, which is a function of the shear stress at the boundary (see Exercise 5.19):

$$u_* = \sqrt{\frac{\tau_0}{\rho}} \qquad (5.153)$$

Here τ_0 is the total mean shear stress at the boundary.

5.13.5. Advanced Turbulence Modeling — the $k - \varepsilon$ Model

In recent years, the modeling of turbulent flows has developed greatly, especially with the advent of high-speed computer workstations and efficient numerical techniques for solving differential equations. One of the most useful turbulence models suitable for numerical solution is called the $k - \varepsilon$ *model*. In addition to the Reynolds-averaged continuity and Navier–Stokes equations, this approach requires writing balance equations for the *turbulent kinetic energy*, k, and the *turbulent energy dissipation*, ε. The turbulent kinetic energy is defined as

$$k = \tfrac{1}{2}\overline{u_i' u_i'} = \tfrac{1}{2}\overline{u_1'^2 + u_2'^2 + u_3'^2} \qquad (5.154)$$

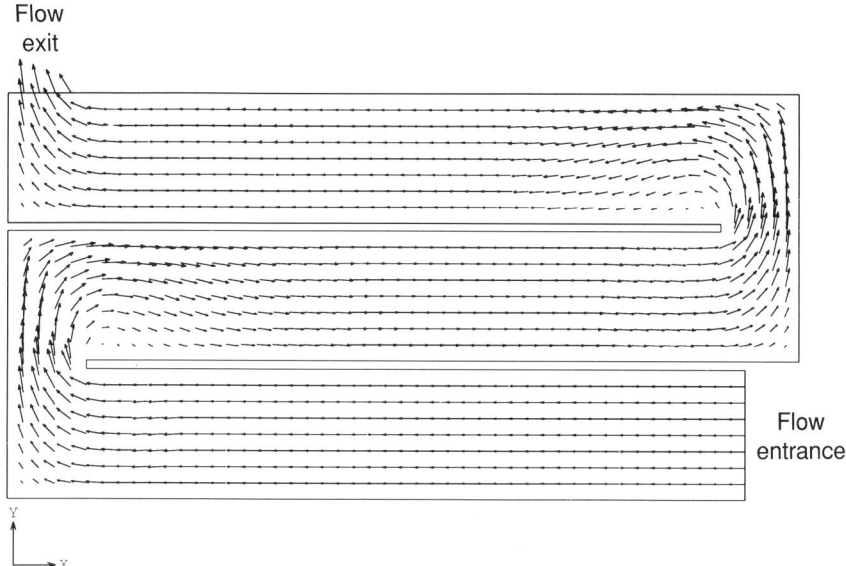

Flow
exit

Y

X

Flow
entrance

Figure 5.21. Simulation of mean velocity vector field in a two-dimensional, serpentine channel using the $k - \varepsilon$ model. Calculations were made with the FIDAP software package (Fluid Dynamics International, Evanston, Illinois).

and the turbulent energy dissipation is defined as

$$\varepsilon = \frac{v}{2} \overline{\left(\frac{\partial u_i'}{\partial x_j} + \frac{\partial u_j'}{\partial x_i} \right)^2} \tag{5.155}$$

Note that since Eq. 5.155 has two repeated indices, there are actually nine terms that contribute to ε (six of which are unique). The $k - \varepsilon$ model also uses a new estimate for the eddy viscosity, which includes k and ε:

$$\mu_t = C_{k-\varepsilon} \frac{k^2}{\varepsilon} \tag{5.156}$$

where $C_{k-\varepsilon}$ is a constant (Rodi, 1984). In Figure 5.21, results from a numerical solution of the governing equations of the $k - \varepsilon$ model are shown for flow through a sinuous, two-dimensional duct. Advanced turbulence modeling is just beginning to see widespread use in environmental modeling.

5.13.6. Isotropic Turbulence and Kolmogoroff's Universal Equilibrium Range

Figure 5.19*a* compares the streamwise and cross-stream turbulent velocity fluctuations in a rectangular channel. Note that near the wall, the streamwise, root-mean-square velocity fluctuation is considerably greater than the corresponding cross-stream component. This characteristic is referred to as the nonisotropy of the turbulence. As we move farther from the wall, the values become somewhat closer to each other. We say the turbulence becomes more *isotropic* near the center of the channel. In general, all shear-flow turbulence is nonisotropic. In certain aspects of turbulence theory, it is nonetheless useful to assume isotropic turbulence. In this case, we assume the following:

$$\overline{u_1'^2} = \overline{u_2'^2} = \overline{u_3'^2} \tag{5.157}$$

In a famous paper, Taylor (1935) derived other important statistics for isotropic turbulence. One of these results is used in Exercise 5.20 to determine the turbulent coagulation rate of small particles in liquids and gases.

The Russian scientist A. N. Kolmogoroff made a famous argument about isotropy and the small scales of turbulence (Hinze, 1975). Recall our earlier observation that turbulent motions exist over a wide range of scales, and that the kinetic energy flow in turbulence is from large- to small-scale motions. Since most of the kinetic energy ends up in the small-scale motions, it must also be dissipated to heat by the small-scale motions. Kolmogoroff argued that at sufficiently high Reynolds numbers, the smallest-scale turbulence loses memory of the exact details of the large-scale motion. In a sense, the small-scale turbulence becomes statistically independent of the boundary conditions and the large-scale motions. Even if the large-scale motions are not isotropic, the small-scale motions may well be isotropic. Then Kolmogoroff hypothesized that since the small-scale turbulence was statistically independent, it could only depend on a limited number of parameters; he argued that those parameters could only be the viscosity, v, and the turbulent energy dissipation rate, ε. Considering the dimensions of v and ε, he further argued that the length scale of the isotropic small-scale motions must be

$$\eta = \left(\frac{v^3}{\varepsilon}\right)^{1/4} \tag{5.158}$$

where η is known as the *Kolmogoroff microscale*. With this same approach, the velocity scale of the small-scale motions must be given by

$$u = (v\varepsilon)^{1/4} \tag{5.159}$$

Note that with these two definitions, the Reynolds number in this "universal equilibrium range" of turbulence is

$$\text{Re} = \frac{\eta u}{v} = 1 \tag{5.160}$$

Although the approximations in Kolmogoroff's theory seem rather extreme, his estimates have been very useful in theoretical and experimental studies of turbulence and turbulent transport.

5.13.7. Flow Through Cylindrical Conduits

Those students who have completed an introductory course in fluid mechanics have undoubtedly seen the famous *Moody diagram*, which relates the pipe-friction factor in cylindrical conduits to the pipe roughness and Reynolds number (Figure 5.22). The y axis in the Moody diagram is the *pipe-friction factor* or the *Fanning friction factor*. The friction factor is given by

$$f_f = \frac{-\tau_w}{(1/2)\rho u_{z,\text{ave}}^2} = -\frac{(d\bar{p}/dz)a}{\rho u_{z,\text{ave}}^2} \tag{5.161}$$

where τ_w is the shear stress at the pipe wall (Section 5.7), $d\bar{p}/dz$ is the static-pressure gradient, a is the pipe radius, and $u_{z,\text{ave}}$ is the average flow velocity.* Most of the f_f curves on the Moody diagram were constructed from careful experiments in which the static-pressure gradient (head loss) and mean velocity were substituted into Eq. 5.161. A number of different curves in the Moody diagram correspond to different values of the *relative pipe roughness*, e/D, where e is the *roughness height* and D is the pipe diameter ($D = 2a$). In the original experimental work of determining the curves for different relative roughness, e was actually the diameter of sand grains that were carefully glued to the inside of the pipe. Over the years, e has been considered the height of any regular roughness elements along the pipe surface.

We now make some general observations about the Moody diagram. Note first that for low Reynolds numbers (Re < 2000), the friction factor is represented by a straight line. Recall that fluid flow in a pipe at low Reynolds number was given by the Poiseuille equation (Sections 5.6 and 5.7). Substituting the average velocity from Eq. 5.27 into Eq. 5.161 yields,

$$f_f = -\frac{(d\bar{p}/dz)a}{\rho[-1/8\mu)(dp/dz)a^2]u_{z,\text{ave}}} = \frac{16}{\text{Re}} \tag{5.162}$$

where Re is the Reynolds number based on the average flow velocity:

$$\text{Re} = \frac{\rho u_{z,\text{ave}} D}{\mu} \tag{5.163}$$

Thus, the straight line at low Reynolds number is for (laminar) Poiseuille flow. Next note the curve labeled "smooth" which is the lower bound for all curves at higher Reynolds numbers. This curve represents turbulent flow in smooth

---

*The Fanning friction factor and the skin-friction coefficient (Section 5.12) are the same quantity.

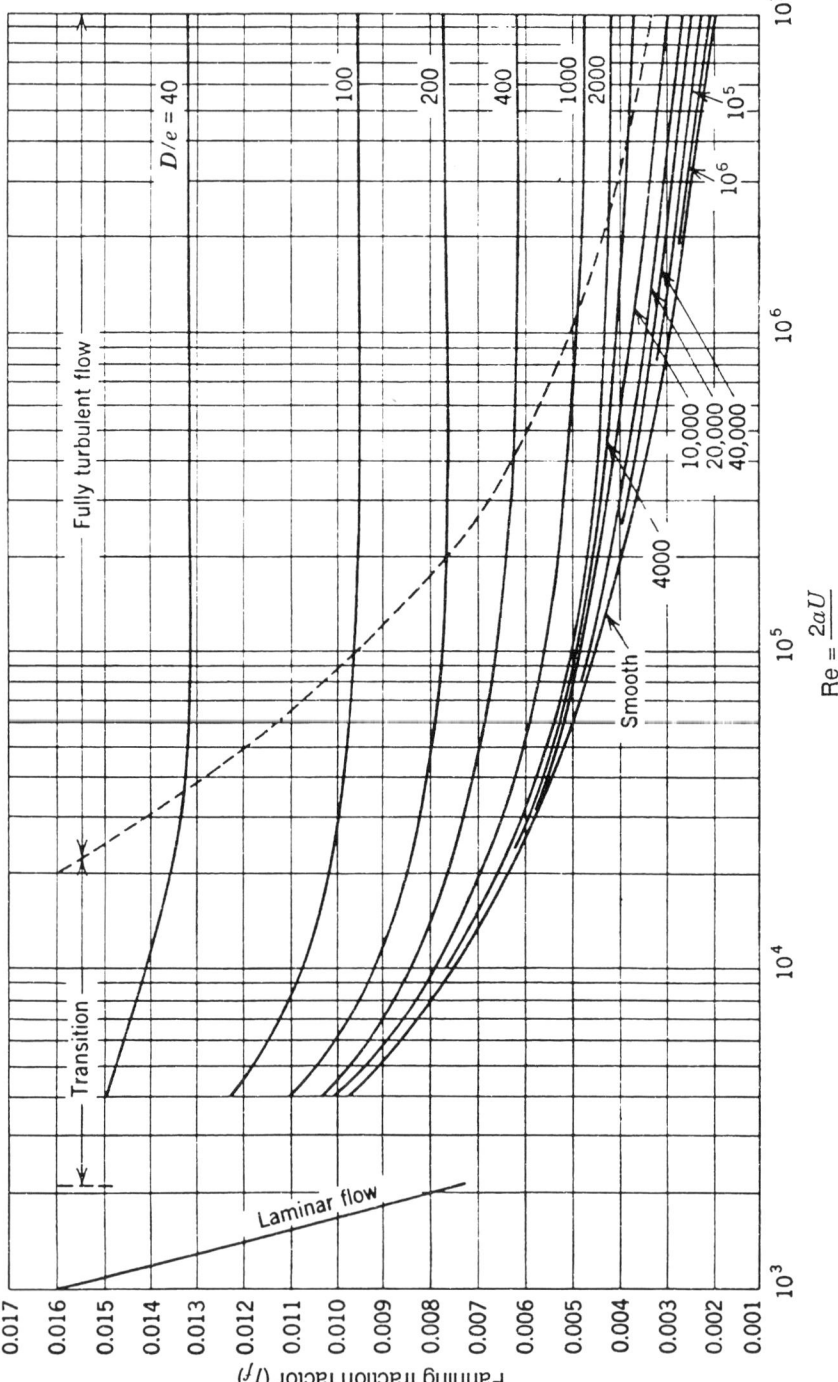

Figure 5.22. The Moody diagram. *Source:* Reproduced from Welty, Wicks, and Wilson, *Fundamentals of Momentum, Heat, and Mass Transport*, p. 209, © 1976. Reprinted by permission of John Wiley & Sons, Inc.

pipes. Now as the relative roughness increase from zero (smooth tube), a family of curves describes the turbulent friction factors. Note also that the curves for nonzero relative roughness asymptotically approach the smooth tube curve as the Reynolds number decreases. This suggests that for low enough Reynolds numbers, even though the pipe surface is rough, the friction factor is given by the smooth tube limit. To understand this behavior, we have to understand more about the structure of turbulence and turbulent boundary layers in pipes and along other surfaces. Recall in our discussion of Reynolds stresses (Section 5.13.3), that near the pipe wall or other boundary in turbulent flows, the small eddies contributing to the Reynolds stresses were eventually dampened out by the effects of viscosity. At a certain distance from the wall, we can imagine that viscous stresses completely outweigh the turbulent Reynolds stresses. This leads to essentially a very thin viscous layer right next to the pipe wall. This region is called the *viscous sublayer*. It is a measurable feature of all turbulence along smooth surfaces. Getting back to our problem, why do the f_f curves for rough pipes eventually approach the smooth-tube curve at low Reynolds numbers? The answer is that the thickness of the viscous sublayer increases as the Reynolds number decreases, just as the thickness of regular laminar bounday layers increases as Reynolds number decreases (Eq. 5.121). When the Reynolds number is sufficiently low, the roughness elements become *buried* in the viscous sublayer. At this point, the turbulence no longer feels the effect of the roughness elements and the friction-factor line melts into the curve for smooth tubes. As far as the turbulence is concerned, the pipe is hydraulically smooth.

The early turbulence researchers attempted to quantify the relationship of the roughness heights to the friction factor by defining a new type of Reynolds number based on the roughness element height and the friction velocity (Schlichting, 1979). The *hydraulically smooth* regime is one in which the *roughness Reynolds number* is between the following limits:

$$0 \leqslant \frac{eu_*}{v} \leqslant 5 \tag{5.164}$$

where u_* is the friction velocity based on the mean wall shear (Eq. 5.153). When the roughness Reynolds number obeys Eq. 5.164, the roughness elements are considered totally buried in the viscous sublayer. The *transition regime* pertains to the following range:

$$5 \leqslant \frac{eu_*}{v} \leqslant 70 \tag{5.165}$$

In the transition regime, we can imagine that the roughness elements are beginning to stick their heads out of the viscous sublayer into the overlying turbulent flow. The mean flow starts to feel the extra wall drag caused by the form drag of the roughness elements. In the Moody diagram, this is the region where the f curves start to diverge from the smooth-tube curve. Finally, the *completely rough* regime pertains to the range

$$\frac{eu_*}{v} \geq 70 \tag{5.166}$$

In the completely rough regime, the viscous sublayer disappears, and the roughness elements have their maximum drag effect on the turbulent flow. The completely rough regime corresponds to the flat part of the f curves for different relative roughness values; hence, in the completely rough regime, the turbulence becomes independent of the Reynolds number (fully turbulent flow).

5.13.8. Turbulent Boundary Layers and Universal Velocity Laws

If we return to the smooth flat plate of Figure 5.13, and consider either higher velocities and/or distances further downstream from the leading edge of the plate, we will generally find that the boundary layer becomes turbulent. This situation is shown in Figure 5.23.

As we noted in our qualitative discussion of turbulence, turbulent boundary layers are more effective than laminar boundary layers in transporting heat, mass, and momentum; this is the reason the turbulent (momentum) boundary layer is fatter than the laminar boundary layer. Turbulent boundary layers are ubiquitous in the environment, and include the boundary layers along the earth's surface and in most water channels. The transition to a turbulent boundary layer over a smooth surface can be predicted with the special

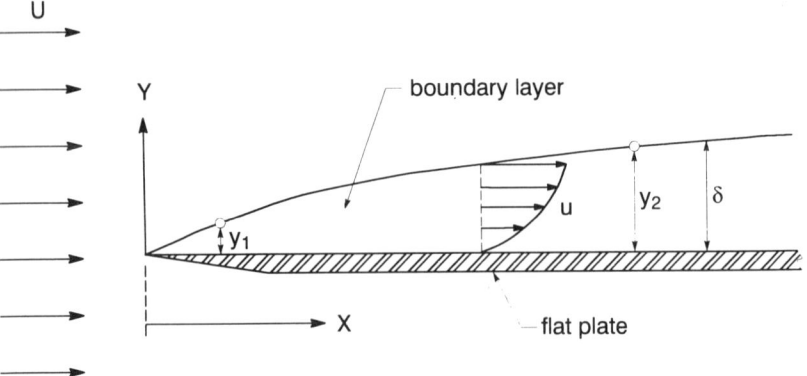

Figure 5.23. Development of turbulent boundary layer above a smooth plate.

boundary-layer Reynolds number, Eq. 5.122 (Welty et al., 1976):

1. Laminar boundary layer $\qquad\qquad\qquad Re_x < 2 \times 10^5$
2. Transitional boundary layer $\qquad 2 \times 10^5 < Re_x < 3 \times 10^6$
3. Turbulent boundary layer $\qquad\qquad\quad Re_x > 3 \times 10^6$

As explained in the previous section, another interesting aspect of turbulent boundary layers is that they are often underlain by a much thinner *viscous sublayer*, which is itself a boundary layer. The viscous sublayer is an inescapable result of the viscosity of real fluids and the necessity of matching the zero-velocity condition at real surfaces.

Quite a bit of effort has gone into characterizing velocities in turbulent boundary layers. Figure 5.24 is a plot showing the vertical variation of experimentally measured mean velocities in the turbulent boundary layers (including the viscous sublayer) above a smooth plate. In the figure, dimensionless distances and velocities are used. The dimensionless velocity is

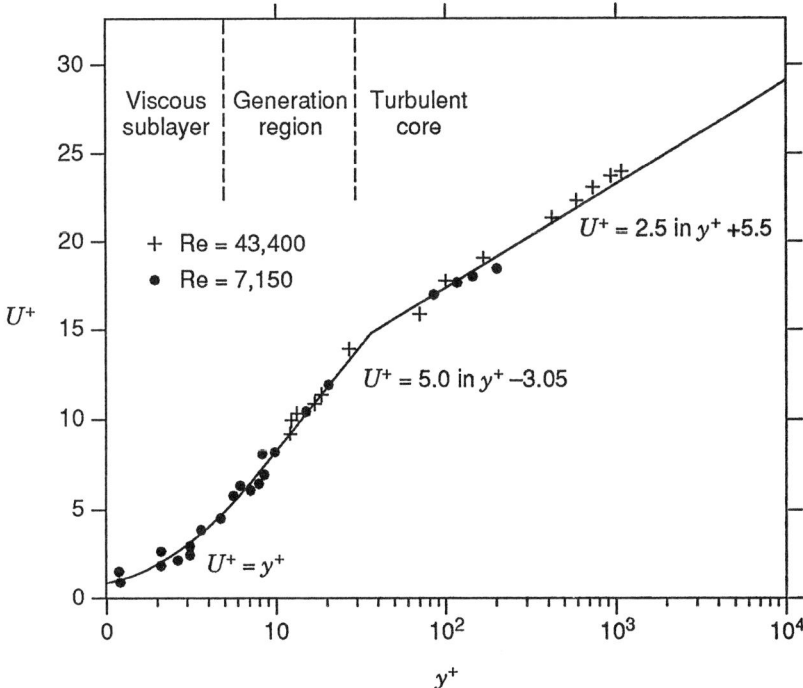

Figure 5.24. Velocities above a smooth flat plate showing universal velocity laws. *Source:* Reproduced from Brodkey and Hershey, *Transport Phenomena: A Unified Approach*, McGraw-Hill Companies, © 1988. Reprinted with permission of The McGraw-Hill Companies.

given by

$$U^+ = \frac{\bar{U}_z}{u_*} \tag{5.167}$$

and

$$y^+ = \frac{y u_*}{\nu} \tag{5.168}$$

where y is measured from the plate surface and u_* is the friction velocity given earlier (Eq. 5.153). Also shown in the figure are some smooth curves which are used to fit the velocity data over the viscous sublayer the "turbulent generation" region (between the viscous sublayer and turbulent layer), and the turbulent region itself. The curves shown are known as *universal velocity laws*, since they should fit all turbulent-boundary-layer data (pipes and flat plates) when velocity and distance are properly nondimensionalized as shown above. For the *viscous sublayer*, the universal law is given by

$$U^+ = y^+ \quad \text{for} \quad y^+ \leqslant 5 \tag{5.169}$$

For the *generation region*, the law is

$$U^+ = 5.0 \ln y^+ - 3.05 \quad \text{for} \quad 5 < y^+ < 30 \tag{5.170}$$

and for the *turbulent core* region, the law is

$$U^+ = \frac{1}{\kappa} \ln y^+ + 5.5 \quad \text{for} \quad y^+ > 30 \tag{5.171}$$

where κ is called the *von Kármán constant*, and usually has a value of 0.4. Equations 5.170 and 5.171 are also known as "logarithmic laws." Exercise 5.19 gives a demonstration of how such a law can be derived using real data and the mixing-length model.

As the reader can appreciate, most surfaces in the environment are not smooth. For example, consider the beds of rivers and other water-flow channels, and the surface of the earth. When the turbulence is completely rough (Section 5.13.7), the viscous sublayer can usually not be discerned. The velocity distribution in the turbulent region can then be given by a modified universal logarithmic law:

$$U^+ = \frac{1}{\kappa} \ln \frac{y}{z_0} \quad \text{for} \quad y \geqslant z_0 \tag{5.172}$$

Table 5.2. Roughness Lengths for Various Surfaces

Surface	z_0 (m)
Very smooth (e.g., ice)	10^{-5}
Snow	10^{-3}
Smooth sea	10^{-3}
Level desert	10^{-3}
Lawn	10^{-2}
Uncut grass	0.05
Fully grown root crops	0.1
Tree covered	1
Low-density residential	2
Central business district	5–10

Source: Seinfeld, *Atmospheric Chemistry and Physics of Air Pollution*, p. 495, © 1986. Reprinted by permission of John Wiley & Sons, Inc.

The term z_0 is called the *roughness length* for the surface. The roughness length is usually approximated as $z_0 = e/30$. Table 5.2 presents roughness lengths for a number of environmental surfaces of interest in atmospheric boundary layers.

5.14. SUMMARY

In this chapter, we used simplified forms of the Navier–Stokes equations to solve numerous types of flow problems. In unidirectional viscous flows, the equation of motion took a simplified form, which led to the development of the Poiseuille equation. The critical concept of the boundary layer was also developed in this chapter. Boundary layers exist in all real laminar and turbulent flows, and are a result of the necessity of matching the zero-velocity condition at solid surfaces. For very low Reynolds number flows, two other important environmental problems were attacked: the computation of drag on a sphere and the flow through porous media.

Inviscid flows are flows in which the viscosity is zero, or equivalently, flows in which the effects of viscosity are minimal. Although all real fluids in the environment have a finite viscosity, regions of many high Reynolds number flows (especially outside of boundary layers) behave essentially as if they were inviscid. Potential functions and stream functions were used to describe inviscid flows. The most important equation to come out of the analysis of inviscid flows was the Bernoulli equation. Finally, the chapter ended with a discussion of turbulence, which is the most common state of flow in the environment. We reviewed elements of the statistical approach to turbulence and discussed the closure of the Navier–Stokes equations in turbulent flows and turbulent boundary layers.

In this chapter we surveyed some of the important topics in fluid mechanics and turbulence, especially those relevant to environmental energy and mass-transport problems. Although the chapter contains many equations, we have literally just scratched the surface of the vast subject of fluid mechanics. Students may wish to follow up on some of our material, especially since the presentation was made as streamlined as possible. The reference section lists some of the best books written on the subject.

EXERCISES

5.1. An isothermal atmosphere is one in which temperature does not change. Use Eq. 5.8 and the ideal gas law (see Example 4.4) to derive the following equation for the ratio of pressure at two points z_1 and z_2 in the isothermal atmosphere:

$$\frac{p_2}{p_1} = \exp\left(-\frac{gm_a(z_2 - z_1)}{RT}\right) \qquad (5.159)$$

where m_a is the molar mass of the gas.

5.2. Perform the gradient operation on Eq. 5.19 and derive Eq. 5.20. Hint: Use the definition of the gradient operation from Appendix II.

5.3. Derive the velocity distribution shown in Figure 5.6a. There should be plenty of hints in Section 5.6. Also note that there is no pressure gradient, dp/dx, in this problem.

5.4. Derive the velocity distribution shown in Figure 5.6b. There should be plenty of hints in Section 5.6. Also note that there is no pressure gradient, dp/dx, in this problem.

5.5. In cylindrical coordinates (r, z, θ), the axial (z) component of the Navier–Stokes equations is (see Appendix III)

$$\rho\left(\frac{\partial u_z}{\partial t} + u_r \frac{\partial u_z}{\partial r} + \frac{u_\theta}{r}\frac{\partial u_z}{\partial \theta} + u_z \frac{\partial u_z}{\partial z}\right)$$

$$= -\frac{\partial P}{\partial z} + \rho g_z + \mu\left[\frac{1}{r}\frac{\partial}{\partial r}\left(r\frac{\partial u_z}{\partial r}\right) + \frac{1}{r^2}\frac{\partial^2 u_z}{\partial \theta^2} + \frac{\partial^2 u_z}{\partial z^2}\right]$$

(a) Show that the equation above reduces to

$$\frac{dP}{dz} = \frac{\mu}{r}\frac{d}{dr}\left(r\frac{du_z}{dr}\right)$$

for steady uniform flow parallel to the z axis (with no free surface). Explain specifically why each term either stays or vanishes.

(b) Solve the simplified equation for laminar flow in a capillary or narrow pipe. The solution should yield the famous Poiseuille equation of fluid mechanics,

$$u_z = -\frac{1}{4\mu}\frac{dP}{dz}(a^2 - r^2)$$

where a is the pipe radius. Hint: the required boundary conditions for the two integrations are quite similar to those used in Section 5.6.

(c) What is the average velocity in the pipe?

(d) Express u_z in terms of the average velocity.

5.6. Imagine a flat membrane sheet with perfect cylindrical pores of radius a.

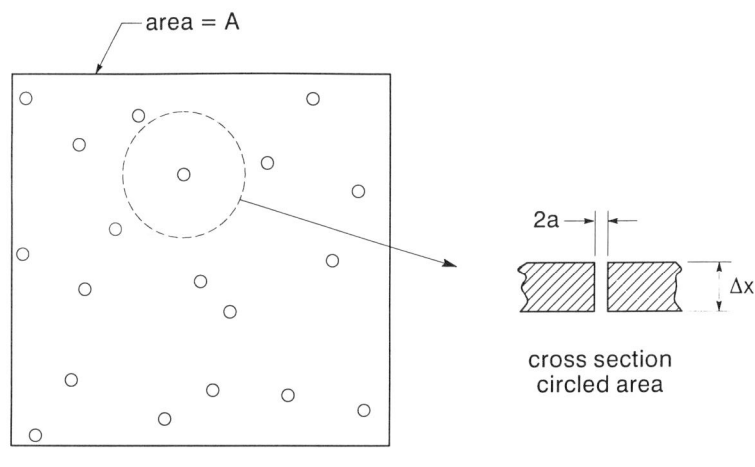

area = A

2a

Δx

cross section
circled area

top view of membrane

Exercise 5.6

Using Eq. 5.27 for average velocity in a small pore, show that the flow through the membrane sheet is given by

$$Q_p = \frac{\epsilon a^2 A}{8\mu}\frac{\Delta p}{\Delta x}$$

where Q_p is the permeate flow (the total flow through the membrane), Δp is the transmembrane pressure drop, Δx is the membrane thickness, and ϵ is the surface porosity of the membrane (i.e., the fraction of membrane surface composed of pores).

5.7. Solve for the velocity profile shown in Figure 5.6d. The θ component of the Navier–Stokes equations in cylindrical coordinates is (Appendix III)

$$\rho\left(\frac{\partial u_\theta}{\partial t} + u_r\frac{\partial u_\theta}{\partial r} + \frac{u_\theta}{r}\frac{\partial u_\theta}{\partial \theta} + \frac{u_r u_\theta}{r} + u_z\frac{\partial u_\theta}{\partial z}\right)$$

$$= -\frac{1}{r}\frac{\partial p}{\partial \theta} + \mu\left[\frac{\partial}{\partial r}\left(\frac{1}{r}\frac{\partial}{\partial r}(ru_\theta)\right) + \frac{1}{r^2}\frac{\partial^2 u_\theta}{\partial \theta^2} + \frac{2}{r^2}\frac{\partial u_r}{\partial \theta} + \frac{\partial^2 u_\theta}{\partial z^2}\right] + \rho g_\theta$$

(a) Show that for the circular streamline flow of Figure 5.6*d*, the equation above reduces to

$$0 = \frac{\partial}{\partial r}\left[\frac{1}{r}\frac{\partial}{\partial r}(ru_\theta)\right]$$

where $g_\theta = 0$.

(b) Integrate the equation above and show that u_θ is given by:

$$u_\theta = \omega R_0\left[\frac{(R_i/r) - (r/R_i)}{(R_i/R_0) - (R_0/R_i)}\right] = \frac{\omega R_0^2}{(R_0^2 - R_i^2)}\left(\frac{r^2 - R_i^2}{r}\right)$$

where the boundary conditions are $u_\theta = 0$ at $r = R_i$ and $u_\theta = \omega R_0$ at $r = R_0$.

5.8. An expanded picture of Figure 5.1*a* is shown below. Simplify the *x* component of the Navier–Stokes equations and show that the steady-velocity profile is given by

$$u = \frac{\rho g}{\mu}\sin\theta y\left(H - \frac{y}{2}\right)$$

Hint: What must the gradient of velocity in the *y* direction be at $y = H$?

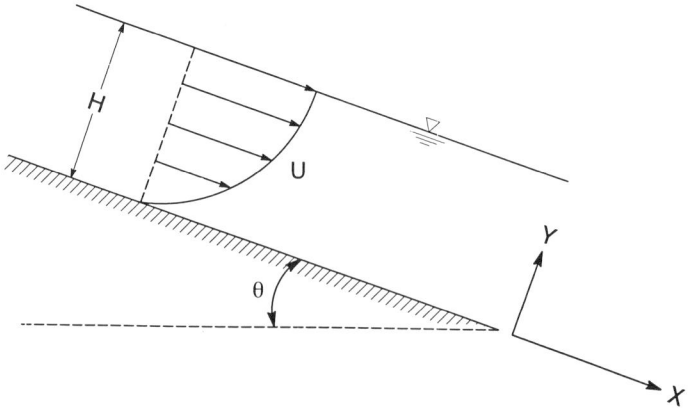

Exercise 5.8

5.9. In cylindrical coordinates, the $r\theta$ component of viscous stress is given by

$$\tau_{r\theta} = \mu \left[r \frac{\partial}{\partial r} \left(\frac{u_\theta}{r} \right) + \frac{1}{r} \frac{\partial u_r}{\partial \theta} \right]$$

(a) Using the result of Exercise 5.7, show that the $r\theta$ component of shear stress in the circular streamline flow of Figure 5.6d is

$$\tau_{r\theta} = 2\mu\omega R_0^2 \left(\frac{1}{r^2} \right) \left[\frac{(R_i/R_0)^2}{1 - (R_i/R_0)^2} \right]$$

(b) The circular couette device of Figure 5.6d is called a viscometer when used to measure viscosity. Show that the force exerted by the fluid on the outer boundary is equal to

$$\text{Force} = 4\pi\mu H\omega R_0 \left[\frac{(R_i/R_0)^2}{1 - (R_i/R_0)^2} \right]$$

where H is the height of the viscometer. Knowing the force and the equation above, we have a neat way of measuring the viscosity of fluids.

5.10. Derive the nondimensional form of the Navier–Stokes equation (Eq. 5.45). Gravity can be ignored. It may help to perform the substitutions on the tensorial form of the Navier–Stokes equation (Eq. 5.4). Note in Eq. 5.45 that the grad operator also has a tilde — it's the dimensionless grad operator. To see this, recall the following:

$$\nabla = \frac{\partial}{\partial x_1} \mathbf{i} + \frac{\partial}{\partial x_2} \mathbf{j} + \frac{\partial}{\partial x_3} \mathbf{k}$$

Using one of the relationships expressed in Eq. 5.44,

$$\tilde{x}_1 = \frac{x_1}{L}; \quad \tilde{x}_2 = \frac{x_2}{L} \quad \text{and} \quad \tilde{x}_3 = \frac{x_3}{L}$$

Substituting into the first equation, we find:

$$\nabla = \frac{1}{L} \frac{\partial}{\partial \tilde{x}_1} \mathbf{i} + \frac{1}{L} \frac{\partial}{\partial \tilde{x}_2} \mathbf{j} + \frac{1}{L} \frac{\partial}{\partial \tilde{x}_3} \mathbf{k}$$

Rearranging we find:

$$L\nabla = \frac{\partial}{\partial \tilde{x}_1} \mathbf{i} + \frac{\partial}{\partial \tilde{x}_2} \mathbf{j} + \frac{\partial}{\partial \tilde{x}_3} \mathbf{k}$$

or finally:

$$L\nabla = \tilde{\nabla}$$

5.11. Use Eq. 5.54 to plot the pressure distribution $p - p_\infty$ around a sphere at low Reynolds number. As suggested in Figure 5.2b, plot positive $p - p_\infty$ as arrows with their heads pointing toward the surface of the sphere, and negative $p - p_\infty$ using arrows with the heads pointing toward the center of the sphere.

5.12. Is Darcy's law a solution to the transformed creeping-flow equations, that is, the Laplace equation (Eq. 5.51)? (It is supposed to be.) Hint: You need to consider that Darcy's law, presented in this chapter (Eq. 5.63), can be generalized to three dimensions:

$$u_1 = -\frac{k}{\mu}\frac{dP}{dx_1}$$

$$u_2 = -\frac{k}{\mu}\frac{dP}{dx_2}$$

$$u_3 = -\frac{k}{\mu}\frac{dP}{dx_3}$$

5.13. A problem relevant to flow near injection wells in porous media is the so-called ideal source. The flow patterns near an ideal two-dimensional source are shown below. Because of conservation of mass, the flow through any concentric circle is a constant, which can be given as

$$q_r = 2\pi r u_r$$

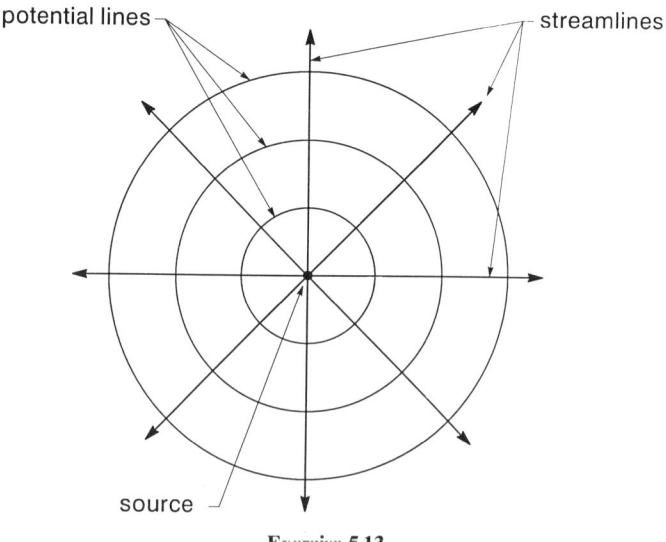

Exercise 5.13

Note from this definition that q_r is the unit flow (i.e., per unit depth into the page). Traditionally, the strength of the source is characterized by a constant m called the *source strength*:

$$2\pi m = q_r = 2\pi r u_r$$

(a) In two-dimensional cylindrical coordinates appropriate to this problem, u_r is given by

$$u_r = \frac{1}{r}\frac{\partial \psi}{\partial \theta}$$

Show that the stream function for the two-dimensional source is given by

$$\psi = m\theta$$

(b) In cylindrical coordinates, u_r is also given by

$$u_r = -\frac{\partial \phi}{\partial r}$$

Show that the potential function for the two-dimensional source is $\phi = -m \ln r$.

5.14. At this point, we have two equations to compute the average velocity or flow in a pipe — the Bernoulli and Poiseuille equations. However, the assumptions of these models are quite different: the Poiseuille equation includes viscous effects and the Bernoulli equation does not. This exercise is designed to compare the predictions of the two equations (or models).

(a) Shown on following page is a tank filled with water discharging through a smooth pipe to the atmosphere. What are the predicted flows (m^3/s) according to the Bernoulli and Poiseuille equations for pipe diameters of 0.1, 1, and 10 mm? Make a figure comparing the two predicted discharges as a function of the pipe diameter. For the purpose of applying these equations, you can assume no entrance losses at the pipe entrance. Also recall that pressure is related to hydrostatic head (H) by Eq. 5.11:

$$P = \rho g H$$

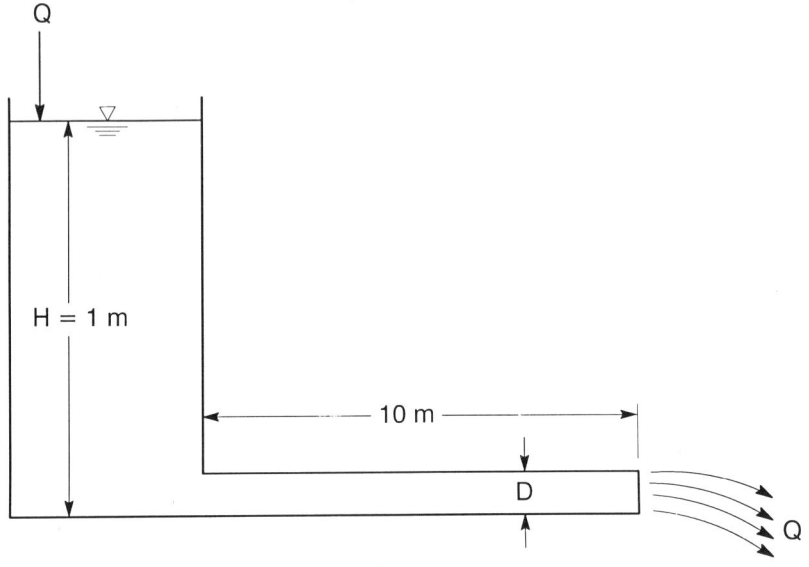

Exercise 5.14

(b) Which equation is correct? Hints: Remember that the Poiseuille equation assumes laminar flow and completely developed flow. According to Welty et al. (1976), the entrance length L_e required for the fully developed laminar velocity profile to develop is

$$\frac{L_e}{D} = 0.0575 \, Re$$

Remember also that laminar flows start to become turbulent in smooth pipes around $Re = 2000$.

(Comment: The Bernoulli equation should overpredict flow rates as the tube diameter becomes smaller. Nevertheless, the Bernoulli equation can be used in this situation if sufficient "major" and "minor" energy losses are accounted for. These corrections can be found in fluid mechanics and engineering reference books.)

5.15. The larva of the cadisfly *Macronema* builds the structure shown on following page to circulate water through its internal environment (Vogel, 1982). Why does this work? Some authors have likened the hydraulic arrangement of this organism to one of the pitot tubes analyzed in Example 5.5.

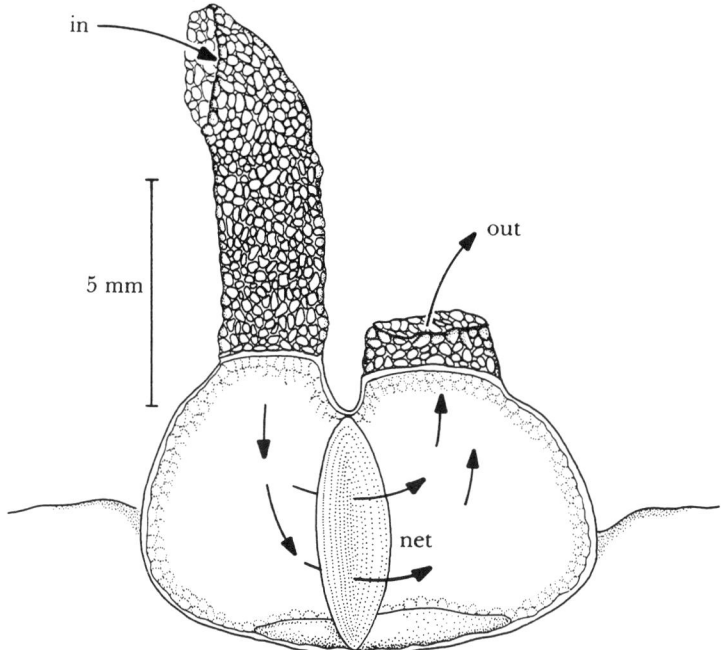

in

out

5 mm

net

Exercise 5.15. *Source*: Reproduced from Vogel, *Life's Devices*, © 1988 by Princeton University Press. Reproduced by permission of Princeton University Press.

5.16. Viscous shear stresses in fluids are in general given by the product of viscosity and a velocity gradient term. In the two-dimensional flow over a flat plate, the shear stress at the surface of the plate is formally calculated with the following equation:

$$\tau_{xy} = \mu \left(\frac{\partial u}{\partial y} + \frac{\partial v}{\partial x} \right)$$

The x and y direction are as we have imagined in Section 5.12. However, because there can be no y component of velocity (hence no gradient in v along the plate surface) at the surface of the plate, the equation above simplifies to:

$$\tau_{xy} = \mu \left(\frac{\partial u}{\partial y} \right)$$

(a) Referring to Section 5.12, show that

$$\frac{\partial u}{\partial y} = \frac{U}{4} \left(\frac{U}{vx} \right)^{1/2} f''$$

Hint: You need to understand how to perform the types of derivatives calculated in Eqs. 5.114–5.118. To get you halfway there,
consider these steps leading to Eq. 5.114:

$$u = \frac{\partial \psi}{\partial y} = \frac{\partial}{\partial y} [\sqrt{vUx} \, f(\eta)]$$

or

$$u = \sqrt{vUx} \, \frac{\partial}{\partial y} f(\eta) = \sqrt{vUx} \, f' \, \frac{\partial \eta}{\partial y}$$

or

$$u = \sqrt{vUx} \, f' \, \frac{1}{2\sqrt{vx/U}} = \frac{U}{2} f'$$

where $f' = \partial f / \partial \eta$. Now keep going.

(b) Using the results of Table 5.1, compute and plot the surface shear
stress along a 10-cm flat plate in a laminar air flow of $U = 10 \, \text{m/s}$.
The kinematic viscosity of air can be taken as $v = 1.5 \times 10^{-5} \, \text{m}^2/\text{s}$.

(c) What implications do your results have for erosion of deposited
particles on the flat plate? Where would erosion be the most
intense — at the beginning or the end of the plate?

5.17. Apply the Reynolds averaging procedure to Eq. 5.4 and develop Eq.
5.141. In addition to the velocity decomposition, you will need the
pressure decomposition:

$$p = \bar{p} + p'$$

Assume that there is no free surface, so that you can drop the gravity
term and use the modified pressure, p. You can assume that the order of
time integration and differentiation can be interchanged. (This can be
proved using Liebnitz' rule.)

5.18. In Section 5.13.2 we proved that the mean velocities must obey the
differential continuity equation. Prove that the fluctuating velocities must
also satisfy the continuity equation.

5.19. Consider turbulent flow in the pipe shown on p. 223, where U characterizes the average velocity distribution in the streamwise (x) direction.

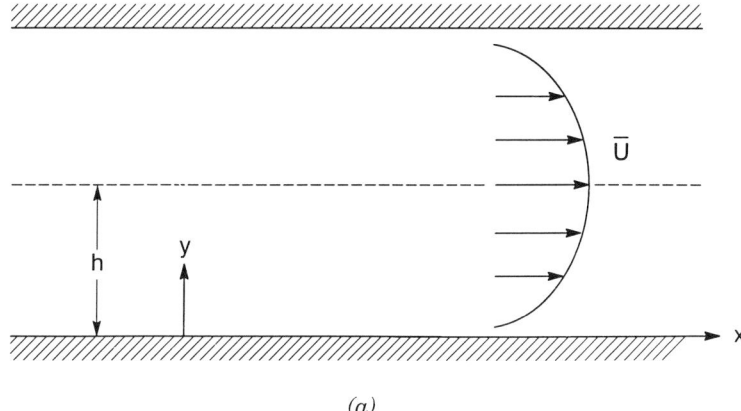

(a)

Exercise 5.19(a)

The xy component of the Reynolds stress is given by

$$\tau^t_{xy} = \rho \nu_t \frac{\partial \bar{U}}{\partial y}$$

where the eddy diffusivity of momentum is given by

$$\nu_t = \kappa u_* y$$

Here u_* is the so-called "friction velocity" (see Section 5.13.4):

$$u_* = \sqrt{\frac{\tau_0}{\rho}}$$

(a) Assuming that τ^t_{xy} is constant across the pipe section and equal to τ_0, develop the famous universal velocity distribution:

$$\bar{U} = \frac{u_*}{\kappa} \ln y + \text{constant}$$

(b) Also develop the so-called "velocity-defect" law:

$$\frac{\bar{U}_{max} - \bar{U}}{u_*} = -\frac{1}{\kappa} \ln \frac{y}{h}$$

where U_{max} is the centerline velocity. Use this equation and the following data plot to determine the value of κ in the defect-law equation.

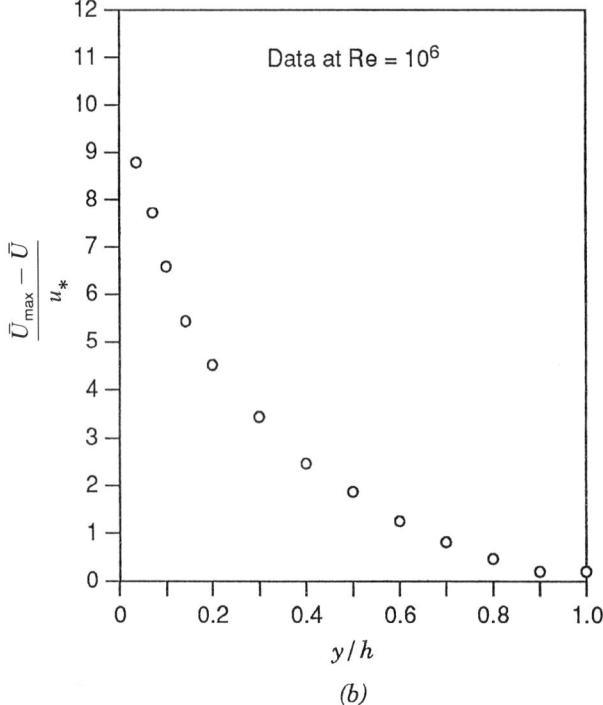

(b)

Exercise 5.19(b). *Source*: Reproduced from Welty, Wicks, and Wilson, *Fundamentals of Momentum, Heat, and Mass Transport*, p. 194, © 1976. Reprinted by permission of John Wiley & Sons, Inc.

(c) Do you see any problems with these equations at $y = 0$ and $y = h$?

5.20. Professor G. I. Taylor showed that for isotropic turbulence, the variance of $\partial u'/\partial x$ is given by

$$\overline{\left(\frac{\partial u'}{\partial x}\right)^2} = \frac{\varepsilon}{15\nu}$$

where ε is the unit mass energy-dissipation rate, ν is the kinematic viscosity, and u' is the fluctuating velocity component. If the statistical distribution of $\partial u'/\partial x$ is assumed to be normal, and if the average value of $\partial u'/\partial x$ is zero, show that the statistical distribution of the *absolute value* of $\partial u'/\partial x$ is given by

$$P\left(\left|\frac{\partial u'}{\partial x}\right|\right) = \frac{2}{\sqrt{\frac{2\pi\varepsilon}{15v}}} \exp\left(-\frac{1}{2}\left(\frac{15v}{\varepsilon}\right)\left(\frac{\partial u}{\partial x}\right)^2\right)$$

Then show that the average value of the absolute value of $\partial u'/\partial x$ is given by

$$\overline{\left|\frac{\partial u'}{\partial x}\right|} = \left(\frac{2\varepsilon}{15\pi v}\right)^{1/2}$$

Saffman and Turner (1956) estimated the flux of class j particles into a collision sphere of radius $r_i + r_j$ with

$$\frac{1}{2}\,n_j\,\iint_{\substack{\text{entire}\\\text{sphere}}}(r_i + r_j)\overline{\left|\frac{\partial u'}{\partial x}\right|}\,dS$$

Finally, using these last two equations, show that the beta function for isotropic turbulence can be estimated by (recall Section 3.7.2)

$$\beta(r_i, r_j)_t = 1.3(r_i + r_j)^3\left(\frac{\varepsilon}{v}\right)^{1/2}$$

REFERENCES

Batchelor, G. K., *An Introduction to Fluid Dynamics*, Cambridge University Press, 1967 (available in paperback).

Bird, R. B., Stewart, W. E., and Lightfoot, E. N., *Transport Phenomena*, Wiley, New York, 1960.

Boyce, W. E. and DiPrima, R. C., *Elementary Differential Equations and Boundary Value Problems*, Wiley, New York, 1969.

Brodkey, R. S., *Transport Phenomena: A Unified Approach*, McGraw-Hill, New York, 1988.

Corrsin, S., "Turbulent Flow," *Am. Sci.*, **40**, 3:300–325 (1961).

Currie, I. G., *Fundamental Mechanics of Fluids*, McGraw-Hill, New York, 1974.

Grant, W. and Madsen, O., "Combined Wave and Current Interactions with a Rough Bottom," *J. Geophys. Res.*, **84**, C4:1798–1808 (1979).

Hinze, J. O., *Turbulence*, 2nd ed., McGraw-Hill, New York, 1975.

Lighthill, J., *An Informal Introduction to Theoretical Fluid Mechanics*, Clarendon Press, Oxford, 1986.

Liu, Z. C., Landreth, C. C., Adrian, R. J., and Hanratty, T. J., "High Resolution Measurement of Turbulent Structure in a Channel with Particle Image Velocimetry," *Exp. Fluids*, **10**, 310–312 (1991).

Panton, R. L., *Incompressible Flow*, Wiley, New York, 1984.

Rodi, W., *Turbulence Models and Their Applications in Hydraulics*, International Association for Hydraulic Research, Rotterdamseweg, Netherlands, 1984.

Saffman, P. G. and Turner, J. S., "On the Collision of Drops in Turbulent Clouds," *J. Fluid Mech.*, **1**, 16:16–30 (1956).

Saleh, S., Thovert, J. F., and Adler, P. M., "Measurement of Two-Dimensional Velocity Fields in Porous Media by Particle Image Displacement Velocimetry," *Exp. Fluids*, **12**, 210–212 (1992).

Schlichting, H., *Boundary Layer Theory*, McGraw-Hill, New York, 1979.

Seinfeld, J. H., *Atmospheric Chemistry and Physics of Air Pollution Control*, Wiley-Interscience, New York, 1986.

Streeter, V. L., *Fluid Mechanics*, McGraw-Hill, New York, 1971.

Taylor, G. I., "Statistical Theory of Turbulence," *Proc. Royal Soc.*, **151**, 421 (1935).

Tennekes, H. and Lumley, J. L., *A First Course in Turbulence*, MIT Press, Cambridge, MA, 1972.

Vogel, S., *Life's Devices*, Princeton University Press, Princeton, NJ, 1982.

Welty, J. R., Wicks, C. E., and Wilson, R. E., *Fundamentals of Momentum, Heat, and Mass Transport*, Wiley, New York, 1976.

BIBLIOGRAPHY

Bear, J., *Dynamics of Fluids in Porous Media*, Dover Publications, New York, 1988 (original publication in 1972).

Corrsin, S., Unpublished lecture notes on turbulence, The Johns Hopkins University, Baltimore, MD, 1980–1981 (notes taken by author).

Freeze, R. A. and Cherry, J. A., *Groundwater*, Prentice-Hall, Englewood Cliffs, NJ, 1979.

Frisch, U. and Orzag, S. A., "Turbulence: Challenges for Theory and Experiment," *Phys. Today*, **43**, 1:24–33 (1990).

Fung, Y. C., *A First Course in Continuum Mechanics*, Prentice-Hall, Englewood Cliffs, NJ, 1977.

Happel, J. and Brenner, H., *Low Reynolds Number Hydrodynamics*, Martinus Nijhoff, Boston, MA, 1983.

Perry, R. H., Green, D. W., and Maloney, J. O., *Perry's Chemical Engineers' Handbook*, McGraw-Hill, New York, 1984.

Skelland, A. H. P., *Diffusional Mass Transfer*, Wiley-Interscience, New York, 1974.

6

DIFFUSIVE MASS TRANSPORT

6.1. INTRODUCTION

In previous chapters, we hinted that diffusion can be an important mass-transport mechanism for gases, molecules, ions, and small particles. Figure 6.1 shows a closed system with two different gases at the same pressure (white and black circles) separated by a partition. When the partition is removed, a spontaneous process of mixing of the two gases occurs. The sharp concentration profiles of the two gases are rounded off, and after enough time has passed, the concentration of gases in the system are homogenized. The process which has homogenized the concentrations of the gases is known as *molecular diffusion*. Note that the situation in the top picture before the removal of the partition is relatively ordered compared with the last picture. Diffusion appears to be a process that destroys order. Recalling that the second law of thermodynamics predicts a decrease in order of the universe, we can also say that the mutual diffusion of the gases in Figure 6.1 is a thermodynamically favored direction of change. In fact, the two gases will never return to the segregated condition of the first picture (unless we do some work on the system). In other words, "unmixing" is not thermodynamically favored in the environment.

In this chapter we develop the mathematics of *ordinary diffusion*, which is diffusion caused by a species-concentration gradient (Bird et al., 1960). The mathematics of ordinary diffusion are connected to several problems in environmental engineering and science. As in previous chapters, the reader is encouraged to seek answers for some simple integrals in mathematics handbooks.

We do not cover, however, diffusion caused by temperature gradients (thermal diffusion and the so-called Soret effect) or forced diffusion, an example of which is particle or ionic drift in electrical fields (see Example 2.2, Chapter 2 for a taste of this kind of diffusion). Diffusion caused by pressure gradients (*pressure diffusion*) is only covered lightly in an exercise on ultracentrifugation (Exercise 6.2).

227

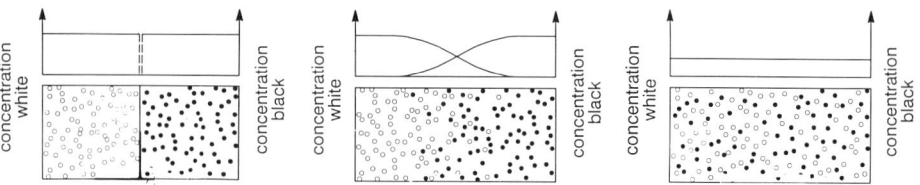

Figure 6.1. Illustration of equimolar diffusion of "white" and "black" gases in a closed system. Time is progressing from left to right.

6.2. THERMODYNAMICS OF DIFFUSION

Imagine the local chemical potential of some chemical species in an ideal solution, as shown in Figure 6.2. In general, spatial variations are not linear, but since we are examining a differential element, we can assume that the variation in chemical potential is linear. Following the presentation of Atkins (1978), we note that because chemical potential is a thermodynamic state function, at constant temperature and pressure, the differential work required to change the chemical potential of 1 mol of solute from $\mu(x)$ *to* $\mu(x + dx)$ is given by

$$dw = \mu(x + dx) - \mu(x) \tag{6.1}$$

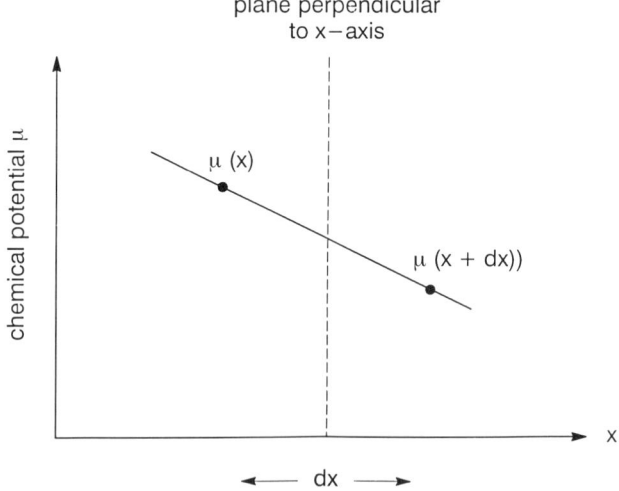

Figure 6.2. One-dimensional diffusion problem, showing chemical potential variations in a small region.

Or, since we are considering a differential region:

$$dw = \left[\mu(x) + \left(\frac{d\mu(x)}{dx} \right) dx \right] - \mu(x) \tag{6.2}$$

Simplifying, we find that

$$dw = \left(\frac{d\mu}{dx} \right) dx \tag{6.3}$$

From mechanics theory, you will remember that work is equal to a force acting through a distance, or $dw = -F\,dx$. Imagining that the chemical potential has been changed by some kind of general force, we can set

$$-F\,dx = \frac{d\mu}{dx} dx \tag{6.4}$$

hence,

$$F = -\frac{d\mu}{dx} \tag{6.5}$$

Recall that for an ideal solution, the chemical potential is expressed as

$$\mu = \mu_0 + RT \ln c \tag{6.6}$$

where R is the gas constant, T is the absolute temperature, c is the solute concentration (mol/L^3), and μ_0 is the chemical potential at standard temperature and pressure. Substituting Eq. 6.6 into Eq. 6.5, we find that

$$F = -RT \frac{d}{dx}(\ln c) \tag{6.7}$$

Taking the derivative:

$$F = -RT \frac{1}{c} \frac{dc}{dx} \tag{6.8}$$

Recall that the mysterious force F was acting on 1 mol of solute, hence the units of F are force/mol. If we multiply F by the molar concentration c (mol/L^3), we arrive at the unit volume force acting on the solute. The flux of solute (moles per unit area per time) passing through the dotted surface in Figure 6.2 should then be proportional to the product of F and c. Hence, we conclude that

$$J \propto cF = -RT \frac{dc}{dx} \tag{6.9}$$

This simple equation predicts that the flux of matter in the region of a concentration gradient dc/dx should have a linear dependence on the concentration gradient. The equation also predicts that a positive flux (and/or mass transport) should be in the direction of decreasing concentration (i.e., J will be positive if dc/dx is negative). Equation 6.9 says that mass transport due to molecular diffusion should occur from left to right, or "down the concentration gradient" in Figure 6.2. Returning to Figure 6.1, we also see that the direction of transport of the two gases (to the right for the white gas, to the left for the black gas) is also predicted by Eq. 6.9.

6.3. FICK'S FIRST LAW AND GENERAL DIFFUSIVE TRANSPORT

Consider a gas that is being produced at the bottom of a tube that is open at the other end (Figure 6.3). The gas is being produced by evaporation of a solid. (It could be an evaporating liquid if the tube were standing right side up.) The tube is open on the right to a constant-pressure atmosphere of some pure gas, which is indicated by the open circles. Since only two pure gases are involved, this is called a *binary* system. As expected from our previous discussions, the concentration of the black gas is highest at the surface of the evaporating black solid, and decreases moving away from the surface. Note also that we could plot the concentration of the background or ambient gas (open circles). Based on the figure, such a plot would also have an exponential shape, where the

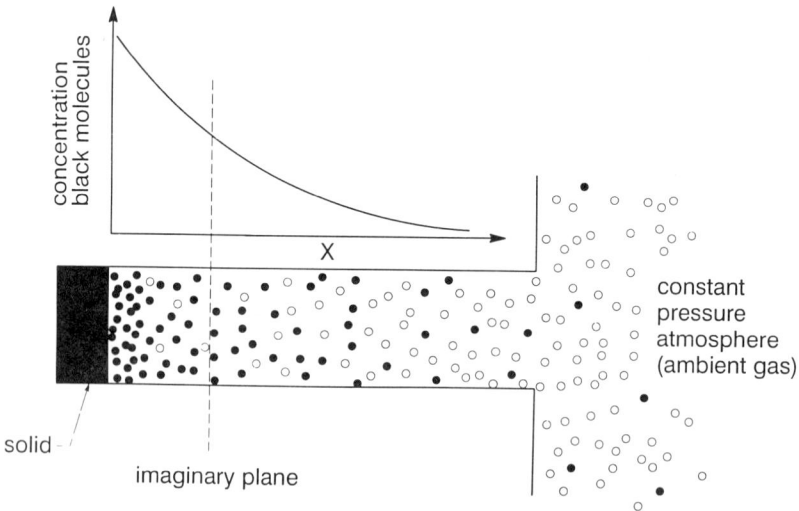

Figure 6.3. Simple one-dimensional binary gas diffusion: Diffusion from an evaporating solid in a tube open to a pure constant-pressure atmosphere.

highest concentration would be at the open end of the tube. The ambient gas is in effect trying to diffuse *into* the tube. However, the black gas is trying to push it out of the tube.

Note also that the concentration profile shown could be either the steady-state profile (achieved after some amount of time) or a snapshot of local concentration prior to the achievement of steady state.

Consider the fixed coordinates of the imaginary fixed plane in Figure 6.3. In the steady-state case, there would obviously be a net transport of black gas out of the tube relative to the fixed plane, although the white-gas transport relative to the fixed plane would be zero. (At steady state, the white-gas is stuck, in a sense, since it cannot pass through the bottom of the tube.) In the nonsteady-state case, there would be transport of both black and white gases in the tube relative to the fixed plane, as the two gases try to achieve the equilibrium condition. To understand better the relative movement of the two gases, it is helpful to define the average gas velocity in the system. Traditionally, two types of average velocities have been considered in diffusion problems (Bird et al., 1960). The *mass average velocity* is defined as

$$\mathbf{u} = \frac{\sum_{i=1}^{m} \rho_i \mathbf{u}_i}{\sum_{i=1}^{m} \rho_i} = \frac{\sum_{i=1}^{m} \mathbf{n}_i}{\rho} \tag{6.10}$$

Here ρ_i is the mass concentration (M/L^3) of the ith component and $\mathbf{u}_i$ is the velocity of the ith component.* The term $\mathbf{n}_i$ is the mass flux of component i relative to fixed or stationary coordinates (e.g., the imaginary plane in Figure 6.3). For the simple binary system shown in Figure 6.3, there are just two components, so $m = 2$. The mass-average velocity described by Eq. 6.10 is not a new concept—it must be the normal fluid velocity we have considered all along in this book. The *molar-average velocity* is defined as

$$\mathbf{u}^* = \frac{\sum_{i=1}^{m} c_i \mathbf{u}_i}{\sum_{i=1}^{m} c_i} = \frac{\sum_{i=1}^{m} \mathbf{N}_i}{c} \tag{6.11}$$

where c_i is the molar concentration (mol/L^3) of the ith component and $\mathbf{N}_i$ is the molar flux of component i relative to fixed or stationary coordinates.

With these definitions, we are in a better position to describe the relative transport of the two gases in Figure 6.3. For example, the *mass flux* of the ith gas component *relative to coordinates moving at the mass average velocity* is

*Actually, since the gas molecules are also in thermal motion, the $\mathbf{u}_i$ values are the average velocities of each gas component.

simply

$$\mathbf{j}_i = \rho_i(\mathbf{u}_i - \mathbf{u}) \tag{6.12}$$

while the *molar flux* of the ith gas component *relative to coordinates moving at the molar average velocity* is

$$\mathbf{J}_i^* = c_i(\mathbf{u}_i - \mathbf{u}^*) \tag{6.13}$$

Note that the *relative* fluxes of the two gases in the example above will in general be different. In fact, they will have opposite signs. The relative flux of the black gas will be positive; the relative flux of the white gas will be negative. (However, the *relative molar flux* of the white gas will exactly equal the negative of the relative molar flux of the black gas—see Exercise 6.3.)

For binary systems, *Fick's first law* states that the diffusive *molar flux* of component A *relative to coordinates moving with the local molar average velocity* is*

$$\mathbf{J}_A^* = -cD_{AB}\nabla x_A \tag{6.14}$$

Here c is the total fluid molar concentration, D_{AB} is the *diffusion coefficient of component A in component B*, and x_A is the *mole fraction* of component A (Section 4.5.1):[†]

$$x_A = \frac{c_A}{c} \tag{4.32}$$

As found in Section 6.2, diffusion occurs "down the concentration gradient." Note also that the diffusive flux is a *vector* quantity; in general, it will have components in all three spatial directions.

In terms of mass quantities, Fick's first law for the diffusive *mass flux* of component A *relative to coordinates moving with the local mass average velocity* is

$$\mathbf{j}_A = -\rho D_{AB}\nabla \omega_A \tag{6.15}$$

where ρ is the total fluid density and ω_A is the *mass fraction* of component A:

$$\omega_A = \frac{\rho_A}{\rho} \tag{6.16}$$

---

*The left-hand side of Eq. 6.14 is obviously a vector quantity, but is the right-hand side a vector? Yes it is, since ∇x_A is a vector (Appendix II).

[†]In Chapter 4, x_A was reserved for the mole fraction in the liquid phase (Eq. 4.32), while y_A was reserved for the mole fraction in the gas phase (Eq. 4.30). Rather than present equations in terms of both x_A and y_A in this chapter, we use only x_A. But the reader should note that y_A can be substituted wherever x_A appears, as long as total concentration c is also replaced by the total pressure p.

The diffusion coefficient D_{AB} is exactly the same in both Eqs. 6.14 and 6.15. Consideration of the dimensions of the terms of either Eq. 6.14 or Eq. 6.15, reveals that the dimensions of the diffusion coefficient are always L^2/T.

We next seek to develop the form of Frick's first law for diffusive transport relative to fixed coordinates, for example, the imaginary fixed plane in Figure 6.3. In physics, if we know the nature of a transfer process relative to moving coordinates (e.g., the two forms of Fick's first law given above), then to find the overall transfer relative to *fixed or stationary* coordinates, all we have to do is add in the transfer associated with the moving coordinates (often called the "convective" mass transfer for mass average quantities). For example, consider the early stages of diffusion (i.e., prior to steady state) in Figure 6.3. As the black gas (A) begins to form at the solid surface, it clearly starts to move away from the bottom of the tube owing to diffusion. But the white gas (B) must be moving to the right to make room for the black gas. In effect, the whole system $A + B$ must be moving to the right. The two forms of Fick's law derived above are for diffusive mass transport relative to this overall "bulk" (mass average or molar average) motion. For a binary system, the *molar average* bulk flow of the system must be the sum of $\mathbf{N}_A$, the molar flux of component A, and $\mathbf{N}_B$, the molar flux of component B. Considering only component A, Fick's first law for *molar flux relative to fixed coordinates* is

$$\mathbf{N}_A = x_A(\mathbf{N}_A + \mathbf{N}_B) + \mathbf{J}_A^* = x_A(c\mathbf{u}^*) - cD_{AB}\nabla x_A = c_A\mathbf{u}^* - cD_{AB}\nabla x_A \qquad (6.17)$$

where we have made use of Eqs. 6.11, 6.14, and 4.32. Using similar reasoning, it follows that Fick's first law for *mass flux relative to fixed coordinates* is

$$\mathbf{n}_A = \omega_A(\mathbf{n}_A + \mathbf{n}_B) + \mathbf{j}_A = \omega_A(\rho\mathbf{u}) - \rho D_{AB}\nabla\omega_A = \rho_A\mathbf{u} - \rho D_{AB}\nabla\omega_A \qquad (6.18)$$

where $\mathbf{n}_A + \mathbf{n}_B$ is the *mass average* bulk flux, and where we have made use of Eqs. 6.10, 6.15, and 6.16. As before, Eqs. 6.17 and 6.18 are vector equations. The first terms to the right of the equal signs are the vector components of flux due to the molar or mass average bulk motion of the fluid, whereas the second terms on the right-hand sides of the equal signs are the vector flux components resulting from diffusion relative to the bulk motion.

Having developed the most general (and complicated) form of Fick's first law, let's consider simpler forms of Eqs. 6.17 and 6.18 which result if the molar and mass-average velocities are zero, that is, there is no molar- or mass-average bulk flow. This in fact is the main type of problem considered in this chapter. For example, $\mathbf{u}$ is usually zero in liquid-diffusion problems when there is no impressed flow due to gravity, viscous forces, or pressure gradients (Chapter 5) and when the concentration of the diffusing substance is low. In gases $\mathbf{u}^*$ is zero for binary-system, *equimolar-diffusion* problems. In equimolar diffusion, there is no impressed flow, and the same number of moles of A are moving in one direction as are moles of B moving in the opposite direction. In any event, if

$u^* = 0,$

$$N_A = J_A^* = -cD_{AB}\nabla x_A \qquad (6.19)$$

and if $u = 0$,

$$n_A = j_A = -\rho D_{AB}\nabla \omega_A \qquad (6.20)$$

Another common simplification usually arises when u or u^* are zero and c or ρ (the *total* molar and mass concentrations) are constant. Then c and ρ cancel with the denominators of ω_A and x_A in Eqs. 6.19 and 6.20. We conclude then that:*

$$N_A = J_A^* = -D_{AB}\nabla c_A \qquad (6.21)$$

and

$$n_A = j_A = -D_{AB}\nabla \rho_A \qquad (6.22)$$

Finally, note that all the formulas in this section assume that the diffusion coefficient is constant, regardless of the local concentration of the diffusing substance. Although this is a very common assumption in environmental science, it is important to note that the diffusion coefficient can change with local composition, temperature, and pressure (for gases). Some of these variations can be accounted for in the formulas found in Section 6.4.

6.4.   THE DIFFUSION COEFFICIENT

Table 6.1 lists the measured diffusion coefficients of various chemicals in air. A more complete list is available in Thibodeaux (1979), which also gives some valuable references for diffusion coefficients in air, water, and various other liquids and gases. Bird et al. (1960) and Reid and Sherwood (1966) discuss corrections of diffusion coefficients for different temperatures and pressures. Reid and Sherwood (1966) also provide leads to different methods of measuring diffusion coefficients.

Often a reliable measurement is not available, and one may need to estimate the diffusion coefficient. The kinetic theory of gases is the starting point for estimating the mutual diffusion coefficients in binary-gas systems; these theories as well as empirical correlations for estimating diffusion coefficients are presented in Bird et al. (1960), Reid and Sherwood (1966), and Lyman et al.

---

*The reader might wonder why we cannot cancel c and ρ in Eqs. 6.14 and 6.15 (or Eqs. 6.17 and 6.18). The reason is that c and ρ are in the denominators of the gradient term (recall the definitions of x and ω). If c and ρ are not constant, we must consider the effect of gradients in total concentration and density and we maintain the more general form of Eqs. 6.14 and 6.15 (and Eqs. 6.17 and 6.18).

Table 6.1. Diffusion Coefficients of Various Chemicals in Air at 1 atm Pressure

Chemical	Temperature (°C)	D_{AB}(cm²/s)
Ammonia	0	0.216
	25	0.28
Benzene	0	0.077
	25	0.088
Carbon dioxide	0	0.138
	25	0.164
Chloroform	0	0.091
Formic acid	25	0.159
Hydrogen	0	0.611
	25	0.410
Methane	0	0.16
Nitrogen	0	0.13
Oxygen	0	0.178
	25	0.206
Toluene	30	0.088
Water	0	0.220
	25	0.256

Source: Thibodeaux, *Chemodynamics: Environmental Movement of Chemicals in Air, Water, and Soil*, © 1979. pp. 458–461, Reprinted by permission of John Wiley & Sons, Inc.

(1982). It is also sometimes possible to estimate the diffusion coefficient of one compound, based on the diffusion coefficient of a similar compound; for example, using Graham's law (see Thibodeaux, 1979). One of the useful empirical approaches for estimating the diffusion of gases in air is called the *FSG method*. The formula is given as (Lyman et al., 1982)

$$D_{AB} = \frac{10^{-3}T^{1.75}\sqrt{M_r}}{p(V_A^{1/3} + V_B^{1/3})^2} \tag{6.23}$$

In this equation, T is the absolute temperature (K), p is the pressure (atm), V_A is the molar volume of the gas and V_B is the molar volume of the air (20.1 cm³/mol), and M_r is a function of the molar mass of the gas and air:

$$M_r = \frac{m_A + m_B}{m_A m_B} \tag{6.24}$$

Here m_A is the molar mass of the gas and m_B is the molar mass of the air (28.97 g/mol). Estimates of the molar volumes of specific gases may require data on atomic structure. For example, Table 6.2 provides data which can be used to estimate V_A for a number of organic gases.

Table 6.2. Atomic Data Used in Calculating V_A

Atom or Group	ΔV_A (cm^3/mol)
C	16.5
H	1.98
O	5.48
Cl	(5.69)
S	(19.5)
Aromatic and heterocyclic rings	-20.2

Source: Reprinted with permission from Lyman, Reehl, and Rosenblatt, *Handbook of Chemical Property Estimation Methods* (1982). Copyright 1982, American Chemical Society.

Equations like Eq. 6.23 provide estimates of diffusion coefficients, and were developed by minimizing the error between the measured and predicted diffusion coefficients of a wide range of gases. Lyman et al. (1982) mention that the FSG method is most accurate for nonpolar gases at low to moderate temperature, and that it provides estimates accurate to within $\pm 5\%$ for aromatics, alkanes, and ketones.

Example 6.1. Estimate the diffusion coefficient of methane (CH_4) in air at (a) 0°C and (b) 20°C.

SOLUTION. (a) The molar mass of methane is equal to $m_B = 12 + 4(1) = 16$ g/mol. Therefore M_r is

$$M_r = \frac{m_A + m_B}{m_A \times m_B} = \frac{28.97 + 16}{28.97 \times 16} = 0.097$$

Using the data in Table 6.1, $V_B = 16.5 + 4 \times 1.98 = 24.4$ cm^3/mol. Therefore, from Eq. 6.23:

$$D_{AB} = \frac{10^{-3}(273)^{1.75}\sqrt{0.097}}{1(20.1^{1/3} + 24.4^{1/3})^2} = 0.18 \frac{\text{cm}^2}{\text{s}}$$

This value can be compared with the measured value shown in Table 6.1.

(b) For this part, everything on the right-hand side of Eq. 6.23 remains the same except the temperature; hence, the equation can be manipulated to yield

$$\frac{D_{AB}^{0°C}}{273^{1.75}} = \frac{D_{AB}^{20°C}}{293^{1.75}}$$

or

$$D_{AB}^{20^\circ C} = \frac{293^{1.75}}{273^{1.75}} D_{AB}^{0^\circ C} = 0.20 \frac{cm^2}{s}$$

The *Stokes-Einstein* equation can be used to estimate diffusion coefficients of large spherical ions, molecules, and particles in water and other liquids, as well as the diffusion coefficient of spherical aerosol particle whose diameters are significantly greater than the air mean-free path:

$$D_{AB} = \frac{kT}{f} = \frac{kT}{3\pi\mu d} \tag{6.25}$$

where D_{AB} refers to the diffusion coefficient of species A in fluid B, f is the friction factor (Section 2.2), k is the Boltzman constant (Appendix I), T is the absolute temperature, μ is the absolute viscosity, and d is the ionic or particle diameter. Note that the friction factor in Eq. 6.25 derives from the Stokes drag on a sphere (low Reynolds-number drag, Section 2.2), that is,

$$F_D = (3\pi\mu d)V = fV \tag{6.26}$$

where V is the velocity of the sphere. In air, we can correct for slip by letting $f = (3\pi\mu d)/C_c$ (Section 2.2). Note the inverse dependence of the diffusion coefficient on particle diameter in Eq. 6.25. As noted in earlier chapters, as the particle size decreases, the particles are increasingly subjected to diffusive forces. Exercise 6.1 uses the Stokes–Einstein relationship to develop a method of using centrifugation and diffusion data to calculate molecular mass.

Care must be taken in the use of Eq. 6.25. For example, for solvated ions and other molecules, the diameter d is the effective *hydrodynamic diameter*. This means that the hydration of certain molecules is great enough to change the effective diameter of the molecule. There are also some notable failures of Eq. 6.25. For example, it severely underestimates the diffusion coefficient of the hydrogen ion, which is a small ion (Atkins, 1978). Table 6.3 lists the measured diffusion coefficients of some chemicals in water. Modern methods of estimating diffusion coefficients in liquids are reviewed in Bird et al. (1960), Reid and Sherwood (1966), and Lyman et al. (1982).

An important empirical formula for estimating diffusion coefficients in water and other solvents is the *Wilke–Chang equation* (e.g., Bird et al., 1960):

$$D_{AB} = 7.4 \times 10^{-8} \frac{(\psi_B m_B)^{1/2} T}{\mu V_A^{0.6}} \tag{6.27}$$

Here m_B is the molar mass of the solvent, T is the absolute temperature (K), μ is the absolute viscosity (g-cm^{-1}-s^{-1}), V_A is the molar volume of the solute (cm^3/mol), and ψ_B is the association parameter ($\psi_B = 2.6$ for water).

Table 6.3. Diffusion Coefficients of Various Chemicals in Water

Chemical	Temperature (°C)	$D_{AB} \times 10^5$ (cm²/s)[a]
Acetic acid	25	0.88
Ammonia	20	1.76
Bromine	20	1.2
Caffeine	25	0.63
Carbon dioxide	20	1.77
Chlorine	20	1.22
Glucose	20	0.6
	25	0.69
Hydrogen	25	5.85
	30	5.42
Hydrogen sulfide	20	1.41
Nitric acid	20	2.6
Nitrogen	25	1.9, 2.01
	40	2.83
Oxygen	25	2.5, 2.2
	40	3.33
Phenol	20	0.84
Sodium chloride	20	1.35
Urea	20	1.06

[a]The values are the diffusion coefficients multiplied by 10^5.
Source: Thibodeaux, *Chemodynamics: Environmental Movement of Chemicals in Air, Water, and Soil*, pp. 462–463, © 1979. Reprinted by permission of John Wiley & Sons, Inc.

By comparing Tables 6.1 and 6.3, we find that diffusion coefficients in air are on the order of 10^4 times greater than those in water. This is a direct result of the fundamentally different structures of water and air. The mean-free path between molecular collisions is vastly greater in air than in water; in effect, an air molecule (or contaminant molecule or particle in air) can travel a much greater distance before encountering another air molecule. Also recall that liquids like water have a more ordered structure because of hydrogen bonding and van der Waals forces.

6.5. STEADY-STATE DIFFUSION PROBLEMS WITH NO OVERALL DIFFUSIVE MASS TRANSFER

In certain diffusion problems, it is sometimes assumed that there is no net mass transfer of the diffusing species relative to fixed reference axes at the steady state. When this condition is assumed, a simple solution for concentration profiles sometimes exists. For example, consider the situation above the membrane shown in Figure 6.4. Fluid is passing through the membrane and carrying particles with it toward the membrane surface. Hence, we say there is

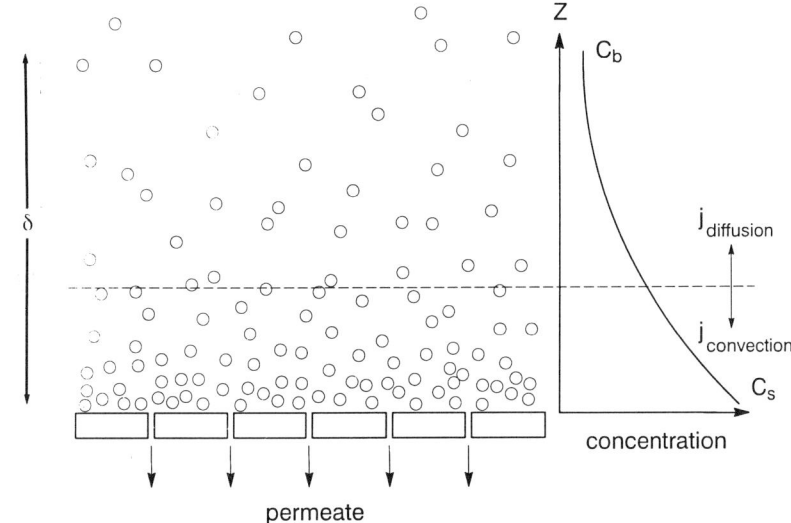

Figure 6.4. Possible concentration profile above a porous membrane during pressure filtration. Molecules are small enough to be under the influence of diffusion, but too large to pass through membrane pores.

a convective flux of particles toward the membrane surface. This is exactly the type of *impressed flow flux* discussed in Section 6.3. If there were no diffusion, particles would just collect in a gel or cake at the membrane surface (recall Exercise 1.12).

However, the particles in this problem are also subject to diffusive forces because of their small size. Hence, the tendency of the particles to pile up against the membrane is counterbalanced by diffusion, which wants to smooth out gradients in concentration. We know from Section 6.3 that in general the total mass flux of a diffusible material is given by

$$\mathbf{n}_A = \rho_A \mathbf{u} - \rho D_{AB} \nabla \omega_A \tag{6.18}$$

Here A refers to the particle phase and B refers to the liquid (and D_{AB} is the diffusion coefficient for particles in water). However, if the mass flux of particles at any plane like that shown in the figure is assumed to be zero,* $\mathbf{n}_A$ is zero,

---

*The student might make a legitimate objection here: If clean fluid is passing through the membrane, the mass of retained material on top of the membrane must be increasing. How can there be no net mass flux across the imaginary plane? This is indeed a problem using the approach described in this section; more correct models of membrane filtration consider solution of the complete convective-diffusion equation (Chapter 7) along with more realistic boundary conditions at the membrane surface.

and we conclude that

$$\rho_A u_z = D_{AB} \frac{d\rho_A}{dz} \tag{6.28}$$

where we have assumed that ρ is constant, and where u_z is the fluid velocity in the z direction. Equation 6.28 can be separated and easily integrated for $\rho_A = C_g$ at $z = 0$ and $\rho_A = C_b$ at $z = \delta$ to yield an equation well known in membrane science:

$$\ln\left(\frac{C_g}{C_b}\right) = (u_{-z}) \frac{\delta}{D_{AB}} \tag{6.29}$$

This parameter δ is the thickness of the diffuse particle layer above the membrane, or the distance over which the particle concentration is greater than the bulk concentration C_b (Figure 6.4). The term C_g is the maximum particle concentration at the membrane surface. The term u_{-z} is the mass-average velocity ($u_{-z} = -u_z$), which in this case is the so-called *approach velocity* of liquid toward the membrane surface, and which is, in a sense, opposite to the positive z direction assumed in the problem. In membrane terminology, u_{-z} is also called the "membrane-permeate" flow velocity. One known limitation of Eq. 6.29 is the assumption that the diffusion coefficient D_{AB} is a constant. In fact, it is so only for small particle concentrations.

In Chapter 2 we mentioned that in some types of ultracentrifugation operation, the diffusiveness of the centrifuged species must be taken into consideration. Exercise 6.2 uses the approach of this section to develop a new equation for determining molecular or particle mass in ultracentrifugation, an example of *pressure diffusion*.

6.6. STEADY-STATE MASS BALANCES OVER DIFFERENTIAL ELEMENTS

6.6.1. Diffusion of One Gas Through Another Stagnant Gas

Consider the situation depicted in Figure 6.5, where a pure gas B is flowing smoothly by the top of a tube of small diameter and cross-sectional area S. The liquid A at the bottom of the tube is evaporating and diffusing out of the tube into the stream of gas B. We stipulate that gas B is not soluble in liquid A.

Taking a mass balance around the differential element at steady state (i.e., no mass accumulation in the differential element), and following our normal approach of summing mass inputs and mass outputs, we find:

$$0 = N_{A,z}S - N_{A,z+\Delta z}S \tag{6.30}$$

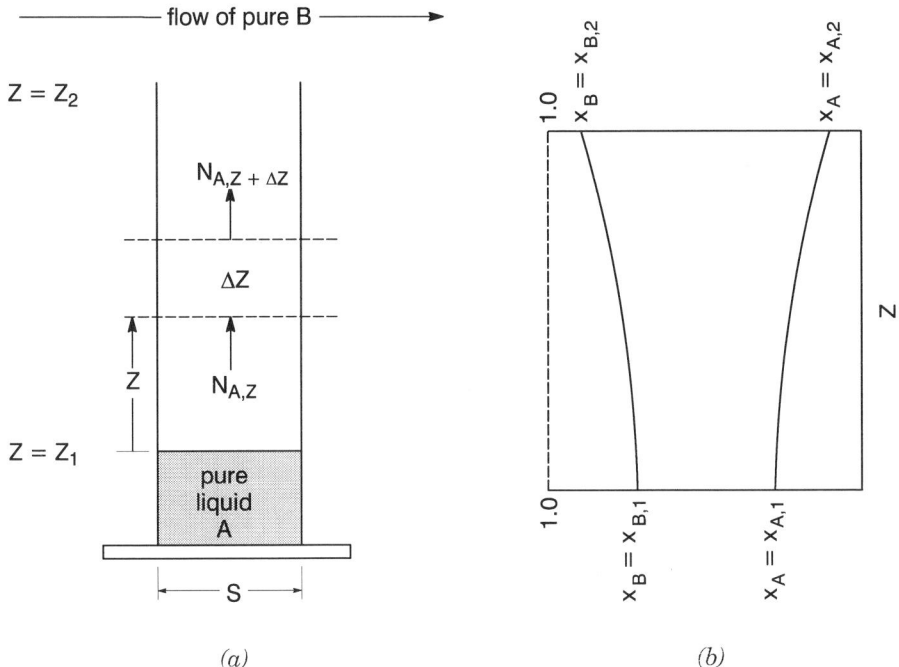

(a)   (b)

Figure 6.5. (a) Classic problem of binary diffusion in a tube. Evaporation of A does not significantly affect the level of liquid A. Here S is cross-sectional area of tube. Positive flux is in the positive z direction. (b) Concentration profiles in the tube.

We now rearrange slightly, divide by Δz, get rid of S, and take the limit as $\Delta z \to 0$:

$$\lim_{\Delta z \to 0} \frac{N_{A,z+\Delta z} - N_{A,z}}{\Delta z} = 0 \qquad (6.31)$$

The left-hand side of Eq. 6.31 is of course the mathematical definition of the differential of $N_{A,z}$; hence,

$$\frac{d}{dz}(N_{A,z}) = 0 \qquad (6.32)$$

Therefore, we conclude that at steady state, $N_{A,z}$ in the tube is constant regardless of height z. Because we stipulated that gas B is not soluble in liquid A, if we use the same type of argument as described above, the transport of gas B in the tube must be zero at steady state (i.e., $N_{B,x} = 0$). This is a classic problem in mass-transfer engineering, known as the *diffusion of gas A through a stagnant gas B.* Since $N_{B,z} = 0$, from Eq. 6.17 we find:

$$N_{A,z} = \frac{-cD_{AB}}{1 - x_A}\frac{dx_A}{dz} \qquad (6.33)$$

Substituting Eq. 6.33 into 6.32 yields

$$\frac{d}{dz}\left(\frac{-cD_{AB}}{1 - x_A}\frac{dx_A}{dz}\right) = 0 \qquad (6.34)$$

Integrating the equation once we find:

$$\frac{-cD_{AB}}{1 - x_A}\frac{dx_A}{dz} = C_1 \qquad (6.35)$$

where C_1 is an integration constant. But if we assume that c and D_{AB} are constants, Eq. 6.35 can be expressed as

$$\frac{1}{1 - x_A}\frac{dx_A}{dz} = C_2 \qquad (6.36)$$

Separating and integrating the equations again we end up with

$$-\ln(1 - x_A) = C_2 z + C_3 \qquad (6.37)$$

From Figure 6.5, the integration limits are $x_A = x_{A,1}$ at $z = z_1$ and $x_A = x_{A,2}$ at $z = z_2$. Substituting into Eq. 6.37, we find, after some effort, that

$$\frac{1 - x_A}{1 - x_{A,1}} = \left(\frac{1 - x_{A,2}}{1 - x_{A,1}}\right)^{z - z_1/z_2 - z_1} \qquad (6.38)$$

Since $x_B = 1 - x_A$, an equivalent expression to Eq. 6.38 is

$$\frac{x_B}{x_{B,1}} = \left(\frac{x_{B,2}}{x_{B,1}}\right)^{z - z_1/z_2 - z_1} \qquad (6.39)$$

It will prove useful later to have an expression for the average concentration of component B in the column:

$$x_{B,\text{ave}} = \frac{\displaystyle\int_{z_1}^{z_2} x_B dz}{\displaystyle\int_{z_1}^{z_2} dz} \qquad (6.40)$$

Substituting Eq. 6.39 into Eq. 6.40 (and evaluating the denominator) results in

$$x_{B,\text{ave}} = \frac{x_{B,1} \displaystyle\int_{z_1}^{z_2} (x_{B,2}/x_{B,1})^{z-z_1/z_2-z_1} dz}{z_2 - z_1} \tag{6.41}$$

Integrating the last equation we find:*

$$x_{B,\text{ave}} = \frac{x_{B,2} - x_{B,1}}{\ln(x_{B,2}/x_{B,1})} = \frac{x_{A,1} - x_{A,2}}{\ln(x_{B,2}/x_{B,1})} \tag{6.42}$$

Now we are in position to calculate $N_{A,z}$ in the column. Recall Eq. 6.33:

$$N_{A,z} = \frac{-cD_{AB}}{1 - x_A} \frac{dx_A}{dz} \tag{6.33}$$

Now set this equation up for integration using the boundary conditions that $x_A = x_{A,1}$ at $z = z_1$, and $x_A = x_{A,2}$ at $z = z_2$:

$$N_{A,z} \int_{z_1}^{z_2} dz = -cD_{AB} \int_{x_{A,1}}^{x_{A,2}} \frac{dx_A}{1 - x_A} \tag{6.43}$$

Evaluating the integral we find:

$$N_{A,z} = \frac{cD_{AB}}{z_2 - z_1} \ln\left(\frac{1 - x_{A,2}}{1 - x_{A,1}}\right) \tag{6.44}$$

and since $x_B = 1 - x_A$,

$$N_{A,z} = \frac{cD_{AB}}{z_2 - z_1} \ln\left(\frac{x_{B,2}}{x_{B,1}}\right) \tag{6.45}$$

Finally, using the result of Eq. 6.42, we arrive at

$$N_{A,z} = \frac{cD_{AB}}{z_2 - z_1} \frac{x_{A,1} - x_{A,2}}{x_{B,\text{ave}}} \tag{6.46}$$

*The integration is made easier if the reader recalls the calculus identity:

$$\int a^u du = \int e^{u\ln a} du = e^{u\ln a}/\ln a = a^u/\ln a \qquad (a > 0 \text{ and } a \neq 1).$$

This result for the mass transfer of pure gas A through stagnant gas B is a famous computation in mass-transfer engineering. A device similar to that shown in Figure 6.5 has been used successfully to measure gas-diffusion coefficients (see Exercise 6.5).

Note that in this model, mass transfer of A through the tube is proportional to the concentration difference at the bottom and top of the column, and is inversely proportional to the height of the column. This simple model has lead to a well-known mass-transfer model (somewhat in disrepute) called the *film theory*, which is described in Chapter 7. In this model, certain mass-transfer operations are imagined to be limited by diffusive transfer through a stationary film, analogous to the height of the tube above the liquid in Figure 6.5a. As noted above, mass transfer is proportional to the concentration difference across the film. The following exercise gives a taste of film-transfer modeling.

Example 6.2. There has been a spill of 55 gal of benzene in a 200-m² retention pond. Being lighter than water, the benzene is floating on the surface of the pond. Estimate the time required for the benzene to evaporate using the approach of this section. Assume air, water, and benzene are maintained at 25°C. Assume that the benzene evaporation occurs through a stagnant air layer of 2.5 mm.

SOLUTION. We would like to use Eq. 6.46 in this problem, where we identify A with the benzene, B with the air, and $z_2 - z_1$ as the thickness of the stagnant air layer (film thickness). If we can calculate the flux of A [i.e., $N_{A,z}$ (mol/time-area)], then dividing this number into the number of moles of benzene per unit area of lake will give us an estimate of the evaporation time.

Calculate c in Eq. 6.46 using the ideal gas law, $PV = nRT$, or

$$c = \frac{n}{V} = \frac{P}{RT}$$

Assuming standard atmospheric pressure (Appendix I), we find:

$$c = \frac{1.013 \times 10^5 \, \text{Pa}}{8.314 \dfrac{\text{J}}{\text{mol K}} 298 \, \text{K}} = \frac{1.013 \times 10^5 \dfrac{\text{N}}{\text{m}^2}}{8.314 \dfrac{\text{Nm}}{\text{mol K}} 298 \, \text{K}} = 40.9 \dfrac{\text{mol}}{\text{m}^3}$$

From Table 4.2, the vapor pressure of benzene is 12,700 Pa. Assuming that the air right at the surface of the benzene is saturated with benzene,

$$x_{A,1} = \frac{1.27 \times 10^4 \, \text{Pa}}{1.013 \times 10^5 \, \text{Pa}} = 0.125$$

and $x_{B,1} = 1 - x_{A,1} = 0.875$. Assuming $x_{B,2} = 1$, we find, for $x_{B,\text{ave}}$:

$$x_{B,\text{ave}} = \frac{x_{B,2} - x_{B,1}}{\ln(x_{B,2}/x_{B,1})} = \frac{1 - 0.875}{\ln(1/0.875)} = 0.936$$

Using $D_{AB} = 0.088 \, \text{cm}^2/\text{s}$ from Table 6.1,

$$N_{A,z} = \frac{(40.9 \, \text{mol/m}^3(0.088 \, \text{cm}^2/\text{s}) \, (\text{m}^2/10^4 \, \text{cm}^2)}{2.5 \, \text{mm} \, (\text{m}/1000 \, \text{mm})} \times \frac{0.125 - 0}{0.936} = 0.018 \frac{\text{mol}}{\text{m}^2\text{s}}$$

For a molar mass of benzene of 78.1 g/mol and a benzene density of 879 g/L, the number of moles of benzene per square meter of pond is

$$\frac{55 \, \text{gal} \, (3.78 \, \text{L/gal})(879 \, \text{g/L})}{(78.1 \, \text{g/mol}) \, 200 \, \text{m}^2} = 11.7 \frac{\text{mol}}{\text{m}^2}$$

Therefore, the time required for evaporation is

$$\text{time} = \frac{11.7 \, \text{mol/m}^2}{1.8 \times 10^{-2} \, \text{mol/m}^2\text{s}} = 1310 \, \text{s} = 21.8 \, \text{min}$$

Note that there were several critical assumptions in this problem, the most important of which was the thickness of the stagnant gas layer. This thickness would definitely be affected by factors like winds and wave action. Unfortunately, there is no good way of estimating film-layer thicknesses, which is one of the main problems of this model! In fact, we show in Chapter 7 that mass-transfer films really do not exist.

6.6.2. Diffusion with Heterogeneous Reaction at the Particle Surface

Thus far, we have avoided going very deeply into reaction kinetics. However, this is an excellent juncture at which to discuss first-order *heterogeneous reactions* and their possible limitation by diffusion. As we will see, heterogeneous reactions occur at the boundary of the diffusion domain and, therefore, the mathematics of diffusion are only affected by an alteration of the boundary condition. Consider the steady-state situation depicted in Figure 6.6. A spherical particle is surrounded by gases A, B, and C. However, at the surface of the particle, gas A is disappearing according to a first-order reaction:

$$A \xrightarrow{k_{A,s}} C \tag{6.47}$$

where $k_{A,s}$ is the first-order heterogeneous reaction-rate constant (L/T) for A at the surface ($r = a$). Hence gas A is diffusing to the surface where it reacts

gases A, B and C

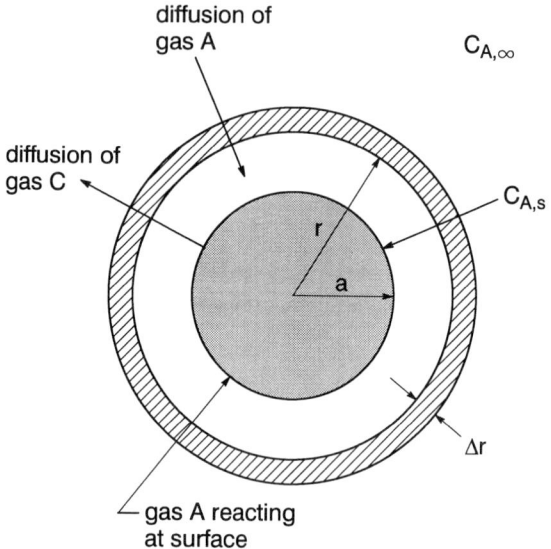

Figure 6.6. Diffusion of gas A toward surface of spherical particle of radius a, where A is disappearing via a first-order reaction. The positive flux is in the positive r direction.

according to Eq. 6.47 and produces gas C, which in turn is diffusing away from the surface of the particle (we assume gas B is inert and stagnant). We could think of this as a catalysis reaction, where the particle surface is the catalyst for reaction 6.47, or we can imagine that A is reacting with and consuming surface molecules to produce gas C.

We can describe the flux of A at the surface as

$$N_{A,s} = -k_{A,s}c_{A,s} \tag{6.48}$$

where $c_{A,s}$ is the concentration of A at the surface ($r = a$), and the negative sign arises because the flux is in the negative r direction (at the surface).* We call this a heterogeneous reaction because it is not really occurring in the external gas phase, but only at the surface of the particle. Heterogeneous reactions are important in environmental science, governing reaction processes as different as the biological consumption of oxygen and nutrients to the consumption of oxygen at the surface of a burning coal particle (see Exercises 6.6 and 6.7). A simple solution to the problem exists if we assume that the reaction at the

---

*We emphasize that $k_{A,s}$ is a first-order *heterogeneous* reaction-rate constant. Because it is defined in terms of a flux at the particle surface, the units must be L/T (Eq. 6.48). This is contrasted with the first-order *homogeneous* reaction-rate constant of Section 6.6.4.

surface is not diffusion limited, but rather is only limited by the reaction itself. In this case even though A is reacting at the surface, it is instantaneously replaced from the rest of solution because either the reaction is very slow or diffusion is infinitely fast. We call this situation *reaction limited*. Then, the flux of A at the surface of the particle is

$$N_{A,s}^{rxn} = -k_{A,s}c_{A,\infty} \tag{6.49}$$

where $c_{A,\infty}$ is the homogeneous phase concentration of A ($r \to \infty$). We assume that, overall, the consumption of A is not great enough to deplete the concentration of A in the rest of the system, that is, $c_{A,\infty}$ is constant.

Determining the flux of A to the surface when diffusion limits the reaction is a little more involved. Examining the differential element in Figure 6.6, a mass balance on diffusing material A yields

$$N_{A,r+\Delta r}4\pi(r + \Delta r)^2 - N_{A,r}4\pi r^2 = 0 \tag{6.50}$$

Expanding and dividing by $r^2\Delta r$ yields

$$\frac{N_{A,r+\Delta r} - N_{A,r}}{\Delta r} + \frac{2}{r}N_{A,r+\Delta r} + \frac{\Delta r}{r^2}N_{A,r+\Delta r} = 0 \tag{6.51}$$

Taking the limit as $\Delta r \to 0$, we find:

$$\frac{dN_{A,r}}{dr} + \frac{2}{r}N_{A,r} = 0 \tag{6.52}$$

However, the reader will note that the equation above is equivalent to

$$\frac{1}{r^2}\frac{d}{dr}(r^2N_{A,r}) = 0 \tag{6.53}$$

or simply

$$\frac{d}{dr}(r^2N_{A,r}) = 0 \tag{6.54}$$

This means that the product of the molar flux of A times the radius squared is constant at any r. Note that the mass transfer across any spherical surface in the region of the particle is given by

$$W_A = 4\pi r^2 N_{A,r} \tag{6.55}$$

This equation and Eq. 6.54 imply that steady-state *mass transfer* at any spherical surface surrounding the particle is a constant. (It is interesting to compare this with the conclusion of Eq. 6.32 regarding *mass flux*.)

Now note that the flux of gas A is given by Eq. 6.17

$$N_{A,r} = x_A(N_{A,r} + N_{B,r} + N_{C,r}) - cD_{AB}\frac{dx_A}{dr} \tag{6.56}$$

However, if the total gas concentration is constant, $N_{C,r}$ must equal the negative of $N_{A,r}$. And since we have assumed gas B is stagnant (i.e., $N_{B,r} = 0$), Eq. 6.56 reduces to

$$N_{A,r} = -cD_{AB}\frac{dx_A}{dr} \tag{6.57}$$

As we stated before, the mass transfer at any spherical surface at r is a constant, W_A. Thus, substituting Eq. 6.57 into Eq. 6.55:

$$W_A = 4\pi r^2 N_{A,r} = -4\pi r^2 cD_{AB}\frac{dx_A}{dr} \tag{6.58}$$

Separating and integrating over the limits that at $r = a$, $x_A = x_{A,s}$ and at $r = \infty$, $x_A = x_{A,\infty}$, we find:*

$$W_A = 4\pi acD_{AB}(x_{A,s} - x_{A,\infty}) \tag{6.59}$$

It is interesting to consider the condition opposite to the reaction limitation case analyzed above, and consider mass transfer under *diffusion-limited* conditions. Here, the reaction is so fast that it totally depletes the A at the surface of the particle; hence, the reaction appears to be limited by the ability of diffusion to deliver reactant A to the surface of the particle. Since the concentration of A at the surface is zero, $x_{A,s}$ must also be zero; hence, we find that mass transfer under diffusion limitation is from Eq. 6.59:

$$W_A^{\text{dif}} = -4\pi acD_{AB}x_{A,\infty} \tag{6.60}$$

When mass transfer is controlled by both reaction and diffusion, we take the following tack: Note that Eq. 6.48 can be rearranged to

$$N_{A,s} = -k_{A,s}cx_{A,s} \tag{6.61}$$

Hence,

$$x_{A,s} = -\frac{N_{A,s}}{k_{A,s}c} \tag{6.62}$$

---

*Equation 6.59 is also a very famous equation in science, and the flux computed from Eq. 6.59 is sometimes called the *Maxwellian flux* (Seinfeld, 1986).

Substitution of Eq. 6.62 into Eq. 6.59 yields

$$W_A = 4\pi a^2 N_{A,s} = 4\pi a c D_{AB} \left(-\frac{N_{A,s}}{k_{A,s}c} - x_{A,\infty} \right) \tag{6.63}$$

Following some rearrangement we find:

$$N_{A,s} = -\left(\frac{D_{AB}k_{A,s}c}{k_{A,s}a + D_{AB}} \right) x_{A,\infty} \tag{6.64}$$

It is also very interesting to take the rearrangement to another level:

$$N_{A,s} = -\left(\frac{D_{AB}k_{A,s}c}{k_{A,s}a + D_{AB}} \right) x_{A,\infty} = -\frac{cx_{A,\infty}}{\dfrac{a}{D_{AB}} + \dfrac{1}{k_{A,s}}} = \frac{-cx_{A,\infty}}{R_d + R_{rxn}} \tag{6.65}$$

The terms R_d and R_{rxn} defined by Eq. 6.65 are the *diffusion* and *reaction* *mass-transfer resistances*, for this problem. The form of the equation above shows that the flux of mass toward the particle is proportional to a driving force (i.e., $cx_{A,\infty}$) and inversely proportional to the sum of two mass-transfer resistances. This is a classic physical model called the *resistances-in-series* model because of its similarity to models for current in electrical circuit analysis. The resistance-in-series model is one of a number of models that help us to understand mass transfer. For example, often one of the two resistances will be much larger than the other. We then say that the larger of the two resistances "controls" mass transfer. This important concept is treated in more depth in Chapter 7.

The following development is another way to understand the limits of diffusion- and reaction-controlled mass transfer for the first-order reaction studied here. From Eqs. 6.49, 6.55, 6.60, and 6.62, the mass transfer in the reaction-controlled limit normalized by the mass transfer in the diffusion-controlled limit is easily calculated as

$$\frac{W_A^{rxn}}{W_A^{dif}} = \frac{4\pi a^2 c k_{A,s} x_{A,\infty}}{4\pi a c D_{AB} x_{A,\infty}} = \frac{a k_{A,s}}{D_{AB}} \tag{6.66}$$

The mass-transfer ratio is a simple function of the parameter grouping $ak_{A,s}/D_{AB}$. This is a famous parameter group in the chemical engineering field known as the *second Damköhler number*.* The reader should demonstrate that the Damköhler number is in fact a dimensionless number. Note that the Damköhler number can be considered the ratio of the characteristic reaction

---

*A more general form of the Damköhler number for reactions other than first order is developed in Exercise 6.18.

time scale and the characteristic diffusion time scale:

$$D_A^{II} = \frac{k_{A,s}}{D_{AB}} = \frac{\dfrac{a^2}{D_{AB}}}{\dfrac{a}{k_{A,s}}} = \frac{\tau_d}{\tau_r} \tag{6.67}$$

Using Eqs. 6.60, 6.63, and 6.64 we find, for the ratio of W_A to W_A^{dif}:

$$\frac{W_A}{W_A^{dif}} = \frac{4\pi a^2 \left(\dfrac{D_{AB}k_{A,s}c}{k_{A,s}a + D_{AB}}\right) x_{A,\infty}}{4\pi a c D_{AB} x_{A,\infty}} = \frac{\dfrac{ak_{A,s}}{D_{AB}}}{1 + \dfrac{ak_{A,s}}{D_{AB}}} = \frac{1}{1 + \dfrac{1}{D_A^{II}}} \tag{6.68}$$

Hence, this dimensionless mass-transfer ratio is also a simple function of just the Damköhler number. In Figure 6.7 we plot Eqs. 6.66 and 6.68 as well as the ratio W_A^{dif}/W_A^{dif} (which is of course equal to unity).

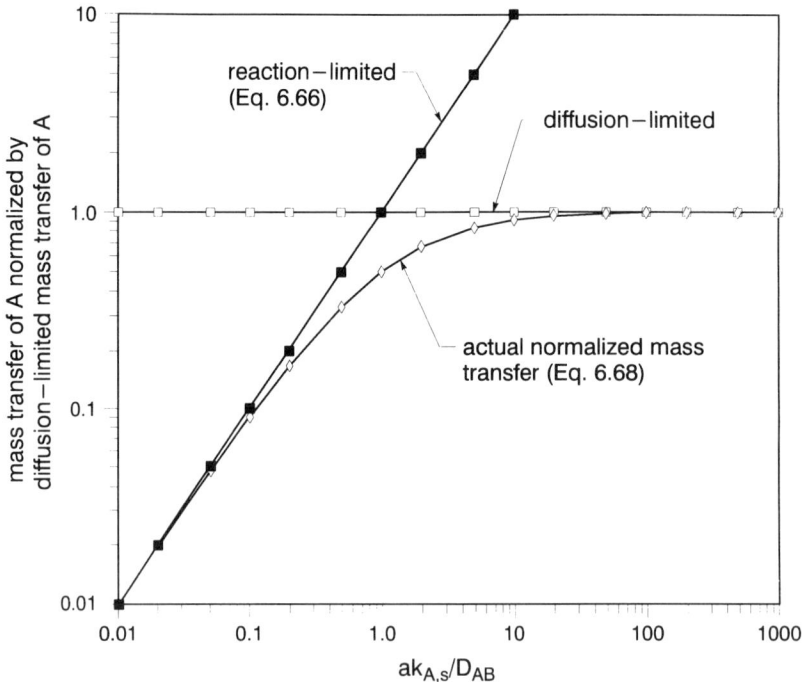

Figure 6.7. Demonstration of reaction- and diffusion-limitation during first-order reaction $A \rightarrow C$ at spherical surface. x axis is the second Damköhler number, $ak_{A,s}/D_{AB}$.

This figure shows well how diffusion and reaction lead to limits in mass transfer in problems where diffusion and reaction both come into play. Note that as the Damköhler number increases above unity, the mass transfer increasingly approaches the diffusion-controlled limit. This makes sense—the Damköhler number is the ratio of the diffusion and reaction time scales. So a large Damköhler number means that the diffusive time scale is greater than the reactive time scale. Large time scales mean slower processes; hence, a Damköhler number greater than unity means that diffusion is slower than reaction. Therefore, the overall process is controlled by diffusion, because, as we showed above, the diffusion resistance is greater than the reaction resistance. However, as the Damköhler number decreases below unity, the mass transfer increasingly approaches the reaction-controlled limit. The smaller Damköhler number means that the reactive time scale is greater than the diffusive time scale; hence, the process is controlled by reaction. We can also imagine in this case that the reaction resistance to mass transfer is greater than the diffusive resistance to mass transfer.

6.6.3. Brownian Coagulation as a Steady, Diffusion-Limited, Heterogeneous Reaction

In Section 3.7.2 we promised to derive the collision-frequency function for Brownian coagulation. Recalling Figure 3.11, imagine that the collision sphere $r_i + r_j$ is equal to a in Figure 6.6. We can think of the diffusion of particles of radius r_j into the collision sphere as being totally analogous to the diffusion of substance A toward the spherical particle of radius a. Any class j particle that arrives at the collision sphere immediately "reacts" with the central class i particle. Hence, the problem involves a heterogeneous reaction, as described in Figure 6.6. In addition, we assume that there is no energy barrier to collision, as discussed in Chapter 3; hence the problem is clearly diffusion-controlled. Then we can consider the mass transfer of class j particles to the diffusion sphere (particles per time) to be given by Eq. 6.60 with $a = r_i + r_j$, $cx_{A,\infty} = c_{A,\infty} = n_j$, and $D_{AB} = D_j$:

$$W = 4\pi D_j(r_i + r_j)n_j \tag{6.69}$$

Here we can think of D_j as the coefficient of diffusion for class j particles in whatever medium the particles are suspended. We must also imagine that the central particle is also experiencing Brownian motion; hence, using the trick developed in Section 3.7.2, we write the collision rate between the class i and j particles as

$$N_{i,j} = 4\pi D_{i,j}(r_i + r_j)n_i n_j \tag{6.70}$$

where $D_{i,j}$ is the *relative diffusion coefficient* for class i and j particles. In Section 6.7.2, we find that the diffusion coefficient for any diffusing species is a function of the mean-square distance traveled by the particle normalized by the time. For example, for the class i particle:

$$D_i = \frac{\overline{x_i^2}}{2t} \tag{6.71}$$

To calculate the relative diffusion coefficient $D_{i,j}$, we simply replace x_i in Eq. 6.71 with the relative particle displacement $x_i - x_j$:

$$D_{i,j} = \frac{\overline{(x_i - x_j)^2}}{2t} \tag{6.72}$$

Expanding under the average bar yields

$$D_{i,j} = \frac{\overline{x_i^2}}{2t} - \frac{\overline{2x_i x_j}}{2t} + \frac{\overline{x_j^2}}{2t} \tag{6.73}$$

The middle term in Eq. 6.73 involves the time average of the product of x_i and x_j, which is essentially the statistical correlation of x_i and x_j. However, the Brownian motion of any two particles (like the class i and j particles) is statistically independent, so the time average of the product of x_i and x_j is also exactly zero. Hence, Eq. 6.73 reduces to

$$D_{i,j} = \frac{\overline{x_i^2}}{2t} + \frac{\overline{x_j^2}}{2t} = D_i + D_j \tag{6.74}$$

Substituting Eq. 6.74 into Eq. 6.70 and assuming that the diffusion coefficients of the class i and j particles D_i and D_j are given by the Stokes–Einstein equation (Eq. 6.25), we find, after some minor algebra, that

$$N_{i,j} = \frac{2kT}{3\mu} \left(\frac{1}{r_i} + \frac{1}{r_j} \right) (r_i + r_j) n_i n_j \tag{6.75}$$

Equation 3.42 immediately falls out of Eq. 6.75, which was our goal.

6.6.4. Diffusion with Homogeneous First-Order Reaction

Figure 6.8 outlines a steady-state diffusion problem consisting of two phases, the lower phase is a liquid B and the upper phase is a gas A. Component A is simultaneously diffusing into B and undergoing a first-order reaction in

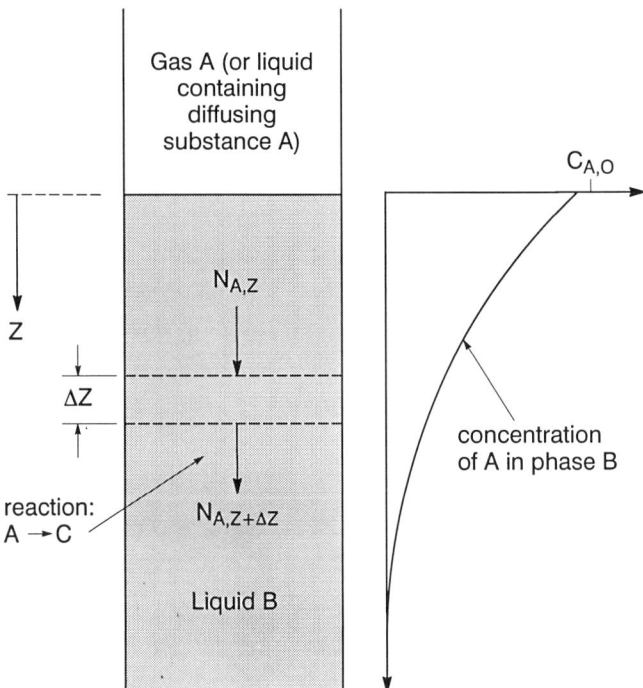

Figure 6.8. Steady-state diffusion of A into phase B where A undergoes first-order, homogeneous, irreversible reaction. Liquid B must be deep enough so that there is no appreciable concentration of A at the bottom of the picture.

phase B:

$$A \xrightarrow{\ k_A\ } C \tag{6.76}$$

where k_A is the homogeneous first-order reaction rate constant (T^{-1}). Note that the situation here is quite different from the last two sections — here the reaction is going on throughout phase B. Therefore, we call this a *homogeneous reaction*. Hence, we write the reaction of A as

$$r_{A,z} = \frac{dc_{A,z}}{dt} = -k_A c_{A,z} \tag{6.77}$$

The upper phase could also be another (lighter) liquid containing some component A which can diffuse into phase B. In this case, we assume that even though A is diffusing into phase B, its concentration remains fixed in phase A

(it is being replenished at the same rate that it is lost to phase B). The situation shown in Figure 6.8 describes a number of mass-transfer situations in the environment, including gas *absorption* in liquids and diffusion of chemicals and nutrients into bottom sediments of natural water bodies and partially penetrated biofilms (see Exercises 6.8, 6.9, 6.10, and 6.12 and Chapter 7). Performing a mass balance on the differential element at steady state yields the following:

$$0 = N_{A,z}S - N_{A,z+\Delta z}S + \Delta V r_{A,z} \tag{6.78}$$

where S is the surface area of the differential element normal to z, and ΔV is the volume of the differential element, $S\Delta z$. Dividing through by $S\Delta z$, substituting for $r_{A,z}$, rearranging slightly, and taking the limit as $\Delta z \to 0$, we find:

$$\frac{dN_{A,z}}{dz} + k_A c_{A,z} = 0 \tag{6.79}$$

If we assume that component A is present in small quantities, we can substitute Eq. 6.21 into Eq. 6.79, which for our one-dimensional problem yields

$$D_{AB}\frac{d^2 c_{A,z}}{dz^2} - k_A c_{A,z} = 0 \tag{6.80}$$

The boundary conditions for Eq. 6.80 are $c_{A,z} = c_{A,0}$ at $z = 0$, and $c_{A,z} = 0$ as $z \to \infty$. The solution to Eq. 6.80 for these boundary conditions is

$$c_{A,z} = c_{A,0} \exp\left(-\frac{z}{\sqrt{\dfrac{D_{AB}}{k_A}}}\right) \tag{6.81}$$

The term $(D_{AB}/k_A)^{1/2}$ in Eq. 6.81 is the *characteristic length scale* or *penetration distance* for this problem, since when depth z is equal to this term, the concentration in phase B has dropped to $\exp(-1) = 0.37$ of the value at the interface. Exercise 6.9 demonstrates the use of Eq. 6.81 in computing the flux of material at the interface of phases A and B.

6.7. FICK'S SECOND LAW AND NONSTEADY-STATE DIFFUSION

6.7.1. Fick's Second Law

When the mass-average velocity (Eq. 6.10) is zero, the diffusion coefficient and density are constant, and there are no reactions, the local diffusive mass flux of some component A is given by Eq. 6.22. A differential region can be considered as shown in Figure 6.9.

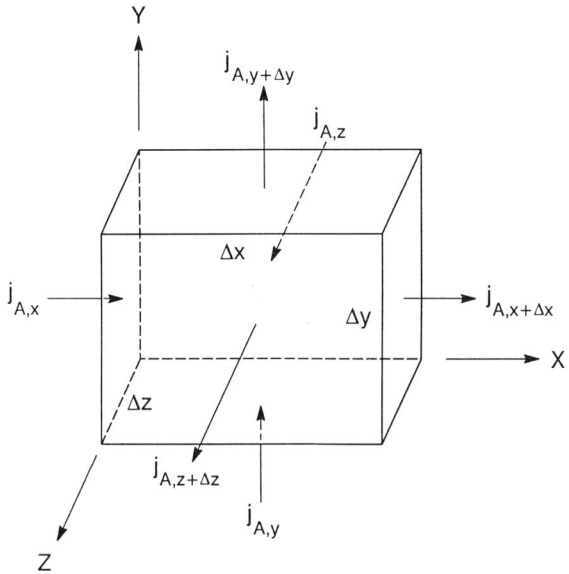

Figure 6.9. Differential element showing diffusive-mass flux components.

Performing a mass balance on species A, we find:

$$\frac{\partial \rho_A}{\partial t} \Delta x \Delta y \Delta z = (j_{A,x} - j_{A,x+\Delta x})\Delta y \Delta z + (j_{A,y} - j_{A,y+\Delta y})\Delta z \Delta x$$

$$+ (j_{A,z} - j_{A,z+\Delta z})\Delta y \Delta x \qquad (6.82)$$

Dividing through by $\Delta x \Delta y \Delta z$ yields

$$\frac{\partial \rho_A}{\partial t} = \frac{j_{A,x} - j_{A,x+\Delta x}}{\Delta x} + \frac{j_{A,y} - j_{A,y+\Delta y}}{\Delta y} + \frac{j_{A,z} - j_{A,z+\Delta z}}{\Delta z} \qquad (6.83)$$

Taking the limit as the differential volume shrinks to zero, we achieve the final form:

$$\frac{\partial \rho_A}{\partial t} + \frac{\partial j_{A,x}}{\partial x} + \frac{\partial j_{A,y}}{\partial y} + \frac{\partial j_{A,z}}{\partial z} = 0 \qquad (6.84)$$

However, since we recall that flux is a vector quantity, the last equation can be expressed as

$$\frac{\partial \rho_A}{\partial t} + \nabla \cdot \mathbf{j}_A = 0 \tag{6.85}$$

Substituting for $\mathbf{j}_A$ from Eq. 6.22, we find:

$$\frac{\partial \rho_A}{\partial t} = D_{AB}\nabla^2 \rho_A \tag{6.86}$$

Dividing through by the molar mass of species A, we find equivalently that

$$\frac{\partial c_A}{\partial t} = D_{AB}\nabla^2 c_A \tag{6.87}$$

Exactly the same equation is derived when the molar-average velocity (Eq. 6.11) is zero (equimolar diffusion), when the diffusion coefficient and total concentration c are constant, and when there are no reactions. Equations 6.86 and 6.87 are known as *Fick's second law*.

6.7.2. Diffusion from a Plane Source

If we imagine a source of diffusible mass A concentrated in an endless, infinitesimally thin plane at $x = 0$, diffusion clearly becomes a one-dimensional problem (Figure 6.10). In addition, since mass spreads out in both the positive and negative x directions, we have no reason to believe this is not also a problem which is symmetrical about $x = 0$. Fick's second law (Eq. 6.87) for this problem becomes

$$\frac{\partial c_A}{\partial t} = D_{AB}\frac{\partial^2 c_A}{\partial x^2} \tag{6.88}$$

It is easy to demonstrate that a particular solution to Eq. 6.88 is (Crank, 1956)

$$c_A = \frac{A}{t^{1/2}}\exp\left(-\frac{x^2}{4D_{AB}t}\right) \tag{6.89}$$

where A is an unknown constant. To determine the value of A, we invoke once again the conservation of mass idea. Even though we expect the mass M_A to spread out symmetrically in the positive and negative x directions, the total mass must remain constant; hence,

$$\int_{-\infty}^{\infty} c_A dx = M_A \qquad (6.90)$$

From inspection of Eq. 6.90, we note that M_A has somewhat unusual units—apparently mol/L^2 if c_A has normal units of mol/L^3. But there is nothing wrong here, we simply interpret M_A as the moles (or sometimes mass) per unit area of the original plane source. A clever substitution simplifies the integration given above. Define a new dimensionless independent variable as

$$\xi^2 = \frac{x^2}{4D_{AB}t} \qquad (6.91)$$

Taking the square root of Eq. 6.91 and differentiating, we find:

$$t^{1/2} = \frac{dx}{d\xi} \frac{1}{\sqrt{4D_{AB}}} \qquad (6.92)$$

Substituting Eq. 6.92 into Eq. 6.89, we find from Eq. 6.90, with the help of a definite integrals table (e.g., Spiegal, 1968):

$$M_A = 2AD_{AB}^{1/2} \int_{-\infty}^{\infty} \exp(-\xi^2) d\xi = 2A(\pi D_{AB})^{1/2} \qquad (6.93)$$

Hence, solving for A and substituting into Eq. 6.89, we get the final result for one-dimensional diffusion from a plane source:

$$c_A = \frac{M_A}{2(\pi D_{AB}t)^{1/2}} \exp\left(-\frac{x^2}{4D_{AB}t}\right) \qquad (6.94)$$

The ratio c_A/M_A from Eq. 6.94 is plotted in Figure 6.10, where each curve represents a different value of the product $D_{AB}t$. Note that for relatively short times (e.g., $D_{AB}t = 0.01$), the material has not diffused very far, and almost resembles the plane source itself. For longer times, we see, as expected, that the material spreads more and more into the medium B. As noted in our Figure 6.1, diffusion tends to smooth sharp concentration gradients.

As usual, we will try to squeeze a little more information from our analysis. Here, we seek information on the statistics of the diffusion process, where distance x is considered to be a random variable. First, for the symmetrical

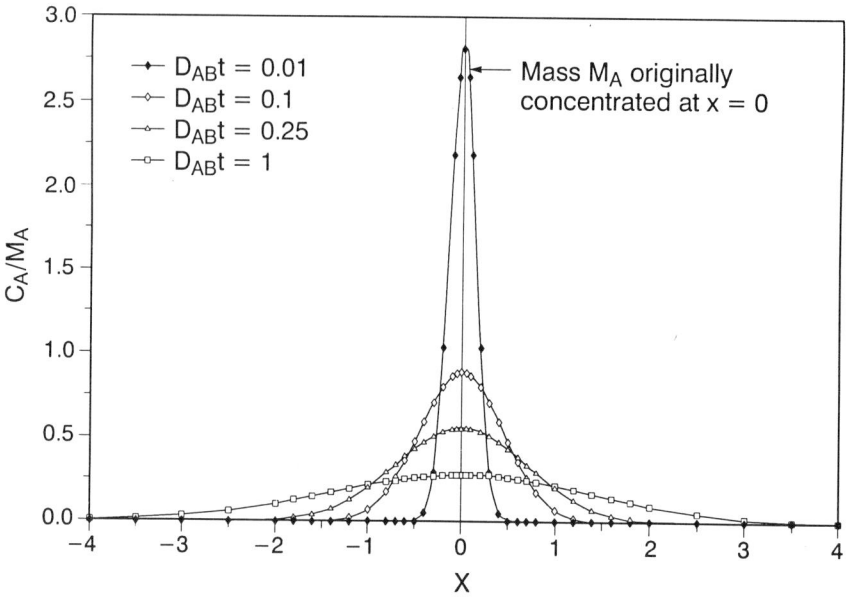

Figure 6.10. Diffusion from a plane source. The different curves are for different values of the product $D_{AB}t$.

diffusion problem considered above, we note that the average diffusion distance at any time is zero. Hence, the "first moment" or average of the random variable x is zero. But this is not very satisfying, since we know that the material is spreading out. The correct parameter to describe the spreading out in space is the second moment, the variance of x, which is called the *mean-squared distance*, $\overline{x^2}$. To calculate $\overline{x^2}$, we resort to a probability approach. Note that the *proportion* of molecules in a differential region of thickness dx at any time t and any distance x is

$$\left\{ \begin{array}{c} \text{proportion of molecules} \\ \text{in region of thickness} \\ dx \text{ at some } x \text{ and some } t \end{array} \right\} = \frac{c_A(x,t)dx}{M_A} = p_A(x,t)dx \qquad (6.95)$$

The function $p_A(x,t)$ is the probability density function of the random variable x. We note immediately from the definition that

$$p_A(x,t) = \frac{c_A(x,t)}{M_A} = \frac{1}{2(\pi D_{AB}t)^{1/2}} \exp\left(-\frac{x^2}{4D_{AB}t} \right) \qquad (6.96)$$

In probability and statistics, the variance of a random variable x is calculated as

$$\overline{x^2} = \int_{-\infty}^{\infty} p_A(x, t)x^2 dx = 2 \int_{0}^{\infty} p_A(x, t)x^2 dx \qquad (6.97)$$

where the last step in Eq. 6.97 is allowed because $p_A(x, t)$ is symmetrical. From Exercise 6.11, we find, with the help of an integrals table:

$$\overline{x^2} = 2D_{AB}t \qquad (6.98)$$

which is a very elegant answer indeed.

In Section 5.8 we noted that a viscous disturbance from a moving boundary would progress into an otherwise still fluid in proportion to the square root of time (Eq. 5.41). By taking the square root of each side of Eq. 6.98, we see that the standard deviation of x also progresses as the square root of time:

$$s = \sqrt{\overline{x^2}} = \sqrt{2D_{AB}}\sqrt{t} \qquad (6.99)$$

We can think of s as a characteristic diffusion distance. Hence, for diffusion from a plane source, the characteristic diffusion distance progresses in a manner similar to the progression of a viscous disturbance. As we mentioned in Chapter 5, this has led researchers to characterize the progression of the viscous disturbance as the "diffusion of momentum." In fact, the analogy is exact, since a mathematically identical differential equation governs both processes (compare Eqs. 5.36 and 6.88).

Example 6.3. Compare the efficiency of one-dimensional diffusive mass transport of ammonia in air at 25°C in (a) a basketball auditorium (characteristic diffusion distance ≈ 100 m), (b) the inside of a closed car (characteristic diffusion distance ≈ 1 m), and (c) the inside of a closed matchbox (characteristic diffusion distance ≈ 0.01 m).

SOLUTION. We imagine the plane source of ammonia at the middle of the rectangular-shaped system. From Eq. 6.99, the characteristic diffusion distance of the ammonia is

$$s = \sqrt{2D_{AB}t}$$

From Table 6.1, the diffusion coefficient of ammonia in air at 25°C is $0.28\,\text{cm}^2/\text{s}$. Hence, plotting t versus s, we find the following:

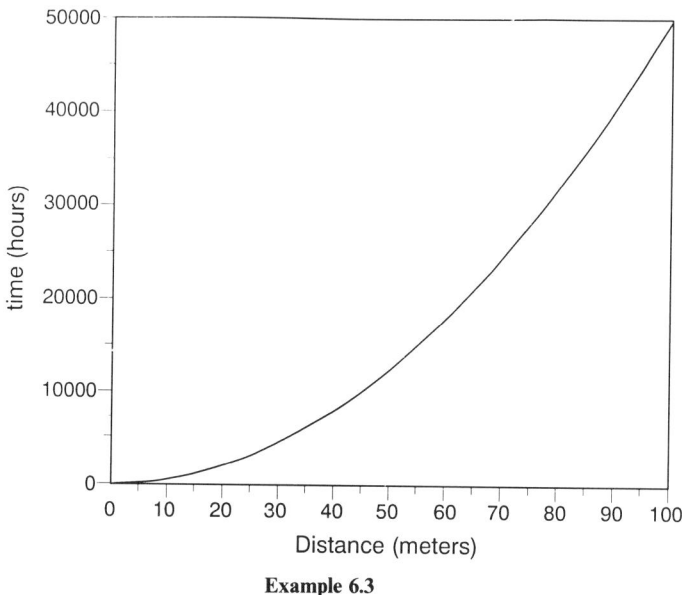

Example 6.3

The plot suggests that as the characteristic diffusion distance grows, the amount of diffusion time required increases rather profoundly. For the specific diffusion distances of this problem:

$$\text{basketball court: } t = \frac{s^2}{2D_{AB}} = \frac{[100\,\text{m}\,(100\,\text{cm/m})]^2}{2 \times 0.28\,\text{cm}^2/\text{s}} \frac{\text{hr}}{3600\,\text{s}} = 50{,}000\,\text{hr}$$

$$\text{car: } t = \frac{s^2}{2D_{AB}} = \frac{[1\,\text{m}\,(100\,\text{cm/m})]^2}{2 \times 0.28\,\text{cm}^2/\text{s}} \frac{\text{hr}}{3600\,\text{s}} = 5\,\text{hr}$$

$$\text{matchbox: } t = \frac{s^2}{2D_{AB}} = \frac{[0.01\,\text{m}\,(100\,\text{cm/m})]^2}{2 \times 0.28\,\text{cm}^2/\text{s}} \frac{\text{hr}}{3600\,\text{s}} = 5.0 \times 10^{-4}\,\text{hr} \ \text{or} \ 1.8\,\text{s}$$

These calculations suggest that although diffusive mass transport can be particularly effective (fast) over small distances, it may be particularly ineffective (slow) over larger distances. But we have to perform a reality check here: If someone opens a bottle of ammonia in a closed car, does it really take 5 h for someone else in the car to smell the ammonia? Ask the same question if someone were smoking in the car (since cigarette smoke has an even lower diffusion coefficient than ammonia). Experience suggests that something is wrong with our analysis. The problem is that we have just ignored some important physics at work in most problems like these. The thing we've ignored is convective mass transport. In basketball auditoriums filled with screaming fans and ventilation systems, or in moving cars with breathing human beings and small drafts, convective mass transport caused by fluid movement is much more effective than diffusive mass transport. However, in very small regions of the environment like the pores of sediments, concentration boundary layers (Chapter 7), or the inside of small rain drops, it may be difficult or impossible to set up a significant fluid flow; hence, diffusive mass transport will greatly dominate convective mass transport. This topic is considered in more detail in Chapter 7.

6.7.3. Diffusion in a Semi-Infinite Medium with Constant Boundary Concentration

A one-dimensional diffusion problem which describes certain gas absorption and other environmental problems is shown in Figure 6.11 (see Exercise 6.13). Here, a source of diffusible matter extends to negative infinity.

The concentration throughout this region (and, most importantly, at the interface $x = 0$) is unchanging in time, $c_A = c_{A,0}$. The material in this region is allowed to diffuse in the positive x direction; hence, the domain of this problem

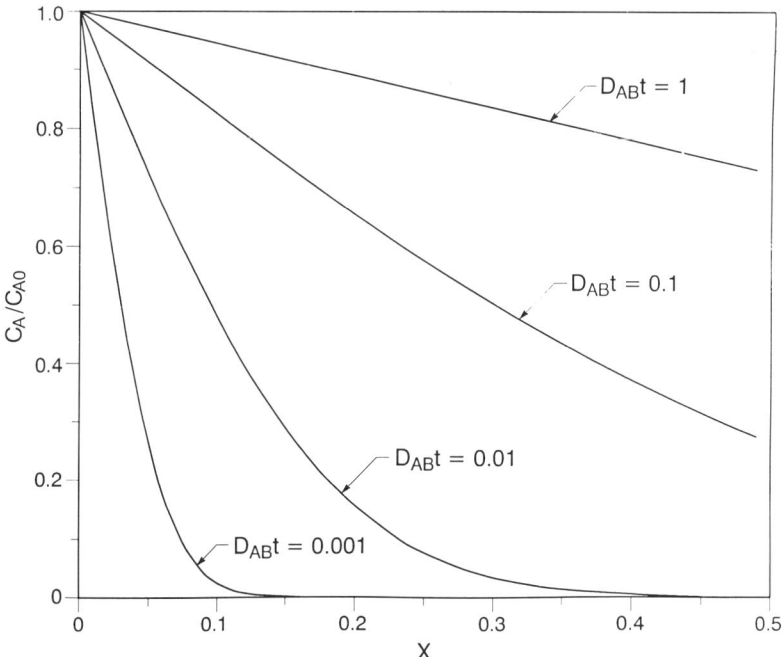

Figure 6.11. One-dimensional diffusion in positive x direction from plane source of constant concentration (Eq. 6.100). The curves indicate different values of the product $D_{AB}t$.

is called semi-infinite. The governing differential equation is still Eq. 6.88. The problem can be solved with Laplace transforms (Crank, 1956) or with the aid of dimensional analysis (Fischer et al., 1979). The answer is

$$c_A = c_{A,0}\,\text{erfc}\left(\frac{x}{2\sqrt{D_{AB}t}}\right) \tag{6.100}$$

Here erfc is the complimentary error function:

$$\text{erfc(argument)} = 1 - \text{erf(argument)} \tag{6.101}$$

The error function, which is given by

$$\text{erf(argument)} = \frac{2}{\sqrt{\pi}} \int_0^{\text{argument}} \exp(-u^2)\,du \tag{6.102}$$

is closely related to the area under the standard normal curve, and values are provided in tabular form in mathematics handbooks (e.g., Spiegal, 1968), or can easily be evaluated numerically with modern spreadsheets and equation-solving software. Equation 6.100 is plotted in Figure 6.11 for different values of the product $D_{AB}t$. The curves in Figure 6.11 show how the diffusing mass spreads out in the x direction over time. Once again, we see the tendency of diffusion to smooth sharp concentration gradients.

Note that in both Eqs. 6.94 and 6.100, distance x in the argument of the exponential or error function is normalized by $(D_{AB}t)^{1/2}$. This term can be thought of as the characteristic penetration distance of diffusing mass A into the medium B.* Thus, once again in a diffusion problem, we find that the characteristic diffusion distance (penetration distance here) increases in proportion to the square root of time.

Finally, at a plane normal to x and situated at $x = 0$, we could imagine measuring the flux of material through the plane caused by diffusion. We would expect this flux to be greatest at very small times (recall the steep concentration gradients at small times in Figure 6.11). For larger times, the flux in material would be expected to decrease. Applying Fick's first law to Eq. 6.100, we find that the flux of material across a plane at $x = 0$ is given by (see Exercise 6.13b):

$$J_{x=0} = \left(\frac{D_{AB}}{\pi t}\right)^{1/2} c_{A,0} \tag{6.103}$$

So we see that the intuitive prediction is verified.

---

*Some scientists say the term $(D_{AB}t)^{1/2}$ "scales" the independent variable x. This means that $(D_{AB}t)^{1/2}$ is the scale over which significant changes in x occur.

Exercise 6.15 investigates the problem of this section with the added complexity of a homogeneous, irreversible first-order reaction. We will see that the reaction leads to an even higher flux of material, for example, when gas absorption is accompanied by an irreversible reaction.

6.7.4. Diffusion and Partitioning into a Homogeneous Sphere

The final nonsteady diffusion problem we consider in this chapter is diffusion in spherical coordinates—specifically, the problem of radial diffusion in a sphere. This is an important diffusion geometry in environmental science, and has application in diffusion, partitioning, and adsorption of contaminants in sediment particles, rain drops, and activated carbon grains (see Exercises 6.16 and 6.17). When the diffusion coefficient and total concentration c are constant, and when there are no reactions, Fick's second law in spherical coordinates for diffusion in the radial direction only is (Appendix III):

$$\frac{\partial c_A}{\partial t} = D_{AB}\left(\frac{\partial^2 c_A}{\partial r^2} + \frac{2}{r}\frac{\partial c_A}{\partial r}\right) \tag{6.104}$$

Crank (1956) provides several solutions to this equation for situations of environmental interest. We focus here on one of those solutions—radial diffusion of component A from a perfectly mixed external phase of volume V into a homogeneous particle of radius a. In this problem, the concentration at the surface of the particle ($r = a$) is assumed to be the concentration in the well-mixed external solution. The concentration of A in the external phase is c_{A0} at $t = 0$. Hence, as time progresses and more A diffuses into the sphere from the external volume V, the concentration of A in V must decrease. There are two important boundary conditions for solution of this problem. The first is that at $t = 0$, the concentration of the diffusing substance within the sphere is zero or a constant, that is,

$$c_A = \text{constant} \quad \text{for } 0 \leqslant r \leqslant a \quad \text{and} \quad t = 0 \tag{6.105}$$

The second boundary condition is a new type of boundary condition in this book, which is related to the fact that the change in mass of A in the external solution must be balanced by the mass transport into the particle, or

$$V\frac{\partial c_A}{\partial t}\Big|_{r=a} = -4\pi a^2 D_{AB}\frac{\partial c_A}{\partial r}\Big|_{r=a} \tag{6.106}$$

The solution to this problem is Crank (1956):

$$\frac{M_{A,t}}{M_{A,\infty}} = 1 - \sum_{n=1}^{\infty} \frac{6\alpha(\alpha+1)\exp\left[-\dfrac{D_{AB}q_n^2 t}{a^2}\right]}{9 + 9\alpha + q_n^2\alpha^2} \tag{6.107}$$

Here $M_{A,t}$ is the total mass of A adsorbed in the sphere at time t and $M_{A,\infty}$ is the total mass of A adsorbed in the sphere at equilibrium with the external fluid concentration (i.e., $t \to \infty$). Note that in the argument of the exponential term, D_{AB} and the particle or sphere radius can be rearranged as a time scale:

$$\tau_d = \frac{a^2}{D_{AB}} \tag{6.108}$$

We call τ_d the *characteristic diffusion time scale* for the particle.

From a mass balance on solute in the system, the left-hand side of Eq. 6.107 can also be expressed as

$$\frac{M_{A,t}}{M_{A,\infty}} = \frac{c_{A,0} - c_{A,t}}{c_{A,0} - c_{A,\infty}} \tag{6.109}$$

where $c_{A,t}$ is the external phase concentration of A at time t, and $c_{A,\infty}$ is the external phase concentration of A at equilibrium. The parameter α in Eq. 6.107 is the equilibrium ratio of mass of solute in the solution to mass of solute in the sphere:

$$\alpha = \frac{3V}{4\pi a^3 K_p} = \frac{c_{A,\infty}}{c_{A,0} - c_{A,\infty}} \tag{6.110}$$

where K_p is the equilibrium partition coefficient for A:

$$K_p = \frac{S_{A,\infty}}{c_{A,\infty}} \tag{6.111}$$

where $S_{A,\infty}$ is the equilibrium concentration of A in the sphere. The q_n are the nonzero roots of the relation

$$\tan q_n = \frac{3q_n}{3 + \alpha q_n^2} \tag{6.112}$$

Hence, knowing a, D_{AB}, V, and K_p, we can evaluate Eq. 6.107 to the desired accuracy. It is important to note again that this solution assumes that the diffusion coefficient in the sphere is indeed a constant. Also, the solution requires that the simple linear partitioning implied by Eq. 6.111 is accurate. A minimum condition to satisfy these two requirements is that the spherical particle is internally homogeneous. If the partitioning is thought of as adsorption, the adsorption isotherm should be linear, and the adsorption should be reversible and instantaneous.*

---

*Exercise 6.16 suggests that the solution may even work in some cases where the adsorption is nonlinear.

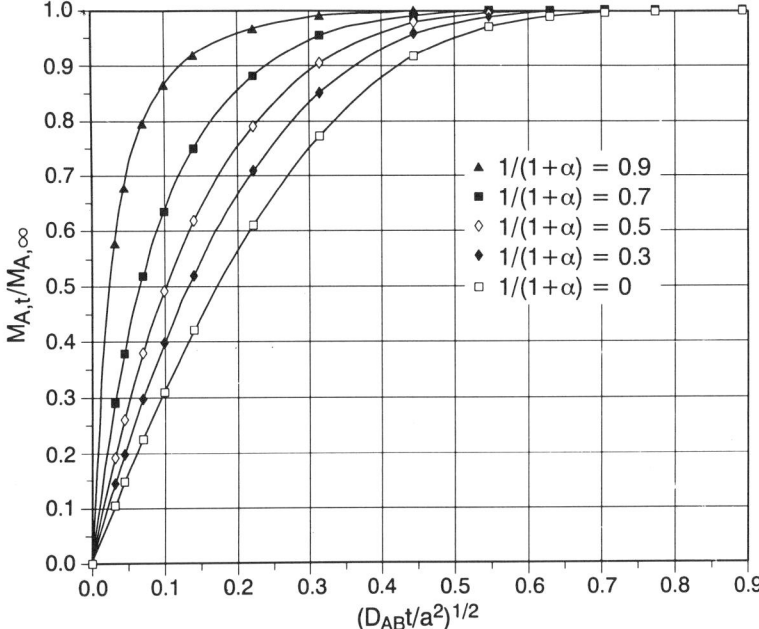

Figure 6.12. Diffusion and partitioning of solute A in a sphere from completely external mixed phase. The curves represent different fractional uptake in the sphere.

In Figure 6.12, the solution expressed by Eq. 6.107 is plotted for several values of the so-called fractional uptake, which is the fraction of mass A (originally in V at time $= 0$) partitioned into the sphere at equilibrium. The fractional uptake can be expressed as a simple function of α from Eq. 6.110. To see this, take a mass balance on solute:

$$V c_{A,0} = M_{A,\infty} + V c_{A,\infty} \tag{6.113}$$

Substituting from Eq. 6.111:

$$V c_{A,0} = M_{A,\infty} + \frac{V}{K_p} S_{A,\infty} \tag{6.114}$$

However, $S_{A,\infty}$ can be defined as

$$S_{A,\infty} = \frac{M_{A,\infty}}{(4/3)\pi a^3} \tag{6.115}$$

Substituting into Eq. 6.113, using the result of Eq. 6.110, and rearranging, the final form for the *fractional uptake* is achieved:

$$\frac{M_{A,\infty}}{Vc_{A,0}} = \frac{1}{1+\alpha} \qquad (6.116)$$

The curves in Figure 6.12 represent various values of the fractional uptake. They were determined through the following procedure:

1. For a selected value of $1/(1 + \alpha)$ in Eq. 116, a series of q_n (usually $n = 1$ to 20) were determined from solution of Eq. 6.112. An equation solver was used to solve Eq. 6.112, utilizing successive seed-solution values suggested by a plot of the function.
2. For the selected α and τ, Eq. 6.107 was evaluated to around 20 terms.*

It is interesting to note that as the fractional uptake of A increases, the kinetics of the diffusion and partitioning increase. Exercise 6.17 uses the results of this section to predict the sorption of contaminants in suspended river sediments. A related problem, diffusion into a sphere with constant surface concentration, is examined in Exercise 6.19.

6.8.   EFFECTIVE DIFFUSION COEFFICIENTS IN POROUS MEDIA

At several points in this chapter, the diffusion of solutes or gases in porous media was mentioned as an important environmental transport mechanism. Examples of such transport processes might be the diffusion of gaseous contaminants in activated carbon fibers or particles (an extremely porous solid material), the diffusion of aqueous contaminants in river or lake sediments (often a particulate deposit with a lot of interstitial water), and even the diffusion of biocides and nutrients through biofilms. But in these applications, there is a problem in using the equations of this chapter. Recall that nearly all the developments of this chapter imagined that diffusion occurred in a homogeneous phase with constant diffusion coefficient; for example, diffusion of contaminants in pure air or water (recall Examples 6.2 and 6.3). But in some of the important examples mentioned above, although the diffusion may well be occurring in an air or water phase, this phase may be a very irregular continuous region within a "solid" material (like activated carbon or sediments). In fact, when data for diffusion of aqueous materials in activated carbon particles are analyzed, they suggest molecular diffusion coefficients many orders of magnitude less than the diffusion coefficient in pure water (Adham et al., 1991). So our intuition is correct: Diffusing through a porous solid containing air or water must be slower than diffusion in a pure air or

---

*The solution for small values of time can converge very slowly, and more terms than indicated may be required for acceptable accuracy.

water phase. There are two reasons for the lower effective-diffusion coefficient. First, consider that during diffusion through the porous media, the diffusing material can adsorb to the internal surface of the porous medium. Intuition tells us here that this phenomenon will tend to slow or retard the mass transfer through the porous medium. Second, because of the irregular and tortuous nature of the internal diffusion path in the porous medium, it takes solutes or gases longer to work their way through.

To see how Fick's second law might be modified for diffusion in porous media, consider Figure 6.13. A two-dimensional slice of a section of a porous medium is shown in Figure 6.13a. The porosity of the three-dimensional porous medium is simply

$$\epsilon = \frac{\text{pore volume}}{\text{total volume}} \tag{6.117}$$

Note that the diffusing component A is indicated by the little dots, some of which are attached to the surface of the porous medium grains; they can be considered temporarily adsorbed to the grain surface. The rest of component A is diffusing through the continuous pore space between the grains. Figure 6.13b suggests an approximation of the porous medium diffusion. The tortuous diffusion path of Figure 6.13a is straightened out, and a regular porous medium with the same porosity is suggested. As in the previous sketch, some of the species A are adsorbed and some are diffusing. We can perform a mass balance in analogy to Section 6.7.1:

$$\frac{\partial}{\partial t}(\rho_A \epsilon \Delta x \Delta y \Delta z + S_A \rho_b \Delta x \Delta y \Delta z)$$
$$= (j_{A,x} - j_{A,x+\Delta x})\epsilon \Delta y \Delta z + (j_{A,y} - j_{A,y+\Delta y})\epsilon \Delta z \Delta x \tag{6.118}$$
$$+ (j_{A,z} - j_{A,z+\Delta z})\epsilon \Delta y \Delta x$$

Some new variables appear in Eq. 6.118: ρ_A is the mass concentration of species A in the interstitial fluid, S_A is the mass of solute A adsorbed per unit mass solid, and ρ_b is the bulk density of solid (mass solid per total volume porous media). In contrast to Eq. 6.82, the accumulation term in Eq. 6.118 has two terms: the first term is the mass of species A in the interstitial fluid, and the second term is the mass of species A associated with the solid phase through adsorption. The sum of these two terms accounts for all A in the differential element. The right-hand side of Eq. 6.118 looks quite similar to Eq. 6.82, except for the appearance of ϵ. The flux of A in the interstitial fluid is partially blocked by the solid material; hence, the flux is proportional to ϵ. Dividing through Eq. 6.118 by $\Delta x \Delta y \Delta z$ and taking the limit as the differential element shrinks to

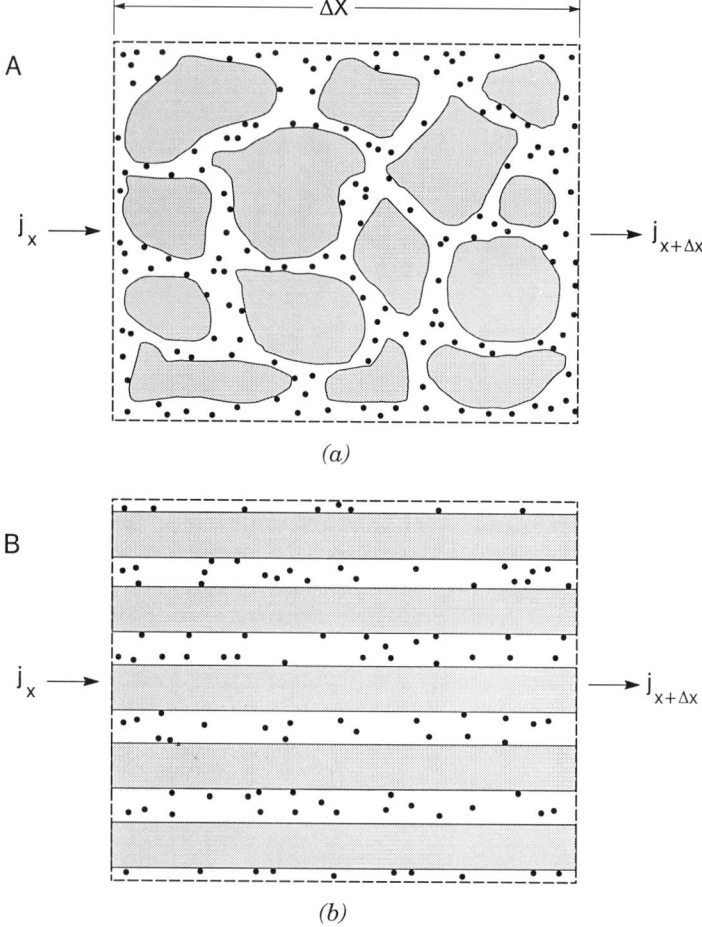

(a)

(b)

Figure 6.13. (a) Two-dimensional slice of a different element of a porous medium, and (b) approximation of a porous medium having the same porosity as the real porous medium.

zero, we find:

$$\frac{\partial}{\partial t}(\rho_A \epsilon + S_A \rho_b) + \epsilon(\nabla \cdot \mathbf{j}_A) = 0 \qquad (6.119)$$

However, we now assume that the adsorption of A on the surface of the solid media is governed by linear partitioning theory (Chapter 4), that is,

$$S_A = k_d \rho_A \qquad (6.120)$$

where k_d is the partition coefficient. Substituting into Eq. 6.119 and rearranging, we find:

$$\frac{\partial \rho_A}{\partial t} + \left[\frac{1}{1 + \frac{k_d \rho_b}{\epsilon}} \right] \nabla \cdot \mathbf{j}_A = \frac{\partial \rho_A}{\partial t} + \frac{1}{R} \nabla \cdot \mathbf{j}_A = 0 \qquad (6.121)$$

The factor R in the equation above,

$$R = 1 + \frac{k_d \rho_b}{\epsilon} \qquad (6.122)$$

is known as the *retardation coefficient*.* R is always greater than or equal to one, so the mathematical effect of the retardation coefficient is to slow diffusion. Substituting Eq. 6.22, we find:

$$\frac{\partial \rho_A}{\partial t} = \frac{D_{AB}}{R} \nabla^2 \rho_A \qquad (6.123)$$

where D_{AB} refers to the coefficient of diffusion of A in the interstitial fluid B. Alternatively, dividing through by the molar mass of species A:

$$\frac{\partial c_A}{\partial t} = \frac{D_{AB}}{R} \nabla^2 c_A \qquad (6.124)$$

where c_A is the molar concentration of A in the interstitial fluid. Finally, note that the ideal diffusion path suggested in Figure 6.13b obviously does not account for the more tortuous diffusion path suggested in the actual porous media. In fact, it might be more realistic to think of the straight channels in Figure 6.13b as being curved or sinusoidal. To account for the additional slowing of diffusion caused by the tortuosity of the diffusion path in porous media, we sometimes add a correction factor to Eq. 6.124:

$$\frac{\partial c_A}{\partial t} = \frac{D_{AB} \tau'}{R} \nabla^2 c_A \qquad (6.125)$$

where τ' is the *tortuosity factor*, which here is assumed to be a constant

---

*An additional mechanism of diffusion called "surface diffusion" has been proposed. It is essentially diffusion of adsorbed species along the solid surface. The retardation coefficient for this situation is different from Eq. 6.122 (see Ball et al., 1991).

throughout the porous medium (see discussion of homogeneous isotropic media in Chapter 8). The tortuosity factor is always less than or equal to unity, so the tortuosity of the diffusion path also slows doen diffusion. Rees et al. (1991) cite values of tortuosity of from 0.2 to 0.6 for soils and 0.5 to 0.9 for sediments. The *effective diffusion coefficient* can be defined as*

$$D_{eff} = \frac{D_{AB}\tau'}{R} \qquad (6.126)$$

Note that if R and τ' are known and constant throughout the diffusion domain, use of the effective diffusion coefficient does not change the mathematics of diffusion (i.e., R and τ' are just constants).

6.9. SUMMARY

This chapter developed the basic mathematics of steady and nonsteady isothermal diffusion and applied this analysis to several problems in environmental transport. Diffusion was shown to be a particularly important mass-transfer mechanism when the characteristic dimensions of the transport were small; for example, in activated carbon grains, liquid drops, and biofilms. Diffusion was also found to be important in problems where the media in which the mass transport was occurring was completely undisturbed; for example, in undisturbed sediments and in a device used to measure gas-diffusion coefficients.

When reactions occur at a spherical diffusion boundary (a heterogeneous reaction), our analysis showed that the overall transport of the diffusing (and reacting) component could be controlled by the reaction, by diffusion, or by reaction and diffusion. For a homogeneous first-order reaction, we showed that the steady-state penetration of a diffusing component into the diffusion domain was a simple function of the reaction-rate constant and the diffusion coefficient. Finally, we showed that the diffusion equation could be modifed to account for linear partitioning in the diffusion domain and for transport through a porous medium. In this regard, the retardation coefficient was developed to compute an effective diffusion coefficient.

At several junctures, we hinted that more information would eventually be required in problems where convective and diffusive transport are present at the same time. For example, in our discussion of film modeling, we admitted that there was no good way to estimate film thicknesses a priori. Chapter 7 concerns itself with these "convective-diffusion" problems and the several approaches that have been developed to solve convective-diffusion problems.

---

*There are actually several definitions of the effective diffusion coefficient. One that is sometimes used in groundwater transport studies is $D_{eff} = D_{AB}\tau$ (Section 8.3).

EXERCISES

6.1. In chapter 2, the determination of particle or molecule size using sedimentation data required the assumption of a spherical particle. It turns out that if diffusion and sedimentation data are both available, molecular or particle mass can be determined without the assumption of a spherical particle. Recall that the Stokes–Einstein equation (Eq. 6.25) can be expressed as

$$D = \frac{kT}{f}$$

In Section 6.4, it was suggested that f was the Stokes friction factor for a sphere. However, f can be the more general particle friction factor (for shapes other than spherical) defined by the particle drag force (Eq. 6.26):

$$F_D = f v_r$$

where v_r is the fluid-particle relative velocity. Using the equations above and the approach of Section 2.11, develop an equation for calculating particle or molecular mass assuming you do not know the particle shape but that you do know the fluid and particle densities, the sedimentation coefficient (S_d), the temperature, and the particle diffusion coefficient (D). Hint: the answer is:

$$m_p = \frac{kT}{D[1 - (\rho_f/\rho_p)]} S_d$$

where all other terms are defined in Section 2.11.

6.2. Often centrifugation is done with an initially homogeneous sample of particles (or molecules) and suspending medium.* Eventually, near the bottom of the centrifuge tube we can no longer ignore the action of diffusion. In analogy to Section 6.5, for a pseudo-steady situation, the sedimentation flux can be considered to be exactly balanced by diffusional flux, or

$$v_r C = D_{AB} \frac{dC}{dr}$$

where C is the particle or molecular concentration. Using this equation and some of the concepts of Exercise 6.1 and Section 2.11, develop an

---

*Recall that earlier we imagined that the particles were initially layered in the suspending medium and more or less drifted together during centrifugation.

equation to determine particle mass in centrifugation knowing only C_1 and C_2 at r_1 and r_2, the centrifuge angular velocity, the particle and suspending liquid densities, and the temperature. Here we need no assumption about particle or molecular shape, and we do not need to know the diffusion coefficient.

6.3. Show using Eqs. 6.11 and 6.13 that the following is true:

$$J_i^* = N_i - x_i \sum_{j=1}^{n} N_j$$

Next, show by summing this equation over $i = 1$ to n, that the sum of the molar diffusion fluxes relative to the molar average velocity is zero:

$$\sum_{i=1}^{n} J_i^* = 0$$

and specifically, for a binary system of A and B:

$$J_A^* = -J_B^*$$

6.4. The molar average flux of component A relative to coordinates moving with the molar average velocity is given by Eq. 6.14:

$$J_A^* = -cD_{AB}\nabla x_A$$

In a similar fashion, we can write the equation for the molar average flux of component B relative to coordinates moving with the molar average velocity as

$$J_B^* = -cD_{BA}\nabla x_B$$

where D_{BA} is interpreted as the diffusion coefficient of B in A. Using the results of Exercise 6.3 and the fact that $x_A + x_B = 1$ in a binary system, prove that

$$D_{AB} = D_{BA}$$

6.5. **(a)** Show using the ideal gas law that Eq. 6.46 can be expressed as

$$N_{A,z} = \frac{\dfrac{pD_{AB}}{RT}}{(z_2 - z_1)x_{B,\text{ave}}}(x_{A,1} - x_{A,2})$$

Here p is the total pressure (atm), and x_A and x_B are interpreted as

$$x_A = \frac{p_A}{p} \qquad x_B = \frac{p_B}{p}$$

where p_A and p_B are the partial pressures of gases A and B. (The numerical subscripts have the same meaning as previously.)

(b) In a study of the diffusion of chloroform in air at 25°C in the device shown in Figure 6.5, it was found at steady state that $N_{A,z} = 1.5 \times 10^{-7}$ mol-cm^{-2}-s^{-1}, where the distance $z_2 - z_1 = 7.7$ cm. Note that the vapor pressure of chloroform is 200 mmHg. From the equation above, calculate the apparent diffusion coefficient of chloroform in air and compare with the value indicated in Table 6.1. It may be of help to have the following form of the universal gas constant:

$$R = 82.06 \frac{\text{atm cm}^3}{\text{mol K}}$$

6.6. In the high-temperature combustion of coal in a power plant (1145 K), the average size of the coal particles is 1.0×10^{-4} m. For complete combustion, the reaction at the surface of the coal particle is

$$C + O_2 = CO_2$$

Assuming that the combustion reaction is controlled by the diffusion of oxygen to the coal particle surface, estimate the instantaneous consumption of oxygen (mol/s) by the average size coal particle. Assume the diffusion coefficient of O_2 in air at this temperature is 1.4×10^{-4} m^2/s (Welty et al., 1976), and that far away from the particle the partial pressure of O_2 is 0.21 atm.

6.7. In Section 6.6.2, we studied the first-order heterogeneous reaction

$$A \xrightarrow{k_{A,s}} C$$

However, another important first-order reaction is

$$A \xrightarrow{k_{A,s}} 2C$$

In this case, for each mole of A that diffuses to the surface of the particle, 2 mol of C must diffuse away. This reaction is relevant to the partial oxidation of surface carbon (e.g., in combustion processes), $2C + O_2 \rightarrow 2CO$.

(a) For the general reaction shown above, complete an analysis similar to that in Section 6.6.2, and show that

$$N_{A,r} = \frac{-cD_{AB}}{1 + x_A} \frac{dx_A}{dr}$$

and

$$W_A = 4\pi a c D_{AB} \ln\left(\frac{1 + x_{A,s}}{1 + x_{A,\infty}}\right)$$

(b) What form does this equation take for the case of diffusion limitation?

6.8. The vertical profile of dissolved oxygen in sediments is measured and analyzed using Eq. 6.81. The penetration distance is found to be 5 cm. If the sediment tortuosity factor is 0.7, determine the first-order reaction-rate constant for the consumption of oxygen. Assume $R = 1$, and $T = 15°C$. Hint: You will need to review the concept of the effective diffusion coefficient in Section 6.8.

6.9. Show using Eq. 6.81 and Fick's first law (Eq. 6.21) that the flux of A into phase B at the interface of A and B (i.e., $z = 0$) is

$$J_A^* = c_{A,0}\sqrt{D_{AB}k_A}$$

Take care to keep track of signs, and make sure that the computed flux is in the correct direction (i.e., down — the positive sense).

6.10. Thibodeaux (1979) describes the problem of accounting for fluxes of radon at sediment–water interfaces. Radon is produced naturally at a constant rate (regardless of depth) in sediments by the decomposition of radium; radon also decays naturally, but as a first-order reaction (i.e., concentration dependent). In the deep sediments, radon production (p, mass per time per unit volume) equals radon decay (d, mass per time unit volume), and there is a constant concentration of radon. However, closer to the sediment–water interface, the radon can escape into the overlying water; hence, the concentration of radon decreases on moving from the deep sediments to the sediment surface.

(a) Consider the differential element in Figure 6.8, and show that at steady state, the differential equation governing radon concentration is

$$D_{eff}\frac{d^2c_A}{dy^2} + p - d = 0$$

where c_A is the radon concentration, and D_{eff} is the effective diffusion coefficient of radon in the sediments. Note: the radon concentration does not look like that sketched in Figure 6.8. How does it look (approximately)?

(b) The first-order decay rate of radon is

$$d = k_A c_A$$

However, as stated above, in the deep sediments the decay rate is equal to the constant production rate; hence, in the deep sediments $(z \to \infty)$,

$$\text{deep decay } d = k_A c_{a,\infty} = p$$

Using these ideas, show that the differential equation governing radon concentration in the sediments can also be expressed as

$$D_{\text{eff}} \frac{d^2 c_A}{dy^2} + k_A (c_{A,\infty} - c_A) = 0$$

(c) For the boundary conditions $c_A = c_{A,\infty}$ as $z \to \infty$ and $c_A = c_{A,0}$ at $z = 0$, the solution to the equation above is (Thibodeaux, 1979 and Section 6.6.4)

$$c_A = c_{A,\infty} - (c_{A,\infty} - c_{A,0}) \exp\left(-\frac{z}{\sqrt{\dfrac{D_{\text{eff}}}{k_A}}}\right)$$

Show using this equation and Fick's first law (Eq. 6.21) that the flux of radon at the sediment–water interface is

$$J_{A,0}^* = -\sqrt{k_A D_{\text{eff}}}\,(c_{A,\infty} - c_{A,0})$$

Hint: Exercise 6.9 should help with this part.

(d) Naturally occurring substances have been used as natural tracers for, among other things, determining diffusion coefficients in places like sediments. Assuming you have measurements of radon concentration with depth in sediments, describe how you would use that data to determine the effective diffusion coefficient of radon.

6.11. Substitute Eq. 6.96 into Eq. 6.97 and evaluate the integral using a definite integrals table.

6.12. A biofilm is a mixed population of microorganisms that are part of a more or less stable thin film. Inside the biofilm, organic substrates are decomposed. The substrate must diffuse from the exterior solution into the biofilm. In certain biofilms along solid surfaces, the consumption of substrate within the biofilm can be assumed to be zero order when the substrate concentration is very high (Characklis, 1990), that is,

$$r_A = -k_A^0 \left(\frac{\text{mol}}{L^3 T}\right)$$

where A stands for the substrate.

(a) Show that the differential equation governing the concentration of substrate within the biofilm is

$$D_{\text{eff}} \frac{d^2 c_A}{dz^2} - k_A^0 = 0$$

(b) Biofilms may be so thin that the substrate never reaches a concentration of zero before $z = L$. This is referred to as a "fully penetrated" biofilm (below). Explain using Fick's first law why the gradient dc_A/dz must be 0 at $z = L$.

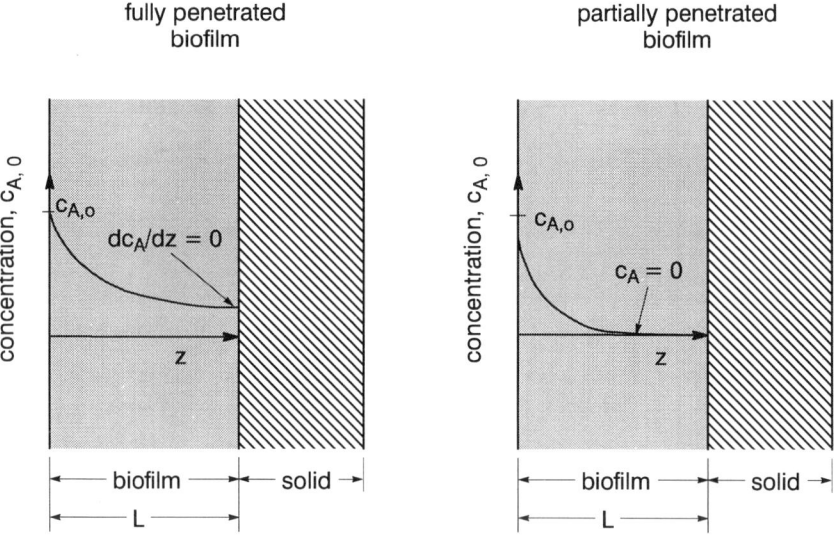

Exercise 6.12 (b).

(c) For the boundary conditions $dc_A/dz = 0$ at $z = L$, and $c_A = c_{A,0}$ at $z = 0$, show that the solution to the differential equation above for the fully penetrated biofilm is

$$\frac{c_A}{c_{A,0}} = 1 - Da^{\text{II}} \left[\frac{z}{L} - \frac{1}{2}\left(\frac{z}{L}\right)^2\right]$$

where Da^{II} is the second Damköhler number for this problem

$$Da^{II} = \frac{k_A^0 L^2}{D_{eff} c_{A,0}}$$

(d) Plot the function $c_A/c_{A,0}$ and confirm that the following figure is obtained for the values of the Damköhler number shown. Does the figure make sense? Explain in terms of the interpretation of the Damköhler as a ratio of diffusive and reactive time scales (recall Section 6.6.2).

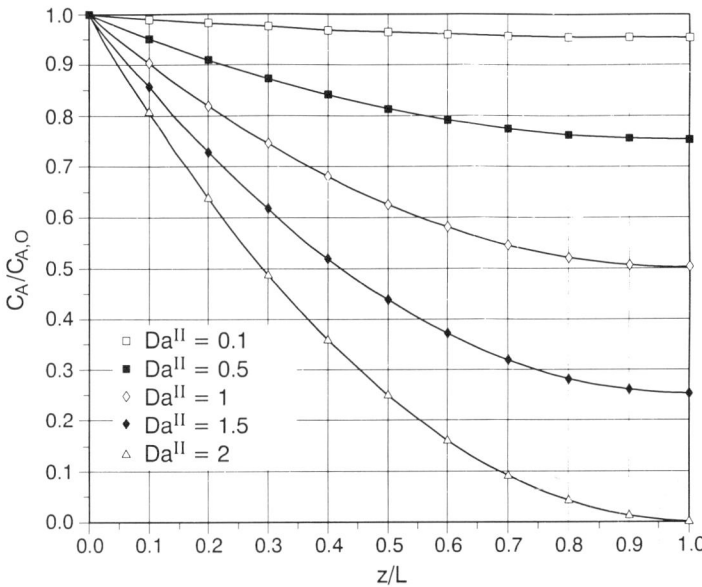

Exercise 6.12 (d). Substrate concentration in fully penetrated biofilm. Curves are for different values of second Damköhler number.

(e) Compute the flux of substrate into the biofilm. Hint: see Exercise 6.9 and 6.10. Answer: $J_A^*|_{z=0} = k_A^0 L$.

6.13. Formica et al. (1988) studied the diffusion of PCBs into sediment from Lake Wedington, Arkansas. Sediment samples were placed in the bottom of a tank, and water containing PCB was gently circulated in the tank and maintained at 8–10°C. The concentration of PCB in the water was maintained constant through periodic addition of more PCB. The profile below shows the PCB concentration in the sediments after 43 days. The PCB concentration at the water–sediment interface was 160 ng/g (that's 160 ng per gram of sediment).

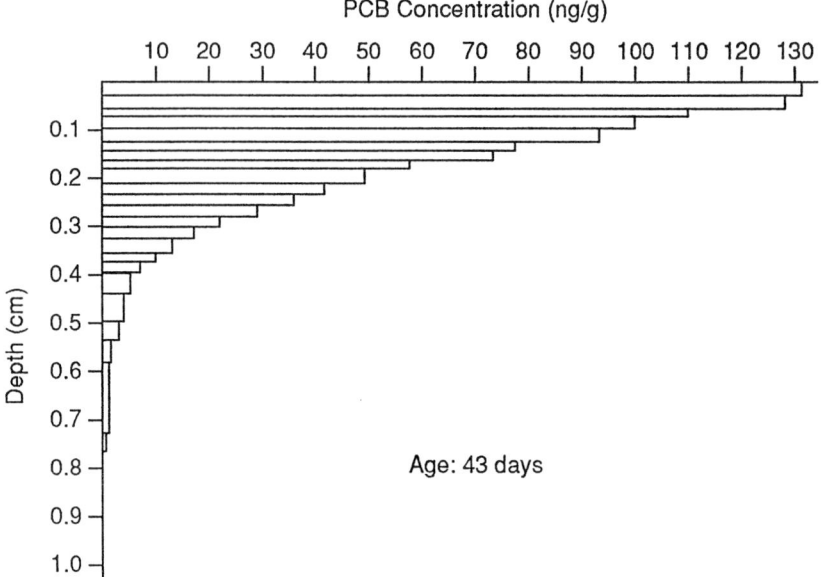

Exercise 6.13. *Source*: Reprinted with permission from Formica, Baron, Thibodeaux, and Valsaraj, *Environ. Sci. Technol.*, **22**, 12, 1437 (1988). Copyright 1988, American Chemical Society.

(a) Using the results of Section 6.7.3, compute the effective diffusion coefficient of PCB in the lake sediments. Present diffusion coefficient in units of square centimeters per second. Compare your concentration profile in the sediments with the data of Formica et al.

(b) Develop an equation to compute the flux of PCB across the sediment–water interface, and estimate the flux at 43 days using the results from part a (recall Eq. 6.103). You will probably need to look up Liebnitz' rule in a calculus text.

6.14. Imagine that the problem of Section 6.7.3 describes absorption of a gas by a liquid. Show that the mass of A absorbed per unit interfacial area is

$$M(t) = c_{A,0} \left(\frac{4 D_{AB} t}{\pi} \right)^{1/2}$$

6.15. Consider the diffusion problem of Section 6.7.3, but with the addition of a homogeneous, first-order reaction in the diffusion domain. In this case, Fick's second law is given as

$$\frac{\partial c_A}{\partial t} = D_{AB} \frac{\partial^2 c_A}{\partial x^2} - k_A c_A$$

For the boundary conditions shown in Figure 6.11, Danckwerts (1970) has shown that the solution for c_A is

$$\frac{c_A}{c_{A,0}} = \frac{1}{2}\exp\left[-x\sqrt{\frac{k_A}{D_{AB}}}\right] * \text{erfc}\left[\frac{x}{2\sqrt{D_{AB}t}} - \sqrt{k_A t}\right]$$

$$+ \frac{1}{2}\exp\left[x\sqrt{\frac{k_A}{D_{AB}}}\right] * \text{erfc}\left[\frac{x}{2\sqrt{D_{AB}t}} + \sqrt{k_A t}\right]$$

(a) Show that the flux of A at $x = 0$ is

$$J_{x=0} = c_{A,0}\sqrt{D_{AB}k_A}\left[\text{erf}\sqrt{k_A t} + \frac{1}{\sqrt{\pi k_A t}}\exp(-k_A t)\right]$$

(b) Show that the mass of A absorbed per unit interfacial area is

$$M(t) = c_{A,0}\sqrt{\frac{D_{AB}}{k_A}}\left[\left(k_A t + \frac{1}{2}\right)\text{erf}(\sqrt{k_A t}) + \sqrt{\frac{k_A t}{\pi}}\exp(-k_A t)\right]$$

(c) Considering the equations above and the answer to Exercise 6.14, what is the effect of the reaction on the amount of A absorbed? The ratio of the mass absorbed with reaction to the amount of A absorbed without reaction is sometimes referred to as the *effectiveness factor*.

6.16. Section 6.7.4 deals with the diffusion and partitioning of a solute from a well-stirred solution into a homogeneous sphere of radius a, the sphere having a uniform diffusion coefficient D_{AB}. Equation 6.107 has been useful in engineering studies of adsorption, and we wish here to see if it can be used to model adsorption from solution by powdered activated carbon (PAC). For example, if a small grain of activated carbon is modeled as a sphere with a uniform effective diffusion coefficient D_{eff} (Section 6.8), the mass $M_{A,t}$ of some solute (adsorbate A) adsorbed from a solution may be modeled with Eq. 6.107 where we substitute D_{eff} for D_{AB}. (This is quite a bold assumption, because we know that activated carbon particles have an inhomogeneous internal microstructure composed of fine pores where, in fact, all the adsorption occurs.) We might hope then that Eq. 6.107 could be used to model the *kinetics* of adsorption by PAC. (This knowledge is critical for the design of real adsorption equipment in environmental control.) In this problem, successful modeling will hopefully yield an effective diffusion coefficient for this carbon and adsorbate.

Some important questions must be asked before we attempt to solve this problem. First, the solution Eq. 6.107 is for a single adsorbing sphere in a well-mixed liquid. However, we cannot run adsorption studies with a single grain of activated carbon; hence, we need to think about the conditions under which Eq. 6.107 might still be useful. If we want to use the results of this section in a typical batch adsorption test, we must find a way to adapt the formulas and parameters of this section for our test.

So consider that each grain of activated carbon is associated with a fraction of the total fluid volume V. Then this small volume V' would be

$$V' = \frac{V}{\text{number of PAC particles in volume } V}$$

(a) Now show that

$$\alpha' = \frac{3V'}{4\pi a^3 K_p} = \frac{c_{A,\infty}}{c_{A,0} - c_{A,\infty}}$$

is the steady-state (equilibrium) ratio of the mass of solute (adsorbate) in the solution to the mass of solute in the sphere; K_p is a partition coefficient which expresses whether the equilibrium distribution of solute favors either the sphere or the solution (see Eq. 6.111).

(b) Prove a slightly different form of Eq. 6.116,

$$\frac{M_{A,\infty}}{V' c_{A,0}} = \frac{1}{1 + \alpha'} = \frac{c_{A,0} - c_{A,\infty}}{c_{A,0}}$$

by writing a steady-state mass balance on the solute.

(c) Now show using a mass balance that

$$M_{A,\infty} = V'(c_{A,0} - c_{A,\infty})$$
$$M_{A,t} = V'(c_{A,0} - c_{A,t})$$

and

$$\frac{M_{A,t}}{M_{A,\infty}} = \frac{c_{A,0} - c_{A,t}}{c_{A,0} - c_{A,\infty}}$$

(d) The following kinetic data were collected in a study of the adsorption of TCP by PAC.

Time (min)	$c_{A,t}/c_{A,0}$
0	1.0
2	0.605
4	0.507
6	0.437
8	0.387
10	0.323
15	0.349
20	0.305
25	0.249
30	0.241
40	0.199
50	0.190
60	0.173
75	0.162
90	0.148
120	0.136
150	0.127
180	0.113
215	0.118

Other important data for this test are:

$c_{A,0} = 450\,\mu g\,\text{TCP/L}$
PAC dose $= 15\,\text{mg/L}$
PAC particle diameter $= 10\,\mu m$
PAC density $= 0.74\,\text{g/cm}^3$
Reactor volume (fluid) $= 2\text{L}$

Using the data and formulas above, and the results of Section 6.7.4, develop a method for determining the effective diffusion coefficient for the PAC. You may estimate $c_{A,\infty}$ by considering the apparent asymptotic concentration from the kinetic data. You will have to develop a procedure to optimize your choice of D_{eff}. This can be done by trial and error in a spreadsheet, or by some kind of optimization routine (also in spreadsheets). Essentially, you need to pick a D_{eff} that tends to result in a best fit of experimental $M_{A,t}/M_{A,\infty}$ values to your calculated $M_{A,t}/M_{A,\infty}$ values. Finally, on a graph, compare your predictions to the measured $M_{A,t}/M_{A,\infty}$. This type of adsorption problem is not normally modeled with a linear isotherm. However, the results of this problem indicate that the linear isotherm assumption works pretty well for modeling the adsorption kinetics.

6.17. Wu and Gschwend (1986) studied the adsorption of hydrophobic organic chemicals on sediment particles from the Charles River in Massachusetts. The following data show the kinetics of adsorption of tetrachlorobenzene on sediments of two different sizes, 232 and 96 μm.

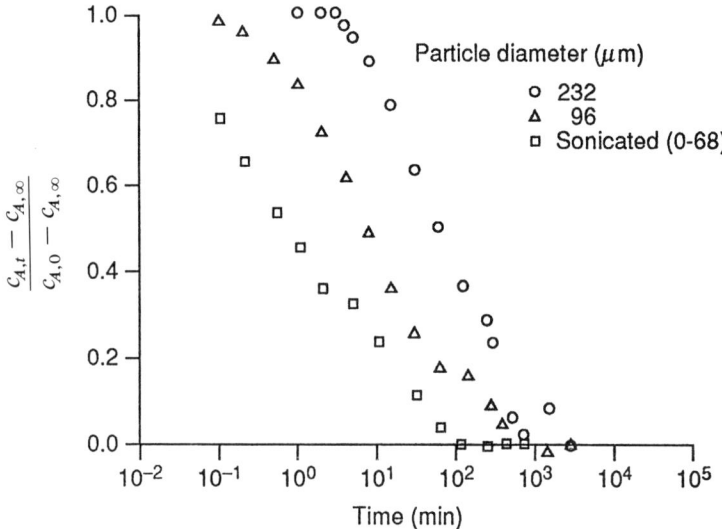

Exercise 6.17. *Source*: Reprinted with permission from Wu and Gschwend, *Environ. Sci. Technol.*, **20**, 7, 721 (1986). Copyright 1986, American Chemical Society.

The following data are also available:

232-μm sediment sample   Mass concentration solids = 442 mg/L
                                Partition coefficient = 1390 cm^3/g

96-μm sediment sample   Mass concentration solids = 1030 mg/L
                               Partition coefficient = 1520 cm^3/g

Using the data given above and the approach of Section 6.7.4, determine the effective diffusion coefficients for the sediment samples. Hints: The y axes of Wu and Gschwend's plots are equal to

$$1 - \frac{c_{A,0} - c_{A,t}}{c_{A,0} - c_{A,\infty}} = 1 - \frac{M_{A,t}}{M_{A,\infty}} = \frac{c_{A,t} - c_{A,\infty}}{c_{A,0} - c_{A,\infty}}$$

You will need to think about the appropriate V in this problem to compute the correct α. For example, consider that each sediment grain is associated with its own subvolume of water and that the sum of all the subvolumes is equal to the total system volume (see Exercise 6.16).

6.18. Show using Fick's first law and the analysis of Section 6.6.2 that at a particle surface where there is a heterogeneous reaction of unknown order n, the following must be true at steady state:

$$4\pi r^2 (k_{A,a} c_{A,a}^n) = 4\pi r^2 D_{AB} \left(\frac{\partial c_A}{\partial r}\right)_{r=a}$$

Next, show through nondimensionalization of the equation with $\tilde{r} = r/a$ and $\tilde{c}_{A,a} = c_{A,a}/C_0$ that the general steady diffusion equation becomes

$$\left(\frac{D_{AB}}{k_{A,a} R C_0^{n-1}}\right) \frac{\partial \tilde{c}_{A,a}}{\partial \tilde{r}} - \tilde{c}_{A,a}^n = 0$$

Note: the term in parentheses is a general form of the second Damköhler number, good for heterogeneous reactions other than first order.

6.19. Another common boundary condition for the type of problem examined in Section 6.7.4 is a constant surface concentration. For example, when adsorption occurs in a particle suspended in a perfectly mixed, continuous reactor, it can be assumed that the concentration of adsorbate is constant in the fluid outside the particle(s). For a constant $c_{A,0}$, Crank (1956) shows that the mass adsorbed over time is given by

$$\frac{M_{A,t}}{M_{A,\infty}} = 1 - \frac{6}{\pi^2} \sum_{n=1}^{\infty} \frac{1}{n^2} \exp\left(-\frac{n^2 \pi^2 t}{\tau_d}\right)$$

where all terms are defined in Section 6.7.4. Comment: note that the adsorption kinetics here are independent of the partition coefficient.

(a) Considering a diffusion coefficient of $1 \times 10^{-10}\,\mathrm{cm}^2/\mathrm{s}$, plot the equation above versus t/τ_d for particles of diameter 10, 1, and $0.1\,\mu\mathrm{m}$. Hence, show the effect of particle size on adsorption kinetics. Hint: it helps to make the plot on semilog paper.
(b) Explain in your own words why particle size affects adsorption as illustrated in your plot.

REFERENCES

Adham, S. S., Snoeyink, V. L., Clark, M. M., and Bersillon, J. L., "Prediction and Verification of the Performance of Powdered Activated Carbon for Removal of Organic Compounds in the PAC/UF Process, "*J. Am. Water Works Assoc.*, **83**, 10 (1991).

Atkins, P. W., *Physical Chemistry*, W. H. Freeman, San Francisco, CA, 1978.

Ball, W. P., and Roberts, P. V., "Long-Term Sorption of Halogenated Organic Chemicals by Aquifer Material: 2. Intraparticle Diffusion," *Env. Sci. Technol.*, **25**, 1237–1249 (1991).

Bird, R. B., Stewart, W. E., and Lightfoot, E. N., *Transport Phenomena*, Wiley, New York, 1960.

Characklis, W. B. and Marshall, K. C., *Biofilms*, Wiley-Interscience, New York, 1990.

Crank, J., *The Mathematics of Diffusion*, Clarendon Press, Oxford, 1956.

Danckwerts, P. V., *Gas–Liquid Reactions*, McGraw-Hill, New York, 1970.

Fisher, H. B., List, E. J., Koh, R. C. Y., Imberger, J., and Brooks, N. H., *Mixing in Inland and Coastal Waters*, Academic, New York, 1979.

Formica, S. J., Baron, J. A., Thibodeaux, L. J., and Valsaraj, K. T., "PCB Transport into Lake Sediments: Conceptual Model and Laboratory Simulation," *Env. Sci. Technol.*, **22**, 1435–1440 (1988).

Lyman, W. J., Reehl, W. F., and Rosenblatt, D. H., *Handbook of Chemical Property Estimation Methods*, American Chemical Society, Washington, DC, 1982.

Rees, K. C. J., Sudicky, E. A., Rao, P. S. C., and Reddy, K. R., "Evaluation of Laboratory Techniques for Measuring Diffusion Coefficients in Sediments," *Env. Sci. Technol.*, **25**, 1605–1611 (1991).

Reid, R. C., and Sherwood, T. K., *The Properties of Gases and Liquids: Their Estimation and Correlation*, McGraw-Hill, New York, 1966.

Seinfeld, J. H., *Atmospheric Chemistry and Physics of Air Pollution Control*, Wiley-Interscience, New York, 1986.

Spiegel, M. R., *Mathematical Handbook of Formulas and Tables (Schaum's Outline Series)*, McGraw-Hill, New York, 1968.

Thibodeaux, L. J., *Chemodynamics: Environmental Movement of Chemicals in Air, Water, and Soil,* Wiley, New York, 1979.

Welty, J. R., Wicks, C. E., and Wilson, R. E., *Fundamentals of Momentum, Heat, and Mass Transport*, Wiley, 1976.

Wu, S., and Gschwend, P. M., "Sorption Kinetics of Hydrophobic Organic Compounds to Natural Sediments and Soils," *Env. Sci. and Technol.*, **20**, 717–725 (1986).

BIBLIOGRAPHY

Brezonik, P. L., *Chemical Kinetics and Process Dynamics in Aquatic Systems*, Lewis Publishers, Boca Raton, FL, 1994.

Flagan, R. C. and Seinfeld, J. H., *Fundamentals of Air Pollution Engineering*, Prentice-Hall, Inc. Englewood Cliffs, NJ, 1988.

Freeze, R. A. and Cherry, J. A., *Groundwater*, Prentice-Hall, Englewood Cliffs, NJ, 1979.

Friedlander, S. K., *Smoke, Dust, and Haze*, Wiley, New York, 1977.

Hiemenz, P. C., *Principles of Colloid and Surface Chemistry*, Marcel Dekker, New York, 1986.

Levich, V. G., *Physicochemical Hydrodynamics*, Prentice-Hall, Englewood Cliffs, NJ, 1962.

Porter, M. C., "Concentration Polarization with Membrane Ultrafiltration," *Ind. Eng. Chem. Prod. Res. Dev.*, **11**, 3:234–248 (1972).

Probstein, R. F., *Physicochemical Hydrodynamics: An Introduction*, Butterworths, Boston, MA, 1989.

Treybal, R. E., *Mass Transfer Operations*, McGraw-Hill, New York, 1968.

7

CONVECTIVE DIFFUSION, DISPERSION, AND MASS TRANSFER

7.1. INTRODUCTION AND SIMPLE EXAMPLE OF CONVECTIVE DIFFUSION

In Chapter 6, we noted that in many mass-transfer problems of environmental interest, we may need to consider the combined effects of convection (or movement with the flow) and diffusion. We suggested that this was especially true at larger characteristic scales, where diffusion by itself might not be very efficient in transporting mass. The purpose of this chapter is to provide the student with a fundamental background in the different aspects of convective diffusion and to introduce approaches to determine mass transfer between solids, liquids, and gases. The chapter also discusses turbulent mass transport, which is so important in all but the smallest-scale environmental-transport phenomena.

The following simple example helps to demonstrate the impact and importance of the combined action of convection and diffusion. Consider the two gas-absorption systems shown in Figure 7.1. In Figure 7.1a, gas A and liquid B are contacted, and the gas is allowed to diffuse into the liquid. Gas A is somehow maintained at the same pressure (concentration); thus, the concentration of A at the liquid surface is a constant. In Figure 7.1b, the gas can also diffuse into the liquid, but fresh liquid is continually being transported at velocity U into the gas container (fresh surface is continually "renewed"). In this system also, the concentration of A in the container is maintained constant. The system shown in Figure 7.1a is essentially the diffusion problem analyzed in Section 6.7.3 (diffusion in a semi-infinite medium with constant boundary concentration) so long as the diffusing substance does not penetrate very far into the liquid.* The solution is

---

*What we mean here is that in the problem discussed in Section 6.7.3, the domain of diffusion was essentially infinite (or semi-infinite, as explained in the development). Clearly, the liquid depth in Figure 7.1a is not infinite. However, so long as the diffusing substance does not penetrate very far into the liquid, it does not "feel" the effect of the bottom of the liquid container.

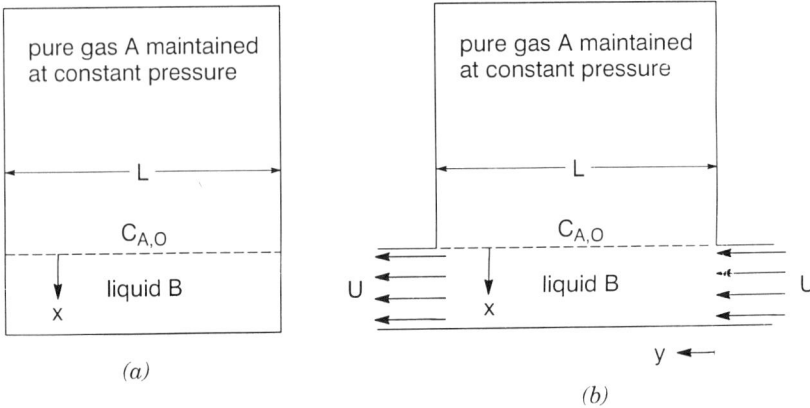

Figure 7.1. Two systems for absorbing a gas into a liquid: (*a*) classic one-dimensional diffusion with a constant boundary concentration, and (*b*) diffusion into a moving liquid stream (surface renewal).

$$c_A = c_{A,0} \, \mathrm{erfc} \left(\frac{x}{2\sqrt{D_{AB}t}} \right) \tag{6.96}$$

where x in the current problem is shown in Figure 7.1, and $c_{A,0}$ is the concentration of diffusing gas at $x = 0$. As demonstrated in Exercise 6.13b, the flux of A into B is

$$J_A = -D_{AB} \left. \frac{dc_A}{dx} \right|_{x=0} = c_{A,0} \sqrt{\frac{D_{AB}}{\pi t}} \tag{7.1}$$

Hence, the total mass transferred per unit width (into the page) of surface between the gas and liquid at any time is obtained by integrating Eq. 7.1 over time (and multiplying by L):

$$\left\{ \begin{array}{l} \text{total mass of } A \text{ transferred} \\ \text{per unit width into page} \end{array} \right\} = 2c_{A,0}L \sqrt{\frac{D_{AB}t}{\pi}} \tag{7.2}$$

Thus, for this system, the mass transferred increases as the square root of time. Now consider the surface-renewal example, Figure 7.1*b*. At $y = 0$, the liquid contains no diffused gas (no chance to be exposed to the gas). It is the same condition as $t = 0$ in the previous system. Fluid at other distances y has had a finite time in which A can diffuse into B. In fact, at steady state, gas A has had time $t = y/U$ in which to diffuse into liquid B at any distance y in Figure 7.1*b*. Hence, the concentration profile in the x direction at any y in Figure 7.1*b* is the same as Eq. 6.96 if we replace t with y/U:

$$c_A = c_{A,0}\,\text{erfc}\left(\frac{x}{2\sqrt{D_{AB}(y/U)}}\right) \tag{7.3}$$

We have the same restriction here as in the previous case, that A does not diffuse very far into B. In analogy to Eq. 7.1, the mass flux of A into the liquid at any point y is

$$J_A(y) = c_{A,0}\sqrt{\frac{D_{AB}U}{\pi y}} \tag{7.4}$$

The average mass flux (per unit width into the page) of A into the liquid is simply the integral of Eq. 7.4 over the distance L times the distance L:

$$\left\{\begin{array}{c}\text{average mass flux}\\\text{of } A \text{ per unit width}\\\text{into the page}\end{array}\right\} = L\left(\frac{\int_0^L J(y)dy}{L}\right) = c_{A,0}\sqrt{\frac{D_{AB}U}{\pi}}\int_0^L y^{-1/2}\,dy$$

$$= 2c_{A,0}\sqrt{\frac{D_{AB}UL}{\pi}} \tag{7.5}$$

The total mass transferred per unit width of gas–liquid interface at steady state is obtained by integrating over time:

$$\left\{\begin{array}{c}\text{total mass of } A \text{ transferred}\\\text{per unit width into page}\end{array}\right\} = 2c_{A,0}\sqrt{\frac{D_{AB}UL}{\pi}}\int_0^t dt' = 2c_{A,0}\sqrt{\frac{D_{AB}ULt^2}{\pi}} \tag{7.6}$$

Hence, for the system in Figure 7.1b, the steady-state mass transfer is proportional to time. The second system here is a very simple example of *convective diffusion*, or diffusion aided by convection. It is a more effective system for transporting mass than the pure diffusion system.

This example shows that for small times, diffusion alone can be a very effective mass-transfer mechanism. For longer times, diffusion aided by convection greatly exceeds pure diffusive mass transfer. This idea is related to a principle developed in Example 6.3 (Chapter 6). There we found that when the diffusion distance was small, the diffusion time scale is small and diffusion was very effective in transporting mass. When the diffusion distance was greater, the diffusion time scale increased greatly (as the square of the diffusion distance) and diffusion was a very ineffective mass-transport mechanism.

Finally, we have been slightly deceptive in this initial demonstration of convective diffusion. First, note that Eq. 7.3 implies that for any x, there is a positive concentration gradient in the positive y direction. Fick's first law would then seem to suggest that there should also be a diffusive mass transport in the negative y direction. We seem to have ignored a mass-transport mechanism. It turns out, though, that our solution is not necessarily wrong. If the velocity U is high enough, we can expect that the mass transport in the y direction due to convection greatly outweighs diffusive mass transport in the negative y direction. We can make a crude analysis of the situation using the idea of characteristic time scales, which has already aided us at several junctures in this book. The time scale for convective mass transport in the y direction is estimated as L/U. The time scale for diffusive mass transport in the y direction is estimated as L^2/D_{AB}. Taking the ratio of these two time scales yields a dimensionless ratio:

$$\frac{\left\{\begin{array}{l}\text{diffusive mass transfer time}\\ \text{scale in streamwise direction}\end{array}\right\}}{\left\{\begin{array}{l}\text{convective mass transfer time}\\ \text{scale in streamwise direction}\end{array}\right\}} = \frac{\dfrac{L^2}{D_{AB}}}{\dfrac{L}{U}} = \frac{LU}{D_{AB}} \tag{7.7}$$

This parameter grouping is very common in convective-diffusion problems and is called the *Peclet number*.

$$Pe = \frac{LU}{D_{AB}} \tag{7.8}$$

Hence when L and/or U are high enough, or when D_{AB} is small enough, the time ratio is large and we can ignore diffusion in the *streamwise* (y) direction. (Note that this does not say that we can ignore diffusive mass transport in the x-direction.) By ignoring streamwise diffusive mass transport, we have effectively said that x-direction diffusive mass transport is balanced by y-direction convective mass transport. This aspect of the problem is reviewed further in Exercise 7.1.

A second possible problem with the simple example given above is that we have assumed that the concentration at the interface was always $c_{A,0}$. However, for high mass-transfer rates, there can also be a concentration gradient in the gas near the interface. We return to this question again in our discussion of mass transfer from the gas phase to liquid droplets (Section 7.9.1).

7.2. THE CONVECTIVE-DIFFUSION EQUATION

Recall from Chapter 6 that the diffusion equation (Fick's second law) was derived by first writing a mass balance equation, and then substituting in a form of Fick's first law with $\mathbf{u} = 0$. We can also write a mass balance equation in terms of the more general flux $\mathbf{n}_A$ of Chapter 6, and substitute into this equation for the $\mathbf{n}_A$ expressed in Eq. 6.18 (i.e., where $\mathbf{u}$ is not equal to zero). This will yield an important new equation. Consider again the differential element of Figure 6.9, and replace all j_A's with their equivalent n_A's. Following the procedure leading to Eq. 6.82, we find that the mass balance yields

$$\frac{\partial \rho_A}{\partial t} \Delta x \Delta y \Delta z = (n_{A,x} - n_{A,x+\Delta x})\Delta y \Delta z + (n_{A,y} - n_{A,y+\Delta y})\Delta z \Delta x$$

$$+ (n_{A,z} - n_{A,z+\Delta z})\Delta y \Delta x + r_A \Delta x \Delta y \Delta z \tag{7.9}$$

Note that we have also added to the analysis the possibility that A is either produced or consumed by a homogeneous chemical reaction, $r_A \Delta x \Delta y \Delta z$.* The units of r_A here are $M\text{-}L^{-3}\text{-}T^{-1}$. Dividing through by $\Delta x \Delta y \Delta z$ and taking the limit as the differential element shrinks to zero, we find, in analogy to Section 6.7.1:

$$\frac{\partial \rho_A}{\partial t} + \nabla \cdot \mathbf{n}_A = r_A \tag{7.10}$$

Substituting Eq. 6.18 for $\mathbf{n}_A$ we find, with slight rearrangement:

$$\frac{\partial \rho_A}{\partial t} + \mathbf{u} \cdot \nabla \rho_A + \rho_A(\nabla \cdot \mathbf{u}) = \nabla \cdot (\rho D_{AB} \nabla \omega_A) + r_A \tag{7.11}$$

If ρ is constant, the fluid is incompressible, that is, $\nabla \cdot \mathbf{u} = 0$ (Eq. 1.24); then, for constant D_{AB},

$$\frac{\partial \rho_A}{\partial t} + \mathbf{u} \cdot \nabla \rho_A = D_{AB} \nabla^2 \rho_A + r_A \tag{7.12}$$

Finally, dividing through by the molar mass of component A, we find:

$$\frac{\partial c_A}{\partial t} + \mathbf{u} \cdot \nabla c_A = D_{AB} \nabla^2 c_A + R_A \tag{7.13}$$

where c_A has the units of mol/L^3 and R_A has units of $\text{mol-}L^{-3}\text{-}T^{-1}$). Equations 7.11–7.13 are different versions of the so-called *convective-diffusion equation* in rectangular coordinates (see Appendix III for a list of equations applicable to other coordinate systems). One of these three equations will usually sufficiently describe problems in liquids where component A is subject to transport by the

---

*By homogeneous phase reaction we mean a reaction throughout the flow domain (where A is present). This is in contrast to heterogeneous reactions (i.e., reactions at surfaces) where the effect of the reaction is not accounted for in the convective-diffusion equation, but rather in the boundary conditions (see also Section 6.6.2).

fluid flow, diffusion, and reaction. Note that the term on the left-hand side of the equation involving the dot product is analogous to the convective-acceleration terms in the Navier–Stokes equation (Chapter 5). It is called the *convective-transport* term. The term on the right-hand side involving the Laplacian is of course the same term occurring in the diffusion equation (Eq. 6.87). It is called the *diffusive-transport* term.

We can complete a similar analysis in terms of N_A and show that in analogy to Eq. 7.11,

$$\frac{\partial c_A}{\partial t} + \mathbf{u}^* \cdot \nabla c_A + c_A(\nabla \cdot \mathbf{u}^*) = \nabla \cdot (c D_{AB} \nabla x_A) + R_A \qquad (7.14)$$

For constant c and D_{AB} we find:

$$\frac{\partial c_A}{\partial t} + \mathbf{u}^* \cdot \nabla c_A + c_A(\nabla \cdot \mathbf{u}^*) = D_{AB} \nabla^2 c_A + R_A \qquad (7.15)$$

Equation 7.15 is another form of the convective-diffusion equation, developed in terms of the molar flux. The following form of Eq. 7.15 is sometimes of use in binary gas systems (Bird et al., 1960, and Exercise 7.4):

$$\frac{\partial c_A}{\partial t} + \mathbf{u}^* \cdot \nabla c_A = D_{AB} \nabla^2 c_A + R_A - \frac{c_A}{c}(R_A + R_B) \qquad (7.16)$$

In Chapter 5, we found that some important insights into the Navier–Stokes equation were discovered through nondimensionalization of the equation. We now attempt the same type of analysis with the convective-diffusion equation. Define the following dimensionless variables:

$$\tilde{u} = \frac{\mathbf{u}}{U}; \qquad \tilde{x} = \frac{\mathbf{x}}{L}; \qquad \tilde{t} = \frac{tU}{L}; \qquad \tilde{c}_A = \frac{c_A}{C_A} \qquad (7.17)$$

where U, L, and C_A are characteristic velocity, length, and concentration scales, respectively. If we ignore any reactions, substitution of the dimensionless variables into Eq. 7.13 yields the following equation (see Exercise 7.5):

$$\frac{\partial \tilde{c}_A}{\partial \tilde{t}} + \tilde{u} \cdot \tilde{\nabla} \tilde{c}_A = \frac{1}{\mathrm{Pe}} \tilde{\nabla}^2 \tilde{c}_A \qquad (7.18)$$

where $\tilde{\nabla}$ is defined in Exercise 5.10, and Pe is the *Peclet number*:

$$\mathrm{Pe} = \frac{LU}{D_{AB}} \qquad (7.8)$$

As pointed out in Section 7.1, the Peclet number can be considered the ratio of diffusive and convective time scales:

$$Pe = \frac{LU}{D_{AB}} = \frac{L^2/D_{AB}}{L/U} = \frac{\tau_d}{\tau_c} \tag{7.19}$$

When the Peclet number is large, certain diffusive-transport terms can sometimes be ignored in comparison to the convective-transport terms; when the Peclet number is small, the convective terms can sometimes be ignored in comparison to certain diffusive terms. In the latter case, we may only need to solve the diffusion equation. The following sections provide further evidence of the importance of the Peclet number in convective-diffusion and mass-transfer problems.

7.3. MASS TRANSPORT IN STEADY LAMINAR FLOW IN A CYLINDRICAL TUBE

Taylor (1953) was interested in how a small slug of miscible tracer* would be transported in another liquid when both liquids were subject to laminar flow in a tube. For example, consider Figure 7.2. At t_0, a thin slug of tracer has been instantaneously placed in the cross section of laminar tube flow. (Poiseuille flow, Section 5.6). In this first case, we assume there is no molecular diffusion; hence, the slug of tracer is simply stretched over time. Concentration within the tracer region is constant, that is, the tracer region is stretched, but the tracer is not diluted. It is a fairly simple matter to compute the coordinates of the tracer profile. Recall from Exercise 5.5 that the velocity profile in the tube can be expressed in terms of the average velocity

$$u_z = 2u_{z,ave} \left[1 - \left(\frac{r}{a} \right)^2 \right] \tag{7.20}$$

Since u_z is equal to dz/dt, we can substitute into Eq. 7.20 for u_z, separate, and integrate. For the case of the "trailing edge" of the tracer region, the initial condition is $z = z_0$ at $t = t_0$; hence,

$$z = z_0 + 2u_{z,ave} \left[1 - \left(\frac{r_t}{a} \right)^2 \right] t \tag{7.21}$$

---

*An ideal tracer is a chemically inert compound that has properties similar to the carrier fluid (especially density and viscosity in liquids). The ideal tracer can easily be distinguished from the carrier fluid by some simple technique. In liquids, typical tracers have been dyes that either fluoresce or adsorb UV light, hydrogen bubbles, radiolabeled compounds, or dilute salts. In gases, gas chromatograph and nondispersive infrared detectors have been used to measure concentrations of a variety of gas tracers (Nauman and Buffham, 1983).

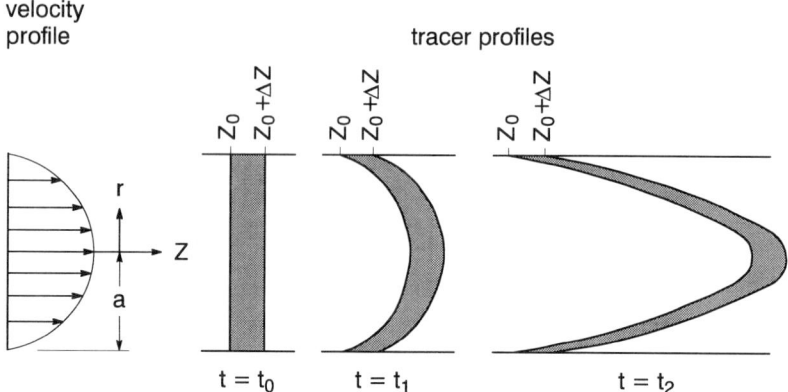

Figure 7.2. Tracer profiles in the absence of diffusion in a small-diameter tube subject to laminar flow.

For the "leading edge" of the tracer region, the boundary condition is $z = z_0 + \Delta z$ at $t = 0$; hence,

$$z = z_0 + \Delta z + 2u_{z,\text{ave}} \left[1 - \left(\frac{r_l}{a} \right)^2 \right] t \tag{7.22}$$

Figure 7.3 shows the slug after some time has passed. If the concentration of tracer is c_0, the radially averaged tracer concentration anywhere in the domain

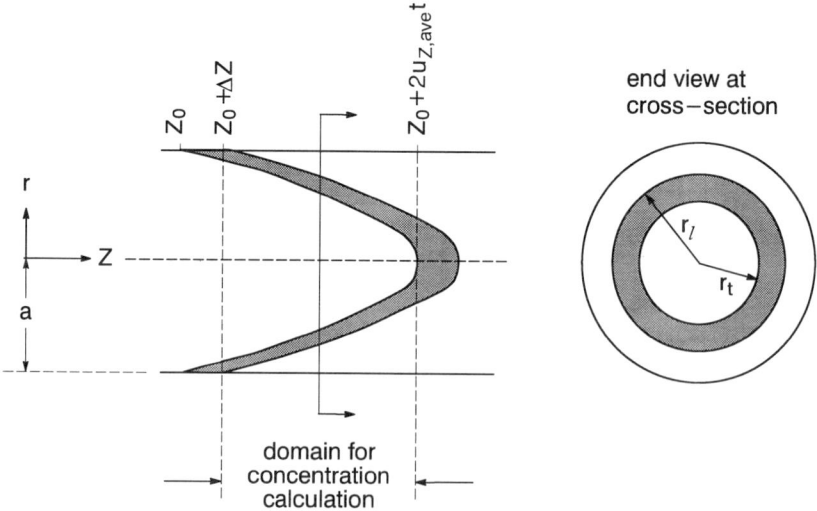

Figure 7.3. Nomenclature for calculation of mean concentration during purely convective tracer transport.

$z_0 + \Delta z < z < z_0 + 2u_{z,ave}t$ is simply the fraction of cross-sectional area occupied by the tracer times c_0,

$$c_{ave} = \frac{c_0(\pi r_l^2 - \pi r_t^2)}{\pi a^2} = c_0 \left[\left(\frac{r_l}{a}\right)^2 - \left(\frac{r_t}{a}\right)^2 \right] \tag{7.23}$$

Rearranging Eqs. 7.21 and 7.22 and substituting into Eq. 7.23, we find, after performing some minor algebra, that in the region $z_0 + \Delta z < z < z_0 + 2u_{z,ave}t$:

$$c_{ave} = c_0 \frac{\Delta z}{2tu_{z,ave}} \tag{7.24}$$

In other words, for a fixed time, the concentration in the stated region is constant; alternatively, the mean concentration in the stated region decreases as time increases (and the tracer is stretched further and further).

There are some consistencies and inconsistencies between this simple model and our experience of mixing tracers in pipes (and the environment). First, consider a lab experiment in which we inject identical slugs of tracer into several pipes of the same diameter but increasing lengths. The flow in all pipes is the same. Experience (or intuition) tells us that the peak concentration measured at the end of the pipe will decrease as the pipe becomes longer. We have the notion that there is greater mixing of the tracer in longer pipes, therefore, the concentration measured at the end of the pipe is lower. Equation 7.24 is consistent with this notion, since for the same $u_{z,ave}$, t would be greater and c_{ave} would be lower in the longer pipes. Next, consider the arrival time of tracer at the end of the pipe. For example, for a pipe of length L, the leading edge of the tracer region reaches the end of the pipe at an arrival time specified by

$$L = z_0 + \Delta z + 2u_{z,ave}t_{arr} \tag{7.25}$$

or rearranging,

$$t_{arr} = \frac{L - (z_0 + \Delta z)}{2u_{z,ave}} \tag{7.26}$$

Equation 7.26 predicts that the arrival time increases linearly with the pipe length. Although it is true that the arrival time of real concentration peaks increases linearly with the pipe length, the actual arrival time of the concentration peak can be considerably later than implied by Eq. 7.26. Another problem with the purely convective solute-transport model is that in some situations, the concentration profile, and particularly the peak concentration, are poorly predicted. For example, the model suggests that for an initially thin tracer slug, the *peak concentration* occurs close to the arrival time estimated above, and that afterward, the concentration at the end of the pipe decreases as a function of t^{-1}. Both of these predictions can be erroneous. Interestingly enough, similar experiments have shown that for sufficiently long times, concentration profiles at the end of the pipe are often bell or Gaussian shaped, just like the curves shown in Figure 6.10, and the tracer slug travels down the pipe at the mean-flow velocity. Recall that the model development shown above assumes there is no molecular diffusion. Hence, it is likely that neglect of diffusion is leading to some of these discrepancies. In fact, Exercise 7.6 shows that the analysis is only accurate when the Peclet number, $2aU/D_{AB}$, is much greater than L/a, where L is the characteristic streamwise length scale over which significant changes in concentration occur.

Taylor (1953) was aware of these phenomena and made a classic analysis of tracer transport in laminar tube flow with diffusion. Before getting into that analysis, consider again the tracer profile of Figure 7.3. As noted above, since there is no diffusion allowed, some of the tracer always remains attached to the wall at $z_0 < z < z_0 + \Delta z$. This is another way to think of the no-slip condition of fluid mechanics (Chapter 5). Probstein (1989) points out that when diffusion is present, the high concentration gradients at the trailing edge of the tracer near $z_0 < z < z_0 + \Delta z$ tend to diffuse mass toward the inside of the pipe. Therefore, mass is transported away from the slower-moving fluid near the wall to the faster-moving fluid closer to the center of the pipe. However, considering the leading edge of the tracer in the faster fluid, high concentration gradients near $z = z_0 + 2u_{z,\text{ave}}t$ tend to transport mass away from the faster fluid to the slower fluid. In effect, diffusion is working against the severe stretching in pure convective transport (above) and tends to consolidate the tracer.

The convective-diffusion equation is given in Section 7.2 as Eq. 7.13. Expanding the dot product and Laplacian terms in cylindrical pipe coordinates results in (Appendix III)

$$\frac{\partial c_A}{\partial t} + u_r \frac{\partial c_A}{\partial r} + u_\theta \frac{1}{r}\frac{\partial c_A}{\partial \theta} + u_z \frac{\partial c_A}{\partial z} = D_{AB}\left[\frac{1}{r}\frac{\partial}{\partial r}\left(r\frac{\partial c_A}{\partial r}\right) + \frac{1}{r^2}\frac{\partial^2 c_A}{\partial \theta^2} + \frac{\partial^2 c_A}{\partial z^2}\right]$$

(7.27)

We have already assumed that the concentration profile is symmetrical about $r = 0$; hence, there can be no variation of concentration with angle θ. Also, there is of course no r or θ component of velocity in Poiseuille flow. Thus, Eq. 7.27 reduces to

$$\frac{\partial c_A}{\partial t} + u_z \frac{\partial c_A}{\partial z} = D_{AB}\left[\frac{1}{r}\frac{\partial}{\partial r}\left(r\frac{\partial c_A}{\partial r}\right) + \frac{\partial^2 c_A}{\partial z^2}\right]$$

(7.28)

Next, Taylor noted that for sufficiently high velocity and/or sufficiently low D_{AB}, the streamwise diffusion term (last term in Eq. 7.28) might be ignored in comparison with streamwise mass transport due to convection (second term on the left-hand side of Eq. 7.28) (also recall discussion in Section 7.1). If this is true, Eq. 7.28 reduces to

$$\frac{\partial c_A}{\partial t} + u_z \frac{\partial c_A}{\partial z} = D_{AB} \left[\frac{1}{r} \frac{\partial}{\partial r} \left(r \frac{\partial c_A}{\partial r} \right) \right] \tag{7.29}$$

It will help to transform the problem to a coordinate system moving with the average fluid velocity $u_{z,ave}$:

$$\tilde{z} = z - u_{z,ave} t \tag{7.30}$$

This means that the streamwise velocity relative to the moving coordinate system is

$$u_{\tilde{z}} = u_z - u_{z,ave} = u_{z,ave} \left[1 - 2 \left(\frac{r}{a} \right)^2 \right] \tag{7.31}$$

Taylor then made some judgments about the problem, which (considering their success) we can imagine were the result of a good mixture of experience, intuition, and genius. Taylor said that in the moving coordinate system, it would be likely that at a long enough time after the tracer injection, the smoothing of *radial* concentration gradients would be fast relative to any changes in concentration due to *axial* convective-mass transport. In other words, any creation of radial concentration gradients by the parabolic velocity field would quickly be smeared out by fast radial diffusion. We might say this situation corresponds to a radial diffusion time scale that is short relative to the axial-convective mass-transfer time scale. Keep in mind that this situation does not arise immediately after tracer injection, but at some time downstream (typically $t \gg a^2/D_{AB}$). There is an important modeling idea that we can call the *principle of time-scale separation*. This means that when the time scales of two simultaneous mass-transfer mechanisms are sufficiently different (separated in time), the overall process can be assumed to be in a local steady state where the fast process balances the slower one. The overall changes are limited by the slower of the two processes. In the current problem, Taylor then reasoned that the fast radial diffusion would quickly adjust to changes caused by the slower axial convective mass transport, and that changes due to axial mass transport would be slow enough for us to consider the process to be at a pseudosteady state.* In this case, we drop the time derivative (nonsteady part) in Eq. 7.29

---

*Note that we have already encountered the pseudosteady-state assumption at several junctures in this book. For example, when we considered the burning of a coal particle in Exercise 6.6, we assumed that the surface reaction was so fast that the process was controlled by diffusion. In other words, there was separation of the diffusion and reaction time scales. We also assumed that the rate of change of particle radius due to the oxidation reaction was slow enough that we could model mass transfer to the particle surface with a steady-state (diffusion-limited) mass-transfer model.

and consider the remaining left-hand side convective term to be a pseudoconstant. Thus, substituting Eq. 7.31 into Eq. 7.29:

$$u_{\bar{z}} \frac{\partial c_A}{\partial \bar{z}} = D_{AB} \left[\frac{1}{r} \frac{\partial}{\partial r} \left(r \frac{\partial c_A}{\partial r} \right) \right] \tag{7.32}$$

This simplified equation is a very interesting equation: It says that in a differential element at some r moving with the mean velocity, the convective mass transport in the streamwise direction (left-hand side of Eq. 7.32) is balanced by radial diffusive mass transport (right-hand side of Eq. 7.32). Equation 7.32 can be integrated fairly easily for the boundary conditions that $\partial c_A / \partial r = 0$ at $r = a$ (i.e., the impermeable boundary assumption) and $c_A = c_{A,0}$ at $r = 0$. Integrating Eq. 7.32 we find:

$$c_A = c_{A,0} + \frac{u_{z,\text{ave}} a^2}{4 D_{AB}} \left[\left(\frac{r}{a} \right)^2 - \frac{1}{2} \left(\frac{r}{a} \right)^4 \right] \frac{\partial c_A}{\partial \bar{z}} \tag{7.33}$$

It is now useful to compute the flux of A in the streamwise direction with respect to the moving coordinate system. We set this up as an integral over a differential ring $2\pi r dr$:

$$\tilde{J}_A = \frac{1}{\pi a^2} \int_0^a c_A u_{\bar{z}} 2\pi r dr \tag{7.34}$$

After a bit of work we find (see Exercise 7.7):

$$\tilde{J}_A = - \frac{u_{z,\text{ave}}^2 a^2}{48 D_{AB}} \frac{\partial c_A}{\partial \bar{z}} \tag{7.35}$$

Levich (1962) shows that when radial diffusion is fast enough, the streamwise gradients in c_A and the radially averaged concentration are the same:

$$\frac{\partial c_A}{\partial \bar{z}} = \frac{\partial c_{A,\text{ave}}}{\partial \bar{z}} \tag{7.36}$$

where $c_{A,\text{ave}}$ is average concentration of A across the pipe section. Thus, Eq. 7.35 can also be written as

$$\tilde{J}_A = - \frac{u_{z,\text{ave}}^2 a^2}{48 D_{AB}} \frac{\partial c_{A,\text{ave}}}{\partial \bar{z}} \tag{7.37}$$

From this equation, we see that Taylor's analysis seems to predict that mass transport relative to the coordinates moving with the mean velocity of the fluid is analogous to Fick's first law of Chapter 6:

$$\tilde{J}_A = -D_{dis} \frac{\partial c_{A,ave}}{\partial \tilde{z}} \tag{7.38}$$

where D_{dis} is the *dispersion coefficient* for convective diffusion in laminar pipe flow:

$$D_{dis} = \frac{u_{z,ave}^2 a^2}{48 D_{AB}} = D_{AB}\left(\frac{Pe^2}{48}\right) \tag{7.39}$$

Taylor (1954b) shows that this model, now called the *Taylor dispersion model*, is valid for a range of Peclet numbers corresponding to (see Exercises 7.8 and 7.9)

$$4\frac{L}{a} \gg Pe \gg 7 \tag{7.40}$$

where L is a characteristic streamwise length scale over which significant changes in concentration occur, and Pe here is the *Peclet number*, $u_{z,ave}a/D_{AB}$. Taylor dispersion is called dispersion because it combines the effects of molecular diffusion and convection.

Example 7.1. Explore the implications of Eq. 7.40 in terms of convective and diffusive time scales. Also determine the relevant velocity limits for application of Taylor's dispersion analysis in a 0.5-mm-diameter pipe for dispersion of (1) ammonia in air and (2) ammonia in water.

SOLUTION. Equation 7.40 indicates that for Taylor's analysis to be appropriate,

$$4\frac{L}{a} \gg Pe \gg 7$$

The left-hand side inequality requires that $Pe \ll 4L/a$, where Pe is the diffusion Peclet number. Therefore, this implies

$$Pe = \frac{\dfrac{a^2}{D_{AB}}}{\dfrac{a}{u_{z,ave}}} \ll \frac{4L}{a} \Rightarrow \frac{\dfrac{a^2}{D_{AB}}}{\dfrac{a}{u_{z,ave}}} \ll 4\frac{L}{a} \Rightarrow \frac{\dfrac{a^2}{D_{AB}}}{\dfrac{L}{u_{z,ave}}} \ll 4$$

Therefore, the last inequality implies that

$$\frac{\dfrac{a^2}{D_{AB}}}{\dfrac{L}{u_{z,\text{ave}}}} \ll 4 \Rightarrow \frac{\left\{\begin{array}{c}\text{radial diffusion}\\\text{time scale}\end{array}\right\}}{\left\{\begin{array}{c}\text{axial convection}\\\text{time scale}\end{array}\right\}} \ll 4$$

This last inequality makes sense in terms of Taylor's analysis; the reader will recall that the pseudosteady-state assumption required that radial diffusion be fast relative to changes caused by axial convection. Rearranging this last inequality we find that the upper limit on velocity implied by Eq. 7.40 is

$$u_{z,\text{ave}} \ll \frac{4LD_{AB}}{a^2}$$

For ammonia diffusion in air, $D_{AB} = 0.28 \text{ cm}^2/\text{s}$; for ammonia diffusion in water, $D_{AB} = 1.76 \times 10^{-5} \text{ cm}^2/\text{s}$ (Tables 6.1 and 6.3); hence,

$$\text{ammonia in air: } u_{z,\text{ave}} \ll \frac{4\left[0.28(\text{cm}^2/\text{s})\,L\right]}{(0.025\,\text{cm})^2} = 1800\,L\left(\frac{\text{cm}}{\text{s}}\right)$$

$$\text{ammonia in water: } u_{z,\text{ave}} \ll \frac{4[1.76 \times 10^{-5}\,(\text{cm}^2/\text{s})\,L]}{(0.025\,\text{cm})^2} = 0.113\,L\left(\frac{\text{cm}}{\text{s}}\right)$$

Considering the right-hand side of the inequality:

$$\text{Pe} = \frac{au_{z,\text{ave}}}{D_{AB}} = \frac{\dfrac{a^2}{D_{AB}}}{\dfrac{a}{u_{z,\text{ave}}}} = \frac{\left\{\begin{array}{c}\text{radial diffusion}\\\text{time scale}\end{array}\right\}}{\left\{\begin{array}{c}\text{convective transport}\\\text{time scale}\end{array}\right\}} \gg 7$$

This inequality ensures that diffusive mass transport is not too great; Taylor realized that diffusion could not be so fast as to swamp out all effects of convection. Hence, the original inequality Eq. 7.40 says that diffusive transport can be neither too big nor too little. Exercise 7.9 shows that this inequality also has implications with respect to D_{AB} and D_{dis}. Rearranging,

$$u_{z,\text{ave}} \gg \frac{7D_{AB}}{a}$$

$$\text{ammonia in air: } u_{z,\text{ave}} \gg \frac{7 \times 0.28 \text{ cm}^2/\text{s}}{0.025\,\text{cm}} = 78.4\,\frac{\text{cm}}{\text{s}}$$

$$\text{ammonia in water: } u_{z,\text{ave}} \gg \frac{7[1.76 \times 10^{-5}\,\text{cm}^2/\text{s}]}{0.025\,\text{cm}} = 0.005\,\frac{\text{cm}}{\text{s}}$$

Since the distance L is not yet specified, it is not possible to pin down the exact velocity limits. But Exercise 7.10 will suggest a way to estimate this and hence prescribe fairly definite limits on velocity for a given tube diameter.

Taylor was so intrigued by the theoretical prediction of Eq. 7.38 that he set out for the laboratory in an effort to confirm his prediction. Note that if dispersion relative to moving coordinates is really analogous to pure molecular diffusion, we should be able to write a mass balance as we did for pure molecular diffusion in Section 6.7.2. For the one-dimensional dispersion process considered here, Eq. 6.84 becomes

$$\frac{\partial c_{A,\text{ave}}}{\partial t} + \frac{\partial \tilde{J}_A}{\partial \tilde{z}} = 0 \qquad (7.41)$$

Substituting from Eq. 7.38 for $\tilde{J}_A$, we find:

$$\frac{\partial c_{A,\text{ave}}}{\partial t} = D_{\text{dis}} \frac{\partial^2 c_{A,\text{ave}}}{\partial \tilde{z}^2} \qquad (7.42)$$

Hence, we find an equation that is completely analogous to Fick's second law (Eq. 6.87). The solution to Eq. 7.42 is analogous to Eq. 6.94:

$$c_{A,\text{ave}} = \frac{M_A}{2(\pi D_{\text{dis}} t)^{1/2}} \exp\left(-\frac{\tilde{z}^2}{4D_{\text{dis}} t}\right) \qquad (7.43)$$

Taylor then reasoned that if a thin slug of tracer could be injected into a laminar pipe flow, at a sufficiently far distance downstream, the concentration profile should resemble the curves for pure molecular diffusion from a plane source (recall Figure 6.10 of Section 6.7.2). Figure 7.4 shows the results of Taylor's (1953) measurements of tracer concentration at three positions in a pipe. Note that the theoretical prediction does seem to be verified: At a sufficiently far distance downstream, the tracer pulse resembles a bell-shaped curve, just like the Gaussian curves of pure one-dimensional molecular diffusion from a plane source. Taylor also plotted (as a dotted line) the concentration of tracer that would be predicted for pure convection, as analyzed earlier in this section. Note how the pure convection model severely underpredicts the actual measured tracer concentration. This leads to the inescapable conclusion that in certain conditions (i.e., Eq. 7.40), molecular diffusion can actually *impede* dispersal of the tracer in laminar pipe flow. Therefore, we have to adjust slightly the truism developed in Chapter 6 that molecular diffusion always *decreases* concentration gradients. Subsequent investigators have further verified Taylor's measurements, but this incredible phenomenon is still quite thrilling. As Taylor observed, "since water moves at twice the mean speed near the centre of the pipe and the patch of colour at the mean speed, the clear water in the middle must approach the colour patch,

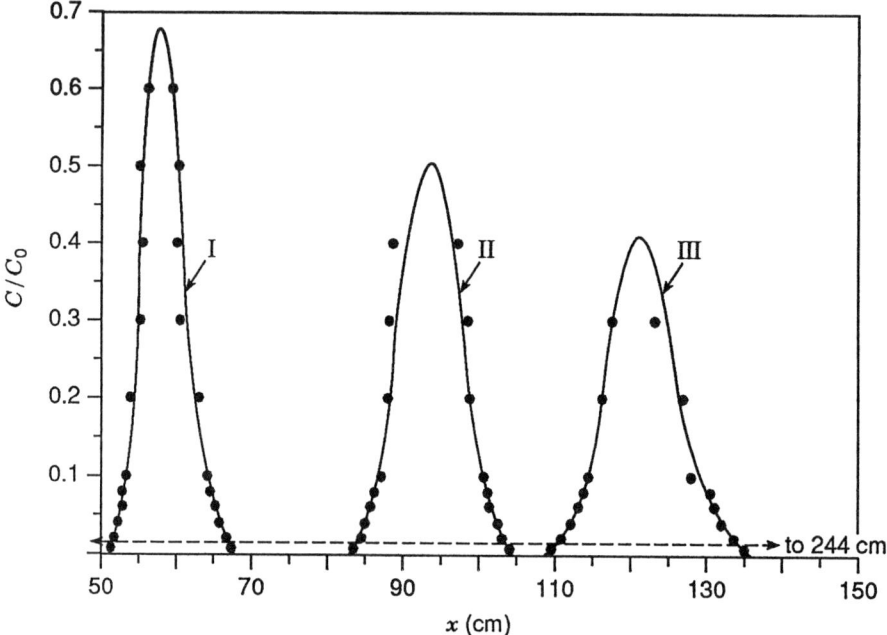

Figure 7.4. $KMnO_4$ concentration measured at three positions in a pipe. The dots are data and the solid lines are convective-diffusion model fits. The dashed curve is the prediction of pure convective dispersion that would obtain at the time corresponding to curve III. *Source*: Reprinted with permission from Taylor, *Proc. Roy. Soc. London*, **A219**, 200 (1953). Copyright (1953), The Royal Society.

absorb colour as it passes into it and then lose colour as it passes out, finally leaving the patch as perfectly clear water." This almost seems like magic!

Our final question in this section is whether experiments like Taylor's could be used to measure the molecular-diffusion coefficient of compounds in liquids or gases. For example, now think of the tracer as a compound with unknown diffusion coefficient. If the concentration profile of the compound can be measured in laminar pipe flow, the dispersion coefficient resulting in the best fit of the dispersion-model solution (Eq. 7.43) to the experimental data should allow calculation of the molecular-diffusion coefficient from Eq. 7.39. This is the subject of Exercise 7.11.

7.4. TAYLOR–ARIS DISPERSION

Even though we violate Eq. 7.40, note that if the Peclet number approaches zero, Eq. 7.39 predicts that the dispersion coefficient approaches zero. This cannot be correct, since molecular diffusion would still lead to axial dispersal of the tracer. Aris (1956) realized that another law was needed to blend

Taylor's analysis of convective diffusion with the pure axial-diffusive mass-transport mechanism at low Peclet numbers.

Because Eq. 7.43 is mathematically identical to Eq. 6.94, the computation of the variance (second moment) of the mean concentration distribution is also identical to the calculation of the variance of the concentration distribution in diffusion from a plane source (Chapter 6). Taking the derivative of Eq. 6.98 and considering D_{AB} in the purely diffusive analysis to correspond to D_{dis} in the Taylor analysis, we find:

$$\frac{1}{2}\frac{d\overline{z^2}}{dt} = D_{AB}\left(\frac{Pe^2}{48}\right) \tag{7.44}$$

Aris considered a less restrictive Peclet number range

$$Pe \ll \frac{4L}{a} \tag{7.45}$$

and showed that the asymptotic limit of the variance of the concentration distribution relative to coordinates moving with the mean velocity is

$$\frac{1}{2}\frac{d\overline{z^2}}{dt} = D_{AB}\left(1 + \frac{Pe^2}{48}\right) \tag{7.46}$$

This implies that a new (less restrictive) estimate for the dispersion coefficient in laminar pipe flow is

$$D_{dis} = D_{AB}\left(1 + \frac{Pe^2}{48}\right) \tag{7.47}$$

This physical law provides a nice mathematical transition between convective diffusion and pure axial diffusion.

7.5. TURBULENT DISPERSION: THE LAGRANGIAN APPROACH

As noted in Chapter 5, over a large section of turbulent flow, momentum transport due to turbulent motions greatly exceeds momentum transport due to viscous diffusion. We will find a completely analogous result in this chapter: mass transport (e.g., of a contaminant or tracer) due to turbulent motions will in general greatly exceed mass transport by molecular diffusion.*

---

*Chemical and mechanical engineering students will probably already have had a course on heat, mass, and momentum transport. Starting in about 1960, with the publication of Bird, Stewart, and Lightfoot's *Transport Phenomena*, students and teachers have realized that there is a high degree of similarity in the transport of heat (energy), mass, and momentum. Developments in one area of transport engineering (for example, heat transport) have often aided developments in another area (mass transport for instance). We only touch lightly on these important analogies in this book, and the reader is referred to one of the excellent transport books listed at the end of this chapter.

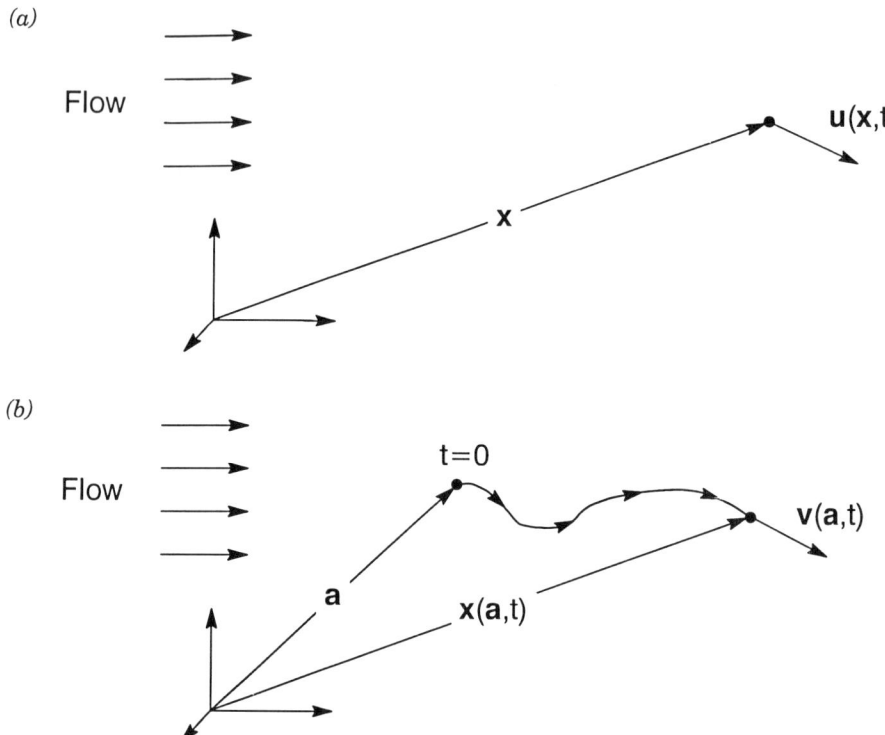

Figure 7.5. Velocity and position specified in (*a*) Eulerian and (*b*) Lagrangian coordinate systems.

We begin our analysis of mass transport in turbulent flows with another classic analysis of Taylor (1922). This theory was the basis for succeeding analyses of turbulent mass transport in chemical and environmental engineering. However, before launching into the Taylor analysis, we need to introduce the concept of Lagrangian and Eulerian coordinate systems. Thus far, our coordinate system has been a *Eulerian* coordinate system. For example, consider the Eulerian coordinate system in the turbulent flow of Figure 7.5*a*. Velocity measurements are made at some fixed place x in space. We can continue to monitor velocity at this fixed point, that is, $u(x, t)$. In the Eulerian system, we can also measure a scalar quantity at x, for example, $c(x, t)$.

Now consider the *Lagrangian* coordinate system of Figure 7.5*b*. Here we have monitored the progress of a small point or particle in the flow. This point can be considered a single tagged fluid molecule or a tracer molecule, as discussed in Section 7.3. We began monitoring the point at position a at time $= 0$. Over time, the point migrates to position $x(a, t)$. This nomenclature means the following: Lagrangian position $x(a, t)$ is the position of a fluid particle at time $= t$ that was released at **a** at time $= 0$. Similarly, we can consider $\mathbf{v}(\mathbf{a}, t)$ as the Lagrangian velocity of a particle at time $= t$ that was released at **a** at time $= 0$. As was the case for Eulerian coordinate systems, we can also describe scalar quantities, for example, concentration, $c(\mathbf{a}, t)$. One of the nice things about Lagrangian analysis is the simple connection between distance of travel **x** and velocity **v**:

$$\mathbf{x}(\mathbf{a}, t) = \mathbf{a} + \int_0^t \mathbf{v}(\mathbf{a}, \tau)d\tau \qquad (7.48)$$

Thus, the final position of the particle is equal to the initial position plus the integral of the Lagrangian velocity over time (i.e., along the path of migration). This equation of course implies that velocity **v** was known continuously along the path of travel of the particle.

The Lagrangian analysis and Eq. 7.48 are powerful ideas: If we know the initial position of a tagged particle, and if we know the Lagrangian velocity field, we can predict the position of the particle at some later time t. The reader sees immediately why Lagrangian statistics are often preferable to Eulerian statistics in pollution-dispersion studies: for a pollutant generated at a specific point, our goal is often to find out how far and where it goes during time t.

The reader will note that we have shown Eulerian and Lagrangian velocities at the same time and place in the (same) flow in Figure 7.5. However, in turbulent flows, it is very difficult to predict where a particle will go in the Eulerian system. Nevertheless, there is a formal connection between Eulerian and Lagrangian velocities:

$$\mathbf{v}(\mathbf{a}, t) = \mathbf{u}[\mathbf{x}(\mathbf{a}, t), t] \qquad (7.49)$$

Equation 7.49 seems to imply that we can also use Eulerian velocity data in diffusion problems. The problem with this approach is seen in the arguments of the Eulerian velocity (right-hand side) of Eq. 7.49. Note that to determine the correct Eulerian velocity, you must know the correct Lagrangian displacement vector, $\mathbf{x}(\mathbf{a}, t)$. Thus, the Lagrangian description is indeed more simple.

Unfortunately (for mankind and students of pollution control), the relevant Lagrangian data are very difficult to collect in the real world. For example, to collect Lagrangian velocity or concentration data, we must "ride" along with the particle and sample its velocity and concentration continuously during time. To obtain significant averages, we have to do this for a large number of particles (see ensemble averaging below).* Historically, this has been difficult

---

*Actually, for stationary, homogeneous turbulence, the ergodic principle lets us substitute an average following a single particle over a long time for ensemble averaging of many different particles. But this is a little more complicated than we need to get in this book (e.g., Bendat and Piersol, 1971).

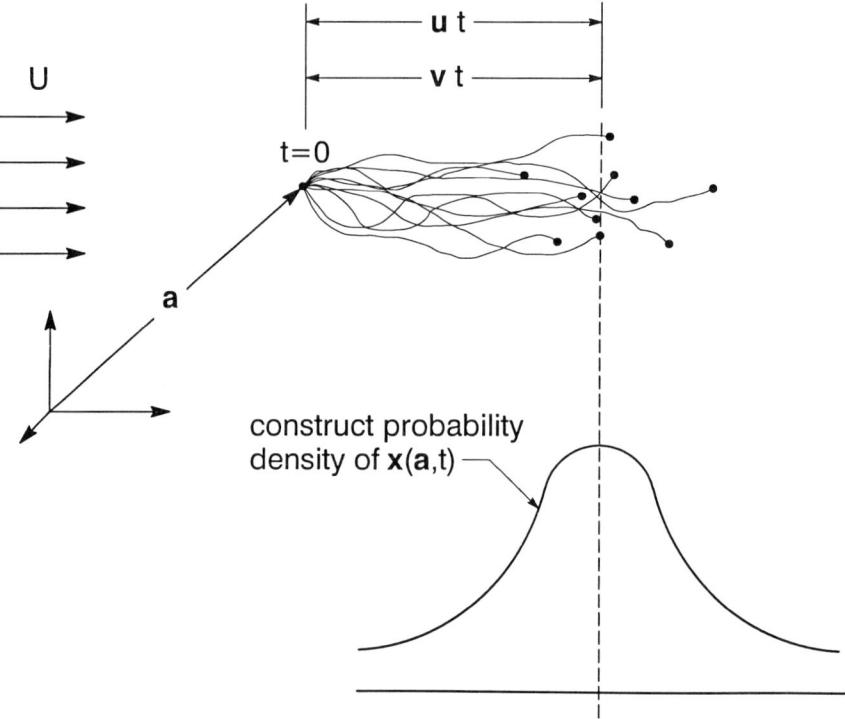

Figure 7.6. Development of the statistics of $x(a, t)$ using the concept of ensemble averaging. The turbulence is stationary and homogeneous.

to accomplish. However, we have developed quite a few tools for accurately measuring Eulerian velocities — our velocity or concentration measuring probe can just be stuck into the flow at different places.

Despite the problem of measuring Lagrangian velocities, scientists have found that much can be learned about diffusion and dispersion problems simply by posing them in the Lagrangian frame of reference. Therefore, we are finally ready to dive into Taylor's analysis.

Consider that a number of particles are released into a *stationary, homogeneous turbulent flow* (Figure 7.6) and that we can track the position of the particles throughout time. In turbulent flows, the term stationary means that the statistical properties of the turbulence at any point in the flow do not vary in time (Chapter 5). In homogeneous turbulence, the statistical properties of the turbulence do not vary along the x, y, or z axes. In reality, stationary, homogeneous turbulent flows do not exist, mainly because most turbulence has a mean velocity gradient (called turbulent shear flows) that feeds energy to the turbulence.* Therefore, for the flow to be stationary, energy must be fed to the turbulence at a constant rate, but there must be a velocity gradient to supply the energy; hence, turbulence cannot at once be stationary and homogeneous. Nevertheless, it has been noted that in a pipe, the flow can be stationary and homogeneous along the direction of flow (i.e., no statistical variations in the streamwise direction). We then say the flow is homogeneous in the streamwise direction. Thus, we can have flows that may approximate stationary, homogeneous turbulence. In any event, if we release enough particles at **a**, and record their position at a certain time t, we can get an idea of the statistics of $x(\mathbf{a}, t)$. In particular, the variance of $x(\mathbf{a}, t)$ will tell us much about the dispersion of contaminants. This type of experiment is sometimes called an *ensemble average*. It means that each time we release a tagged particle into the stationary, homogeneous flow, we can imagine that we are starting a new experiment. We record the particle position at the same time after the start of the new experiment, and from that develop the statistics required for the ensemble of experiments (sometimes called realizations). Taylor (1922) asked us to consider the following integral of ensemble-averaged Lagrangian velocity fluctuations:

$$\int_0^t \langle v_t' v_\tau' \rangle d\tau \qquad (7.50)$$

Here the angles indicate the ensemble-averaging process, and we consider v_t' to be the Lagrangian velocity fluctuation at some time t in the streamwise direction, that is, the streamwise Lagrangian velocity with the streamwise mean velocity subtracted out (Taylor, 1954a):[†]

$$v_t = \bar{V}_t + v_t' \qquad (7.51)$$

Through manipulations of the integral Taylor showed (see also Hinze, 1975; Fischer et al., 1979; and Seinfeld, 1985) that this could also be expressed as follows:

$$\int_0^t \langle v_t' v_\tau' \rangle d\tau = \int_0^t \langle v_t' v_{t+\tau}' \rangle d\tau \qquad (7.52)$$

*For example, recall how radically the average streamwise velocity fluctuations changed in the radial direction in the channel flow shown in Figure 5.19. Nevertheless, we know that for *a particular radial position*, the statistics of the velocity fluctuations in any direction do not change in the streamwise direction. Hence, the flow is homogeneous in the streamwise direction.

[†]It is fairly easy to show that because of conservation of mass, the Lagrangian and Eulerian mean velocities are the same in homogeneous turbulent flows. Interestingly, the mean square Lagrangian and Eulerian velocity fluctuations are also the same in homogeneous turbulence (Tennekes and Lumley, 1972). Unfortunately, there is no simple connection between Eulerian and Lagrangian correlation coefficients.

Consider first integral
in Eq. 7.52:

Consider second integral
in Eq. 7.52:

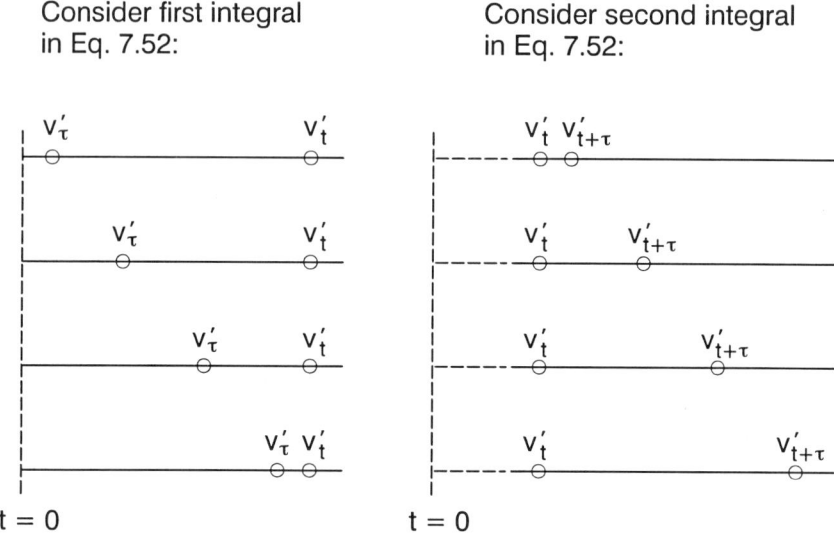

Figure 7.7. Demonstration of equivalence of integrals on either side of Eq. 7.52.

This unlikely looking transformation can be understood by considering two time lines, one of which concerns the first integral, and a second that concerns the second integral in Eq. 7.52. In Figure 7.7 we show the time separation of the fluctuating velocities of the ensemble averages for four different values of τ. Note that as we progress through the integration, the time lag is different at each step in the two integrals. However, when the integrations are finished, it is easy to see that we have covered the same ground, just in reverse order. Since the average is an ensemble average, we can imagine that at each step (actually an infinite number of small steps like those shown in the figure) we have a large number of similar realizations over which to average. Since the turbulence is stationary and homogeneous, the two integrals are identical; it doesn't matter that the second integral appears to be performed at later times because the critical factor is the time separation. Now note that the *Lagrangian autocorrelation coefficient* is formally defined by

$$R_\tau = \frac{\langle v_t' v_{t+\tau}' \rangle}{\langle v_t'^2 \rangle} \tag{7.53}$$

where $\langle v_t'^2 \rangle$ is the mean-square Lagrangian velocity in the streamwise direction. In fact, because we have assumed a stationary homogeneous turbulent flow, $\langle v_t'^2 \rangle$ can be considered a constant. Rearranging Eqs. 7.52 and 7.53, we find:

$$\int_0^t \langle v_t' v_\tau' \rangle d\tau = \langle v_t'^2 \rangle \int_0^t R_\tau d\tau \tag{7.54}$$

Now note that with regard to the integral on the left-hand side of Eq. 7.54, time t can be considered a constant. Therefore, we can take the following step:

$$\int_0^t \langle v_t' v_\tau' \rangle d\tau = \left\langle v_t' \int_0^t v_\tau' d\tau \right\rangle = \langle v_t' x \rangle = \frac{1}{2} \frac{d}{dt} \langle x^2 \rangle \tag{7.55}$$

since, from Eq. 7.48,

$$\mathbf{v} = \frac{d\mathbf{x}}{dt} \tag{7.56}$$

Finally, combining Eqs. 7.54 and 7.55 we can write, after Taylor (1922),

$$\langle x^2 \rangle = 2\langle v_t'^2 \rangle \int_0^t \left(\int_0^t R_\tau d\tau \right) dt' \tag{7.57}$$

The Lagrangian autocorrelation coefficient has the following asymptotic limits. As $\tau \to 0$, the autocorrelation coefficient must approach unity. Hence, for $\tau \to 0$, $R_\tau \to 1$ and the integral inside the parentheses in Eq. 7.57 must approach t. So Eq. 7.57 yields

$$\langle x^2 \rangle = \langle v_t'^2 \rangle t^2 \tag{7.58}$$

Therefore, we conclude that for very small times, the variance of x grows as the square of time.

Now consider the opposite limit, $t \to \infty$. It is safe to imagine that in this limit, the autocorrelation must approach zero: in stationary, homogeneous flows, the velocity of a Lagrangian point must become uncorrelated with its velocity at sufficiently earlier time. This means that the integral inside the parentheses in Eq. 7.57 must converge to a finite value; we call this constant the Lagrangian integral scale:

$$F = \int_0^\infty R_\tau d\tau \tag{7.59}$$

Hence, Eq. 7.57 yields

$$\langle x^2 \rangle = 2\langle v_t'^2 \rangle Ft \tag{7.60}$$

We conclude that for long times, the variance of x increases linearly with time. We can assume here that by "long times" we mean a time long enough that the Lagrangian velocity fluctuations are no longer correlated with themselves (i.e., $t \gg F$). This last result has very important implications in the history of our understanding of turbulent transport. In Chapter 6 we showed that in pure molecular diffusion from a plane source, the variance of the concentration distribution increases in proportion to time. We now find an analogous result for stationary, homogeneous turbulence. As shown in the next section, this observation encouraged researchers to invent an important transport idea, the "eddy diffusivity" of turbulence.

7.6.   TURBULENT DISPERSION: THE EULERIAN APPROACH

We hinted above that practical problems in turbulent transport will probably have to be solved in the Eulerian coordinate system. Recall that for constant ρ and D_{AB} and no reactions, the convective-diffusion equation takes the following form (Section 7.2):

$$\frac{\partial c_A}{\partial t} + \mathbf{u} \cdot \nabla c_A = D_{AB} \nabla^2 c_A \tag{7.61}$$

In turbulent flow, we decompose the velocity at a certain point in space as in Chapter 5:

$$u_i = \bar{U}_i + u_i' \tag{7.62}$$

However, the concentration at a point must also be decomposed:

$$c_A = \bar{C}_A + c_A' \tag{7.63}$$

Substituting Eqs. 7.62 and 7.63 into Eq. 7.61 we find, after Reynolds averaging (see Exercise 7.12):

$$\frac{\partial \bar{C}_A}{\partial t} + \bar{U}_1 \frac{\partial \bar{C}_A}{\partial x_1} + \bar{U}_2 \frac{\partial \bar{C}_A}{\partial x_2} + \bar{U}_3 \frac{\partial \bar{C}_A}{\partial x_3}$$
$$= D_{AB} \left(\frac{\partial^2 \bar{C}_A}{\partial x_1^2} + \frac{\partial^2 \bar{C}_A}{\partial x_2^2} + \frac{\partial^2 \bar{C}_A}{\partial x_3^2} \right) - \left(\frac{\partial}{\partial x_1} \overline{u_1' c_A'} + \frac{\partial}{\partial x_2} \overline{u_2' c_A'} + \frac{\partial}{\partial x_3} \overline{u_3' c_A'} \right) \tag{7.64}$$

As noted earlier, in turbulent flows, the mass transport due to pure molecular diffusion can usually be ignored, in comparison to the turbulent mass transport. Therefore, the diffusion terms in Eq. 7.64 are almost always dropped, leaving the following:*

$$\frac{\partial \bar{C}_A}{\partial t} + \bar{U}_1 \frac{\partial \bar{C}_A}{\partial x_1} + \bar{U}_2 \frac{\partial \bar{C}_A}{\partial x_2} + \bar{U}_3 \frac{\partial \bar{C}_A}{\partial x_3} = -\frac{\partial}{\partial x_1}\overline{u_1' c_A'} - \frac{\partial}{\partial x_2}\overline{u_2' c_A'} - \frac{\partial}{\partial x_3}\overline{u_3' c_A'}$$

$$(7.65)$$

Expressed in tensor form, Eq. 7.65 becomes

$$\frac{\partial \bar{C}_A}{\partial t} + \bar{U}_k \frac{\partial \bar{C}_A}{\partial x_k} = -\frac{\partial \overline{u_j' c_A'}}{\partial x_j} \qquad (7.66)$$

The new terms on the right-hand sides of Eqs. 7.65 and 7.66 involve the correlations between fluctuating velocities and concentrations, and they are analogous to the Reynolds stresses discovered in Chapter 5 when we Reynolds-averaged the Navier–Stokes equations. If the mean velocity field in the turbulence is known, Eqs. 7.65 and 7.66 imply that there are still four unknowns: $\bar{C}_A$, $\overline{u_1' c_A'}$, $\overline{u_2' c_A'}$, and $\overline{u_3' c_A'}$. It appears that we have discovered yet another "closure problem" of turbulence.

In Section 7.5, it was shown that after a sufficient time, the dispersion of particles relative to coordinates moving with the mean velocity in stationary, homogeneous turbulence was analogous to a molecular-diffusion process. In analogy to the development of eddy viscosity in Chapter 5, we introduce here the following approximation of the fluctuating turbulent velocity and concentration correlations:

$$\frac{\partial \bar{C}_A}{\partial t} + \bar{U}_1 \frac{\partial \bar{C}_A}{\partial x_1} + \bar{U}_2 \frac{\partial \bar{C}_A}{\partial x_2} + \bar{U}_3 \frac{\partial \bar{C}_A}{\partial x_3}$$
$$= \frac{\partial}{\partial x_1}\left(\epsilon_{11}\frac{\partial \bar{C}_A}{\partial x_1}\right) + \frac{\partial}{\partial x_2}\left(\epsilon_{22}\frac{\partial \bar{C}_A}{\partial x_2}\right) + \frac{\partial}{\partial x_3}\left(\epsilon_{33}\frac{\partial \bar{C}_A}{\partial x_3}\right) \qquad (7.67)$$

The ϵ_{ii} are known as the *eddy diffusivities* or *turbulent mixing coefficients*. If the eddy diffusivities were constant, they would come out in front of the derivatives and the new eddy-diffusion terms would exactly resemble the pure molecular-diffusion terms in the Reynolds-averaged convective-diffusion equation (Eq. 7.64). Yet the analogy to molecular diffusion must ultimately fail, since the turbulent diffusion is a property of the turbulence, not of the fluid (as suggested by the analogy with molecular diffusion). In fact, the eddy diffusivity is generally not constant throughout a turbulent flow. We see that Eq. 7.67 is not yet necessarily closed, that is, the values of ϵ_{ii} may vary spatially.

---

*There are some turbulent flows where pure diffusive mass transport cannot totally be ignored. For example, in turbulent flows where the Reynolds number is not high and the laminar sublayer is large, mass can in effect get "trapped" or retarded in the laminar sublayer. This is thought to be one reason for the long "tails" in some tracer profiles.

Just as more modeling was required in Chapter 5 to close the Navier–Stokes equations (recall the discussion of eddy viscosity), we apparently need more modeling to close Eq. 7.67. Perhaps the reader will not be surprised to learn that Sir G. I. Taylor (1954a) was one of the first to find a closure of the Eulerian turbulent-transport equations for stationary, homogeneous turbulence in pipe flow. The reader is asked to recall Taylor's (1953) analysis of convective diffusion in steady laminar pipe flow (Section 7.3) because the new turbulent analysis is very similar in approach. First, the turbulent convective-diffusion equation is written in radial coordinates moving with the main average streamwise velocity, and just like the laminar-flow analysis, the radial turbulent transport of mass is balanced by axial-convective mass transport:

$$\bar{U}_{\hat{z}} \frac{\partial \bar{C}_A}{\partial \hat{z}} = \left[\frac{1}{r} \frac{\partial}{\partial r} \left(\epsilon_{rz} r \frac{\partial \bar{C}_A}{\partial r} \right) \right] \tag{7.68}$$

For $\bar{U}_{\hat{z}}$, Taylor cited a "velocity defect" law for turbulent pipe flow (recall Exercise 5.19):

$$\frac{\bar{U}_{z,\max} - \bar{U}}{u_*} = f\left(\frac{r}{a}\right) \tag{7.69}$$

where f stands for function, r is measured from the central axis of the pipe, a is the pipe diameter, and u_* is the friction velocity (also Chapter 5):

$$u_* = \sqrt{\frac{\tau_0}{\rho}} \tag{5.153}$$

Here τ_0 is the mean shear stress at the pipe wall. Recall from Section 5.7 that in pipe flow, the stress at any radial position r is related to τ_0 by

$$\tau = \tau_0 \left(\frac{r}{a}\right) \tag{7.70}$$

Therefore, from the last two equations,

$$\tau = \rho u_*^2 \left(\frac{r}{a}\right) \tag{7.71}$$

However, the most critical step in Taylor's analysis involved the model for ϵ_{rz}, the radial turbulent mixing coefficient. Here Taylor appealed to the *Reynolds analogy*. The idea here is that heat, mass, and momentum are transported in exactly the same way in turbulence. For example, we recall from Chapter 6 that the rate of mass transfer across a cylindrical surface at some r can be given by a gradient transport law:

$$J_A = -\epsilon_{rz} \frac{\partial \bar{C}}{\partial r} \tag{7.72}$$

The Reynolds analogy says that momentum transfer across the same surface will not only be analogous, it will have exactly the same transport coefficient:

$$\tau = -\epsilon_{rz}\rho \frac{\partial \bar{U}}{\partial r} \tag{7.73}$$

Hence, the Reynolds analogy allows us to calculate the turbulent mass-transfer coefficient (the eddy diffusivity) from information on turbulent momentum transport. For example, from Eq. 7.69,

$$\frac{\partial \bar{U}}{\partial r} = \frac{-u_* f'(r/a)}{a} \tag{7.74}$$

where f' is the derivative of f with respect to r/a. Substituting into Eq. 7.73,

$$\tau = \frac{\epsilon_{rz}\rho u_* f'(r/a)}{a} \tag{7.75}$$

Equating Eqs. 7.75 and 7.71, we find:

$$\epsilon_{rz} = \frac{a(r/a)u_*}{f'(r/a)} \tag{7.76}$$

At this point Taylor had all the ingredients to evaluate Eq. 7.68 numerically with tabulated values for $f(r/a)$. As he did for mass transport in laminar pipe flow, he calculated the mass flux across a plane traveling with the mean velocity. The final result is

$$\tilde{J}_A = -10.06 a u_* \frac{\partial \bar{C}_{A,\tilde{z}}}{\partial \tilde{z}} \tag{7.77}$$

As we noticed in Section 7.3, Eq. 7.77 seems to imply that axial mass transport relative to coordinates moving with the mean velocity is analogous to Fick's first law:

$$\tilde{J}_A = -D_{dis}^t \frac{\partial \bar{C}_{A,\tilde{z}}}{\partial \tilde{z}} \tag{7.78}$$

where D_{dis}^t is the *turbulent diffusion coefficient*. Comparison of the last two equations indicates that for stationary pipe flow turbulence,

$$D_{dis}^t = 10.06 a u_* \tag{7.79}$$

As in Section 7.3, this suggests an axial turbulent diffusion equation relative to moving coordinates:

$$\frac{\partial \bar{C}_A}{\partial t} = D^t_{\text{dis}} \frac{\partial^2 \bar{C}_A}{\partial \tilde{z}^2} \qquad (7.80)$$

The solution of Eq. 7.80 is of course the same as the solution of Eq. 7.42. If we go back to regular Eulerian coordinates, we simply add in the convective transport term:

$$\frac{\partial \bar{C}_A}{\partial t} + \bar{U} \frac{\partial \bar{C}_A}{\partial z} = D^t_{\text{dis}} \frac{\partial^2 \bar{C}_A}{\partial z^2} \qquad (7.81)$$

This equation is known as the *one-dimensional dispersion equation*. It is, in effect, the one-dimensional approximation for turbulent convective diffusion in a pipe. Exercise 7.13 shows how to compute the turbulent diffusion coefficient from real tracer dispersion data in turbulent pipe flow. We meet the dispersion equation again in Chapters 8 and 10 when we discuss dispersion in porous media.

Thus, we end our discussion and analysis of turbulent convective diffusion in the Eulerian framework. We have considered only the basic underpinnings of the theory. Over the past 30 years or so, much progress has been made in implementing more detailed, thorough, and widely usable formulations of turbulent diffusion in large-scale air and water systems. The developments we refer to here have been successful in describing all aspects of turbulent mixing and dispersion, considering such problems as buoyancy, initial mixing, initial plume momentum, and scales of mixing. The computation of the dispersion of contaminants in air and water environments has become quite sophisticated and quite successful (perhaps the greatest success story of environmental transport modeling!). In this chapter, the reader has been exposed to the majority of concepts required for understanding these advanced models.

Excellent introductions to advanced modeling of pollutant transport and mixing at large, environmental-length scales are provided by Fischer et al. (1979), Thomann and Miller (1987), and Seinfeld (1985).

7.7. MASS TRANSFER IN LAMINAR FLOW ALONG REACTING OR DISSOLVING SOLID SURFACES

The previous sections in this chapter focused on the convective diffusion and dispersion of tracers, contaminants, and other compounds in liquid and air flows. But what if the substance being convected, diffused, or dispersed undergoes heterogeneous reaction at the solid boundary? From our discussion of heterogeneous reactions in Chapter 6, we can guess that the flux of contaminants will certainly have something to do with the reaction rate, the diffusion coefficient, and the diffusion distance. But the situation in flowing (convective-diffusion) systems must be more complex than the analyses of the last chapter because of the added convective-mass transport.

Consider the transport of reactive contaminants in the environment. For example, particulate contaminants from automobiles and industry are transported effectively in the trachea of the lungs by our breathing. But these contaminants are also (unfortunately) deposited in our lungs when they run into the sticky mucous layer along the internal lung surface. We can imagine that next to the lung surface, the concentration of particles must go to zero if there is no possibility of rebound from the surface. If the contaminant is being continuously replaced by the breathing cycle, there must be some kind of matching between the zero concentration at the lung surface and the higher concentration in the main stream. We call this region the *concentration boundary layer*. Concentration boundary layers are ubiquitous in the environment, and can be anticipated at any surface where there is consumption of a particulate or chemical phase. Similarly, we find concentration boundary layers along surfaces that are producing some compound or dissolving. In this case, the concentration at the surface might be considered the (constant) saturation concentration of whatever is dissolving. Here, there must be matching between the high concentration near the surface and the lower concentration out in the main flow. Concentration boundary layers are important because they describe the concentration profile near surfaces, which in turn determines the rate of diffusion at the surface via Fick's first law.

The reader has probably guessed that there are similarities between concentration boundary layers and viscous boundary layers. As we learned in Chapter 5, even when viscosity can be ignored in the majority of a flow, the no-slip condition requires that in the region of a boundary there must be a smooth viscous boundary layer to match velocities in the outer flow and the solid surface. This is the laminar boundary-region layer, and it is quite analogous to the concentration boundary layer. Momentum boundary layers are important because they describe velocity profiles near surfaces, which in turn determines the shear at the surface via Newton's viscosity relation (Chapter 5).

7.7.1. Flat Plate with Fast Heterogeneous Reaction or Constant Surface Concentration

Our mathematical analysis begins with consideration of the laminar concentration boundary layer along a flat plate, as shown in Figure 7.8. We have shown the concentration in the boundary layer with arrows. Although arrows are normally reserved for vectors (like velocity), we use them here for concentration, since it is difficult to show the magnitude of scalars in any other way. The concentration at the surface of the plate is zero because there is a very fast reaction at the surface, or because the surface is a perfect sink for mass (as at the sticky lung surface).

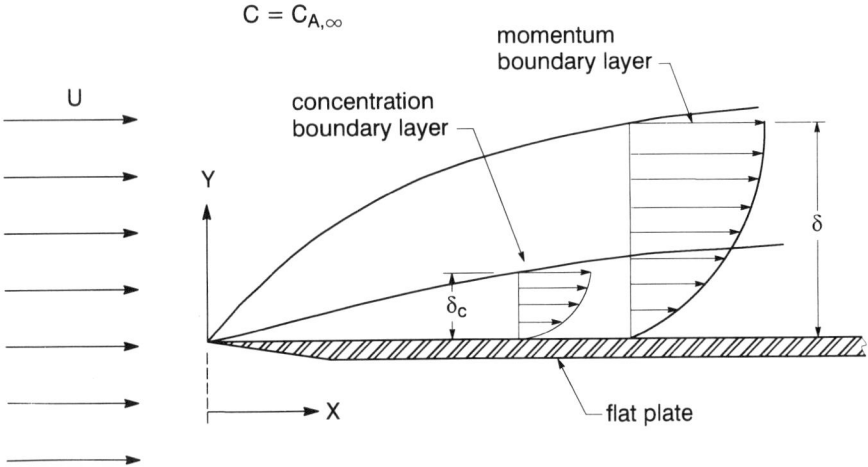

Figure 7.8. Development of laminar momentum and concentration boundary layers along a flat plate in uniform flow.

The concentration boundary layer is indicated by the notation δ_c. Note that the velocity (momentum) boundary layer is drawn thicker than the concentration boundary layer. There is a good mathematical reason for this assumption, as we shall see shortly. It is useful to estimate the thickness of the momentum and concentration boundary layers. Of course, we know from Chapter 5 that the laminar momentum boundary layer thickness is

$$\frac{\delta}{x} = \frac{5}{\sqrt{Re_x}} \qquad (5.121)$$

where Re_x is the boundary-layer Reynolds number:

$$Re_x = \frac{Ux}{\nu} \qquad (5.122)$$

We have noted that in the laminar boundary layer, there was a competition between the main flow, which is trying to sweep the boundary layer downstream and viscosity, which is trying to match the zero-velocity condition at the surface with the free-stream velocity. In the concentration boundary layer, we can imagine an analogous situation in which the main flow is trying to sweep the boundary layer downstream, while molecular diffusion is trying to match the zero concentration at the surface with the free-stream concentration. We can also imagine that the thickness of the concentration boundary layer is sensitive to the Reynolds number, since the Reynolds number ultimately determines the nature of the velocity profile, and the velocity profile obviously has a lot to do with the transport of mass. But what determines the relative thickness of the two boundary layers? It should help to compare the viscous diffusion (τ_v) and mass diffusion (τ_d) time scales:

$$\frac{\tau_d}{\tau_v} = \frac{\dfrac{L^2}{D_{AB}}}{\dfrac{L^2}{\nu}} = \frac{\nu}{D_{AB}} = \text{Sc} = \frac{\text{Pe}}{\text{Re}} \qquad (7.82)$$

where L is an unspecified length scale. The new dimensionless number in Eq. 7.82 is the *Schmidt number*, the ratio of mass- and viscous-diffusion time scales. When the Schmidt number is large (Sc $\gg$ 1), viscous diffusion has a smaller time scale than molecular diffusion, and we will find that the momentum boundary layer is thicker than the diffusion boundary layer. Alternatively, if the Schmidt number is small (Sc $\ll$ 1), molecular diffusion has the smaller time scale and the concentration boundary layer will be thicker than the momentum boundary layer. Note also from Eq. 7.82 that the Schmidt number is the ratio of Peclet and Reynolds numbers; hence, these three dimensionless numbers are not independent of one another, and by specifying any two, the third is automatically known.

Next, we attempt to solve the convective-diffusion equation for the concentration boundary layer. As we found in the laminar momentum boundary-layer solution of Chapter 5, we will have to simplify the governing equation. For a wide plate (into the page), we can consider this a two-dimensional problem; hence, Eq. 7.13 reduces to (with no *homogeneous* reactions),

$$\frac{\partial c_A}{\partial t} + u\frac{\partial c_A}{\partial x} + v\frac{\partial c_A}{\partial y} = D_{AB}\left(\frac{\partial^2 c_A}{\partial x^2} + \frac{\partial^2 c_A}{\partial y^2}\right) \qquad (7.83)$$

We can continue the simplification procedure with an argument similar to the one presented in Section 5.12; taking $c_{A,\infty}$ as the characteristic concentration scale, we find:

$$\frac{\dfrac{\partial^2 c_A}{\partial y^2}}{\dfrac{\partial^2 c_A}{\partial x^2}} = O\left(\frac{\left(\dfrac{c_{A,\infty}}{\delta_c^2}\right)}{\left(\dfrac{c_{A,\infty}}{x^2}\right)}\right) = \left(\frac{x}{\delta_c}\right)^2 \qquad (7.84)$$

We have already hypothesized a thin concentration boundary layer (i.e., Sc $\gg$ 1); therefore, it is clear from Eq. 7.84 that we can ignore the second derivative of concentration in the streamwise (x) direction; hence, for the steady state,

$$u \frac{\partial c_A}{\partial x} + v \frac{\partial c_A}{\partial y} = D_{AB} \frac{\partial^2 c_A}{\partial y^2} \tag{7.85}$$

Now we must figure out how to handle u and v in Eq. 7.85. When the Schmidt number is high (i.e., the concentration boundary layer is thin relative to the momentum boundary layer), we can use truncated series solutions of the boundary-layer velocities which are accurate close to the plate (see Levich, 1962; Friedlander, 1977; or Probstein, 1989);

$$u = 0.332U\eta \tag{7.86}$$

and for the small y-direction velocity:

$$v = 0.083U \left(\frac{v}{xU}\right)^{1/2} \eta^2 \tag{7.87}$$

where η is a dimensionless variable:

$$\eta = \left(\frac{U}{vx}\right)^{1/2} y \tag{7.88}$$

Note that Eqs. 7.86 and 7.87 are increasingly accurate for $\eta \to 0$. Substituting Eqs. 7.86–7.88 into Eq. 7.85 and nondimensionalizing c_A, we find:

$$0.332\eta \frac{\partial \tilde{c}_A}{\partial x} + 0.083 \left(\frac{v}{Ux}\right)^{1/2} \eta^2 \frac{\partial \tilde{c}_A}{\partial y} = \frac{D_{AB}}{U} \frac{\partial^2 \tilde{c}_A}{\partial y^2} \tag{7.89}$$

where $\tilde{c}_A$ is the dimensionless concentration:

$$\tilde{c}_A = \frac{c_A - c_{A,s}}{c_{A,\infty} - c_{A,s}} \tag{7.90}$$

Reflecting on the definition above and Figure 7.8, the boundary conditions for solution of Eq. 7.89 are that $\tilde{c}_A = 1$ at $y = \infty$ and $\tilde{c}_A = 0$ at $y = 0$. Now, following Friedlander (1977), we assume that $\tilde{c}_A$ is only a function of the dimensionless variable η:

$$\tilde{c}_A = g(\eta) \tag{7.91}$$

After some work, we find the following ordinary differential equation (see Exercise 7.14):

$$\frac{d^2 g}{d\eta^2} + 0.083 \left(\frac{v}{D_{AB}}\right) \eta^2 \frac{dg}{d\eta} = 0 \tag{7.92}$$

The solution to Eq. 7.92 is given in Levich (1962):

$$\tilde{c}_A = \frac{(0.22 \mathrm{Sc})^{1/3}}{0.89} \int_0^{\frac{1}{2}\sqrt{\frac{Uy^2}{vx}}} \exp(-0.22 \mathrm{Sc}\, z^3)\, dz \tag{7.93}$$

To calculate the *local flux* of A toward the surface, we use Fick's first law:

$$J_A = -D_{AB} \left. \frac{\partial c_A}{\partial y} \right|_{y=0} = -0.34 \left(\frac{D_{AB}}{x}\right)\left(\frac{xU}{v}\right)^{1/2}\left(\frac{v}{D_{AB}}\right)^{1/3}(c_{A,\infty} - c_{A,s})$$

$$= 0.34 \left(\frac{D_{AB}}{x}\right) \mathrm{Re}_x^{1/2}\, \mathrm{Sc}^{1/3}(c_{A,s} - c_{A,s}) \tag{7.94}$$

Equation 7.94 is a very important result in mass-transfer engineering. Note that the flux of A toward the surface increases with the Reynolds number. This makes sense: as the Reynolds number increases, both the momentum and concentration boundary layers can be expected to become thinner. Recall that the surface reaction is assumed to be infinitely fast, so the consumption of A at the surface is not limited by the reaction, only by the transport of A to the surface (analogous to diffusion-limited heterogeneous reactions analyzed in Chapter 6). Thus, the higher the Reynolds number is, the steeper the concentration gradients and the more effective the mass transfer. Note also that since mass flux is proportional to $x^{-1/2}$, mass flux is highest at small x and decreases in the positive x direction. Finally, we note that Eq. 7.94 says that the flux is also proportional to $\mathrm{Sc}^{1/3}$. Recall from Eq. 7.82 that the Schmidt number is the ratio of the mass- and viscous-diffusion time scales. As the Schmidt number increases, the concentration boundary layer becomes relatively thinner than the viscous boundary layer. Since thin boundary layers mean greater concentration gradients, and greater concentration gradients imply greater mass transfer, it does seem reasonable that the mass flux will increase with increasing Schmidt number.

Many times we are interested in the average mass-transfer rate for the entire plate of length L. Exercise 7.15 shows that the average molar flux of A is

$$J_{A,\mathrm{ave}} = 0.678 \left(\frac{D_{AB}}{L}\right)\left(\frac{LU}{v}\right)^{1/2}\left(\frac{v}{D_{AB}}\right)^{1/3}(c_{A,s} - c_{A,\infty})$$

$$= 0.678 \left(\frac{D_{AB}}{L}\right) \mathrm{Re}_L^{1/2}\, \mathrm{Sc}^{1/3}(c_{A,s} - c_{A,\infty}) \tag{7.95}$$

It is also interesting to reflect on the difference between pure diffusive mass transfer (Chapter 6) and our recent result (Eq. 7.94). Recall from Section 6.6.2 that mass transfer in diffusion-limited conditions was proportional to the

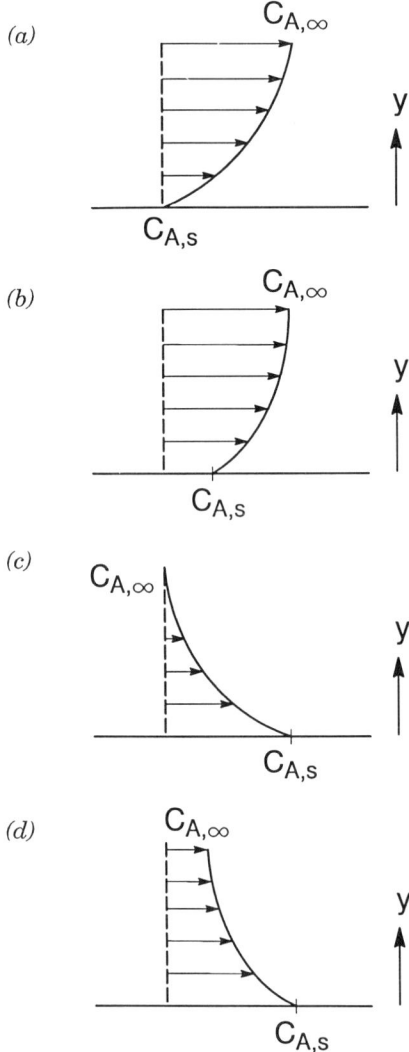

Figure 7.9. Four examples of mass-transfer problems with the same dimensionless boundary conditions.

diffusion coefficient, D_{AB}. However, in this section, where mass transfer is due to convective diffusion, the dependence of mass flux is on $D_{AB}^{2/3}$. Intuitively, we can imagine that the dependence here on D_{AB} is different because the diffusion-length scale in convective diffusion to the flat plate is smaller (and the concentration gradient is higher) than in pure diffusion. Exercise 7.16 uses the analysis of this section to compare aerosol deposition rates on different-size leaves.

It is worthwhile to think about other problems which might have the same solution (Eqs. 7.93–7.95). Consider the four general boundary conditions sketched in Figure 7.9. The situation sketched in Figure 7.9a is of course identical to the problem already considered in this section, laminar flow past a flat plate with a very fast surface reaction; the boundary conditions in terms of the nondimensional concentration are $\tilde{c}_A = 1$ at $y = \infty$ and $\tilde{c}_a = 0$ at $y = 0$. But what if the surface concentration is not zero (i.e., Figure 7.9b)? We quickly realize from Eq. 7.90 that as long as the surface concentration is constant, we get exactly the same dimensionless boundary conditions as the original problem in Figure 7.9a. Apparently, a very different looking problem is sketched in Figure 7.9c. Here, the surface is dissolving or producing some chemical species, such that the concentration of this species is constant along the surface, while the concentration out in the solution is zero. Here again, by substituting into Eq. 7.90, we find that $\tilde{c}_A = 1$ at $y = \infty$ and $\tilde{c}_a = 0$ at $y = 0$. This means that the solution represented by Eqs. 7.93–7.95 will also be appropriate for this problem. Finally, from examination of the problem sketched in Figure 7.9d, we find that the same solution obtains for the situation in which there is surface dissolution and a constant (nonzero) concentration out in the flow. It must be stressed, however, that the solution to the four problems given above can be the same only when all the assumptions of the analysis of the section are true: high Schmidt number (i.e., relatively thin concentration boundary layer) and laminar flow along a flat surface.

Finally, we can estimate the thickness of the concentration boundary layer in Figure 7.8 (from Levich, 1962). Interestingly enough, the estimate arises from an equation that looks much like film theory; we estimate the mass flux as

$$J_A = \frac{D_{AB}}{\delta_c}(c_{A,\infty} - c_{A,0}) \tag{7.96}$$

Since $c_{A,0} = 0$ in the analysis above, we find from Eqs. 7.94 and 7.96 that

$$\frac{\delta_c}{x} = 3\sqrt{\frac{\nu}{Ux}}\left(\frac{D_{AB}}{\nu}\right)^{1/3} \tag{7.97}$$

or incorporating the definition of the laminar momentum boundary layer on a flat plate (Eq. 5.121):

$$\delta_c = 0.6\,Sc^{-1/3}\delta \tag{7.98}$$

For example, in water the viscosity is usually near 0.01 cm^2/s and molecular diffusion coefficients for dissolved species are usually near 10^{-5} cm^2/s; hence, the Schmidt number is around 10^3. Equation 7.97 then suggests that the concentration boundary layer is 6% of the thickness of the momentum boundary layer in laminar flow along the flat plate.

Example 7.2. The derivations above require that the concentration boundary layer be thin relative to the laminar momentum boundary layer. How appropriate is the model for air systems when the diffusing species is (a) CO_2 and (b) 0.01-, 0.1-, and 1.0-μm-diameter spherical aerosol particles and $T = 25°C$.

SOLUTION. For part **a**, we can get the diffusion coefficient of CO_2 in air from Table 6.1, $D_{CO_2\text{-air}} = 0.164\,\text{cm}^2/\text{s}$. Taking the kinematic viscosity of air as $0.156\,\text{cm}^2/\text{s}$, the Schmidt number is $v/D_{CO_2\text{-air}} = 0.951$, and from Eq. 7.97 we get

$$\delta_c = 0.6(0.951)^{-1/3}\delta = 0.610\delta$$

So it seems that the assumption of a relatively thin concentration boundary layer may not be safe for CO_2 reactions with surfaces (and most other gas-diffusion problems in air). For part **b**, we estimate the diffusion coefficient of the spherical aerosol particles with the Stokes–Einstein equation (Eq. 6.25), modified for slip correction as discussed in Section 6.4. The Cunningham slip-correction factors from Table 2.1 are 22.7, 2.91, and 1.168 for the 0.01-, 0.1-, and 1.0-μm-diameter particles, respectively. For dry air, we can take the viscosity as $1.85 \times 10^{-4}\,\text{g-cm}^{-1}\text{-s}^{-1}$. Hence, for the 0.01-$\mu$m-diameter particle, the diffusion coefficient in dry air is

$$D_{0.01\text{-air}} = \frac{kTC_c}{3\pi\mu d} = \frac{1.38 \times 10^{-23}\,(\text{J/K})\,298\,\text{K}\,(\text{kg m}^2/\text{s}^2\,\text{J})\,22.7}{3\pi\,1.85 \times 10^{-4}\,(\text{g/cm s})(\text{kg}/1000\,\text{g})(100\,\text{cm/m})\,1 \times 10^{-8}\,\text{m}}$$

$$= 5.35 \times 10^{-8}\,\frac{\text{m}^2}{\text{s}}$$

In a similar way, we find for the other two aerosols that $D_{0.1\text{-air}} = 6.86 \times 10^{-10}\,\text{m}^2/\text{s}$, $D_{1.0\text{-air}} = 2.76 \times 10^{-11}\,\text{m}^2/\text{s}$. Hence, the concentration boundary layer thicknesses are estimated as

$$0.01\,\mu\text{m: Sc} = \frac{v}{D_{AB}} = \frac{1.56 \times 10^{-5}(\text{m}^2/\text{s})}{5.35 \times 10^{-8}(\text{m}^2/\text{s})} = 292; \qquad \delta_c = 0.6\,\text{Sc}^{-1/3}\delta = 0.090\delta$$

and similarly,

$$0.1\,\mu\text{m: Sc} = 22740; \qquad \delta_c = 0.021\delta$$

and

$$1.0\,\mu\text{m: Sc} = 565217; \qquad \delta_c = 0.0073\delta$$

We conclude that although the model of this section is probably going to be appropriate for water systems, we have to be careful when applying it to air systems (e.g., gas-transport problems). Certainly it looks like it will work with most aerosols because of the relatively high Schmidt numbers in these systems.

7.7.2.   Laminar Tube Flow with Fast Heterogeneous Reaction or Constant Surface Concentration

For laminar flow in a tube where mass is being transferred either to or from the surface of the tube, we must imagine as we did for the flat plate that the thickness of the boundary layer increases downstream. For example, in Figure 7.10a, we indicate a situation in which fluid is entering a pipe with some uniform concentration $c_{A,i}$.

The component reacts very quickly with the tube wall; hence, the concentration of A at the tube wall is zero. However, as we know from our consideration of the flat-plate situation, the A lost through reaction with the surface cannot be replaced instantaneously. Therefore, a concentration boundary layer develops along the inside surface of the pipe. As with the laminar momentum boundary layer, we say the concentration boundary layer is fully developed when the boundary layer reaches the center of the tube. Also, as with the flat plate in the last section, our modeling will require some knowledge of the relative thickness of the momentum and concentration boundary layers.

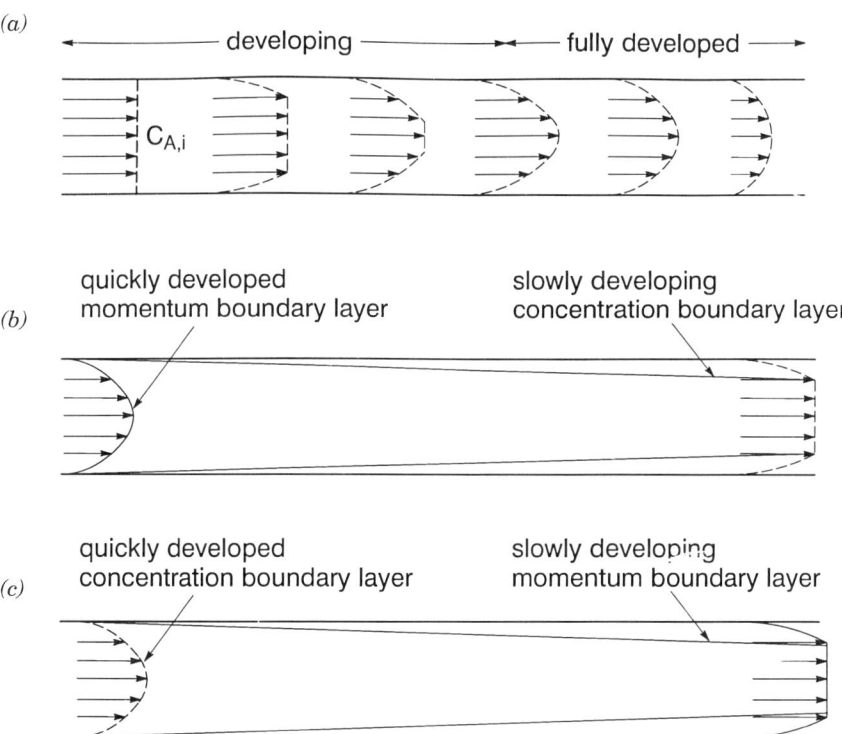

Figure 7.10. Boundary-layer development in laminar flow in a tube with fast surface reaction: (a) growth of concentration boundary layer, (b) high Schmidt number case, and (c) low Schmidt number case.

We can imagine that for the high Schmidt number case, the momentum boundary layer will develop much more quickly than the concentration boundary layer; this is shown in Figure 7.10b. For Schmidt numbers much less than unity, however, the concentration boundary layer will develop more quickly than the momentum boundary layer (Figure 10.7c). For Schmidt numbers near unity, the two boundary layers will grow at about the same rate.

As in the last section, we focus here only on the high Schmidt number case. Therefore, we can expect that, in general, the analysis will pertain to almost all liquid and aerosol mass-transfer problems. The basic differential equation governing mass transfer in this cylindrical coordinate system is the convective-diffusion equation provided earlier — Eq. 7.27. However, for the high Schmidt number case, we assume a fully developed Poiseuille flow; hence, there is no u_r or u_θ velocity component. Since the concentration profile will be assumed symmetrical about $r = 0$, there is also no variation in concentration in the θ direction, hence, no diffusion term related to variation of concentration in the θ direction. Finally, as in the last section, we assume that the streamwise diffusion component is much smaller than the cross-stream component. Thus, for steady-state conditions we arrive again at Eq. 7.29:

$$u_z \frac{\partial c_A}{\partial z} = D_{AB} \left[\frac{1}{r} \frac{\partial}{\partial r} \left(r \frac{\partial c_A}{\partial r} \right) \right] \tag{7.29a}$$

The next step is to nondimensionalize Eq. 7.29 in terms of the wall concentration (zero in the current example) and the uniform inlet concentration, $c_{A,i}$:

$$\tilde{c}_A = \frac{c_A - c_{A,s}}{c_{A,i} - c_{A,s}} \tag{7.99}$$

while the axial coordinate z is transformed in the following way:

$$\tilde{z} = \frac{z/a}{Re\,Sc} \tag{7.100}$$

Finally, r is nondimensionalized with the pipe radius:

$$\tilde{r} = \frac{r}{a} \tag{7.101}$$

The following boundary conditions are required:

$$\tilde{c}_A(\tilde{r}, 0) = 1; \quad \tilde{c}_A(1, \tilde{z}) = 0; \quad \frac{\partial \tilde{c}_A(0, \tilde{z})}{\partial \tilde{r}} = 0; \quad \tilde{c}_A(\tilde{r}, \infty) = 0 \tag{7.102}$$

The first boundary condition concerns the uniform initial concentration at the pipe inlet, the second concerns the wall boundary condition, the third is a result of the assumption of a symmetrical concentration distribution about the central pipe axis, and the fourth states that the ultimate concentration at a sufficiently far distance downstream must be uniform. Assuming Poiseuille flow, substituting Eqs. 7.99–7.101 into Eq. 7.29a, and considering the four boundary conditions, we solve Eq. 7.29 using the method of separation of variables (Skelland, 1974; Freidlander, 1977):

$$\tilde{c}_A(\tilde{r}, \tilde{z}) = \sum_{n=0}^{\infty} C_n R_n(\tilde{r}) \exp(-\lambda_n^2 \tilde{z}) \tag{7.103}$$

In this equation, C_n is an infinite series of constants, R_n represents a function that depends only on the dimensionless coordinate $\tilde{r}$, and λ_n are the eigenvalues.

In the pipe-flow problem under consideration, we are often most interested in the average concentration in the pipe, which we calculate as the classic "mixing cup" concentration:*

$$c_{A,\text{ave}} = \frac{2\pi \int_0^a c_A(r, \tilde{z})u(r)r\,dr}{\pi a^2 u_{\text{ave}}} = \frac{2}{a^2 u_{\text{ave}}} \int_0^a c_A(r, \tilde{z})u(r)r\,dr \tag{7.104}$$

For example, in the removal of aerosol particles along the walls of the trachea or other lung airway, we would typically be interested in total removal along a specific length of pipe, which is easily calculated if we know $c_{A,i}$ and the average exit concentration, $c_{A,\text{ave}}$. Skelland (1974) and Friedlander (1977) also discuss this problem; the solution is

$$\frac{c_{A,i} - c_{A,\text{ave}}}{c_{A,i} - c_{A,s}} = \frac{c_{A,\text{ave}} - c_{A,i}}{c_{A,s} - c_{A,i}} = 1 - 8 \sum_{n=0}^{\infty} \frac{G_n}{\lambda_n^2} \exp(-\lambda_n^2 \tilde{z}) \tag{7.105}$$

where G_n is equal to $(C_n/2)\, R'_n(1)$. Table 7.1 provides values for G_n and λ_n, which are attributed to Sellars et al. (1956).

For $n > 4$, the eigenvalues and constants in Eq. 7.105 can be estimated using the following equations:

$$\lambda_n = 4n + \frac{8}{3} \tag{7.106}$$

*Note that we normally think of the average concentration at any point in a pipe as the spatial average over the cross section, just as was calculated in Section 7.3. Another useful average is the so-called "mixing-cup" average, which as shown above is an average weighted by the local velocity (Levinspiel, 1972). This strange name comes from the idea that the average defined by Eq. 7.104 would be the average concentration in a cup that could somehow catch all the flow in the pipe for a short interval of time.

Table 7.1. The First Five Constants and Eigenvalues for the Evaluation of Eq. 7.104

n	λ_n^2	G_n
0	7.312	0.749
1	44.62	0.544
2	113.8	0.463
3	215.2	0.414
4	348.5	0.382

Source: Friedlander, *Smoke, Dust and Haze*, John Wiley & Sons, New York, p. 75, © 1977. Used with permission of S. K. Friedlander, 1995.

and

$$G_n = 1.01276\lambda_n^{-1/3} \qquad (7.107)$$

With these values we can calculate the mass transfer in the pipe to any desired accuracy. Exercise 7.17 shows how to use these equations for computation of aerosol deposition in model trachea. Note that Eq. 7.104 is valid for any high Schmidt number laminar-flow problem satisfying the same dimensionless boundary conditions as specified in Eq. 7.102, for example, the average concentration of some component dissolving from pipe walls (i.e., $c_{A,s}$ = constant $> c_{A,i}$).

Finally, it is valuable to have an estimate of the distance required to obtain a fully developed concentration boundary layer. Levich (1962) derived the following formula:

$$L_c \simeq \frac{a}{3.32}\,\mathrm{Sc}\,\mathrm{Re} \qquad (7.108)$$

This can be compared with the estimate from Exercise 5.14 for the distance required to achieve fully developed Poiseuille flow:

$$L_e \simeq \frac{a}{8.7}\,\mathrm{Re}$$

It is a simple matter to verify that these two formulas are consistent with our earlier assertion that for high Schmidt number (i.e., $\mathrm{Sc} \gg 1$), we can assume that $L_c \gg L_e$.

Example 7.3. Chlorine is added to drinking water after treatment as a disinfectant. After the water leaves the treatment plant it flows through piping until it arrives at the consumer (e.g., your water tap). We know that the concentration of chlorine decreases as the water flows along the pipe. Although some of the early reactions with chlorine are often with inorganic compounds in the water itself, it is believed that in the farther sections of pipe, chlorine decay is caused by reactions with organic matter in the water and by oxidation reactions along the pipe surface. In a particular section of pipe far from the plant, the initial concentration of chlorine Cl_2 is 0.5 mg/L. The pipe diameter is 5 cm, the average velocity of flow is 1 cm/s, and the temperature is 20°C. Find the maximum decrease in concentration of chlorine in a 75-m pipe using the analysis developed in this section. Assume that chlorine is only consumed at the pipe surface.

SOLUTION. The maximum uptake of chlorine at the pipe walls will occur when the reaction is mass-transport limited, that is, when $c_{A,s} = 0$. Rearranging Eq. 7.105, we find, therefore, that

$$\frac{c_{A,\text{ave}}}{c_{A,i}} = 8 \sum_{n=0}^{\infty} \frac{G_n}{\lambda_n^2} \exp(-\lambda_n^2 \tilde{z})$$

The Reynolds and Schmidt numbers are

$$\text{Re} = \frac{2au_{\text{ave}}}{\nu} = \frac{5\,\text{cm}(1\,\text{cm/s})}{0.01\,\text{cm}^2/\text{s}} = 500 \quad \text{and} \quad \text{Sc} = \frac{\nu}{D_{AB}} = \frac{0.01\,\text{cm}^2/\text{s}}{1.22 \times 10^{-5}\,\text{cm}^2/\text{s}} = 820$$

where the diffusion coefficient of chlorine is from Table 6.3 and the kinematic viscosity is 0.01 cm²/s. Hence, the dimensionless axial coordinate is

$$\tilde{z} = \frac{z/a}{\text{Re Sc}} = \frac{[(75\,\text{m})\,100\,\text{cm/m}]/2.5\,\text{cm}}{500 \times 820} = 7.32 \times 10^{-3}$$

Plugging in the values for λ_n^2 and G_n from Table 7.1, we quickly find that after five terms, the addition of extra terms is causing a change of less than one hundredth of a percent. So we stop after five terms in the summation and say that the answer converges to

$$\frac{c_{A,\text{ave}}}{c_{A,i}} = 0.866$$

This means that the greatest possible change in chlorine concentration in the 75-m pipe due solely to wall reactions is about 13%.

7.7.3. Laminar Flow Past Sphere with Fast Heterogeneous Reaction or Constant Surface Concentration

In many systems in the environment and in pollution-control engineering, there is mass transfer between a spherical particle and the surrounding fluid. For example, biogenic and mineral particles settling in the ocean dissolve over time, and this dissolution affects the ocean chemistry. We can imagine that the concentration of the dissolving component at the particle surface is constant, and decreases out into the fluid. Under certain limiting conditions, fast reactions inside liquid droplets in the atmosphere may cause the reacting and diffusing component concentration to be zero at the particle surface; mass transfer to the particle then occurs across a boundary layer around the droplet. In Chapter 8, we consider the problem of diffusional transport to spherical collectors in a filter.

In this section, we examine the limiting condition of mass transfer to or from a particle in unidirectional laminar flow. The situation is indicated in Figure 7.11a. The viscous flow around the particle will be given by the Stokes flow solution studied in Section 5.9.2. As in the previous two sections, we will imagine that a concentration boundary layer grows around the particle. In the case shown above, that boundary layer starts forming at $\theta = 0$ and grows as we follow the surface around the particle. As before, we also consider the limit of high Schmidt numbers; therefore, the concentration boundary layer will be close enough to the particle surface that simple functions can be assumed for velocity near the particle surface. In normal spherical coordinates, we usually also discuss the azimuthal angle that comes out of the plane; however, since we are assuming Stokes flow, there is no variation in velocity or concentration in the azimuthal direction. The appropriate form of the convective-diffusion equation is, therefore (Appendix III),

$$\frac{\partial c_A}{\partial t} + u_r \frac{\partial c_A}{\partial r} + u_\theta \frac{1}{r} \frac{\partial c_A}{\partial \theta} = D_{AB} \left[\frac{1}{r^2} \frac{\partial}{\partial r} \left(r^2 \frac{\partial c_A}{\partial r} \right) + \frac{1}{r^2 \sin \theta} \frac{\partial}{\partial \theta} \left(\sin \theta \frac{\partial c_A}{\partial \theta} \right) \right]$$

(7.109)

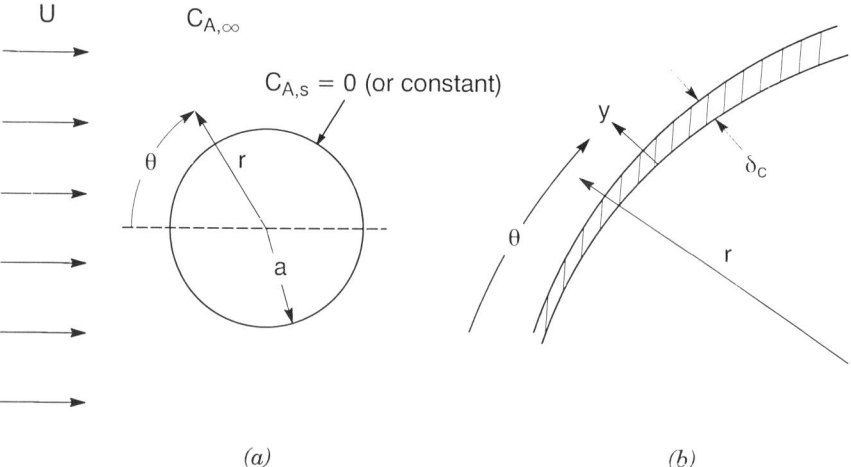

(a)   *(b)*

Figure 7.11. Mass transfer to spherical surface: (*a*) general coordinates and (*b*) small section of concentration boundary layer.

As usual, we will attempt to simplify Eq. 7.109. Note that for the thin concentration boundary layer assumed, we make our normal assumption that diffusive mass transport in the θ direction is small relative to the corresponding term in the r direction. Hence, for steady-state conditions, Eq. 7.109 reduces to

$$u_r \frac{\partial c_A}{\partial r} + u_\theta \frac{1}{r} \frac{\partial c_A}{\partial \theta} = D_{AB} \left[\frac{1}{r^2} \frac{\partial}{\partial r} \left(r^2 \frac{\partial c_A}{\partial r} \right) \right] \qquad (7.110)$$

The next step in the simplification of the problem may surprise the reader somewhat. Note that if the concentration boundary layer is thin relative to the particle radius, the concentration boundary layer does not feel the effect of the curvature of the particle surface. Therefore, if we are careful, it may be possible to imagine the same boundary layer along a flat surface, as suggested in Figure 7.12b. This has been a classic tool in physics; the reader will recall a similar

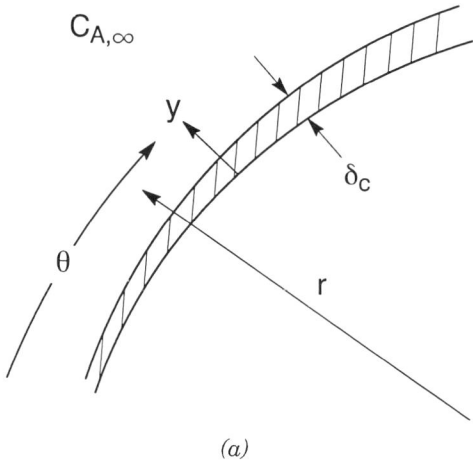

(a)

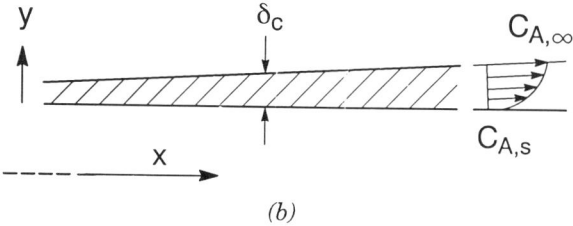

(b)

Figure 7.12. Transformation of (a) thin concentration boundary layer in spherical coordinates (a to b) equivalent thin concentration boundary layer in rectangular coordinates.

simplification in our discussion of the double layer surrounding colloidal particles in Section 3.5.1. Therefore, our governing equation reduces to a familiar equation from Section 7.7.1:

$$u \frac{\partial c_A}{\partial x} + v \frac{\partial c_A}{\partial y} = D_{AB} \frac{\partial^2 c_A}{\partial y^2} \tag{7.85}$$

As usual, we next nondimensionalize concentration, also as in Section 7.7.1:

$$\tilde{c}_A = \frac{c_A - c_{A,s}}{c_{A,\infty} - c_{A,s}} \tag{7.90}$$

where $c_{A,\infty}$ is the concentration of A far from the particle surface and $c_{A,s}$ is the (constant) concentration of A at the particle surface. Hence, the boundary conditions are also those familiar from previous sections, $\tilde{c}_A = 1$ at $y = \infty$ and $\tilde{c}_A = 0$ at $y = 0$. The last physical ingredient in the solution of this problem is the proper functions for the velocities in the boundary layer. For distances not far from the particle surface, u and v can be expressed in terms of a simplified stream function. Substituting these into Eq. 7.85, Levich (1962) showed, after a lengthy and ingenious analysis, that the solution for the given boundary conditions is

$$\frac{c_A - c_{A,s}}{c_{A,\infty} - c_{A,s}} = \frac{1}{1.15} \int_0^z \exp\left(-\frac{4}{9} z^3\right) dz \tag{7.111}$$

where z is given by

$$z = \left(\frac{3U}{4D_{AB}a^2}\right)^{1/3} \frac{y \sin\theta}{\left[\theta - \frac{\sin 2\theta}{2}\right]^{1/3}} \tag{7.112}$$

As the reader recalls, one of the critical assumptions of this analysis is that the boundary layer is thin relative to the particle radius. Actually, although this is quite legitimate for most of the region $\theta < \pi$, as θ approaches π, the boundary-layer thickness approaches something on the order of the particle radius. Hence, the analysis becomes less and less accurate on the back side of the particle. Nevertheless, Levich (1962) points out that by the time θ gets close to π, the boundary layer has become thick enough (mathematically and physically) that there is not much mass transfer anyway. This means that even though the analysis is in error near $\theta = \pi$, the error can probably be ignored.

We are interested in computing the flux of material to the particle surface. As usual, this is obtained from Fick's first law evaluated at the particle surface

$(y = 0)$:

$$J_A = -D_{AB}\left(\frac{\partial c_A}{\partial y}\right)_{y=0} = \frac{-D_{AB}(c_{A,\infty} - c_{A,s})}{1.15}\left(\frac{3U}{4D_{AB}a^2}\right)^{1/3}\frac{\sin\theta}{\left[\theta - \dfrac{\sin 2\theta}{2}\right]^{1/3}}(7.113)$$

The concentration boundary layer thickness is obtained from a "film-theory" type analysis:

$$\delta_c = \frac{D_{AB}(c_{A,\infty} - c_{A,s})}{J_A} = \frac{1.15\left[\theta - \dfrac{\sin 2\theta}{2}\right]^{1/3}}{\sin\theta}\left(\frac{4D_{AB}a^2}{3U}\right)^{1/3} \qquad (7.114)$$

The reader will find that this equation predicts a boundary layer of infinite thickness as $\theta \to \pi$, which is not physically realistic. This is a problem related to the approximations used in the analysis, and is discussed by Levich (1962). Hence, Eq. 7.114 should probably only be applied for $\theta < \pi/2$.

Finally, to get the total mass transfer rate from the particle we need to integrate Eq. 7.113 over the total particle surface. Since there is no variation in flux in the azimuthal angle, we can integrate over the differential strip shown in Fig. 5.9 (see also Eq. 5.56):

$$W_A = -\int_S \mathbf{n}\cdot\mathbf{J}_A\,dS = 2\pi a^2\int_0^\pi J_A\sin\theta\,d\theta$$

$$= \frac{D_{AB}(c_{A,\infty} - c_{A,s})a^{4/3}}{1.15}\left(\frac{3U}{4D_{AB}}\right)^{1/3}2\pi\int_0^\pi\frac{\sin^2\theta}{\left[\theta - \dfrac{\sin 2\theta}{2}\right]^{1/3}}d\theta \qquad (7.115)$$

Evaluating the integral, we find the following:

$$W_A = 7.98a^{4/3}D_{AB}^{2/3}U^{1/3}(c_{A,\infty} - c_{A,s}) = 6.33aD_{AB}\,\mathrm{Pe}^{1/3}(c_{A,\infty} - c_{A,s}) \quad (7.116)$$

where the Peclet number is defined here as

$$\mathrm{Pe} = \frac{2aU}{D_{AB}} \qquad (7.117)$$

The reader recalls that the analysis of this section requires a thin boundary layer, or large Schmidt number. Also recall that the Stokes flow assumed in this section requires a very low Reynolds number; hence, a high Schmidt number also implies a large Peclet number, since $\mathrm{Sc} = \mathrm{Pe}/\mathrm{Re}$. So Eq. 7.116 can be considered to apply for high Schmidt and Peclet numbers only. A later analysis relaxed these restrictions by considering concentration boundary layer curvature at lower Schmidt and Peclet numbers. This equation is given as (Clift et al., 1978)

$$W_A = 2\pi a D_{AB}(0.92 + 0.991\ \mathrm{Pe}^{1/3})(c_{A,\infty} - c_{A,s}) \qquad (7.118)$$

However, even Eq. 7.118 becomes inaccurate as Peclet numbers approach zero. Clift et al. (1978) recommend that the following formula will approximate to within 2% an accurate numerical solution for all Peclet numbers:

$$W_A = 2\pi a D_{AB}[1 + (1 + \mathrm{Pe})^{1/3}](c_{A,\infty} - c_{A,s}) \qquad (7.119)$$

As we have found in the previous three sections, the boundary conditions for a dissolving surface are identical to the fast reaction surface; hence, the equations above can also be used to predict mass transfer from the particle to the fluid when the particle is dissolving or sublimating (assuming no significant temperature effects and constant $c_{A,s}$).

Finally, the reader is encouraged to compare Eq. 7.119 (convective diffusion to a sphere) to the pure diffusion limit analyzed in Section 6.6.2. (Eq. 6.59). Note that as Pe → 0, the predictions of Eqs. 7.119 and 6.59 converge. This of course makes sense, since the case Pe → 0 corresponds to pure diffusion.

The analysis of this section is used in Exercise 7.20 to predict the dissolution rates of biological particles in the ocean.

7.8. MASS-TRANSFER COEFFICIENTS, MODELS, AND CORRELATIONS FOR LAMINAR AND TURBULENT FLOWS

In Section 7.7 we found that for several relatively simple laminar flow problems, classic exact analyses exist for predicting mass transfer to and from surfaces. In many other situations in the environment and in pollution-control engineering, such analyses do not exist. For example, none of the analyses of the previous section can be used to predict the dissolution of solid particles composing a porous medium. Likewise, we do not yet in this book have a way to estimate mass transfer to surfaces in turbulent flows. (Consider, for example, mass transfer of contaminants to or from sediments in turbulent river flows, or transfer of gases to a raindrop in a turbulent cloud.) As we know from our previous discussion of momentum and mass transfer, turbulent flows are much more complicated than laminar flows. In fact, for the turbulent flows (and some complicated laminar flows), we will have to rely mainly on experimentally determined mass-transfer correlations (Section 7.8.2).

The purpose of this section is to develop approaches for modeling these more complex problems of mass transfer at solid surfaces or at surfaces of a second phase.

7.8.1. Mass-Transfer Coefficients and the Sherwood Number

We begin this section with a review of the some of the mass flux expressions developed in Chapters 5 and 6. For example, in Section 6.6.1, we developed steady-state mass flux expressions for one-dimensional diffusion of A through a stagnant column of B:

$$J_A = \frac{D_{AB}}{z_2 - z_1}(c_{A,1} - c_{A,2}) \tag{7.120}$$

Here $z_2 - z_1$ is the thickness of the diffusion layer, $c_{A,2} - c_{A,1}$ is the concentration difference across the stagnant layer, and we have assumed that the molar concentration is constant and the concentration of background gas B is much greater than the diffusing gas A. In Section 6.6.2 we derived a steady-state expression for diffusion of component A to the surface of a particle where there was a heterogeneous reaction:

$$J_A = \frac{W_A}{4\pi a^2} = -\frac{D_{AB}}{a}(c_{A,\infty} - c_{A,s}) \tag{7.121}$$

where a is the particle radius, $c_{A,s}$ is the surface concentration of A, and $c_{A,\infty}$ is the concentration of A far from the surface. In Section 7.1, we developed a simple surface-renewal model for interphase mass transfer of pure gas A into a liquid B:

$$J_A = 2c_{A,0}\sqrt{\frac{D_{AB}U}{\pi L}} \tag{7.5}$$

where U is the liquid velocity, $c_{A,0}$ is the interfacial concentration of A, and L is length of the exposed surface. In Section 7.7.1, the average flux of A to (or from) a flat plate in laminar flow was found to be

$$J_{A,\text{ave}} = -0.678\left(\frac{D_{AB}}{L}\right)\text{Re}_L^{1/2}\,\text{Sc}^{1/3}(c_{A,\infty} - c_{A,s}) \tag{7.95}$$

where L is the plate length. In Section 7.7.3 the total mass flux at the surface of a sphere was found (see Eq. 7.116), to be

$$J_A = \frac{-W_A}{4\pi a^2} = -0.504\frac{D_{AB}}{a}\text{Re}^{1/3}\,\text{Sc}^{1/3}(c_{A,\infty} - c_{A,s}) \tag{7.122}$$

Even though there are important differences between the geometry and physics of these five problems, previous researchers noticed some important similarities between all these (and other) mass-transfer expressions:

1. Flux is proportional to a concentration or concentration difference characteristic of the mass-transfer process.

2. Flux is inversely proportional to some length scale characteristic of the mass-transfer process.

3. Flux is proportional to the diffusion coefficient raised to a power between 0.5 and 1. (Don't forget the D_{AB} in the denominator of the Schmidt number!) In the next section we will find that almost all mass-transfer rate expressions (laminar and turbulent flows) involve the diffusion coefficient raised to a power between 0.5 and 1.

The first of these observations led to the idea of a *mass-transfer coefficient, k_c,* which is defined by the following expression:

$$J_A = k_c c_A^* \tag{7.123}$$

where c_A^* is a characteristic concentration scale or concentration difference, and where the flux is, as usual, always in the direction of the concentration decrease. For example, from Eq. 7.120, the mass-transfer coefficient would be

$$k_c = \frac{D_{AB}}{z_1 - z_2} \tag{7.124}$$

whereas for Eq. 7.95, the mass-transfer coefficient would be

$$k_c = 0.678 \left(\frac{D_{AB}}{L} \right) \mathrm{Re}_L^{1/2} \, \mathrm{Sc}^{1/3} \tag{7.125}$$

It is just as simple to calculate the mass-transfer coefficients for the models represented in Eqs. 7.5, 7.121, and 7.122. Another useful approach is to consider all three observations listed above and the form of the equations discussed. Dimensional analysis suggests that by multiplying J_A by a characteristic length scale, and dividing by the diffusion coefficient and the characteristic concentration scale, we should arrive at a dimensionless parameter group. In fact, this group is called the *Sherwood number:*

$$\mathrm{Sh} = \frac{J_A L}{D_{AB} c_A^*} = \frac{k_c L}{D_{AB}} \tag{7.126}$$

Here L is the characteristic length scale. For example, from Eqs. 7.5, 7.94, and

7.120–7.121, 7.122we find:

$$\left\{\begin{array}{l}\text{pure diffusion of } A \\ \text{through stagnant } B\end{array}\right\}: \text{Sh} = \frac{J_A(z_2 - z_1)}{D_{AB}(c_{A,1} - c_{A,2})} = 1 \qquad (7.127)$$

$$\left\{\begin{array}{l}\text{pure diffusion of } A \\ \text{to surface of sphere}\end{array}\right\}: \text{Sh} = \frac{2aJ_A}{D_{AB}(c_{A,\infty} - c_{A,s})} = 2 \qquad (7.128)$$

$$\left\{\begin{array}{l}\text{diffusion of } A \text{ in simple} \\ \text{surface-renewal example}\end{array}\right\}: \text{Sh} = \frac{J_A L}{D_{AB}c_{A,0}} = \frac{2}{\sqrt{\pi}}\,\text{Pe}^{1/2} = \frac{2}{\sqrt{\pi}}\,\text{Re}^{1/2}\,\text{Sc}^{1/2}$$

$$(7.129)$$

$$\left\{\begin{array}{l}\text{flux of } A \text{ to flat} \\ \text{plate in laminar flow}\end{array}\right\}: \text{Sh} = \frac{J_A L}{D_{AB}(c_{A,\infty} - c_{A,s})} = 0.678\,\text{Re}_L^{1/2}\,\text{Sc}^{1/3}$$

$$(7.130)$$

$$\left\{\begin{array}{l}\text{flux of } A \text{ to sphere in} \\ \text{Stokes flow (high Sc)}\end{array}\right\}: \text{Sh} = \frac{2aJ_A}{D_{AB}(c_{A,\infty} - c_{A,s})} = 1.01\,\text{Pe}^{1/3} = 1.01\,\text{Re}^{1/3}\,\text{Sc}^{1/3}$$

$$(7.131)$$

Note for the two pure-diffusion cases that the Sherwood number is equal to a constant; this will in general be true for all pure-diffusion problems. Note that the Sherwood number for the three convective-diffusion cases is also a constant as long as the Reynolds and Schmidt numbers are the same. In fact, for these five cases, the computation of the Sherwood number adds nothing to our ability to calculate mass-transfer rates. It does give us insight into the general parameters of importance in mass-transfer processes. The real power of the Sherwood number comes into play when we note that for laminar- or turbulent-flow problems with the same geometry, Reynolds number, Schmidt number, and dimensionless boundary conditions, the dimensionless mass-transfer rate (expressed by the Sherwood number) is often a constant. This is an extremely useful technique for "correlating" experimental data and is pursued in the following sections.

7.8.2. Mass-Transfer Models and Analogies

7.8.2.1. Film Theory and Other Models. We have already hinted that with the exception of a few laminar-flow, mass-transfer models (such as those studied in Section 7.7), prediction of mass transfer by exact analysis has been very difficult. This is especially true for mass-transfer in turbulent flows. This problem has led to the study of mass-transfer models and analogies. By far the most widespread mass transfer model has been the *film model* or *film theory*. As mentioned above and in Example 6.1 (Chapter 6), this model was motivated by the elegant steady-state analysis of diffusion of A through a column of stagnant B (where the concentration of A is much less than the concentration of B), as illustrated by Eq. 7.120 and

$$J_A = \frac{D_{AB}}{\delta_f} c_A^*$$ (7.132)

where δ_f is the so-called *film thickness* and c_A^* is a characteristic concentration or concentration difference.* In application to mass transfer at rigid boundaries in turbulent flows, the film model envisions mass transfer caused only by diffusion through a stagnant film at the solid boundary. It is important to fully appreciate the rather severe approximation implied by the film theory in the case of turbulent flows. For example, consider mass transfer to a solid surface with fast heterogeneous reaction in a turbulent flow, as shown in Figure 7.13a. The film theory imagines that the rather complicated mass transfer to the surface of the plate can be modeled by diffusion through a completely stagnant fluid layer near the surface (Figure 7.13b). This has been a traditional way of thinking about mass transport near surfaces and interfaces between two fluids. Nevertheless, several problems with the film theory have been noted. First, as we learned in Chapter 5, when solid surfaces are concerned, only the molecules adjacent to the boundary are motionless (although even these can be ex-changed with other molecules as a result of molecular diffusion). Thus, there really can be no stagnant diffusion film in convective mass-transfer problems near solid surfaces. Second, even if we accept Eq. 7.132, there is no a priori way to predict the film thickness, δ_f. Note further that even if δ_f could be predicted, the dependence of J_A on D_{AB} is generally wrong: in turbulent flows, mass transfer to and from surfaces usually varies as a power of D_{AB} between 0.5 and 1.0. As we stated earlier, even though molecular diffusion of A must be a critical step in the mass-transfer process at some scale of the process (evidently some small scale near the surface), the overall mass transfer is ultimately not well represented by the exact analysis of molecular diffusion in a stagnant fluid in which the mass flux is proportional to D_{AB}. Other possible problems have been noted with the film theory as we have learned more about the complexity of turbulent flows. As stated in Chapter 5, turbulent bursts may be an important source of momentum transfer near surfaces. By analogy, these turbulent bursts are also expected to be significant sources of mass transport near surfaces. Hence, even if there is a thin layer of almost stagnant fluid where diffusion is important, it is periodically (and randomly) interrupted by the bursts. These bursts are expected to enhance the average mass transport. We conclude then that even though the film theory provides a convenient way of thinking about mass transfer, it is a poor fundamental predictive theory. Despite these problems, the film theory has been very useful, particularly in understanding gas–liquid reactions (Danckwerts, 1970).

---

*Note from comparison of Eqs. 7.123 and 7.132 that the film theory yields a mass-transfer coefficient — $k_c = D_{AB}/\delta_f$.

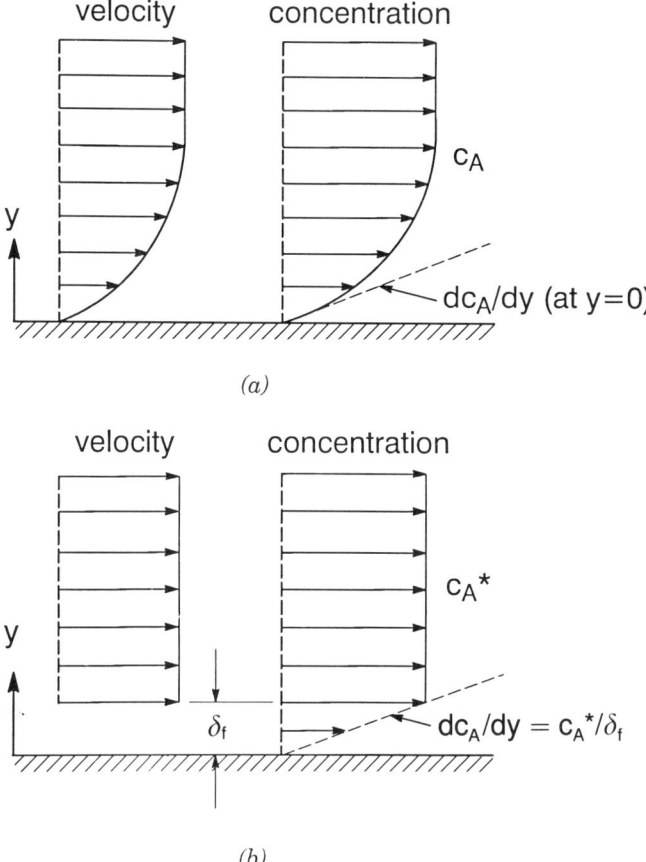

Figure 7.13. Comparison of (a) actual average velocity and concentration profiles in a turbulent boundary layer and (b) approximation of velocity and concentration according to the film theory. Note that the film thickness is generally not the same as the thickness of the concentration boundary layer.

Penetration or *surface-renewal* models of mass-transfer modeling are more appropriate to interphase mass transfer (Section 7.9). In these models, we imagine that fresh fluid is transported to the interface between two fluids (e.g., two immiscible liquids or a liquid and a gas), remains there for a finite amount of time in which pure diffusion occurs, and then leaves the interfacial region (recall the demonstration in Section 7.1). For example, the Higbie (1935) model is based on the assumption of a constant-exposure time of surface elements of one phase to a uniform second phase. The resulting mass-transfer coefficient is

$$k_c = \sqrt{\frac{4D_{AB}}{\pi\theta}} \tag{7.133}$$

Here θ is the exposure time. Unfortunately, it has not generally been possible to relate the exposure time to characteristics of the flow or turbulence. Hence, the Higbie model has not universally been useful in the prediction of mass-transfer rates. As with the film model, it has been useful in predicting the effects of reactions on the rate of interphase mass transfer (Danckwerts, 1970, and Section 7.9.4).

7.8.2.2. Analogies. One of the most useful approaches to solving mass (and heat) transfer-engineering problems (especially in turbulent flows) has its origin in *mass (or heat) transfer analogies.* For example, recall the equation from Chapter 5 for steady-state momentum transfer in a laminar boundary layer along a flat plate:

$$u \frac{\partial u}{\partial x} + v \frac{\partial u}{\partial y} = v \frac{\partial^2 u}{\partial y^2} \tag{5.108}$$

The reader may have noted its similarity to the equation from Section 7.7.1 for steady-state mass transfer in a laminar boundary layer on a flat plate:

$$u \frac{\partial c_A}{\partial x} + v \frac{\partial c_A}{\partial y} = D_{AB} \frac{\partial^2 c_A}{\partial y^2} \tag{7.85}$$

For nonradiative steady-state heat transfer along a flat plate in laminar flow, the following equation is applicable (Treybal, 1968):

$$u \frac{\partial T}{\partial x} + v \frac{\partial T}{\partial y} = \alpha \frac{\partial^2 T}{\partial y^2} \tag{7.134}$$

Here T is temperature and α is called the *thermal diffusivity*:

$$\alpha = \frac{k}{C_p \rho} \tag{7.135}$$

where k is the *thermal conductivity* (energy/T-L-°C), C_p is the specific heat of the medium (energy/M-°C), and ρ is the fluid density. In these three equations, the terms on the left-hand sides represent convective transport of momentum, mass, or energy and the terms on the right-hand sides represent the diffusive transport of momentum, mass, or energy; the coefficients of momentum, mass, and heat diffusivity are v, D_{AB}, and α. It is natural to wonder whether solutions of one problem (either heat, mass, or momentum transport along a flat plate), or experimental data collected in one type of experiment, can be used to solve problems in one of the analogous situations. This may be possible, and one requirement would be that the boundary conditions are the same in the two analogous problems. This would involve a comparison of the following dimensionless variables:

$$\text{dimensionless velocity:} \quad \tilde{u} = \frac{u - u_s}{U - u_s} \tag{7.136}$$

$$\text{dimensionless concentration:} \quad \tilde{c}_A = \frac{c_A - c_{A,s}}{c_{A,\infty} - c_{A,s}} \tag{7.90}$$

$$\text{dimensionless temperature:} \quad \tilde{T} = \frac{T - T_s}{T_\infty - T_s} \tag{7.137}$$

where u_s is the x component of velocity at the surface (equal to zero in the flat-plate problem) and T_s is the surface temperature of the plate.

A classic example of the application of this approach is to use the Blasius solution for the momentum boundary layer of Section 5.12 and Exercise 5.16 to predict mass transfer from (or to) the surface of a flat plate in laminar flow (recall Section 7.7.1 and Figure 7.8). There are two main requirements for making this analogy. First, the Schmidt number must equal unity, or $v = D_{AB}$; this requirement makes Eqs. 5.108 and 7.85 mathematically equivalent. Second, the dimensionless boundary conditions must be identical. This is indeed the case, since by inspection (with $u_s = 0$):

$$\tilde{u} = \frac{u - u_s}{U - u_s} = \frac{u}{U} = \begin{cases} 0 & \text{at } y = 0 \\ 1 & \text{at } y = \infty \end{cases} \qquad \tilde{c}_A = \frac{c_A - c_{A,s}}{c_{A,\infty} - c_{A,s}} = \begin{cases} 0 & \text{at } y = 0 \\ 1 & \text{at } y = \infty \end{cases}$$

Next, using the results of Exercise 5.16 and our nondimensional velocity defined above, note that the gradient in velocity at the surface of the plate can be expressed as

$$\left.\frac{\partial \tilde{u}}{\partial y}\right|_{y=0} = \frac{1}{4}\left(\frac{U}{vx}\right)^{1/2} f''(0) = \frac{1}{4x}\left(\frac{Ux}{v}\right)^{1/2} f''(0) = \frac{1.328}{4x} \operatorname{Re}_x^{1/2} \tag{7.138}$$

The last step in the equation above is based on Table 5.1, where $f'' = 1.328$ at $\eta = 0$. The analogy of mass and momentum transport requires:

$$\left.\frac{\partial \tilde{u}}{\partial y}\right|_{y=0} = \left.\frac{\partial \tilde{c}_A}{\partial y}\right|_{y=0} = \frac{1}{(c_{A,\infty} - c_{A,s})} \left.\frac{\partial c_A}{\partial y}\right|_{y=0} \tag{7.139}$$

Combining Eqs. 7.138 and 7.139, we find:

$$\left.\frac{\partial c_A}{\partial y}\right|_{y=0} = \frac{1.328}{4x} \operatorname{Re}_x^{1/2}(c_{A,\infty} - c_{A,s}) \tag{7.140}$$

Finally, from Fick's first law:

$$J_A = -D_{AB} \frac{\partial c_A}{\partial y}\bigg|_{y=0} = 0.332 \left(\frac{D_{AB}}{x}\right) \mathrm{Re}_x^{1/2}(c_{A,s} - c_{A,\infty}) \qquad (7.141)$$

We can compare this equation with the exact analysis from Section 7.7.1:

$$J_A = -D_{AB} \frac{\partial c_A}{\partial y}\bigg|_{y=0} = 0.34 \left(\frac{D_{AB}}{x}\right) \mathrm{Re}_x^{1/2} \, \mathrm{Sc}^{1/3}(c_{A,s} - c_{A,\infty}) \qquad (7.94)$$

Note that the mass-transfer analogy with momentum transport in the viscous boundary layer along a flat plate results in an equation nearly identical to the exact analysis, as long as the condition Sc = 1 is maintained. Exactly the same approach can be used to calculate heat transfer from a flat plate in laminar flow to the surrounding fluid. The equation analogous to Eq. 7.139 is

$$\frac{\partial \tilde{u}}{\partial y}\bigg|_{y=0} = \frac{\partial \tilde{T}}{\partial y}\bigg|_{y=0} = \frac{1}{(T_\infty - T_s)} \frac{\partial T}{\partial y}\bigg|_{y=0} \qquad (7.142)$$

Fourier's law of heat conduction is

$$q = -k \frac{dT}{dy} \qquad (7.143)$$

where q is the heat flux (energy/T-L^2). An analysis identical to that above then results in

$$q = -\alpha C_p \rho \frac{\partial T}{\partial y}\bigg|_{y=0} = 0.332 C_p \rho \left(\frac{\alpha}{x}\right) \mathrm{Re}_x^{1/2}(T_s - T_\infty) \qquad (7.144)$$

where, as before, we require that $v = \alpha$. The ratio of v and α is called the *Prandtl number*:

$$\mathrm{Pr} = \frac{v}{\alpha} = \frac{\mu}{\rho\alpha} = \frac{\mu C_p}{k} \qquad (7.145)$$

so use of Eq. 7.144 requires that Pr = 1.

The successes of calculations like those shown above emboldened scientists to extend analogies of heat, mass, and momentum to other flow geometries. Before getting to some of these analogies and later experimental correlations, it is a good idea to review the different analogous dimensionless groups and other expressions we have been discussing and will be using. These are presented in Table 7.2.

Table 7.2. Relevant Momentum, Mass, and Heat-Transfer Quantities and Dimensionless Groups[a]

Momentum Transfer	Mass Transfer	Heat Transfer
Dimensionless velocity:	Dimensionless concentration:	Dimensionless temperature:
$\tilde{u} = \dfrac{u - u_s}{U - u_s}$	$\tilde{c}_A = \dfrac{c_A - c_{A,s}}{c_{A,\infty} - c_{A,s}}$	$\tilde{T} = \dfrac{T - T_s}{T_\infty - T_S}$
Momentum flux:	Mass flux:	Heat flux:
$\tau = \mu \dfrac{du}{dz}$	$J_A = -D_{AB} \dfrac{dc_A}{dz}$	$q = -k \dfrac{dT}{dz}$
Reynolds number:	Reynolds number:	Reynolds number:
$\mathrm{Re} = \dfrac{UL}{\nu}$	$\mathrm{Re} = \dfrac{UL}{\nu}$	$\mathrm{Re} = \dfrac{UL}{\nu}$
	Peclet number:	Peclet number (for heat transfer)
	$\mathrm{Pe} = \dfrac{UL}{D_{AB}}$	$\mathrm{Pe} = \dfrac{UL}{\alpha}$
	Schmidt number:	Prandtl number:
	$\mathrm{Sc} = \dfrac{\nu}{D_{AB}}$	$\mathrm{Pr} = \dfrac{\nu}{\alpha}$
	Sherwood number:	Nusselt number:
	$\mathrm{Sh} = \dfrac{k_c L}{D_{AB}}$	$\mathrm{Nu} = \dfrac{hL}{k}$

[a]L is the characteristic distance and h is the convective heat-transfer coefficient.

The first analogy is called the *Reynolds analogy*, to which we were exposed in Section 7.6, and that we will see is a generalized form of the analysis just completed. Consider first the case of mass transfer to or from a flat plate in laminar flow when $\mathrm{Sc} = 1$. For the flat plate, $u_s = 0$. The Reynolds analogy says that the dimensionless velocity and concentration (and temperature) profiles must be the same. Thus, for $\mathrm{Sc} = 1$ (hence, $\mu = \rho D_{AB}$), the Reynolds analogy leads to the following mathematical statement:

$$\mu \frac{\partial}{\partial y}\left(\frac{u}{U}\right)\Bigg|_{y=0} = \rho D_{AB} \frac{\partial}{\partial y}\left(\frac{c_A - c_{A,s}}{c_{A,\infty} - c_{A,s}}\right)\Bigg|_{y=0} \tag{7.146}$$

This can be rearranged to

$$\frac{\mu}{U}\frac{\partial u}{\partial y}\bigg|_{y=0} = \rho D_{AB}\frac{\partial}{\partial y}\left(\frac{c_{A,s}-c_A}{c_{A,s}-c_{A,\infty}}\right)\bigg|_{y=0} \tag{7.147}$$

From Eq. 7.123 and Fick's first law, we can now state the following:

$$J_A = -D_{AB}\frac{\partial c_A}{\partial y}\bigg|_{y=0} = -D_{AB}\frac{\partial}{\partial y}(c_A - c_{A,s})|_{y=0} = k_c(c_{A,s}-c_{A,\infty}) \tag{7.148}$$

Rearranging,

$$k_c = \frac{-D_{AB}(\partial/\partial y)(c_A - c_{A,s})|_{y=0}}{c_{A,s}-c_{A,\infty}} = D_{AB}\frac{\partial}{\partial y}\left(\frac{c_{A,s}-c_A}{c_{A,s}-c_{A,\infty}}\right)\bigg|_{y=0} \tag{7.149}$$

Comparing Eqs. 7.149 and 7.147, we find that the following must be true:

$$\frac{\mu}{U}\frac{du}{dy}\bigg|_{y=0} = \rho k_c \tag{7.150}$$

In Section 5.12 we defined the coefficient of skin friction, C_f, with the following equation:

$$C_f = \frac{2\tau_{xy}}{\rho U^2} = \frac{2\mu(\partial u/\partial y)|_{y=0}}{\rho U^2} \tag{5.125}$$

Comparing this with Eq. 7.150 we find:

$$\frac{k_c}{U} = \frac{C_f}{2} \tag{7.151}$$

Recalling the definition of the Sherwood number in Eq. 7.126, note that Eq. 7.151 can also be expressed as

$$\text{Sh} = \frac{k_c L}{D_{AB}} = \frac{C_f UL}{2D_{AB}} = \frac{C_f}{2}\text{Pe} \tag{7.152}$$

By exactly the same kind of analysis, using the analogous quantities, we will find that the parallel result for convective heat transfer is

$$\frac{h}{\rho U C_p} = \frac{C_f}{2} \tag{7.153}$$

where h is the heat-transfer coefficient (Table 7.2). Or, using the definition of the Nusselt number from Table 7.2:

$$\text{Nu} = \frac{hL}{k} = \frac{C_f}{2} \frac{\rho C_p}{k} UL = \frac{C_f}{2} \frac{UL}{\alpha} = \frac{C_f}{2} \text{Pe} \qquad (7.154)$$

Note that the Peclet number here is the Peclet number for heat transfer, also defined in Table 7.2. Equations 7.152 and 7.154 are the Reynolds analogy between momentum and mass and heat transport in laminar flow along a flat plate. An alternate presentation of these results comes from rearranging Eqs. 7.151 and 7.153:

$$\frac{k_c}{U} = \frac{C_f}{2} = \frac{h}{\rho U C_p} \qquad (7.155)$$

which is also known as the Reynolds analogy. Caution must be used when applying these equations however. First, recall that they were derived for Schmidt and Prandtl numbers of one. Second, they were derived for a condition in which the total fluid drag was characterizable by the skin-friction coefficient. Although this is accurate for flows like the flat plate considered here, or flows in pipes and other straight ducts, recall that flow past immersed bodies usually results in a component of drag related to unequal pressure distribution (called the form drag, Section 5.9.2), which can be a significant part of the total fluid drag force, especially in the extreme case of flow separation. In these later cases, the full Reynolds analogy (Eq. 7.155) cannot be correct, since C_f does not completely characterize momentum transport.

Regarding the problem of restriction to Schmidt and Prandtl numbers of one, three other equations have been developed using the analogy between momentum and mass transport (or momentum and heat transport). Colburn (1933) and Chilton and Colburn (1934) found that the following improvement of the Reynolds analogy adequately matched experimental heat- and mass-transfer data in laminar and turbulent flows:

$$\frac{k_c}{U} \text{Sc}^{2/3} = \frac{C_f}{2} = \frac{h}{\rho U C_p} \text{Pr}^{2/3} \qquad (7.156)$$

This is called the *Chilton–Colburn analogy*, and was found to be adequate for $0.5 < \text{Pr} < 50$ and $0.6 < \text{Sc} < 2500$, and for a variety of flow shapes, as long as there was no form drag present. When form drag is present, it is typically found that the Chilton–Colburn analogy between heat and mass transfer is still valid, or

$$\frac{k_c}{U} \text{Sc}^{2/3} = \frac{h}{\rho U C_p} \text{Pr}^{2/3} \qquad (7.157)$$

This result is a very powerful idea. For example, imagine that because of the geometry, environment, method of measurement, or any other reason that it is more accurate (or easy) to measure the heat-transfer coefficient than the mass-transfer coefficient (this actually has been true in many cases). Then the Chilton–Colburn analogy says that under certain conditions, the heat-transfer data can be used to estimate the mass-transfer coefficient.

Two other analogies have also been developed (Skelland, 1974; Welty et al. 1976). The first is called the *Prandtl analogy*, which was developed through consideration of mass transport in the turbulent core and viscous sublayer of shear flows. The Prandtl analogy results in

$$\text{Sh} = \frac{C_f}{2} \frac{\text{Pe}}{1 + 5\sqrt{C_f/2}\,(\text{Sc} - 1)} \tag{7.158}$$

The *von Kármán analogy* also considered momentum transport in the buffer zone of turbulence. This resulted in the following equation:

$$\text{Sh} = \frac{C_f}{2} \frac{\text{Pe}}{1 + 5\sqrt{C_f/2}\,(\text{Sc} - 1) + \ln[(1 + 5\,\text{Sc})/6]} \tag{7.159}$$

Note that both of these equations reduce to the Reynolds analogy for Sc = 1. Analogous equations can be written for heat transfer, by substitution of Pr for Sc in the two equations above. As was the case for the Reynolds analogy, Eqs. 7.158 and 7.159 are also only accurate for flows were fluid drag is only caused by skin friction.

It is worthwhile summarizing the basic requirements of mass, heat, and momentum analogies.

Analogies Between Mass (or Heat) and Momentum

1. The dimensionless boundary conditions must be identical.

2. No form drag can be present, that is, there can be no drag component related to differential pressure distribution on the surface of the object (Chapter 5). Therefore, momentum transfer data for flows around spheres and other so-called "bluff bodies" cannot be used to form an analogy with mass or heat transfer around the same objects. This is not a problem in pipes, ducts, and flat plates, since here there is usually only drag due to shear forces at the surface (skin friction).

3. The velocity profile for mass (or heat) transfer must be the same as when there is no mass (or heat) transfer. In other words, the magnitude of mass or heat transfer cannot be so large as to affect the velocity field.

4. In turbulent flows, the eddy diffusivity of the turbulence must equal the turbulent mass (or heat) diffusivity.

Analogies Between Mass and Heat Transfer   Besides the requirement of identical dimensionless boundary conditions, identical velocity fields, and the same turbulent mass and heat diffusivity suggested above, the heat- and mass-transfer analogies also require that there can be no significant heat transfer during mass transfer, or mass transfer during heat transfer. For example, viscous dissipation of energy cannot be so large as to significantly increase the temperature of the fluid. Also in the mass-transfer case, there can be no mass transfer due to pressure or thermal diffusion.

Example 7.4.   Assume you have no available experimentally determined data for mass transfer for laminar flow in a pipe (i.e., pipe flow for $Re < 2300$). Use the Chilton–Colburn analogy to estimate the Sherwood number, the mass-transfer coefficient, and the mass flux J_A.

SOLUTION.   From the Chilton–Colburn analogy between mass and momentum transfer, we note that

$$\frac{k_c}{U} Sc^{2/3} = \frac{C_f}{2}$$

Recall from Chapter 5 that the friction factor in a pipe in laminar flow is given by

$$C_f = f_f = \frac{16}{Re}$$

Substituting into the first equation, we find:

$$k_c = 8 \, Re^{-1} \, Sc^{-2/3} U$$

The Sherwood number is calculated as

$$Sh = \frac{k_c d}{D_{AB}} = 8 \, Re^{-1} \, Sc^{-2/3} \frac{Ud}{D_{AB}} = 8 \, Re^{-1} \, Sc^{-2/3} \, Pe = 8 \, Sc^{1/3}$$

where we have made use of the identity $Pe = Sc \, Re$. The mass-transfer coefficient can then be reexpressed as

$$k_c = \frac{Sh \, D_{AB}}{d} = 8 \frac{D_{AB}}{d} Sc^{1/3}$$

Finally, the flux J_A can be expressed as

$$J_A = \frac{Sh \, D_{AB}}{d} c_A^* = k_c c_A^* = 8 \frac{D_{AB}}{d} Sc^{1/3} c_A^*$$

These equations are estimated from the Chilton–Colburn analogy. They have not been tested against experimental data and must be used with caution.

7.8.3. Mass-Transfer Correlations and Calculations

In many situations, mass-transfer models and equations from the preceding section cannot be used to predict heat- or mass-transfer coefficients (e.g., C_f may not be known for the entire surface). In this case, experimental data on heat or mass transfer must be collected. The term "mass-transfer correlation" refers to the efficient organization of these experimental mass-transfer data in dimensionless form. The form of the dimensionless equations is often suggested by dimensional analysis or exact analysis of related heat- or mass-transfer situations. For the case of the mass-transfer correlations, these data are typically collected in "wetted-wall-column" experiments, where dry air is blown across wetted walls, or in pipes or other surfaces made of a material that slowly dissolves in the moving fluid. As suggested by the Chilton–Colburn analogy discussed above, often the mass-transfer expression can be calculated from the analogous heat-transfer expression by substitution of the correct dimensionless group.

7.8.3.1. Flat Plate. From Eqs. 7.94 and 7.126, the Sherwood number for the local mass transfer from (or to) a smooth flat plate in laminar flow is easily calculated to be

$$Sh = 0.34\, Re_x^{1/2}\, Sc^{1/3} \tag{7.160}$$

At a Reynolds number between roughly 3 to 5×10^5, the laminar boundary layer becomes turbulent. Through experimentation, the Sherwood number for the flat plate with turbulent boundary layer is found to be (Welty et al., 1976)

$$Sh = 0.0292\, Re_x^{0.8}\, Sc^{1/3} \tag{7.161}$$

Example 7.5. A layer of benzene covers a piece of flooring, 2 m long by 1 m wide. A fan is blowing air parallel to the surface (along the length) at a velocity of 4 m/s. What is the rate of evaporation of benzene if the boundary layer above the benzene is turbulent?

SOLUTION. The first step in this problem is to calculate the average flux of benzene into the air. The local Sherwood number is given by Eq. 7.161, which can be expressed in terms of a flux as

$$J_A = \frac{-D_{AB}(c_{A,\infty} - c_{A,s})\, Sh}{x} = \frac{0.0292 D_{AB}}{x}\, Re_x^{0.8}\, Sc^{1/3}(c_{A,s} - c_{A,\infty})$$

The average flux per unit area of flooring is calculated by integrating over the plate of length L:

$$J_{A,ave} = \frac{\int_0^L J_A dx}{L} = \frac{0.0292}{L} D_{AB} \, Sc^{1/3}(c_{A,s} - c_{A,\infty}) \left(\frac{U}{\nu}\right)^{0.8} \int_0^L \frac{x^{0.8}}{x} dx$$

or completing the integration:

$$J_{A,ave} = \frac{0.0365}{L} D_{AB}(c_{A,s} - c_{A,\infty}) \, Sc^{1/3} \, Re_L^{0.8}$$

The vapor pressure of pure benzene at 25°C is 12,700 Pa (Table 4.2). From Example 6.1, the saturation concentration of benzene at the interface is

$$c_{A,s} = \frac{n}{V} = \frac{P}{RT} = \frac{12,700 \, Pa}{8.314 \, (J/K \, mol) \, 298 \, K} = 5.13 \, \frac{mol}{m^3}$$

Using a dry-air kinematic viscosity of 1.56×10^{-5} m^2/s and a benzene diffusivity of 0.088 cm^2/s (Table 6.1), the Schmidt number is

$$Sc = \frac{\nu}{D_{AB}} = \frac{1.56 \times 10^{-5} \, m^2/s}{0.088(cm^2/s) \times (m^2/100^2 \, cm^2)} = 1.77$$

The Reynolds number is

$$Re_L = \frac{4(m/s)2 \, m}{1.56 \times 10^{-5}(m^2/s)} = 5.13 \times 10^5$$

Plugging into the equation for $J_{A,ave}$, and assuming $c_{A,\infty} = 0$, we find that the unit width flux of benzene into the air is

$$J_{A,ave} = \frac{0.0365}{2 \, m} \left(0.088 \, \frac{cm^2}{s} \times \frac{m^2}{100^2 \, cm^2}\right)\left(5.13 \, \frac{mol}{m^3}\right)(1.77)^{1/3}(5.13 \times 10^5)^{0.8}$$

$$= 3.69 \times 10^{-2} \frac{mol}{m^2 - s}$$

Therefore, the rate of mass transfer from the benzene surface is

$$W_{A,ave} = J_{A,ave}(2 \, m^2) = 0.0737 \, \frac{mol}{s}$$

7.8.3.2. Turbulent Flow in Pipes.   In a smooth pipe, fully developed turbulence can be anticipated after an entry length L_e given by (Brodkey and

Hershey, 1988)

$$\frac{L_e}{2a} = 0.693 \, \text{Re}^{1/4} \tag{7.162}$$

Fully developed turbulence in rough pipes will be expected at an even shorter L_e. For turbulent flow in a pipe, two correlations based on experimental data are available. The first is due to Gilliard and Sherwood (1934), who studied the evaporation of nine different liquids in a wetted-wall column. It is appropriate for gas flows because of the experimental conditions and applicable Schmidt number range:

$$\text{Sh} = \frac{2ak_c}{D_{AB}} \frac{P_{B,lm}}{P}$$

$$= 0.023 \, \text{Re}^{0.83} \, \text{Sc}^{0.44} \quad \text{for } 2000 < \text{Re} < 35{,}000 \quad \text{and} \quad 0.6 < \text{Sc} < 2.5 \tag{7.163}$$

Here a is the pipe radius, P is the total pressure, and $p_{B,lm}$ is the log mean concentration of the carrier gas B:

$$p_{B,lm} = \frac{p_B(\text{bulk phase}) - p_B(\text{interface})}{\ln[p_B(\text{bulk phase})] - \ln[p_B(\text{interface})]} \tag{7.164}$$

Linton and Sherwood (1950) examined a higher range of Schmidt numbers by measuring the dissolution of pipes whose inside surfaces were coated with sparingly soluble solids like benzoic acid and found the following correlation, which is appropriate for liquids:

$$\text{Sh} = 0.023 \, \text{Re}^{0.83} \, \text{Sc}^{1/3} \quad \text{for } 2000 < \text{Re} < 70{,}000 \quad \text{and} \quad 1000 < \text{Sc} < 2260 \tag{7.165}$$

Brodkey and Hershey (1988) recommend the following correlation of Harriot and Hamilton (1965) for smooth pipe flow, which they claim is the most up to date:

$$\text{Sh} = \frac{2ak_c}{D_{AB}} = 0.0096 \, \text{Re}^{0.913} \, \text{Sc}^{0.346}$$

$$\text{for } 10{,}000 < \text{Re} < 100{,}000 \quad \text{and} \quad 432 < \text{Sc} < 97{,}600 \tag{7.166}$$

where k_c is the average liquid-phase mass-transfer coefficient.

Example 7.6. The addition of chlorine to drinking water after treatment was discussed in Example 7.3. In a particular section of pipe far from the plant, the initial concentration of chlorine Cl_2 is 0.5 mg/L. The pipe diameter is 20 cm, the velocity of flow is 0.5 m/s, and the temperature is 20°C. Find the maximum decrease in concentration of chlorine in a 200-m long pipe using the analysis developed in this section. Assume that chlorine is only consumed at the pipe surface, and that the chlorine concentration at the pipe surface is zero.

SOLUTION. Equation 7.166 gives the Sherwood number for turbulent flow in a pipe. The average mass-transfer coefficient is obtained by rearrangement:

$$k_c = \frac{D_{AB}\,\text{Sh}}{2a} = \frac{D_{AB}(0.0096)\,\text{Re}^{0.913}\,\text{Sc}^{0.346}}{2a}$$

The Reynolds and Schmidt numbers are

$$\text{Re} = \frac{2au_{ave}}{v} = \frac{20\ \text{cm}[0.5(\text{m/s})]\,100\ \text{cm/m}}{0.01\ \text{cm}^2/\text{s}} = 100{,}000$$

$$\text{Sc} = \frac{v}{D_{AB}} = \frac{0.01\ \text{cm}^2/\text{s}}{1.22 \times 10^{-5}\ \text{cm}^2/\text{s}} = 820$$

Plugging into the equation for k_c, we find:

$$k_c = \frac{1.22 \times 10^{-5}(\text{cm}^2/\text{s})(0.0096)(10^5)^{0.913}820^{0.346}}{2(10\ \text{cm})} = 2.2 \times 10^{-3}\ \frac{\text{cm}}{\text{s}}$$

From Eq. 7.123, the flux to the pipe wall can then be given by

$$J_A = k_c c_A^*$$

The next step is to somehow account for the change in chlorine concentration in the pipe as the water flows through the pipe. Consider the differential section of pipe shown in the following figure.

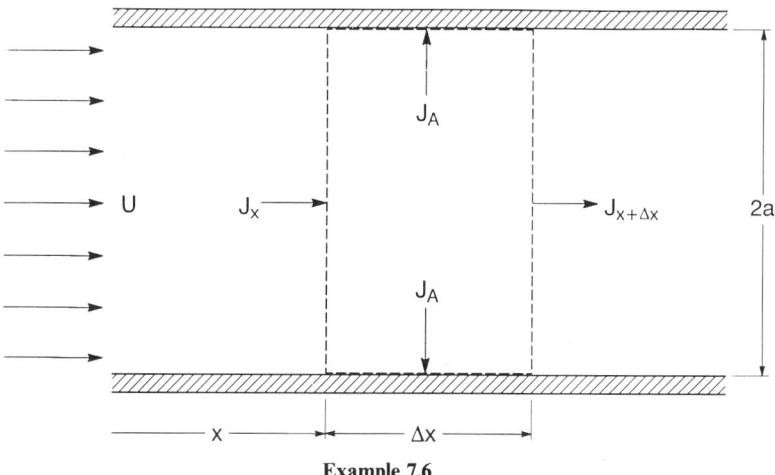

Example 7.6

If plug-flow conditions and uniform concentration in the radial direction are assumed, we can perform the following mass balance around the pipe section:

$$V \frac{\partial c_x}{\partial t} = J_x A - J_{x+\Delta x} A - 2\pi a \Delta x J_A$$

where A is the cross-sectional area of the pipe ($A = \pi a^2$), and V is the volume of the element. As usual, the term on the left of the equal sign is the rate of change of mass in the differential volume, the first term on the right is the mass flow into the differential element, the second term on the right is the mass flow out of the differential element, and the last term on the right is the mass flow out of the differential element from mass transfer to the wall. If we let $J_x = U c_x$ and $J_{x+\Delta x} = U(c_x + \partial c_x/\partial x \, dx)$, then dividing through by $V = \pi a^2 \Delta x$ and rearranging, we find the following one-dimensional conservation equation:

$$\frac{\partial c_x}{\partial t} + U \frac{\partial c_x}{\partial x} = -\frac{2J_A}{a}$$

Now letting $c_A^* = c_x$ in the previous mass-transfer expression, and considering only steady-state conditions, we find the following equation:

$$U \frac{dc_x}{dx} = -\frac{2k_c c_x}{a}$$

For the initial condition $c_x = c_{x=0}$ at $x = 0$ (i.e., the head of the pipe), this integrates to

$$c_x = c_{x=0} \exp\left(-\frac{2k_c}{aU} x\right)$$

Substituting in for the given values, we find:

$$c_{x=200\,m} = c_{x=0} \exp\left[-\frac{2(2.2 \times 10^{-5}(\text{m/s})]\,200\,\text{m}}{0.1\,\text{m}[0.5(\text{m/s})]}\right] = 0.84 c_{x=0}$$

This indicates that the maximum decrease in chlorine concentration under the given turbulent-flow conditions will be expected to be 16% or $c_{x=200\,m} = 0.42$ mg/L.

7.8.3.3. Spheres.

In Section 7.7.3, we discussed the flux to or from a spherical surface under the effect of convective–diffusive mass transport in Stokes flow. To calculate the flux of A under these conditions, we divide Eq.

7.118 by the surface area of a sphere:

$$J_A = \frac{W_A}{4\pi a^2} = \frac{2\pi D_{AB}a(0.92 + 0.991\,\mathrm{Pe}^{1/3})(c_{A,\infty} - c_{A,s})}{4\pi a^2}$$

$$= \frac{D_{AB}(0.92 + 0.991\,\mathrm{Pe}^{1/3})(c_{A,\infty} - c_{A,s})}{2a} \qquad (7.167)$$

The Sherwood number is then

$$\mathrm{Sh} = \frac{2aJ_A}{D_{AB}(c_{A,\infty} - c_{A,s})} = 0.92 + 0.991\,\mathrm{Pe}^{1/3} \qquad (7.168)$$

These equations will only be accurate up to small Reynolds numbers, not much greater than $\mathrm{Re} = 1$ (i.e., Stokes flow). For greater Reynolds numbers, the fluid-flow patterns diverge from those of ideal Stokes flow, and we have to again resort to experimentally determined correlations. Welty et al. (1976) review a number of these. Garner and Suckling (1958) determined the following correlation for liquids:

$$\mathrm{Sh} = 2.0 + 0.95\,\mathrm{Re}^{1/2}\,\mathrm{Sc}^{1/3} \qquad \text{for } 100 < \mathrm{Re} < 700 \quad \text{and} \quad 1200 < \mathrm{Sc} < 1525$$

$$(7.169)$$

where Re is the particle Reynolds number (Eq. 2.1). Fröessling (1938) determined the following correlation for gases:

$$\mathrm{Sh} = 2.0 + 0.552\,\mathrm{Re}^{1/2}\,\mathrm{Sc}^{1/3} \qquad \text{for } 2 < \mathrm{Re} < 800 \quad \text{and} \quad 0.6 < \mathrm{Sc} < 2.7$$

$$(7.170)$$

An additional complication sometimes arises in mass- (and heat-) transfer problems which is related to *natural convection*. For example, when a surface is hotter than the surrounding medium, heat transfer can occur in the absence of forced convection (i.e., a forced flow). Because the fluid at the surface is less dense than the surrounding fluid, it may rise; this sets up a natural convective flow which can definitely result in heat transfer from the surface. The same possibility exists in mass transfer, because the density of the fluid at the surface can be changed by the dissolution of the surface or other changes in concentration. We have ignored this possibility in the preceding sections, because we can often assume that the forced-convection effect is much greater than the natural-convection effect. The importance of natural convective effects can be estimated by calculating the *Grashof number*, which for a sphere is given by

$$\mathrm{Gr} = \frac{(2a)^3 \rho g \Delta \rho_A}{\mu^2} \qquad (7.171)$$

where $\Delta\rho_A$ is the difference in density of the fluid at the surface and in the bulk fluid, and g is the acceleration of gravity. Welty et al. (1976) indicate that the two correlations above will be accurate in the absence of natural-convection effects, or when the following condition holds:

$$Re \geqslant 0.4\, Gr^{1/2}\, Sc^{-1/6} \tag{7.172}$$

When Eq. 7.172 is not satisfied, correlations are available to consider higher Grashof number conditions (e.g., Steinberger and Treybal, 1960).

7.8.3.4. Packed Beds. In a number of situations in the environment and in pollution-control engineering, there can be a flow through assemblages of particles, for example, in a fixed or "packed" bed. In these cases, we may be interested in mass transfer between the fluid and the assemblage of particles. Although data for all types of particles are incomplete, several correlations are available. Wilson and Geankoplis (1966) provided the following equations for liquids in fixed beds of spherical particles with bed porosity (Eq. 6.117) between 0.35 and 0.75:

$$Sh = \frac{1.09}{\epsilon}\, Sc^{1/3} \qquad \text{for } 0.0016 < Re_{pb} < 55 \quad \text{and} \quad 165 < Sc < 70{,}600 \tag{7.173}$$

and

$$Sh = \frac{0.25}{\epsilon}\, Re_{pb}^{0.69}\, Sc^{1/3} \qquad \text{for } 55 < Re_{pb} < 1500 \quad \text{and} \quad 165 < Sc < 10690$$

$$\tag{7.174}$$

Here Re_{pb} is the Reynolds number for the packed bed,

$$Re_{pb} = \frac{2aU_{sup}}{\nu} \tag{7.175}$$

where U_{sup} is the superficial bed velocity, which is the total flow through the bed divided by the bed cross-sectional area. For mass transfer in packed (and fluidized) beds of spherical and nonspherical particles in both gas and liquid flows, the following correlation of Dwidevi and Upadhyay (1977) is recommended in Fogler (1992):

$$Sh = (0.765\, Re_{pb}^{0.18} + 0.365\, Re_{pb}^{0.614})\, \frac{Sc^{1/3}}{\epsilon}$$

$$\text{for } Re_{pb} > 10 \text{ (gases)} \quad \text{and} \quad Re_{pb} > 0.01 \text{ (liquids)} \tag{7.176}$$

For nonspherical particles, the equivalent particle radius can be estimated using the following formula:

$$a = \frac{1}{2}\sqrt{\frac{A_p}{\pi}} \qquad (7.177)$$

where A_p is the particle surface area.

Example 7.7. A packed bed is composed of spherical particles of 0.5 cm diameter; the bed porosity is 0.47, and the specific surface area of the porous material is $M = 3.24\,\mathrm{cm}^{-1}$. Water passes through the bed with a superficial velocity of 0.8 m/min, and a compound dissolved in the water reacts quickly at the particle surface. If the diffusion coefficient of the compound is $1 \times 10^{-5}\,\mathrm{cm}^2/\mathrm{s}$, what is the percent removal of the compound through a 1-m deep bed?

SOLUTION. It is helpful to first construct a picture of the bed.

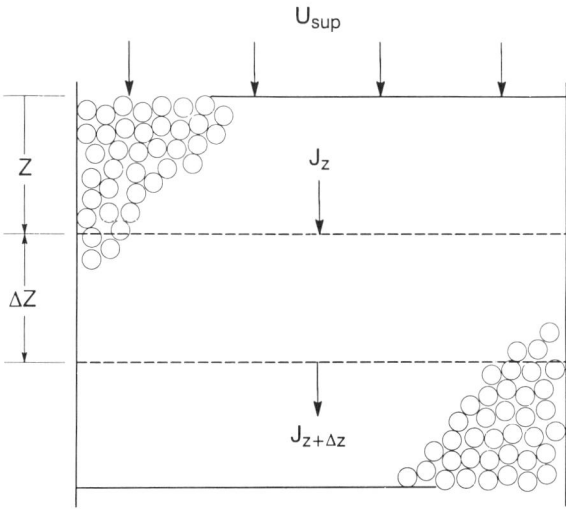

Example 7.7

As usual, we perform a mass balance around the differential element:

$$V\frac{dc_z}{dt} = J_z A - J_{z+\Delta z} A - Sk_c c_z$$

where V is the element volume, A is the plane-surface area of the side of the differential element, and S is the surface area of all spheres in the differential element. Now we let $J_z = Uc_z$ and $J_{z+\Delta z} = U(c_z + \partial c_z/\partial z\, dz)$, and dividing through by V we arrive at

$$\frac{\partial c_z}{\partial t} = -U_{sup}\frac{\partial c_z}{\partial z} - \frac{S}{V}k_c c_z$$

In this equation, S/V is called the *specific surface* in the bed — the surface area of the particles per bulk volume of bed. Considering steady-state conditions, and giving the specific surface the name S_0, we find:

$$\frac{dc_x}{dx} = -\frac{S_0 k_c}{U_{sup}}c_z$$

Separating and integrating with the initial condition $c_z = c_{z=0}$ at $z = 0$, we find:

$$\frac{c_z}{c_{z=0}} = \exp\left(-\frac{S_0 k_c}{U_{sup}}z\right)$$

We can choose between Eqs. 7.173 and 7.174 to provide the value of k_c; Re_{pb} is calculated from Eq. 7.175:

$$Re_{pb} = \frac{2aU_{sup}}{\nu} = \frac{2(0.25\text{ cm})80(\text{cm/min}) \times (\text{min}/60\text{ s})}{0.01(\text{cm}^2/\text{s})} = 66.7$$

The Schmidt number is calculated as

$$Sc = \frac{\nu}{D_{AB}} = \frac{0.01(\text{cm}^2/\text{s})}{1 \times 10^{-5}(\text{cm}^2/\text{s})} = 1000$$

Then we can calculate the mass transfer coefficient using Eq. 7.174:

$$k_c = \frac{Sh\, D_{AB}}{2a} = \frac{0.25\, D_{AB}}{2\varepsilon}\frac{Re_{pb}^{0.69}\, Sc^{1/3}}{a}$$

$$= \frac{0.25}{2 \times 0.47}\frac{1 \times 10^{-5}(\text{cm}^2/\text{s})}{0.25\text{ cm}}(66.7^{0.69})1000^{1/3} = 1.9 \times 10^{-3}\frac{\text{cm}}{\text{s}}$$

Therefore, the ratio of final to initial concentrations is

$$\frac{c_z}{c_{z=0}} = \exp\left(-\frac{S_0 k_c}{U_{sup}}z\right) = \exp\left\{-\frac{3.24\text{ cm}^{-1}[1.9 \times 10^{-3}(\text{cm/s})]}{0.833(\text{cm/s})}100\text{ cm}\right\} = 0.48$$

So the percent removal is 52%.

7.9. INTERPHASE MASS TRANSPORT AND RESISTANCE MODELS

The term interphase mass transfer simply means the transfer of a component between two or more phases in contact with each other. The component being transferred can undergo reactions in one or both of the phases, or it can be conservative (i.e., nonreactive). The phases are usually defined so that we can imagine an interfacial region between the phases. Actually, in a simplified way, we have already considered mass transfer between phases in contact with one another at numerous junctures in this book.

In Chapter 4 we considered the equilibrium adsorption or partitioning of a component between two phases. For example, we looked at the adsorption of contaminants like TCE and TCP on activated carbon, the partitioning of DDT between air and water, and the partitioning of TCE between octanol and water. We did not talk about kinetics in Chapter 4, but obviously there was a net mass transfer between the phases prior to the establishment of the equilibrium condition.

We considered steady-state and nonsteady-state diffusion of components in and between different phases in Chapter 6. For example, we looked at the diffusion of benzene through a stagnant air column (interphase transfer from the pure benzene through a benzene-air phase and into the pure air phase). We also examined the transfer of PCBs from the water column (phase) to the sediment phase, as well as diffusion of tetrachlorobenzene from a water phase into a sediment-particle phase. Note in all these problems, the interfacial concentrations were well defined — either zero or a constant. The constant interfacial concentration was set by a saturation value, an equilibrium vapor pressure, a heterogeneous reaction rate, or simply by the assumption (not always explicitly stated) that there was no mass transfer limitation in the reservoir supplying the diffusing constituent.

In this chapter, although we have considered the added complexity of convective-mass transport working with diffusive-mass transport, we have still maintained the basic assumptions of Chapter 6, that is, a constant or zero interfacial concentration and mass-transport limitation in only a single phase. In certain mass-transfer operations, we will have to consider mass-transfer limitation in more than one phase, and/or the effect of reactions on overall mass flux.

7.9.1. Nonsteady-State Mass Transfer to Particle

The implications of mass-transfer limitation in two phases can be seen by returning to an important mass-transfer problem, that between an external phase (gas or liquid) into a particle. This problem was addressed in Chapter 6, where we considered diffusion into a sphere from a well-mixed external phase (Section 6.7.4). The governing partial differential equation for the particle phase was

$$\frac{\partial c_A}{\partial t} = D_{AB}\left(\frac{\partial^2 c_A}{\partial r^2} + \frac{2}{r}\frac{\partial c_A}{\partial r}\right) \tag{6.104}$$

where D_{AB} was the effective diffusion coefficient for porous particles. The boundary conditions for Eq. 6.104 were

$$\frac{\partial c_A}{\partial t} = 0 \qquad \text{at } r = 0$$

$$c_A = 0 \text{ (or a constant)} \qquad \text{at } t = 0;\ 0 < r < a \tag{6.105}$$

Another required boundary condition concerns the surface concentration, which was set to the value in the homogeneously mixed external fluid. As noted in the discussion above, there was no mass-transfer resistance allowed for in the external fluid. If there is limitation, it can easily be incorporated into a new type of boundary condition, called a *flux boundary condition*. This is derived simply by noting that the flux of mass into the particle at its surface (Fick's first law) must equal the mass transfer from the external solution by convective diffusion:

$$J_A = -D_{AB}\frac{\partial c_A}{\partial r}\bigg|_{r=a} = k_c(c_{A,s} - c_{A,\infty}) \tag{7.178}$$

The concentration profile in the particle and the external fluid at a certain time might then resemble that shown in Figure 7.14a. Crank (1956) provides a solution for this problem, given as

$$\frac{M_t}{M_\infty} = 1 - \sum_{n=1}^{\infty}\frac{6\,\text{Bi}^2\exp[-\beta_n^2(D_{AB}t/a^2)]}{\beta_n^2[\beta_n^2 + \text{Bi}(\text{Bi} - 1)]} \tag{7.179}$$

In the equation above, M_∞ is the mass transferred to the particle at equilibrium, M_t is the mass transferred to the particle at some time t, and the β_n are the roots of the following relation:

$$\beta_n \cot \beta_n + \text{Bi} - 1 = 0 \tag{7.180}$$

where Bi is a new dimensionless parameter known as the *Biot number*, which

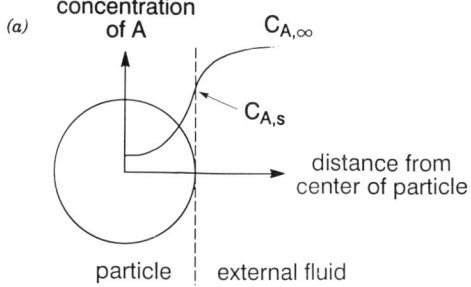

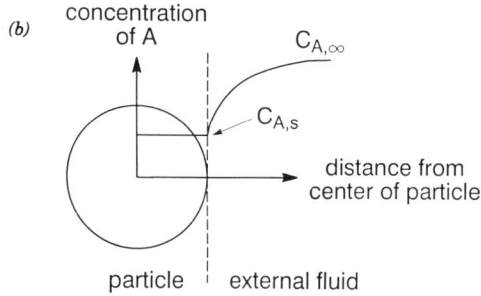

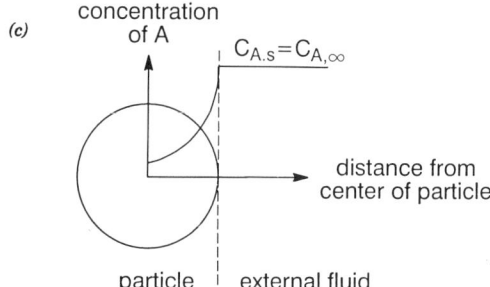

Figure 7.14. Concentration profile in and around particle. The particle is porous and saturated with the external fluid: (*a*) mass-transfer limitation in both phases; (*b*) limitation in external phase; and (*c*) limitation in internal phase.

is named after the analogous parameter from heat-transfer studies:

$$\text{Bi} = \frac{ak_c}{D_{AB}} = \frac{\dfrac{a^2}{D_{AB}}}{\dfrac{a}{k_c}} = \left\{ \frac{\text{internal diffusion time scale}}{\text{external convective-diffusive time scale}} \right\} \qquad (7.181)$$

As indicated, the Biot number can be considered the ratio of the time scale for diffusion inside the particle, to the time scale of convective diffusion in the external fluid surrounding the particle. When the Biot number is small, we might expect the external-phase mass transfer to play the biggest role (i.e., internal-phase mass transfer is relatively fast; see Figure 7.14b); if the Biot number is large, internal-phase mass transfer will control the overall rate (i.e., external phase mass transfer is relatively fast; see Figure 7.14c).

Example 7.8. Calculate the Biot numbers for mass transfer to a 100-μm-diameter particle, where the particle Reynolds number is 150. Consider two cases: (a) mass transfer of a chlorobenzene compound to a sediment particle in water, and (b) mass transfer of SO_2 to a water droplet from air.

SOLUTION. For part a, you are given the following physical/chemical data on the chlorobenzene compound: diffusion coefficient in water $= 0.8 \times 10^{-5}$ cm^2/s, effective diffusion coefficient in the sediment particle $= 2.0 \times 10^{-10}$ cm^2/s, and kinematic viscosity of water $= 0.01$ cm^2/s. The Schmidt number in the external fluid is calculated as Sc $= v/D_{AD} = 0.01/0.8 \times 10^{-5} = 1250$. The Sherwood number for mass transfer to a spherical particle in water is estimated with Eq. 7.169:

$$Sh = 2.0 + 0.95\, Re^{1/2}\, Sc^{1/3}$$

Using the definition of the Sherwood number, we find the mass-transfer coefficient through substitution of the known diffusion coefficient, particle radius, and Reynolds and Schmidt numbers:

$$k_c = \frac{D_{AB}\, Sh}{2a} = \frac{0.8 \times 10^{-5}(cm^2/s)}{(2)50 \times 10^{-4}\, cm}[2.0 + (0.95)(150^{1/2})(1250^{1/3})] = 0.10\,\frac{cm}{s}$$

The Biot number is then calculated as

$$Bi = \frac{ak_c}{D_{AB}} = \frac{(50 \times 10^{-4}\, cm)[0.10(cm/s)]}{2 \times 10^{-10}(cm^2/s)} = 2.55 \times 10^6$$

Note that we needed two diffusion coefficients for the calculation—the diffusion coefficient in pure water (to calculate the Sherwood number for mass transfer from the external pure-water phase), and the effective diffusion coefficient inside the particle (to calculate the Biot number).

For part b, you are given the following physical data on the SO_2: diffusion coefficient in air $= 0.126$ cm^2/s, diffusion coefficient in water $= 1.8 \times 10^{-5}$ cm^2/s, and kinematic viscosity of dry air $= 0.148$ cm^2/s. The Schmidt number in the external fluid (air) is, therefore, Sc $= v/D_{AB} = 0.148/0.126 = 1.17$. The mass-transfer coefficient is calculated as above, but we use Eq. 7.170 instead of

Eq. 7.169:

$$k_c = \frac{D_{AB}\,\text{Sh}}{2a} = \frac{0.126(\text{cm}^2/\text{s})}{(2)50 \times 10^{-4}\,\text{cm}}\,[2.0 + (0.552)(150^{1/2})(1.17^{1/3})] = 115\,\frac{\text{cm}}{\text{s}}$$

The Biot number for part b is then calculated as in part a:

$$\text{Bi} = \frac{ak_c}{D_{AB}} = \frac{(50 \times 10^{-4}\,\text{cm})[115(\text{cm}/\text{s})]}{1.8 \times 10^{-5}(\text{cm}^2/\text{s})} = 3.12 \times 10^4$$

These are indeed very large Biot numbers, and they suggest that in the two given problems, we do not have to worry about mass-transfer limitation up to the surface of the particle (Figure 7.14c). In other words, internal diffusion is so slow that contaminant concentration in the external fluid is homogeneous and, therefore, there are no gradients in concentration surrounding the particle.

7.9.2. Steady-State Interphase Mass Transfer with Homogeneous Reaction in One Phase

In numerous situations in environmental science, there is mass transfer between two adjacent phases, with a constant flux determined by conditions in one of the phases. This situation differs from the problem examined in the last section, where the flux changed over time. For example, consider pipe flow where a component in the main stream diffuses into the wall, where it undergoes first-order reaction. This could pertain to our previous problem of chlorine transport in turbulent pipe flow, where the chlorine is consumed in a biofilm along the wall by a first-order reaction within the biofilm. This is also quite similar to the situation depicted in Section 6.6.4, where we considered gas diffusion into a stagnant liquid phase. Consider the situation shown in Figure 7.15. The reactant concentration in the main flow is assumed to be constant. Between the main turbulent flow and the biofilm, there is a mass transfer resistance due to convective diffusion, which we can imagine occurs through a liquid film. This is not a *film model* per se, since here we will not attempt to predict the thickness of the film. But it is appealing to think of a fictitious liquid film through which pure diffusion occurs, so many people call this a film model (see the discussion in Section 7.8.2.1). After passing through the liquid phase, the reactant diffuses into the biofilm, where it is consumed by a first-order reaction. We make the important assumption here that the reaction always goes to completion before the wall is reached; hence, the reactant concentration goes to zero before the wall is reached. As noted in Chapter 6, this is called a "partially penetrated biofilm" (see the figure in Exercise 6.12).

We first model transport through the liquid film, which can be described as usual with a mass-transfer coefficient:

$$J_A = k_c(c_{A,\infty} - c_{A,z=0}) \tag{7.182}$$

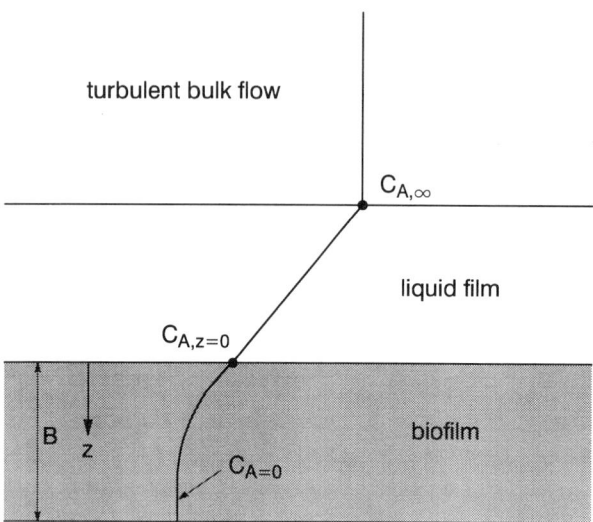

Figure 7.15. Mass transfer from flowing stream into biofilm at wall.

Note that this flux is a constant through the liquid film. Note also that the flux out of the liquid film must equal the flux into the biofilm. From Exercise 6.9, we know that for the partially penetrated biofilm with a first-order homogeneous reaction, this flux into the biofilm is a constant given by

$$J_A = c_{A,z=0}\sqrt{D_{AB}k_A} \qquad (7.183)$$

where D_{AB} is the diffusion coefficient of contaminant in the biofilm and k_A is the first-order homogeneous reaction-rate constant in the biofilm. Note that Eq. 7.183 can be rearranged to yield

$$c_{A,z=0} = \frac{J_A}{\sqrt{D_{AB}k_A}} \qquad (7.184)$$

We can substitute Eq. 7.184 into Eq. 7.182, and after a bit of rearranging we find the following:

$$J_A = \left[\frac{1}{(1/k_c) + (1/\sqrt{D_{AB}k_A})} \right] c_{A,\infty} \qquad (7.185)$$

A rather pleasing result is obtained if we look closely at the terms in the denominator of the expression in brackets. Note first that they are all fundamental constants of this problem: k_c is related to mass transfer through the imaginary liquid film, and D_{AB} and k_A are related to mass transfer in the biofilm. Let's make the following definitions:

$$R_f = \frac{1}{k_c} \tag{7.186}$$

:and

$$R_b = \frac{1}{\sqrt{D_{AB}k_A}} \tag{7.187}$$

Using these definitions, we can reexpress Eq. 7.185 as

$$J_A = \left(\frac{1}{R_f + R_b}\right) c_{A,\infty} \tag{7.188}$$

We call R_f and R_b the *mass-transfer resistances* of the liquid film and biofilm. We make all these transformations because of a classic (and useful) result from electrical engineering that has to do with potential drop across a circuit with two resistances in series. Ohm's law states that the voltage drop is given by

$$V = IR \tag{7.189}$$

where R is the electrical resistance and I is the current. Now consider the circuit shown in Figure 7.16. For the resistances in series, $R = R_1 + R_2$; thus, substituting into Eq. 7.189 and rearranging,

$$I = \left(\frac{1}{R_1 + R_2}\right) V_{12} \tag{7.190}$$

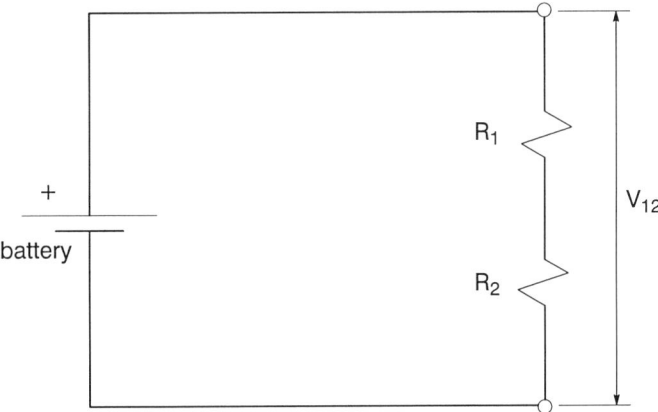

Figure 7.16. Electrical analog for two mass-transfer resistances in series.

Note the obvious similarity between this equation and Eq. 7.188. In fact, the mass flux J_A is analogous to current I, the concentration c_A is analogous to the electrical potential V_{12}, and the sum of the mass-transfer resistances $R_f + R_b$ is analogous to the sum of the electrical resistances $R_1 + R_2$. Therefore, we call the mass-transfer model of this section a *resistance-in-series model*. It has been an extremely useful concept in environmental modeling. Section 7.9.3 will show that it is even possible to have mass-transfer situations with more than two resistances in series. The main usefulness of the electrical analogy is quite simple: Just as in electrical theory, when one of the resistances is much greater than the other, the mass-transfer mechanism represented by the larger resistance can be said to *control* the overall mass transport. This can be critical information for environmental modelers.

7.9.3. Steady-State Interphase Mass Transfer with Heterogeneous Reaction

A slightly more complex relative of the problem considered in the last section is shown in Figure 7.17, where a reacting component is transferred in turbulent flow through a liquid film, a biofilm, and finally to a reacting surface (hence the name heterogeneous reaction).*

There is a reaction in this problem, but unlike the problem of the last section, it is a heterogeneous reaction occurring at the pipe wall. (We assume there is no reaction in the biofilm in this case.) Thus, the liquid film and biofilm can be considered to be mass-transfer resistances in series, which may limit mass transfer to the reacting surface. It is simpler to analyze this problem working from the reacting surface back toward the liquid film. Within the biofilm, the mass flux of reacting compound can be described by Fick's first law:

$$J_A = -D_{AB,b} \frac{dc_A}{dz} \tag{7.191}$$

where $D_{AB,b}$ is the effective diffusion coefficient in the biofilm. Separating and integrating from the limits $c_A = c_{A,z=0}$ at $z = 0$, and $c_A = c_{A,z=B}$ at $z = B$, we find:

$$J_A = \frac{D_{AB,b}}{B}(c_{A,z=0} - c_{A,z=B}) \tag{7.192}$$

If we assume the surface reaction is a first-order reaction occurring at $z = B$, the flux at the wall must also be equal to

$$J_A = k_{A,B} c_{A,z=B} \tag{7.193}$$

---

*The analysis is not necessarily restricted to liquid systems with biofilms. It could apply to any turbulent-flow problem near a wall, where the physicochemical and boundary conditions are the same.

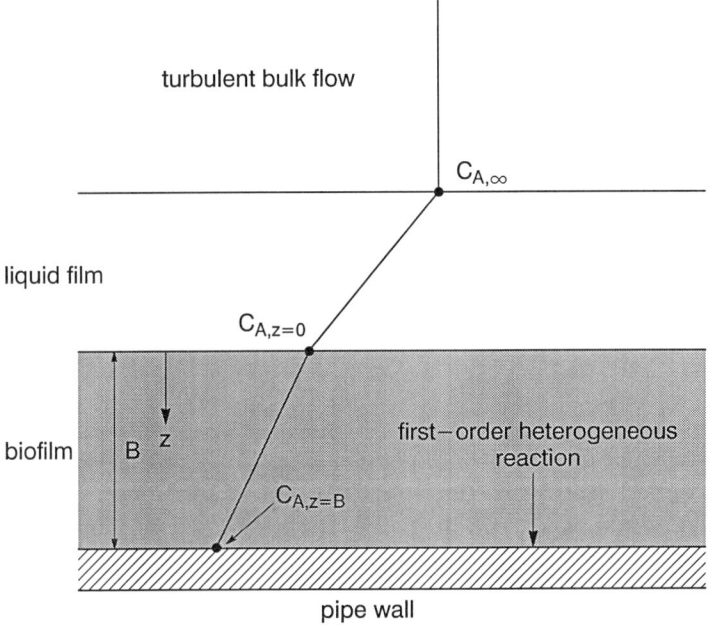

Figure 7.17. Mass transfer with liquid film, biofilm, and surface reaction.

where $k_{A,B}$ is the first-order heterogeneous reaction-rate constant at $z = B$ (Section 6.6.2).

Substituting Eq. 7.193 into Eq. 7.192 and rearranging, we find:

$$J_A = \left[\frac{1}{(B/D_{AB,b}) + (1/k_{A,B})} \right] c_{A,z=0} \qquad (7.194)$$

This is of course a two resistance-in-series model. Hence, we define a new biofilm mass transfer resistance as

$$R_b = \frac{B}{D_{AB}} \qquad (7.195)$$

and the surface reaction resistance as

$$R_s = \frac{1}{k_{A,B}} \qquad (7.196)$$

The flux through the liquid film will be modeled as usual with a mass-transfer coefficient:

$$J_A = k_c(c_{A,\infty} - c_{A,z=0}) \qquad (7.182)$$

Substituting this equation into Eq. 7.194 and rearranging, we quickly discover that the mass-transfer situation shown in Figure 7.17 leads to a three resistance-in-series model:

$$J_A = \left(\frac{1}{R_f + R_b + R_s} \right) c_{A,\infty} \qquad (7.197)$$

where R_f was given earlier by Eq. 7.186. As we have been stressing in this section, it is sometimes possible to evaluate the resistances and find one (large) resistance that controls mass transfer. This can simplify both modeling and experimental programs.

However, it may not always be possible to estimate all the resistances based on available chemical and physical data. For example, in many cases the surface-reactivity resistance is unknown. This is to be expected, because surface reactivities depend on a number of physicochemical conditions like the surface material, surface roughness, surface chemistry, surface age, and previous reaction/deposition history. Even if all the resistance are not known, we can still use the resistance-in-series model. For example, Wu et al. (1992) were interested in the dry deposition of gaseous SO_2 and particulate SO_4^{2-} on various surfaces. Both SO_2 and SO_4^{2-} are important pollutants in acid deposition, and Wu et al. were interested in the dry deposition on statues (in an effort to understand corrosion). They studied in detail deposition on the General George Meade equestrian statue in the Gettysburg National Military Park (Figure 7.18). Dry deposition to different parts of the statue were modeled with a series-resistance model (analogous to Eq. 7.197):

$$J_A = \frac{1}{r_a(z) + r_b + r_c} c(z) \qquad (7.198)$$

Here J_A is the flux of SO_2 or particulate SO_4^{2-} to the surface, $r_a(z)$ is an "aerodynamic resistance," r_b is a film resistance called the "boundary-layer" resistance, r_c is the surface-reactivity resistance, and $c(z)$ is the particulate or gas concentration at a reference height z normal to the statue surface. In their model, r_a represents a hypothetical resistance to mass transfer from the atmosphere if there were no film resistance (r_b) or reaction resistance (r_c).* The inverse of the sum of the three resistance terms has the units of velocity, and is, in fact, called the "deposition velocity" by air-pollution control scientists.

Estimation of the different resistances by Wu et al. showed a clever use of data and models. First, r_b was estimated using the Prandtl analogy, discussed in Section 7.8.2.2. Note that we calculate a mass-transfer coefficient from

---

*Although the model of Wu et al. (1992) is technically not an interphase mass-transport model (since the transfer is essentially through a single phase—the air), it fits very well in this section, since the various resistances can be imagined to represent different phases.

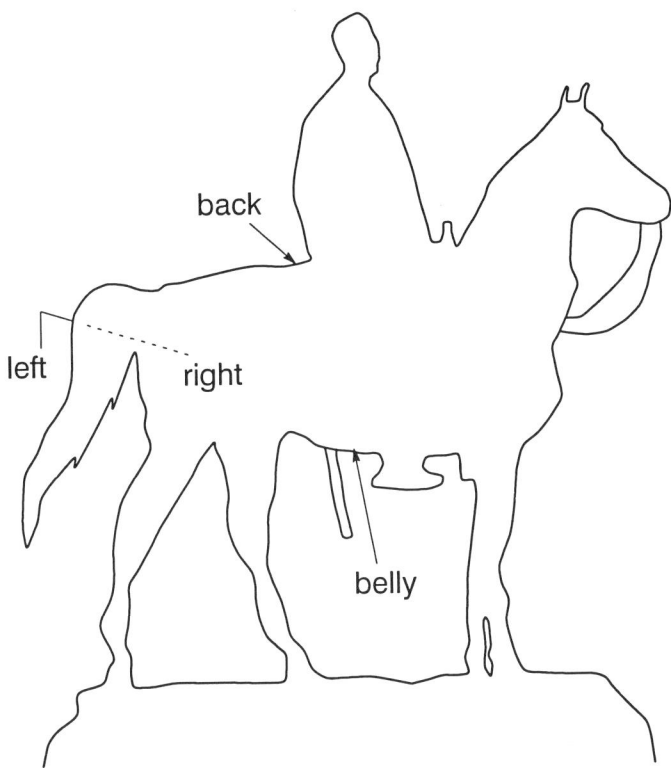

Figure 7.18. General George Meade equestrian statue, showing study locations. *Source*: Used by permission of the publisher from Wu, Davidson, Dolske and Sherwood, *Aerosol Sci. Technol.*, **16**, 68–76. Copyright 1994 by Elsevier Science Inc.

Eq. 7.158 as

$$k_c = \frac{\text{Sh}\, D_{AB}}{L} = \frac{D_{AB}}{L}\left(\frac{C_f}{2}\right)\frac{UL/D_{AB}}{1+5\sqrt{C_f/2}\,(\text{Sc}-1)} = \frac{C_f U}{2}\frac{1}{1+5\sqrt{C_f/2}\,(\text{Sc}-1)}$$

(7.199)

where L is the plate (surface) length, and where the relation $\text{Pe} = UL/D_{AB}$ has been incorporated. The mass-transfer resistance for the boundary layer is then simply $r_b = 1/k_c$. Two types of sampler surfaces were placed at the different locations indicated in Figure 7.18 to measure contaminant flux to the surface. Greased and bare mylar surfaces were used to collect particulate SO_4^{2-} and a special reactive filter was used to collect SO_2; these measurements yielded the total fluxes of SO_2 and SO_4^{2-} to the different parts of the statue. In the case of SO_2, the reaction with the special filter was assumed to be

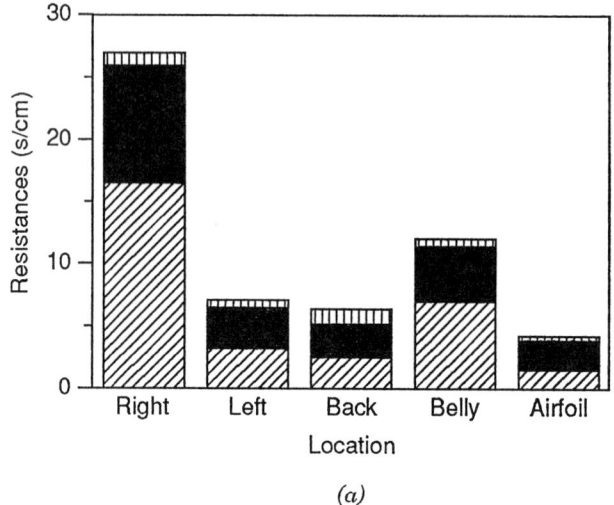

(a)

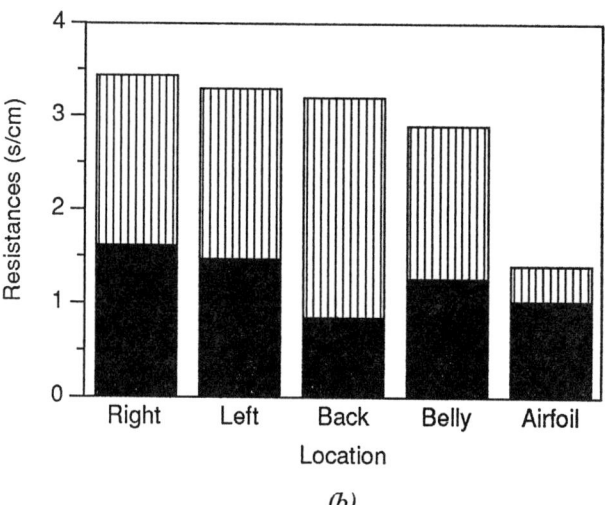

(b)

Figure 7.19. Average resistances for (*a*) SO_4^{2-} and (*b*) SO_2 dry deposition. Vertical stripes = aerodynamic resistance: shading = film resistance; diagonal stripes = surface resistance. *Source*: Used by permission of the publisher from Wu, Davidson, Dolske and Sherwood, *Aerosol Sci. Technol.*, **16**, 68–76. Copyright 1994 by Elsevier Science Inc.

instantaneous; hence, there was no r_c. Since r_b could be calculated as shown above (Prandtl analogy), that left only the aerodynamic resistance as an unknown for the SO_2 case. The aerodynamic resistance was assumed to be the same for both SO_2 and SO_4^{2-}. Then for the SO_4^{2-} case, since r_b was calculated from the Prandtl analogy, and r_a was known from the SO_2 experiments, that left only one unknown, the surface reactivity resistance, r_c. In this way, Wu et al. (1992) were able to determine the mechanisms of mass-transfer limitation for SO_4^{2-} and SO_2. Figure 7.19 shows the results of some of their analyses. From Figure 7.19a, note that the total resistance to mass transfer can differ at different locations on the statue, which indicates that deposition to various parts of the statue is different as well. Also, the relative values of the three resistances vary at any location, with boundary layer (or film) resistance usually less than or equal to the surface-reactivity resistance. The aerodynamic resistance was relatively small in all cases for particulate SO_4^{2-} deposition. By contrast, the aerodynamic resistance was a large percentage of the total resistance for gaseous SO_2 deposition (Figure 7.19b). Wu et al. noted that their estimates of SO_4^{2-} resistances were in the same range as other resistance measurements for deposition to crops and vegetation.

7.9.4. Two-Film Models, Limiting Forms, and Gas Absorption

A number of mass-transfer processes in environmental science and pollution-control engineering involve the contact of gas and liquid phases, and mass transfer of one or more compounds between the phases. These mass-transfer processes can be accompanied by reactions. For example, in the treatment of gas fumes from combustion of fossil fuels, the gas is passed through a packed tower in which the gas is contacted with a stream of water. Gas-phase pollutants like SO_2 can effectively be transferred to the liquid phase, making the gas safer for release to the environment ("gas scrubbing," e.g., Flagan and Seinfeld, 1988). We now know that analogous processes occur in the atmosphere where gases are transferred into raindrops and other hydrated aerosols ("scavenging"). After absorption in the liquid phase, the dissolved gases can undergo further reactions (e.g., Schwartz and Freiberg, 1981; Graedel and Weschler, 1981). In water and wastewater treatment, gases like ozone and oxygen are bubbled into the water phase to oxidize organic and inorganic compounds and to provide oxygen for biological processes or for "stripping" volatile organic compounds from the water (e.g., Pontius, 1990). As in many of the processes considered in this book, the mechanism of mass transfer in some of the cases mentioned above is complicated, and simplified models have been used in the analysis of experimental results, environmental transport modeling, and treatment-process design. One of the classic models of these gas–liquid interphase mass-transport processes has been the two-film model. In Figure 7.20 we show an idealized interface between a gas and a liquid. The concentration of A in the bulk gas and imaginary gas film are expressed in terms of partial pressures, while the concentration in the liquid and imaginary liquid film are expressed in terms of molar concentration. (The apparent discontinuity at the interface is an artifact of the difference in concentration units.) The gas and liquid phases outside of the films are assumed to be well mixed; hence, the concentrations there are constant. At the interface itself, $p_{A,i}$ is assumed to be in equilibrium with $c_{A,i}$. Using the concept of the mass-transfer coefficients, we can express the transfer of A between the gas and liquid phases as

$$J_A = k_G(p_{A,\infty} - p_{A,i}) = k_L(c_{A,i} - c_{A,\infty}) \qquad (7.200)$$

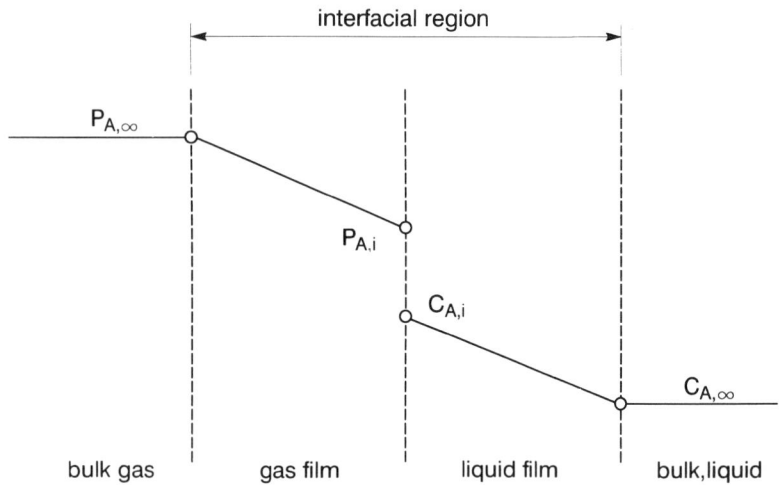

Figure 7.20. Two-film model of transfer of A from gas to liquid phase.

Here k_G and k_L are the gas- and liquid-phase mass-transfer coefficients respectively, and the subscript i refers to values at the interface. Considering Figure 7.20 and the resistance-model discussion of the previous section, it seems reasonable to expect that the present model can be interpreted as two mass-transfer resistances in series. To see this we need to recall Henry's law from Chapter 4. Recall that Henry's law is a linear relation between the partial pressure of a gas and its equilibrium aqueous concentration. Since $c_{A,i}$ and $p_{A,i}$ are assumed to be in equilibrium, Henry's law yields

$$p_{A,i} = Hc_{A,i} \tag{7.201}$$

where H is the Henry's law constant. Substituting into the middle term in Eq. 7.200 we find:

$$J_A = k_G(p_{A,\infty} - Hc_{A,i}) \tag{7.202}$$

But rearranging the right-hand term in Eq. 7.200 yields

$$c_{A,i} = \frac{J_A}{k_L} + c_{A,\infty} \tag{7.203}$$

Substituting this into Eq. 7.202 and rearranging (recall Section 7.9.2), we quickly find that J_A can be expressed as

$$J_A = \left(\frac{1}{(1/k_G) + (H/k_L)}\right)(p_{A,\infty} - Hc_{A,\infty}) \tag{7.204}$$

We then define the *gas-side resistance* as

$$R_G = \frac{1}{k_G} \tag{7.205}$$

and the *liquid-side resistance* as

$$R_L = \frac{H}{k_L} \tag{7.206}$$

Substituting into Eq. 7.204 we find:

$$J_A = \left(\frac{1}{R_G + R_L}\right)(p_{A,\infty} - Hc_{A,\infty}) \tag{7.207}$$

which shows that the two-film model does indeed boil down to a resistances-in-series model. As we noted in the previous section, one of the two resistance R_G or R_L might be much bigger than the other; hence, it may be expected to control mass transfer at the gas–liquid interface. We then say that, mathematically, Eq. 7.207 takes a "limiting form."

The bad news is that there really are no liquid or gas films. Thus, we have to consider the model above an approximation. But even if the two film-model were correct, we have no way of determining $c_{A,i}$ and $p_{A,i}$. Therefore, it is common to simplify the model even further as follows:

$$J_A = K_G(p_{A,\infty} - p_A^*) = K_L(c_A^* - c_{A,\infty}) \tag{7.208}$$

where K_G and K_L are the new gas- and liquid-phase mass-transfer coefficients, respectively (sometimes called "overall mass transfer coefficients"). The concentrations p_A^* and c_A^* are also new variables and require some discussion. We define them as follows:

$$c_A^* = \begin{cases} \text{aqueous concentration of } A, \text{ which} \\ \text{would be in equilibrium with } p_{A,\infty} \end{cases} \tag{7.209}$$

$$p_A^* = \begin{cases} \text{partial pressure of } A, \text{ which would} \\ \text{be in equilibrium with } c_{A,\infty} \end{cases} \tag{7.210}$$

Since p_a^* and c_A^* can be evaluated with Henry's law, Eq. 7.208 can be very useful in environmental problems, especially when it takes a limiting form.

One of the most common limiting forms of Eq. 7.208 occurs when gas-side resistance is negligible. This means that on the gas side of the interface, mass transfer of the transferable component is much quicker. This situation is not difficult to imagine, since we already know that diffusion coefficients in air are orders of magnitude greater than in water. In any event, in this case we should be able to represent gas–liquid mass transfer by the right-hand side of Eq. 7.208. Using Henry's law and the definition in Eq. 7.209, this can be transformed further to

$$J_A = K_L(c_A^* - c_{A,\infty}) = K_L \left(\frac{p_{A,\infty}}{H} - c_{A,\infty} \right) \tag{7.211}$$

To see how this equation has been used, consider the situation shown in Figure 7.21. Here an infinite liquid phase is being mixed with uniformly spaced mixing impellers. We imagine the setup is such that the mixing creates homogeneous turbulence below the gas–liquid interface. Even though the turbulence is homogeneous, we assume that it does not disrupt the interface (e.g., no waves or splashing). We can write a mass balance on the dotted control volume in Figure 7.21, resulting in

$$V \frac{dc_{A,\infty}}{dt} = J_A A = A K_L(c_A^* - c_{A,\infty}) \tag{7.212}$$

gas, $P_{A,\infty}$

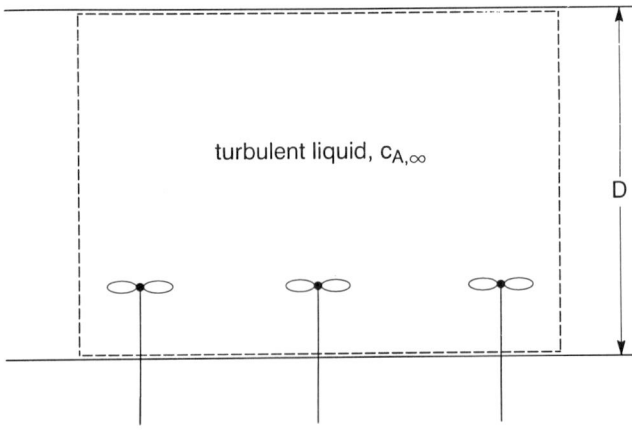

turbulent liquid, $c_{A,\infty}$

D

Figure 7.21. Transport of some component A from gas phase to turbulent liquid.

Here A is the interfacial area and V is the volume of the control volume ($V = AD$). Dividing through by V, separating, and integrating, we find:

$$\left(\frac{c_A^* - c_{A,\infty}}{c_A^* - c_{A,\infty}^0}\right) = \exp\left(-\frac{K_L}{D}t\right) \tag{7.213}$$

where $c_{A,\infty}^0$ is the concentration of the transferable component in the liquid phase at time $= 0$. In integrating Eq. 7.212, we have also made the assumption that the gas phase represents an infinite source or sink for the transferable component (i.e., $p_{A,\infty}$ is constant). Equation 7.213 can be used in two basic ways. First, if K_L is not known, we can collect data from an experiment like that shown above and fit the data to Eq. 7.213, thereby determining K_L (see Exercise 7.26). Alternatively, if K_L is known from theoretical considerations or from previous experiments, we can use Eq. 7.213 to predict the rate of exchange of gas between liquid and gas phases. In fact, Eq. 7.213 works for gas absorption from the gas phase to the liquid, or gas desorption from the liquid into the gas phase.

Lamont and Scott (1970) determined an expression for K_L based on consideration of surface-renewal theory discussed in Section 7.8.2.1. Their analysis suggests that

$$K_L = c_1(\varepsilon v)^{1/4} \, Sc^{-1/2} \tag{7.214}$$

where ε is the unit mass-energy dissipation rate discussed in Chapter 5, v is the kinematic viscosity of the liquid, and c_1 is a constant with a theoretical value of 0.4. Asher and Pankow (1986) have tested the theory described above in a laboratory device having turbulence properties like those suggested in Figure 7.21. They verified Eq. 7.214, although the numerical constant c_1 depended on the state and cleanliness of the liquid surface. Example 7.9 uses some of the results of Asher and Pankow.

Example 7.9. A mixed-water basin 50 cm deep initially contains no CO_2. The pH has been adjusted to 5.0 by the addition of a mineral acid (HCl). How long does it take to achieve 99% saturation of CO_2 if the K_L is given by Eq. 7.214 and $(\varepsilon v)^{1/4} = 0.143$ cm/s?

SOLUTION. Asher and Pankow (1986) found that Eq. 7.214 was valid for values of $(\varepsilon v)^{1/4} \, Sc^{-1/2} > 30 \times 10^{-4}$ cm/s for a carefully cleaned water surface. For this surface condition, the constant c_1 found by Asher and Pankow was 1.4. From Table 6.3, the diffusivity of CO_2 in water at 20°C is 1.77×10^{-5} cm²/s. Hence, the Schmidt number is

$$Sc = \frac{v}{D_{AB}} = \frac{0.01 \text{ cm}^2/\text{s}}{1.77 \times 10^{-5} \text{ cm}^2/\text{s}} = 565$$

We can calculate

$$(\varepsilon v)^{1/4} \, Sc^{-1/2} = 60 \times 10^{-4} \, \frac{cm}{s}$$

which means that the correlation of Asher and Pankow can safely be used. Then the value of K_L is given by Eq. 7.214 with $c_1 = 1.4$:

$$K_L = 1.4(\varepsilon v)^{1/4} \, Sc^{-1/2} = (1.4)60 \times 10^{-4} \, cm/s = 8.40 \times 10^{-3} \, cm/s$$

At pH 5.0 and below, the dissolved CO_2 is essentially undissociated and we do not have to worry about conversion of CO_2 to other carbonate species. If we desire 99% saturation of CO_2, for $c^0_{A,\infty} = 0$, the left-hand side of Eq. 7.213 reduces to 0.01. Therefore,

$$\exp\left(-\frac{K_L}{D}t\right) = 0.01 \Rightarrow t = \frac{4.61D}{K_L} = \frac{4.61(50 \, cm)}{8.40 \times 10^{-3} \, cm/s} = 2.74 \times 10^4 \, s = 7.62 \, h$$

In laboratory experiments, we often let systems exposed to the atmosphere come to equilibrium "overnight." The result of this exercise shows that this is a reasonable amount of time when mixing is present.

Exercise 7.26 compares this problem (a very clean surface) to CO_2 transport across a dirty surface. When the surface is dirty or contains an organic surface film, mass transfer of gases can be slowed. In effect, we have added another mass-transfer resistance. Many natural waters contain organic films of natural surface-active compounds (even rain droplets in the atmosphere). This can significantly slow the transfer of gases to and from the water body, which in turn, affects the chemistry of the system.

A similar limiting form of Eq. 7.208 obtains when gas is bubbled into a liquid, with or without mixing of the liquid. For example, imagine the situation shown in Figure 7.22a. If we examine a small section of any one of the bubbles (Figure 7.22b), we can also imagine gas transfer through an effective liquid mass-transfer film (i.e., no mass-transfer limitation in gas phase). If we perform a mass balance on the dissolved phase gas, we again end up with Eq. 7.212:

$$V\frac{dc_{A,\infty}}{dt} = J_A A = AK_L(c_A^* - c_{A,\infty}) \tag{7.212}$$

However, there is an important difference between the bubble system and the flat interface in Figure 7.21: we do not know the total gas–liquid interfacial surface area, A, in the bubble system. Therefore, since we don't know A, we cannot divide V by A in this system. The integrated form of Eq. 7.212 is,

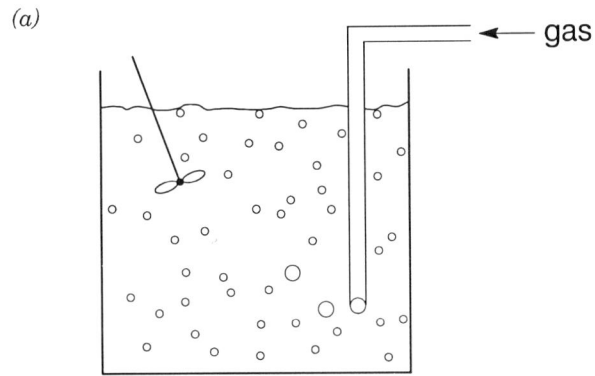

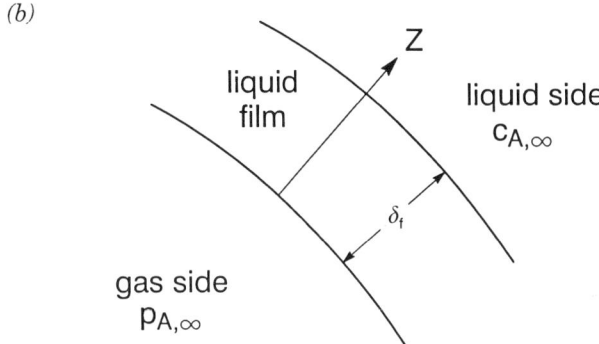

Figure 7.22. Gas transfer to a liquid using bubbling plus mixing: (a) process schematic and (b) gas transfer across a small element of the bubble surface.

therefore,

$$\left(\frac{c_A^* - c_{A,\infty}}{c_A^* - c_{A,\infty}^0}\right) = \exp\left(-K_L \frac{A}{V} t\right) \tag{7.215}$$

In much environmental-engineering literature, the constants in the argument of the exponential term are grouped together as a new constant:

$$K_L a = \frac{A}{V} K_L \tag{7.216}$$

Although this admittedly complex mass-transfer problem has boiled down to a useful practical equation, we have paid a price because of all of our approximations and simplifications: The parameter K_La is a function of almost every variable we can imagine. For example, in the simple gas-transfer system shown in Figure 7.22a, we can expect K_La to be a function of the following variables:

1. *Mixing energy or unit mass energy dissipation rate, ε.* The power applied to the mixer influences the intensity of the turbulence, the mixing, and perhaps even the bubble size. The mixing and turbulence in turn influence K_L (as was demonstrated by Eq. 7.214). Bubble size influences the interfacial area A.

2. *Gas flow rate.* Obviously if more gas is pumped into the system and more gas bubbles are circulating in the fluid, there is greater interfacial area. The gas flow itself can induce mixing, which is important in systems with no separate mechanical mixing.

3. *System size.* Obviously, for the same gas-flow rate and turbulent mixing, we would expect a smaller K_La in a larger system (i.e., greater V).

4. *Diffusion coefficient.* We know that the diffusion coefficient is important in all mass-transfer processes we have considered. It is important here as well, and we expect K_La to change with the diffusion coefficient. We therefore expect that K_La is also a function of temperature (because D_{AB} is a function of temperature).

5. *Chemistry.* As we noted above, certain surface-active agents can form surface films. These compounds could very well concentrate at the bubble surface, hence affecting the mass-transfer coefficient and even the bubble size. Dissolved salts and other solids can also affect the diffusion coefficient.

Therefore, if the sensitivity of K_La to any of the variables listed above is desired, the other variables have to be carefully controlled.

7.10. SUMMARY

In this chapter we began with a development of the convective-diffusion equation and applied this equation to the problem of dispersion of a tracer slug in laminar pipe flow. For a certain range of the Peclet number, the slug of tracer was predicted to disperse relative to a moving coordinate system in a manner similar to pure molecular diffusion from a plane source; the dispersion coefficient was found to play a role similar to the molecular-diffusion coefficient. This diffusion-like characteristic was also noted in the Lagrangian analysis of dispersion in turbulent flows, which showed that the variance of a system of tracer molecules increased in time very much like the variance of tracer molecules in molecular diffusion. These observations led the way for the development of the turbulent diffusion coefficient and the one-dimensional turbulent dispersion equation in Section 7.6.

The middle section of the chapter focused on convective-diffusion problems in which reactions occurred at one flow boundary, for example, at the surface of a flat plate, along the inside of a pipe, and on the surface of a sphere. This led to the development of the concentration boundary layer, which, combined with Fick's first law, allowed us to compute mass flux at the various surfaces.

In many mass-transfer problems in laminar and turbulent flows, exact analytical analysis is difficult or impossible. This led researchers to develop various mass-transfer models, analogies, and correlations. These approaches were used in several problems of mass transfer along flat plates, in turbulent pipe flow, and in systems of particles. The chapter ended with a discussion of interphase mass transfer, and showed that many complex problems could be boiled down to a type of model in which the different components of mass transfer are looked at as mass-transfer resistances.

EXERCISES

7.1. Draw a control volume around the liquid element of Figure 7.1*b* as shown below.

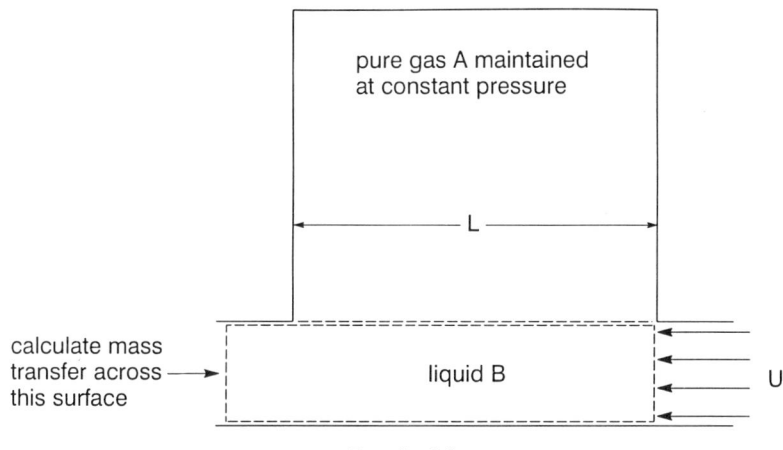

Exercise 7.1

What is the convective mass-transfer rate (M/T) of A out of the control volume on the left-hand side? Hint: Consider mass-transfer rate into the control volume.

7.2. Estimate the rate of mass transfer into the flowing stream at steady state if the gas in Figure 7.1*b* is replaced by a solid, where the interfacial concentration of A is the solubility of A in B, $c_{A,0}$. You can assume the solid is dissolving slowly enough that the system geometry is not significantly changed. Note: for the interested reader, an exact solution for dissolution into a "falling film" (i.e., with a more realistic velocity profile obeying the no-slip condition) is given in Bird et al. (1960, p. 551).

7.3. Considering a binary system composed of A and B, note that an equation completely analogous to Eq. 7.10 can be written for component B:

$$\frac{\partial \rho_B}{\partial t} + \nabla \cdot \mathbf{n}_B = r_B$$

From Eq. 6.10 we know that

$$\mathbf{n}_A + \mathbf{n}_B = \rho \mathbf{u}$$

and from the conservation of mass

$$r_A = -r_B$$

Show using the equations discussed that

$$\nabla \cdot \mathbf{u} = 0$$

7.4. The sum of molar reaction rates, $R_A + R_B$, is not necessarily zero in a binary system (i. e., mass is always conserved but "moles" may not be).

(a) Hence, show that

$$\frac{\partial c}{\partial t} + \nabla \cdot c\mathbf{u}^* = R_A + R_B$$

(b) Consider the form of this equation for $c = $ constant and, as a result, show that Eq. 7.16 is correct.

(c) Why do we say that in general $\nabla \cdot \mathbf{u}^* \neq 0$ for a constant molar density system?

7.5. Derive Eq. 7.18 following instructions in the text. It is quite straightforward.

7.6. The purely convective transport analysis of Section 7.3 can only be true if the radial-diffusion time scale is much greater than the axial-convection time scale. Show that this is equivalent to the restriction $\text{Pe} \gg L/a$, where L is the characteristic axial-length scale over which significant changes in concentration occur.

7.7. Substitute Eqs. 7.31 and 7.33 into Eq. 7.34 and integrate them, hence verifying Eq. 7.35. Hint: the integral is a little long, but gives a satisfying answer.

7.8. The purpose of this exercise is to show how the left-hand side of Eq. 7.40 is derived.

(a) First, show that the radially averaged concentration in Taylor's analysis is

$$c_{A,\text{ave}} = c_0 + \frac{1}{3} \frac{U a^2}{4 D_{AB}} \frac{\partial c_A}{\partial \tilde{z}}$$

where c_0 is the centerline concentration (i.e., $r = 0$).

(b) Therefore, show that the concentration of A can be expressed in terms of the radially averaged concentration:

$$c_A = c_{A,\text{ave}} + \frac{a^2 U}{4 D_{AB}} \frac{\partial c_{A,\text{ave}}}{\partial \tilde{z}} \left(-\frac{1}{3} + \frac{r^2}{a^2} - \frac{1}{2} \frac{r^4}{a^4} \right)$$

(c) Finally, show that for Eq. 7.36 to be true, the following must be true:

$$\frac{\partial c_{A,\text{ave}}}{\partial \tilde{z}} \gg \frac{a^2 U}{4 D_{AB}} \frac{\partial^2 c_{A,\text{ave}}}{\partial \tilde{z}^2}$$

which in turn requires that

$$\frac{4 L D_{AB}}{a^2 U} \gg 1$$

and

$$4 \frac{L}{a} \gg \text{Pe}$$

7.9. Consider that another condition for the applicability of the Taylor dispersion analysis is that the axial-diffusion time scale must be much greater than the time scale characterizing axial dispersion; hence, verify the right-hand side of Eq. 7.40.

7.10. In Example 7.1, we were able to narrow in on the limits of applicability of Taylor's analysis of dispersion in laminar pipe flow. The purpose of this problem is to see if Taylor's own (1953) experiments on permanganate dispersion in laminar pipe flow fit the limits of the analysis which he derived later (Taylor, 1954b).

(a) Assuming a diffusion coefficient of $1 \times 10^{-5} \, \text{cm}^2/\text{s}$, use the data provided in Exercise 7.11 to determine if $\text{Pe} \gg 7$.

(b) The reader will recall that we ran into a dead end in Example 7.1 when we had to come up with a value for L. Let's approximate L

with four times the standard deviation of the tracer cloud, that is, $L = 4\sigma$. We are in effect saying that the most significant concentration changes occur over a scale of 4σ. Is $Pe \ll 4L/a$ for curves II and III in Figure 7.4?

7.11. Studies like Taylor's (1953) measurements of dispersion in laminar flow can be used to estimate diffusion coefficients. Taylor fit the Gaussian law (Eq. 7.43) to curves II and II in Figure 7.4 and found that the variance, $2D_{dis}t$ was given by

$$2D_{dis}t_{II} = \sigma_{II}^2 = 13.4 \text{ cm}^2$$

and

$$2D_{dis}t_{III} = \sigma_{III}^2 = 21.8 \text{ cm}^2$$

(a) Show that

$$D_{dis} = \frac{1}{2(t_{III} - t_{II})}(\sigma_{III}^2 - \sigma_{II}^2)$$

(b) Taylor measured $t_{III} - t_{II}$ as 330 s. The pipe diameter was 0.0504 cm. If $u_{z,ave} = 0.0865$ cm/s, calculate the diffusion coefficient of permanganate indicated by Taylor's experiments. (Note: as mentioned in Chapter 6, Taylor found that the diffusion coefficient could depend on tracer concentration.)

7.12. Derive Eq. 7.64 from the instructions in the text. Hint: don't forget the tricks used in Section 5.13.2.

7.13. Taylor (1954a) discussed some measurements of the dispersion of a conductive tracer in turbulent pipe flow. The mean pipe flow was $\bar{U} = 105$ cm/s and the pipe diameter was 20 in. At 103 s after injection the variance of the tracer was $2D_{dis}^t t = 4.49 \times 10^5$ cm^2. Compute the apparent numerical coefficient in Eq. 7.79, if, at the Reynolds number of the experiments, the value of $\bar{U}/u_* = 26.0$. Why might the calculated value of the numerical coefficient be bigger (or smaller) than the value derived in Section 7.6 of 10.06?

7.14. Derive Eq. 7.92. This problem makes good use of the chain rule of calculus. For example,

$$\frac{\partial \tilde{c}_A}{\partial y} = \frac{dg}{d\eta}\frac{\partial \eta}{\partial y} = \frac{dg}{d\eta}\left(\frac{U}{vx}\right)^{1/2}$$

7.15. Derive Eq. 7.95. It may help to consider the following defining integral:

$$J_{A,\text{ave}} = \frac{1}{L} \int_0^L J_A dx$$

where L is the plate length.

7.16. Imagine that the flat plate in Figure 7.8 is a perfectly flat leaf, and that there is a laminar flow of air of 0.1 m/s along the leaf surface.

(a) Compute the average aerosol flux ($J_{A,\text{ave}}$) to leaves of 3 and 10 cm length. The aerosol particles are 1.0 μm in diameter, and the air temperature is 25°C. Assume that small hairs on the surface of the leaves make the surfaces perfect sinks for the aerosol particles, yet do not affect the fluid mechanics on the leaf surface. Explain your results in a qualitative sense, invoking Fick's first law to explain why average deposition is greater on the smaller leaf.

(b) Often the rate of deposition of aerosols to surfaces decreases over time. Make an argument(s) why this might be. Often the deposition rate is found to increase just after a rain storm. Why might this be?

7.17. (a) Calculate the percent removal of 1.0-, 0.1-, and 0.01-μm-diameter aerosol particles in a model trachea pipe of 20 cm length and 2 mm diameter. Assume $U = 0.5$ m/s and $T = 25$°C. Be sure to state how and where you decided to truncate the summation in Eq. 7.105.

(b) On the basis of the results of part **a**, what kinds of particles are most likely to be deposited in the lungs because of convective diffusion — small or large particles?

7.18. Your boss at Earth, Wind, and Fire Environmental Services wants you to design an aerosol sizer based on the analysis of Section 7.7.2. The equipment available to construct the device is just pipes, pumps, and an accurate particle-concentration detector. You can assume that the inside pipe surface is a perfect sink for the aerosol particles and that the aerosol particles are of a single size. Design a system for measuring aerosol size in the range of 0.1 to 10 μm. Note any assumptions made, and don't forget that flow in the pipe must stay laminar. Also check to make sure that Sc $\gg$ 1 for all particles measured.

7.19. Regarding Example 7.3, explain (a) how an increase or decrease in a affects $c_{A,\text{ave}}/c_{A,i}$ when U is a constant and (b) how an increase or decrease in U affects $c_{A,\text{ave}}/c_{A,i}$ when a is a constant. Do these predictions make sense? Explain in words or with words and equations.

7.20. In some parts of the ocean, oxygen concentrations are observed to attain minimum values in certain horizontal layers, sometimes as deep as 700 m (Csanady, 1986). One explanation for this is that large biological particles sinking with their Stokes velocity slow down as they dissolve, and tend to congregate at the same depth where their final dissolution or oxidation consumes much oxygen.

(a) Using Eq. 7.116, derive a function for particle size as a function of time. This is considered the high Peclet number limit.

(b) Using the expression for the high Peclet number limit, calculate the change in time of the radius of a 200-μm-diameter organic particle initially at the surface of the ocean. Assume that there is no turbulence in the ocean and that the particle sinks at its Stokes velocity. Assume further that the oxidation of this particle can be represented by the equation

$$COH_2 + O_2 = CO_2 + H_2O$$

and that dissolution of the particle is limited by the oxygen diffusion rate, whose coefficient of diffusion can be estimated as $D_{AB} \simeq 2 \times 10^{-5}\,cm^2/s$; Csanady (1986) estimates the oxygen deficiency normalized by the particle density as

$$\frac{c_{A,s} - c_{A,\infty}}{\rho_p} = 10^{-6}$$

The following value is also available:

$$\frac{2(\rho_p - \rho_f)g}{9\mu} = 2 \times 10^5\,m^{-1}\,s^{-1}$$

(c) Using the results of part **b**, determine how far the particle will have settled before its radius decreases to 50 μm.

7.21. Correlations for mass transfer in rough pipes or tubes seem to be nonexistent. In Chapter 5, the friction factor for fully rough turbulent flow in a tube was given by

$$\frac{1}{\sqrt{C_f}} = 4.0 \log_{10} \frac{D}{e} + 2.28$$

where e/D is the roughness parameter.

(a) Using the Chilton–Colburn analogy, show that $Sh = 4.73 \times 10^{-3}$ $Re\,Sc^{1/3}$ when $D/e = 100$. Hint: recall Example 7.4.

(b) We expect the mass transfer at a rough surface in turbulent flow to be greater than that at a smooth surface. Compare your Sherwood number result for the case of $Sc = 1000$ to the smooth-tube correlation of Linton and Sherwood in Section 7.8.3.2 if $Re = 4 \times 10^5$. Does your equation conform to our expectation?

7.22. Reconsider the piece of benzene-covered flooring discussed in Example 7.5. At the leading edge of the section, the boundary layer is laminar and transforms to a turbulent boundary layer at $Re_x = 4 \times 10^5$. What is the maximum average benzene evaporation rate per unit width of flooring? Hint: compute the mass-transfer coefficient (or J_A) for the laminar and turbulent sections, and average over the floor length $L = 2$ m.

7.23. Reconsider Example 7.6 with the added complexity of a homogeneous first-order reaction in the liquid phase. In other words, consider the combined effects of convection, surface reaction, and homogeneous reaction in the pipe.

(a) Show that the appropriate partial differential equation is

$$\frac{\partial c_x}{\partial t} + U \frac{\partial c_x}{\partial x} = \frac{-2J_A}{a} + r_A$$

where r_A is the homogeneous liquid phase reaction:

$$r_A = \frac{\partial c_x}{\partial t} = -k_{A,l}c_x$$

and, as in Example 7.6,

$$J_A = k_c c_x$$

(b) For the initial condition $c_x = c_{x=0}$ at $x = 0$, show that at steady state, the integrated form of the differential equation above is

$$c_x = c_{x=0} \exp\left[-\left(\frac{2k_c}{a} + k_{A,l}\right)\frac{x}{U}\right]$$

(c) Consider the same section of pipe and the same flow conditions described in Example 7.6. A sample of water was taken from the head of the pipe and the homogeneous phase first-order reaction-rate constant for chlorine was determined to be 1.6×10^{-2} min^{-1}. If the initial chlorine concentration is 6.5 mg/L, determine the consumption of chlorine at a point 175 m downstream of the head of the pipe using the equation developed in part **b**.

7.24. Using the analysis of the fully penetrated biofilm with zero-order homogeneous reaction (Exercise 6.12), complete an analysis parallel to that of Section 7.9.2.

(a) Show that

$$J_A = \left(\frac{1}{(1/k_c) + (1/k_A^0 L)}\right) c_{A,\infty}$$

or equivalently,

$$J_A = \left(\frac{1}{R_f + R_{fpb}} \right) c_{A,\infty}$$

where R_f is given by Eq. 7.186 and where R_{fpb} is the mass-transfer resistance for the fully penetrated biofilm:

$$R_{fpb} = \frac{1}{k_A^0 L}$$

(b) From the data of Exercise 7.23, which concerns a section of piping in Paris, France, the laminar-flow mass-transfer coefficient is estimated to be 5.48×10^{-3} cm/min. If the biofilm thickness is $100 \, \mu$m, determine the required k_A^0 for the fully penetrated biofilm such that the diffusional and reaction mass-transfer resistance are identical.

7.25. Nitrates, NO_3^-, are thought to be biologically degraded in anaerobic sediment layers in lakes and wetlands. Shown below is a hypothetical profile of nitrate concentration in a aerobic–anaerobic sediment system.

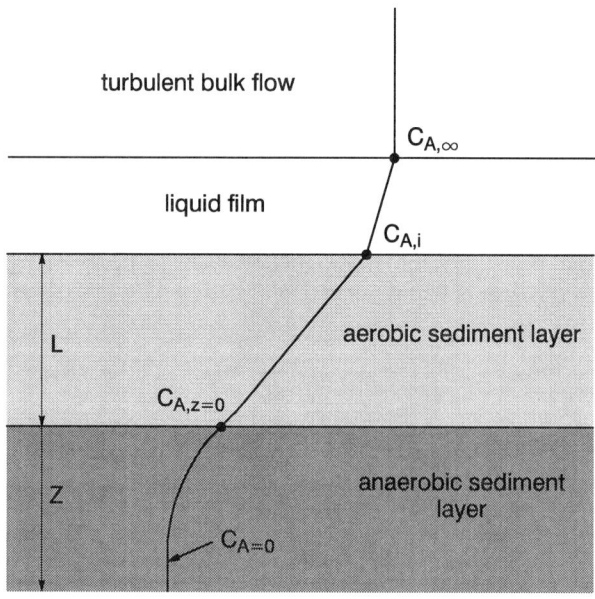

Exercise 7.25

Nitrate is not consumed in the aerobic layer; it simply diffuses through the aerobic layer on its way to the anaerobic layer, much like the diffusion through the biofilm in Figure 7.17. Once in the anaerobic layer, the nitrate diffuses and reacts, where the reaction is a first-order homogeneous reaction. Hence, mass transport in the anaerobic layer is like that shown in the biofilm of Figure 7.15. Show that the flux into the sediments can be given by

$$J_A = \left(\frac{1}{(L/D_{AB}^{aer}) + (1/\sqrt{D_{AB}^{an} k_A}) + (1/(k_c))} \right) C_{A,\infty}$$

where D_{AB}^{aer} and D_{AB}^{an} are the diffusion coefficients for nitrate in the aerobic and anaerobic layers, k_A is the first-order homogeneous reaction rate constant, and k_c is defined in Sections 7.9.2 and 7.9.3.

7.26. The complete data set of Asher and Pankow (1986) for four different water surface conditions is shown below.

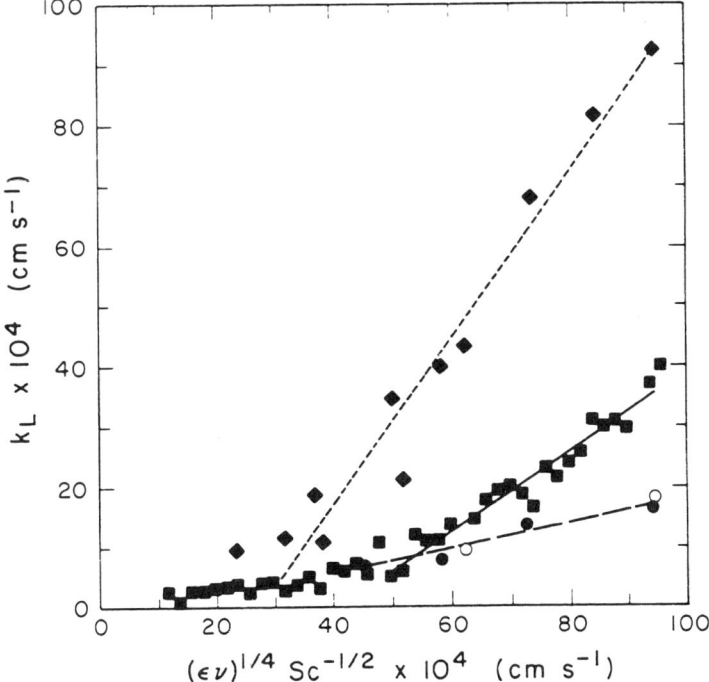

Exercise 7.26. *Source*: Reprinted with permission from Asher and Pankow, *Tellus*, **38B**, 314 (1986). Copyright 1986, Munksgäard International Publishers Inc.

The straight-line fits were made using Eq. 7.214. The four conditions are the following:

1. Solid diamonds: surface cleaning by contact of surface with rayon cloth followed by surface vacuuming with a micropipet.
2. Solid squares: surface cleaning by contact of surface with additive-free lens paper.
3. Solid circles: 1-octadecanol surface monolayer.
4. Open circles: average of four measurements of uncleaned water surface.

(a) Determine the slopes of lines through the four data sets, and determine the effect of surface cleanliness and impurities of the liquid mass-transfer coefficient. Interpret your results in terms of the resistances-in-series model. Note: as seen from the plot, the data from the experiments corresponding to the closed and open circles could not be discriminated.

(b) Using the system described in Example 7.9, compute the time required to achieve 99% saturation of CO_2 for the systems represented by solid squares, solid circles, and open circles.

REFERENCES

Aris, R., "On the Dispersion of a Solute in a Fluid Flowing Through a Tube," *Proc. R. Soc. London Ser. A*, **235**, 67–77 (1956).

Asher, W. E. and Pankow, J. F., "The Interaction of Mechanically Generated Turbulence and Interfacial Films with a Liquid Phase Controlled Gas/Liquid Transport Process," *Tellus*, **38B**, 305–318 (1986).

Bird, R. B., Stewart, W. E., and Lightfoot, E. N., *Transport Phenomena*, Wiley, New York, 1960.

Brodkey, R. S. and Hershey, H. C., *Transport Phenomena: A Unified Approach*, McGraw-Hill, New York, 1988.

Chilton, T. H. and Colburn, A. P., *Ind. Eng. Chem.*, **26**, 1183 (1934).

Clift, R., Grace, J. R., and Weber, M. E., *Bubbles, Drops, and Particles*, Academic, New York, 1978.

Colburn, A. P., *Trans. A.I.Ch.E.*, **29**, 174–210 (1933).

Crank, J., *The Mathematics of Diffusion*, Oxford University Press, 1956.

Csanady, G. T., "Mass Transfer to and from Small Particles in the Sea," *Limnol. Oceanogr.*, **31**, 2:237–248 (1986).

Danckwerts, P. V., *Gas–Liquid Reactions*, McGraw-Hill, New York, 1970.

Dwidevi, P. N. and Upadhyay, S. N., *Ind. Eng. Chem. Proc. Des. Dev.*, **16**, 157 (1977).

Fischer, H. B., List, E. J., Koh, R. C. Y., Imberger, J., and Brooks, N. H., *Mixing in Inland and Coastal Waters*, Academic, New York, 1979.

Flagan, R. C. and Seinfeld, J. H., *Fundamentals of Air Pollution Engineering*, Prentice-Hall, Englewood Cliffs, NJ, 1988.

Fogler, H. S., *Elements of Chemical Reaction Engineering*, Prentice-Hall, Englewood Cliffs, NJ, 1992.

Friedlander, S. K., *Smoke, Dust, and Haze*, Wiley, New York, 1977.

Garner, F. H. and Suckling, R. D., "Mass Transfer from a Soluble Solid Sphere," *A.I.Ch.E.J.*, **4**, 1, 1114–1124, 1958.

Gilliard, E. R. and Sherwood, T. K., *Ind. Eng. Chem.*, **26**, 516 (1934).

Graedel, T. E. and Weschler, C. J., "Chemistry within Aqueous Atmospheric Aerosols and Raindrops," *Rev. Geophys. Space Phys.*, **19**, 4:505–539 (1981).

Harriot, P. and Hamilton, R. M., "Solid-Liquid Mass Transfer in Turbulent Pipe Flow," *Chem. Eng. Sci.*, **20**, 1073 (1965).

Higbie, R., *Trans. A.I.Ch.E.*, **35**, 365 (1935).

Hinze, J. O., *Turbulence*, 2nd ed., McGraw-Hill, New York, 1975.

Lamont, J. C. and Scott, D. S., "An Eddy Cell Model of Mass Transfer into the Surface of a Turbulent Liquid," *A.I.Ch.E.J.*, **16**, 513–519 (1970).

Levich, V. G., *Physicochemical Hydrodynamics*, Prentice-Hall, Englewood Cliffs, NJ, 1962.

Levinspiel, O., *Chemical Reaction Engineering*, 2nd ed., Wiley, New York, 1972.

Linton, W. H. and Sherwood, T. K., *Chem. Eng. Prog.*, **46**, 258 (1950).

Nauman, E. B. and Buffham, B. A., *Mixing in Continuous Flow Systems*, Wiley-Interscience, New York, 1983.

Pontius, F. W., Ed., *Water Quality and Treatment*, 4th ed., McGraw-Hill, New York, 1990.

Probstein, R. F., *Physicochemical Hydrodynamics: An Introduction*, Butterworths, Boston, MA, 1989.

Seinfeld, J. H., *Atmospheric Chemistry and Physics of Air Pollution Control*, Wiley-Interscience, New York, 1985.

Skelland, A. H. P., *Diffusional Mass Transfer*, Wiley-Interscience, New York, 1974.

Schwartz, S. E. and Freiberg, J. E., "Mass Transport Limitation to the Rate of Reaction of Gases in Liquid Droplets: Application to Oxidation of SO_2 in Aqueous Solutions," *Atmos. Env.*, **15**, 7:1129–1144 (1981).

Steinberger, R. L. and Treybal, R. E., "Mass Transfer from a Solid Soluble Sphere to a Flowing Liquid Stream, *A.I.Ch.E.J.*, **6**, 227 (1960).

Taylor, G. I., "Diffusion by Continuous Movements," *Proc. London Math. Soc. A*, **20**, 196–211 (1922).

Taylor, G. I., "Dispersion of Soluble Matter in Solvent Flowing Slowly Through a Pipe," *Proc. R. Soc. London Ser. A*, **219**, 186–203 (1953).

Taylor, G. I., "The Dispersion of Matter in Turbulent Flow Through a Pipe," *Proc. R. Soc. London Ser. A*, **223**, 446–468 (1954a).

Taylor, G. I., "Conditions Under Which Dispersion of a Solute in a Stream of Solvent Can Be Used to Measure Molecular Diffusion," *Proc. R. Soc. London Ser. A*, **255**, 473–477 (1954b).

Thomann, R. V. and Miller, J. A., *Principles of Surface Water Quality Modeling and Control*, Harper and Row, New York, 1987.

Treybal, R. E., *Mass Transfer Operations*, McGraw-Hill, New York, 1968.

Welty, J. R., Wicks, C. E., and Wilson, R. E., *Fundamentals of Momentum, Heat, and Mass Transport*, Wiley, New York, 1976.

Wilson, E. J. and Geankoplis, C. J., "Liquid Mass Transfer at Very Low Reynolds Numbers in Packed Beds," *Ind. Eng. Chem. Fund.*, **5**, 1:9–14 (1966).

Wu, Y. L., Davidson, C. I., Dolske, D. A., and Sherwood, S. I., "Dry Deposition of Atmospheric Contaminants: The Relative Importance of Aerodynamic, Boundary Layer, and Surface Resistances," *Aerosol Sci. Technol.*, **16**, 65–81 (1992).

BIBLIOGRAPHY

Bendat, J. S. and Piersol, A. G., *Random Data: Analysis and Measurement Procedures*, Wiley-Interscience, New York, 1971.

Sellars, J. R., Tribus, M., and Klein, J. S., *Trans. Am. Soc. Mech. Eng.*, **78**, 441 (1956).

Tennekes, H. and Lumley, J. L., *A First Course in Turbulence*, The MIT Press, Cambridge, MA, 1972.

8

FILTRATION AND MASS TRANSPORT IN POROUS MEDIA

8.1. INTRODUCTION

This chapter concerns transport through porous media, and two classical subjects of great environmental importance: *dispersion* of compounds and contaminants (gaseous or liquid) in porous media, and the collision and attachment of particulate contaminants (hydrosols or aerosols) at the surface of individual grains or fibers of the porous media, which we call *filtration*. Our definition of *porous media* is broad: We consider movement through soils (groundwater transport), packed filtration beds, air filters, and chromatography columns. These subjects form an area of considerable activity in environmental engineering and science. By necessity, our review is fairly streamlined. In general, we consider one-dimensional systems and *saturated media*, that is, porous media in which the carrier fluid (the liquids or gases passing through the porous media) completely fill the interstitial region between the porous media grains or fibers. Also, since the characteristic distances separating the individual grains or fibers of the porous media are quite small, we generally restrict discussion to laminar flows; as in Chapter 7, we also assume that these velocity fields are known (i.e., there is very little discussion of fluid mechanics in this chapter).

There are many similarities, and some important differences, between porous media transport and the transport processes in homogeneous media discussed in the last two chapters. Recall that for laminar flows in a tube, we found that for certain restrictions on the Peclet number, a thin slug of tracer would eventually broaden into a more or less Gaussian-shaped concentration profile (Figure 8.1*a*). Likewise, we found that for turbulent pipe flow, an initially concentrated slug would eventually broaden into a Gaussian-shaped concentration profile (Figure 8.1*b*). In this chapter, we again find that in certain porous media, an initially concentrated slug of tracer will eventually be transported as a Gaussian plume (Figure 8.1*c*). This would suggest that there are similarities between the three mass-transport situations in Figure 8.1. We know that in turbulent flows, small-scale velocity fluctuations contribute

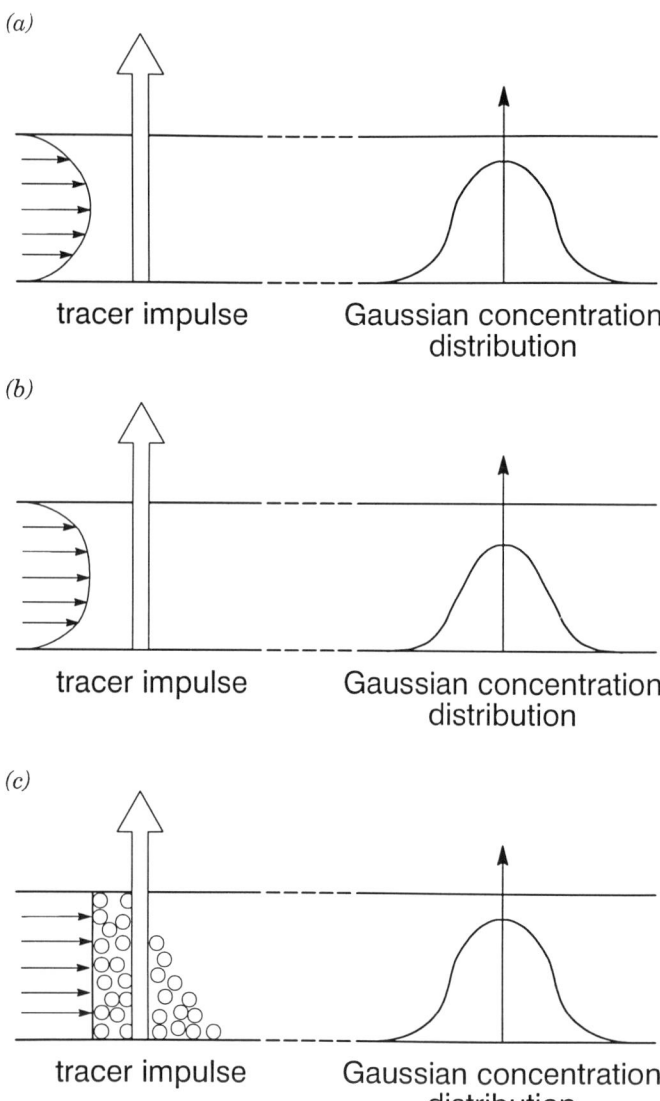

Figure 8.1. Ideal tracer dispersion in (*a*) laminar tube flow, (*b*) turbulent pipe flow, and (*c*) one-dimensional porous media flow.

greatly to the dispersion of a compound, above and beyond the natural molecular diffusion of the compound. If we took a snapshot of a turbulent flow and could visualize all the instantaneous motions, we would see a variety of swirling motions of different scales (Figure 5.16). Because of Taylor's success in modeling turbulent dispersion in a pipe, we modeled the turbulent transport as a Fickian-like gradient-diffusion process. By contrast, in porous media, we are usually dealing with laminar flows. Molecular diffusion is definitely an important transport mechanism, but in many cases, it cannot account for the dispersion observed in real porous media. There is another mechanism at work here, which is generally referred to as *mechanical dispersion*. The key to understanding mechanical dispersion lies in the nature of the flow between the grains of the porous media. As shown in Chapter 5 (Figure 5.10), detailed measurements of flow on the scale of individual porous media grains show a complicated flow pattern: flow is generally curvilinear, with streamline division around individual grains. This results in a division and mixing of small packets of fluid as they work themselves through the porous media. Since the layout of the porous media is generally random, we can think of the mixing in porous media as caused by small-scale (spatially random) motions. The similarity of these motions to the random curvilinear eddy motion in turbulence has been noted by many investigators. Although the range of scales of motion in a homogeneous porous media is much more limited than in a turbulent flow, we can in fact think of a mechanical dispersion in porous media related to the small-scale random motions. In fact, we will see shortly that this mechanical dispersion is classically described in a manner similar to turbulent dispersion.

The chapter ends with a discussion of the basic mechanisms of particle capture by filter grains and fibers, and the effect these deposits have on energy (head) loss through the filter. Particle capture by filter media can be considered a type of heterogeneous reaction, and some of the techniques developed in the last two chapters can be applied to filtration problems.

8.2. POROSITY, VELOCITY, AND POROUS MEDIA CONTINUA

Some basic definitions in this introduction to porous media transport are essential. For a sample of porous media, we define *porosity* as the interstitial or void volume divided by the total sample volume:

$$\epsilon = \frac{V_v}{V_T} \tag{8.1}$$

Another useful parameter describing porous media is the *solid fraction*, which is defined as the ratio of the solid volume to the total volume of a sample:

$$s = \frac{V_s}{V_T} = \frac{V_T - V_v}{V_T} = \frac{V_T - \epsilon V_T}{V_T} = 1 - \epsilon \tag{8.2}$$

Table 8.1 lists typical values of porosity for some ideal and natural porous media. There are a few complications in the simple picture of porosity. Consider the section of porous media in Figure 6.13a. When we consider the

Table 8.1. Porosity of Some Unconsolidated Media

Type of Deposit	Porosity (%)
Uniform spheres (cubical packing)	47.6
Uniform spheres (rhombohedral packing)	26.0
Peat soil	60–80
Soils	50–60
Gravel	25–40
Sand	25–50
Silt	35–50
Clay	40–70

Source: Bear, *Dynamics of Fluids in Porous Media*, p. 46, © 1988. Reprinted by permission of Dover Publications, New York; Freeze and Cherry, *Groundwater*, p. 37, © 1979. Reprinted by permission of Prentice Hall, Upper Saddle River, New Jersey.

flux of some component into the side of the element, we are interested in the *areal porosity*, or the fraction of the plane normal to flow which is void or open. Is this porosity value different from that defined in Eq. 8.1, which was based on a bulk sample? We can in fact assume they are the same, the proof being given by Bear (1988). Another concern arises in some porous media, where the effective flow path may actually occur through a somewhat more limited region of the void space. We then may need to discuss an *effective porosity* value. For example, the grains of some porous media have surface pores that may be considered "dead-end pores" with respect to convective transport through the media. However, in the overall scheme of mass transport and partitioning, these pore spaces are by no means unimportant. In some types of chromatography, these pores are in fact a major reason for chromatographic separation; this is discussed further in Section 8.6. Nevertheless, in the initial part of this chapter, we assume that the porosity defined by Eq. 8.1 is all the information we need.

As has generally been the case in this book, we are less interested in detailed descriptions of individual molecular motions, and more in the overall macroscopic motion of fluids and components dissolved or transported in the fluids. For single-phase flow in nonporous media, we developed the concept of a continuum in Chapter 1. This was based on consideration of molecular spacing in the fluid (gas or liquid). We noted that if our scale of resolution was too small, our sensor would pick up molecular scale fluctuations in density. At a sufficiently larger scale of resolution, these wild fluctuations died out and we arrived at a convenient scale over which to define density (and velocity). The continuum hypothesis was that we could always define density in this way, so that local density value changed smoothly. In other words, we could have smoothly varying "point values" of quantities like density and velocity, which allowed the application of classical approaches from continuum mechanics. We have the same goal in this chapter, to be able to calculate local fluid velocities,

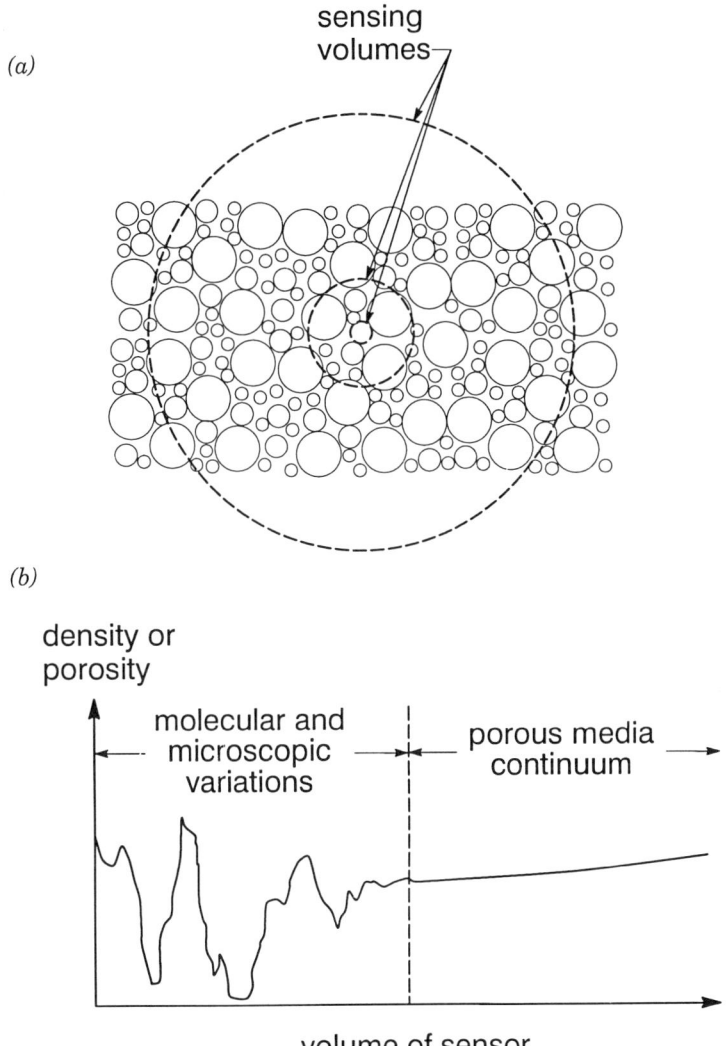

Figure 8.2. Representation of (*a*) porous media structure and density-measuring sensor volume, and (*b*) variations of density as a function of measurement scale.

densities, and fluxes. However, there is an added (but not too severe) complication that is related to the presence of the porous media itself. Consider a somewhat ideal porous media composed of spherical particles, as shown in Figure 8.2*a*. The dotted circles are the effective sensing volume of our density-measuring instrument. When the sensing volume is very small, the output of the density-measuring equipment is quite wild and ragged, as shown in Figure 8.2*b*. However, as the sensor volume grows, the variations in the density signal die out as our instrument seems to converge to a stable density value. The range of wild variations is called the range of *molecular and microscopic effects*: this range corresponds to changes caused by molecular scale variations and microscopic variations of the order of the size of the porous media grains. The smoothly varying density range will simply be called the *porous media continuum*. In most of porous media transport study (and in this book), we usually assume that the porous media continuum can be defined. Hence, when we talk about a density or porosity, we usually assume it is an average value over a region at least as big as that corresponding to the onset of smooth density or porosity shown in Figure 8.2*b*; it can equivalently be considered an average over many pores. This continuum averaging volume goes by a special name in groundwater transport studies, the *representative elementary volume* (REV).

Finally, we arrive at the important consideration of the appropriate velocities, density, and concentration for consideration in porous media-transport problems. One definition of velocity is the average velocity at a continuum point:

$$u = \frac{Q}{A} \tag{8.3}$$

where Q is the flow in the REV and A is the area of the plane normal to the local flow direction. This velocity is usually known as the *specific discharge*. Because the fluid is actually only moving through the pores, it must on average be traveling at a higher rate than the specific discharge. Freeze and Cherry (1979) define this as the *average linear velocity*:

$$v = \frac{u}{\epsilon} \tag{8.4}$$

The average linear velocity will in general have three components corresponding to the three cartesian axes. Therefore, it can also be expressed in vector form:

$$\mathbf{v} = v_1\mathbf{i} + v_2\mathbf{j} + v_3\mathbf{k} \tag{8.5}$$

where $\mathbf{i}$, $\mathbf{j}$, and $\mathbf{k}$ are the unit vectors along the x_1, x_2, and x_3 axes. The REV can also be used to identify the appropriate average density and concentration. We therefore identify ρ_A as the average density of component A in the REV, and c_A as the average molar concentration of A in the REV, where as usual $c_A = \rho'_A$ (molar mass of A).

8.3. COEFFICIENTS OF MECHANICAL, MOLECULAR, AND HYDRODYNAMIC DISPERSION

We suggested in Section 8.1 that the mechanical dispersion of dissolved components is analogous to turbulent transport in a homogeneous medium. Like our development of Reynolds averaging in Chapters 5 and 7, we define density and velocity as the sum of the average value in the REV and the fluctuation from the average:

$$P_A = \rho_A + \rho_A' \tag{8.6}$$

and

$$V_i = v_i + v_i' \tag{8.7}$$

Here P_A and V_i are the total density and velocity, the primed quantities are the fluctuations, and the subscript i refers to the ith coordinate axis. The average flux of A in the REV is the average of the product of the total concentration and velocity.* Expanding the terms under the average, as in Sections 5.13.2 and 7.6, we find:

$$
\begin{aligned}
n_{A,i} = \overline{P_A V_i} = \overline{(\rho_A + \rho_A')(v_i + v_i')} &= \overline{\rho_A v_i} + \overline{\rho_A v_i'} + \overline{\rho_A' v_i} + \overline{\rho_A' v_i'} \\
&= \rho_A v_i + \rho_A \overline{v_i'} + \overline{\rho_A'} v_i + \overline{\rho_A' v_i'}
\end{aligned}
\tag{8.8}
$$

However, as we know, the averages of the single primed quantities are by definition zero:

$$\overline{v_i'} = \overline{\rho_A'} = 0 \tag{8.9}$$

Hence, Eq. 8.8 reduces to

$$n_{A,i} = \rho_A v_i + \overline{\rho_A' v_i'} \tag{8.10}$$

The last term in Eq. 8.10 involves the spatial average of the fluctuating concentration and velocity. Therefore, as we found in Chapters 5 and 7, there appears also to be a "closure problem" in porous-media transport, because the last term in Eq. 8.10 is not known a priori. Researchers in this area have attempted to close the equations of porous-media transport with an approach very similar to that examined in Chapters 5 and 7, by assuming that the last

---

*Note that our average here is a spatial average over the REV at some point in the porous medium. This is in contrast to the average used in Chapters 5 and 7, which was a time average. Nevertheless, the basic techniques of Reynolds averaging are the same, whether a time or spatial average is involved.

term in Eq. 8.10 is proportional to the local gradient of the average concentration in the REV:

$$\overline{\rho'_A v'_i} = -D_{i,j}\frac{\partial \rho_A}{\partial x_j} \tag{8.11}$$

(As usual in this book, tensor notation implies summation over the repeated index j.) Combining this with Eq. 8.10, we arrive at

$$n_{A,i} = \rho'_A v_i - D_{i,j}\frac{\partial \rho_A}{\partial x_j} \tag{8.12}$$

This can also be equivalently expressed in terms of molar concentration:

$$N_{A,i} = c_A v_i - D_{i,j}\frac{\partial c_A}{\partial x_j} \tag{8.13}$$

where c_A is the average concentration in the REV. The coefficient D is called the *coefficient of mechanical dispersion* and is, in general, a second-order tensor:

$$D_{i,j} = \begin{pmatrix} D_{1,1} & D_{1,2} & D_{1,3} \\ D_{2,1} & D_{2,2} & D_{2,3} \\ D_{3,1} & D_{3,2} & D_{3,3} \end{pmatrix} \tag{8.14}$$

The transport of component A can also be caused by molecular diffusion in the pores of the porous material. We define the *hydrodynamic dispersion* as the sum of the mechanical and effective molecular diffusion coefficients. For a homogeneous and isotropic porous medium, i.e., where the statistical properties of the medium are invariant with changes in direction or translation of axes, the hydrodynamic dispersion is simply

$$D_{i,j}^{(h)} = D_{i,j} + D_{\text{eff}} \tag{8.15}$$

Here h stands for the hydrodynamic dispersion and D_{eff} is the effective diffusivity,

$$D_{\text{eff}} = D_{AB}\tau' \tag{8.16}$$

Here, τ' is the tortuosity (Section 6.8), and D_{AB} is the molecular diffusion coefficient in pure water. The general flux expression to replace Eq. 8.12 would be

$$n_{A,i} = \rho_A v_i - D_{i,j}^{(h)}\frac{\partial \rho_A}{\partial x_j} \tag{8.17}$$

In a homogeneous and isotropic porous medium, the porous medium properties do not vary in space or in any direction. In this case, there are only six unique components of the hydrodynamic dispersion tensor: $D_{11}^{(h)}$, $D_{22}^{(h)}$, $D_{33}^{(h)}$, $D_{21}^{(h)}$ $(= D_{12}^{(h)})$, $D_{13}^{(h)}$ $(= D_{31}^{(h)})$, and $D_{23}^{(h)}$ $(= D_{32}^{(h)})$. This is called a *symmetrical tensor*. A further simplification of any symmetrical tensor obtains if the local coordinate axes are rotated to align with the *principal axes* of the tensor;* this yields a simple diagonal tensor:

$$\tilde{D}_{i,j}^{(h)} = \begin{pmatrix} \tilde{D}_{1,1}^{(h)} & 0 & 0 \\ 0 & \tilde{D}_{2,2}^{(h)} & 0 \\ 0 & 0 & \tilde{D}_{3,3}^{(h)} \end{pmatrix} = \begin{pmatrix} \tilde{D}_{1,1} + D_{\text{eff}} & 0 & 0 \\ 0 & \tilde{D}_{2,2} + D_{\text{eff}} & 0 \\ 0 & 0 & \tilde{D}_{3,3} + D_{\text{eff}} \end{pmatrix} \quad (8.18)$$

where the tilde indicates that this is the principal component form (i.e., in general $D_{1,1}^{(h)} \neq \tilde{D}_{1,1}^{(h)}$, etc.).

Bear (1979) states that for a homogeneous and isotropic medium, the principal component form is achieved simply by aligning one of the axes (e.g., the x_1 axis) with the direction of the local average, linear-velocity vector defined earlier. In this case, Bear shows that the principle components of the mechanical-dispersion tensor can be expressed as simple functions of the linear-velocity magnitude:

$$\tilde{D}_{i,j} = \begin{pmatrix} a_L|\mathbf{v}| & 0 & 0 \\ 0 & a_T|\mathbf{v}| & 0 \\ 0 & 0 & a_T|\mathbf{v}| \end{pmatrix} \quad (8.19)$$

The parameters a_L and a_T are called the *coefficients of longitudinal and transversal dispersivity*, and a_L is often 3–10 times greater than a_T in groundwater-transport problems. The longitudinal direction refers to the direction of $\mathbf{v}$, and the transversal directions are normal to the direction of $\mathbf{v}$. Therefore, we write the principal components of Eq. 8.19 as

$$\tilde{D}_{1,1}^{(h)} = \tilde{D}_{1,1} + D_{\text{eff}} = a_L|\mathbf{v}| + D_{\text{eff}} \quad (8.20)$$

$$\tilde{D}_{2,2}^{(h)} = \tilde{D}_{2,2} + D_{\text{eff}} = a_T|\mathbf{v}| + D_{\text{eff}} \quad (8.21)$$

$$\tilde{D}_{3,3}^{(h)} = \tilde{D}_{3,3} + D_{\text{eff}} = a_T|\mathbf{v}| + D_{\text{eff}} \quad (8.22)$$

---

*A property of every symmetrical cartesian tensor is that the coordinate system can be rotated to a unique set of directions such that the tensor collapses to its "principal-component" form. In this form, all off-diagonal terms are zero; the remaining transformed diagonal terms are called the principal components.

8.4.   POROUS MEDIA DISPERSION EQUATION
IN A HOMOGENEOUS ISOTROPIC MEDIUM

The development of the equation for dispersion in porous media is based on conservation of mass in a representative elementary volume of porous media. Referring to Section 6.8 and Figure 6.13, we find that Eq. 6.118 will be appropriate:

$$\frac{\partial}{\partial t} (\rho_A \epsilon \Delta x \Delta y \Delta z + S_A \rho_b \Delta x \Delta y \Delta z)$$

$$= (\mathbf{n}_{A,x} - \mathbf{n}_{A,x+\Delta x})\epsilon \Delta y \Delta z + (\mathbf{n}_{A,y} - \mathbf{n}_{A,y+\Delta y})\epsilon \Delta z \Delta x + (\mathbf{n}_{A,z} - \mathbf{n}_{A,z+\Delta z})\epsilon \Delta y \Delta x$$

$$(6.118)$$

As the reader will recall from Chapter 6, S_A is the mass of solute A adsorbed per unit mass of solid, and ρ_b is the bulk density of solid (mass of solid per total volume porous media). The mass accumulation in Eq. 6.118 (left-hand side) has two terms: the first is the mass of species A in the interstitial fluid, and the second is the mass of species A associated with the solid phase through adsorption. The sum of these two terms accounts for all A in the differential element. The right-hand side of Eq. 6.118 takes into consideration the blockage of the flow path by multiplication by ϵ. Dividing through Eq. 6.118 by $\Delta x \Delta y \Delta z$ and taking the limit as the differential element shrinks to zero, we find, as in Section 6.9:

$$\frac{\partial}{\partial t} (\rho_A \epsilon + S_A \rho_b) + \epsilon(\nabla \cdot \mathbf{n}_A) = 0 \qquad (6.119)$$

where ϵ is assumed to be constant in space. The left-hand side will also be handled as in Section 6.9, by the assumption of a linear isotherm relation between ρ_A and S_A:

$$S_A = k_d \rho_A \qquad (6.120)$$

where k_d is the partition coefficient. Substituting into Eq. 6.112 and rearranging, we again find Eq. 6.114:

$$\frac{\partial \rho_A}{\partial t} + \left[\frac{1}{1 + \dfrac{k_d \rho_b}{\epsilon}} \right] \nabla \cdot \mathbf{n}_A = \frac{\partial \rho_A}{\partial t} + \frac{1}{R} \nabla \cdot \mathbf{n}_A = 0 \qquad (6.121)$$

Here R is the same retardation coefficient found in Chapter 6:

$$R = 1 + \frac{k_d \rho_b}{\epsilon} \qquad (6.122)$$

For $\mathbf{n}_A$ we first use the most general form (i.e., Eq. 8.17). Therefore, substituting Eq. 8.17 into Eq. 6.121, we find:

$$
\begin{aligned}
&\frac{\partial \rho_A}{\partial t} + \frac{1}{R}\left\{\rho_A\left(\frac{\partial v_1}{\partial x_1} + \frac{\partial v_2}{\partial x_2} + \frac{\partial v_3}{\partial x_3}\right) + \left(v_1\frac{\partial \rho_A}{\partial x_1} + v_2\frac{\partial \rho_A}{\partial x_2} + v_3\frac{\partial \rho_A}{\partial x_3}\right)\right\} \\
&- \frac{1}{R}\frac{\partial}{\partial x_1}\left(D_{1,1}^{(h)}\frac{\partial \rho_A}{\partial x_1} + D_{1,2}^{(h)}\frac{\partial \rho_A}{\partial x_2} + D_{1,3}^{(h)}\frac{\partial \rho_A}{\partial x_3}\right) \\
&- \frac{1}{R}\frac{\partial}{\partial x_2}\left(D_{2,1}^{(h)}\frac{\partial \rho_A}{\partial x_1} + D_{2,2}^{(h)}\frac{\partial \rho_A}{\partial x_2} + D_{2,3}^{(h)}\frac{\partial \rho_A}{\partial x_3}\right) \\
&- \frac{1}{R}\frac{\partial}{\partial x_3}\left(D_{3,1}^{(h)}\frac{\partial \rho_A}{\partial x_1} + D_{3,2}^{(h)}\frac{\partial \rho_A}{\partial x_2} + D_{3,3}^{(h)}\frac{\partial \rho_A}{\partial x_3}\right) = 0
\end{aligned}
\qquad (8.23)
$$

However, recall that the sum of the first series of terms in parentheses on the left-hand side of Eq. 8.23 is zero for an incompressible fluid (Eq. 1.24); then if there are no sources or sinks; hence, the porous media dispersion equation reduces to

$$
\begin{aligned}
&\frac{\partial \rho_A}{\partial t} + \frac{1}{R}\left(v_1\frac{\partial \rho_A}{\partial x_1} + v_2\frac{\partial \rho_A}{\partial x_2} + v_3\frac{\partial \rho_A}{\partial x_3}\right) \\
&- \frac{1}{R}\frac{\partial}{\partial x_1}\left(D_{1,1}^{(h)}\frac{\partial \rho_A}{\partial x_1} + D_{1,2}^{(h)}\frac{\partial \rho_A}{\partial x_2} + D_{1,3}^{(h)}\frac{\partial \rho_A}{\partial x_3}\right) \\
&- \frac{1}{R}\frac{\partial}{\partial x_2}\left(D_{2,1}^{(h)}\frac{\partial \rho_A}{\partial x_1} + D_{2,2}^{(h)}\frac{\partial \rho_A}{\partial x_2} + D_{2,3}^{(h)}\frac{\partial \rho_A}{\partial x_3}\right) \\
&- \frac{1}{R}\frac{\partial}{\partial x_3}\left(D_{3,1}^{(h)}\frac{\partial \rho_A}{\partial x_1} + D_{3,2}^{(h)}\frac{\partial \rho_A}{\partial x_2} + D_{3,3}^{(h)}\frac{\partial \rho_A}{\partial x_3}\right) = 0
\end{aligned}
\qquad (8.24)
$$

Another simplification arises for a homogeneous and isotropic porous medium if we assume that the coordinate axes are aligned with the principal axes, and specifically, if the x_1 axis is aligned with $\mathbf{v}_1$ (recall discussion above). In this case, $\mathbf{v}_2 = \mathbf{v}_3 = 0$, and all the off-diagonal hydrodynamic dispersion coefficients are zero; hence,

$$
\frac{\partial \rho_A}{\partial t} + \frac{1}{R}\mathbf{v}_1\frac{\partial \rho_A}{\partial x_1} = \frac{1}{R}\frac{\partial}{\partial x_1}\left(\tilde{D}_{1,1}^{(h)}\frac{\partial \rho_A}{\partial x_1}\right) + \frac{1}{R}\frac{\partial}{\partial x_2}\left(\tilde{D}_{2,2}^{(h)}\frac{\partial \rho_A}{\partial x_2}\right) + \frac{1}{R}\frac{\partial}{\partial x_3}\left(\tilde{D}_{3,3}^{(h)}\frac{\partial \rho_A}{\partial x_3}\right)
$$

$$(8.25)$$

A final simplification of the porous-media dispersion equation comes about if we consider only one-dimensional flow where v_1 is constant. In the one-dimensional case where the x_1 axis is aligned with the average flow direction, $\tilde{D}^{(h)}_{2,2} = \tilde{D}^{(h)}_{3,3} = 0$ and $\tilde{D}^{(h)}_{1,1} =$ a constant; hence Eq. 8.25 reduces to

$$\frac{\partial \rho_A}{\partial t} + \frac{1}{R}\, v_1\, \frac{\partial \rho_A}{\partial x_1} = \frac{1}{R}\, \tilde{D}^{(h)}_{1,1}\, \frac{\partial^2 \rho_A}{\partial x_1^2} \tag{8.26}$$

8.5. SOLUTION OF THE DISPERSION EQUATION IN AN INFINITE ONE-DIMENSIONAL MEDIUM

Our first solution of the one-dimensional dispersion equation will be for the case where there is no dispersion at all. Although this seems like a pointless exercise, we find that it demonstrates an important effect of retardation. When there is no dispersion, Eq. 8.26 simplifies to

$$\frac{\partial \rho_A}{\partial t} + \frac{v_1}{R}\, \frac{\partial \rho_A}{\partial x_1} = 0 \tag{8.27}$$

Now imagine an infinite one-dimensional porous medium. For simplicity, the reader can think of a tube filled with porous media extending to infinity in the positive and negative x_1 directions. At time $= 0$ and at $x_1 = 0$, we inject an infinitesimally thin slug of tracer at $x_1 = 0$. As pointed out in the development, this tracer can partition into or adsorb to the solid phase in the porous medium. As explained in Chapter 10, this ideal type of impulse is called the Dirac function. As also shown in Chapter 10, the solution to equations like Eq. 8.27 for this initial condition is quite simple:

$$\rho_A(t) = \frac{M}{Q}\, \delta(t - \bar{t}) \tag{8.28}$$

Here δ is the Dirac function, M is the total mass injected in the Dirac pulse, and Q is the flow rate through the porous medium. In Chapter 10, $\bar{t}$ is the mean detention time in the system; $\bar{t} = x/v_1$, where x is distance in the x_1 direction. We need a slightly modified form of $\bar{t}$ in this problem, since v_1 is divided by R in Eq. 8.27:

$$\bar{t} = \frac{x_1}{v_1/R} \tag{8.29}$$

The Dirac function has a value of zero for any value of its argument other than zero; when the argument is zero, the Dirac function extends to infinity. Now

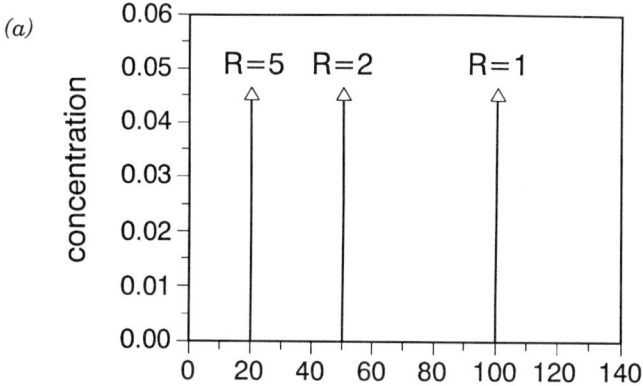

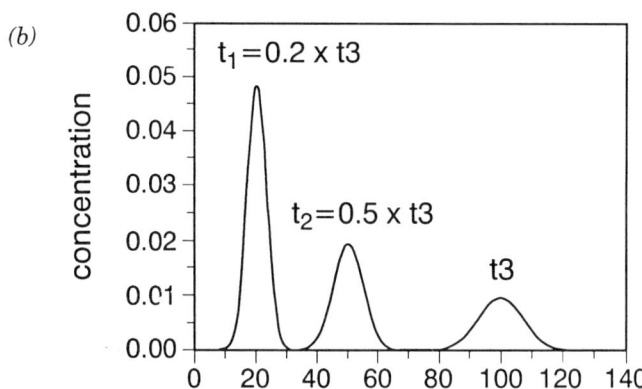

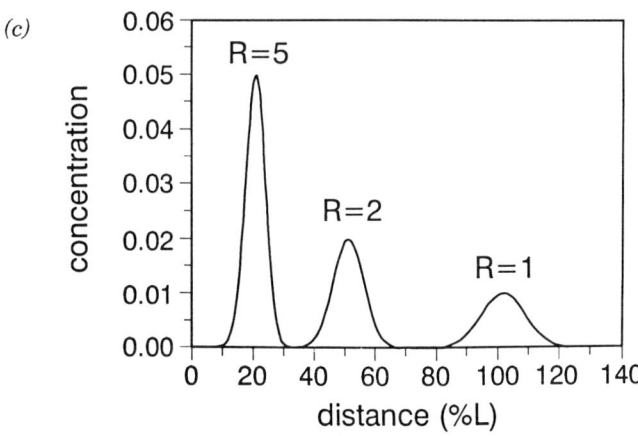

Figure 8.3. Dispersion of tracer for (a) no dispersion, (b) different times and $R = 1$, and (c) same time and different R values.

consider the form of Eq. 8.28 for a fixed distance L and time $t = L/v_1$:

$$\rho_A = \frac{M}{Q} \, \delta\left(\frac{L}{v_1} - \frac{x_1 R}{v_1}\right) \tag{8.30}$$

We now ask where along the x_1 axis the Dirac pulse will occur for different values of R at time $= L/v_1$. For example, for $R = 1$, Eq. 8.30 says the Dirac pulse occurs at $x_1 = L$. For $R = 2$, the pulse occurs at $x_1 = \frac{1}{2}L$, while for $R = 5$, the Dirac pulse occurs at $x_1 = 1/5\, L$. These solutions to the one-dimensional dispersion equation (without dispersion) are plotted in Figure 8.3a. Note that as the retardation factor increases, the concentration pulse (represented by the spike Dirac function) is more and more retarded. This demonstrates a very important principal: even when there is little or no dispersion, partitioning into the fixed solid phase of the porous medium "slows down" the movement of the adsorbing tracer through the medium. This slowing down is known as *retardation*.

We next consider the solution of the one-dimensional dispersion equation when there is dispersion. In a one-dimensional, homogeneous, and isotropic porous medium with constant v_1, the appropriate dispersion equation is given by Eq. 8.26, which simplifies even further if we switch to a moving coordinate system given by

$$x_1' = x_1 - \frac{v_1}{R} t \tag{8.31}$$

In this case Eq. 8.26 reduces to

$$\frac{\partial \rho_A}{\partial t} = \frac{\tilde{D}_{1,1}^{(h)}}{R} \frac{\partial^2 \rho_A}{\partial x_1'^2} \tag{8.32}$$

This transformation is essentially identical to those made for dispersion in laminar tube flow (see Eq. 7.43) and turbulent tube flow (see Eq. 7.80). In the present problem we imagine again an infinitesimally thin slug of tracer of mass M injected at $x_1 = x_1' = 0$ at $t = 0$. In fact, we already have the solution to this problem, which was given in Sections 7.3 and 7.6:

$$\rho_A(x_1, t) = \frac{1}{R} \frac{M/\epsilon}{\{4\pi[\tilde{D}_{1,1}^{(h)}/R]t\}^{1/2}} \exp\left\{-\frac{[x_1 - (v_1/R)t]^2}{4[\tilde{D}_{1,1}^{(h)}/R]t}\right\} \tag{8.33}$$

Figure 8.3b is a plot of Eq. 8.33 for no adsorption (i.e., $R = 1$) and three different times, $t_3 = 2t_2 = 5t_1$. As predicted, in the introduction to this chapter, the concentration profile is a Gaussian curve which broadens with time as the mass of tracer is convected in the x_1 direction.

Example 8.1. The following data (concentration variances) were collected during a study of the one-dimensional dispersion of a compound in a porous soil where the retardation coefficient was unity:

$$t = 10 \, \text{days:} \ \overline{x^2} = 189 \, \text{m}^2$$

$$t = 50 \, \text{days:} \ \overline{x^2} = 1100 \, \text{m}^2$$

$$t = 100 \, \text{days:} \ \overline{x^2} = 1990 \, \text{m}^2$$

Determine the longitudinal dispersion coefficient, a_L, if $\mathbf{v} = 1 \, \text{m/day}$ and the molecular diffusion coefficient of the pure compound is $D_{AB} = 1.5 \times 10^{-5}$ cm^2/s. Assume a tortuosity of $\tau' = 0.5$.

SOLUTION. By comparison of Eq. 8.33 with the solution for the variance of a Gaussian distribution in Section 6.7.2, we note that

$$\overline{x^2} = 2 \frac{\tilde{D}^{(h)}_{1,1}}{R} t$$

Therefore, if we plot the concentration variance versus time and we find an approximate straight line, then for $R = 1$ the slope of that line is $2 \times \tilde{D}^{(h)}_{1,1}$.

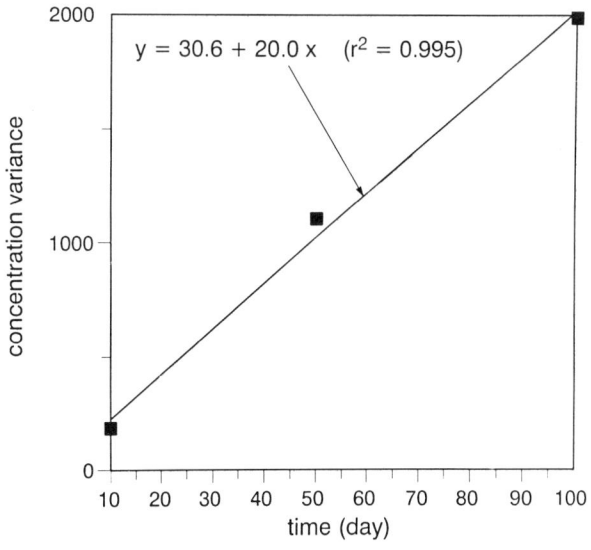

$y = 30.6 + 20.0 \, x \quad (r^2 = 0.995)$

concentration variance

time (day)

Example 8.1

The straight-line fit to the experimental data seems reasonable; hence, from the slope of the line, we find

$$2\tilde{D}_{1,1}^{(h)} = 19.9 \; \frac{m^2}{day} \Rightarrow \tilde{D}_{1,1}^{(h)} = 10.0 \; \frac{m^2}{day}$$

From the definition of the longitudinal dispersion coefficient, Eq. 8.20, we note that

$$\tilde{D}_{1,1}^{(h)} = \tilde{D}_{1,1} + D_{eff} = a_L |v| + D_{eff} \tag{8.20}$$

or,

$$\tilde{D}_{1,1}^{(h)} = a_L |v| + \tau' D_{AB}$$

Substituting into this equation,

$$10.0 \; \frac{m^2}{day} = a_L \left(1 \; \frac{m}{day}\right) + (0.5)1.5 \times 10^{-5} \; \frac{cm^2}{s} \left(\frac{m^2}{10^4 \, cm^2}\right) \left(\frac{8.64 \times 10^4 \, s}{day}\right)$$

$$= a_L \left(1 \; \frac{m}{day}\right) + 6.48 \times 10^{-5} \; \frac{m^2}{day}$$

Solving for a_L we find

$$a_L = 10 \; m$$

Note that for this example, the molecular-diffusion component of the hydrodynamic dispersion is very small.

We next attempt to demonstrate an important relationship between the retardation coefficient and the displacement of the peak concentration of the tracer. Note that the peak concentration from Eq. 8.33 will always occur when the argument of the exponential term is zero:

$$\frac{\left[x_c - \dfrac{v_1}{R} t\right]^2}{4 \dfrac{\tilde{D}_{1,1}^{(h)}}{R} t} = 0 \quad \text{or when } x_c - \frac{v_1}{R} t = 0 \tag{8.34}$$

where we call the position of the peak in maximum concentration x_c.

Rearranging, the peak of the concentration curve will occur when

$$\frac{v_1}{R} = \frac{x_c}{t} = v_c \tag{8.35}$$

The term $v_c = x_c/t$ can be called the velocity of the concentration peak. Therefore, the relationship between the average linear velocity v_1 and v_c is simply

$$\frac{v_1}{v_c} = R \tag{8.36}$$

Figure 8.3c demonstrates this relationship. When $R = 2$, the peak concentration has traveled only half the distance of the $R = 1$ curve. Comparing Figures

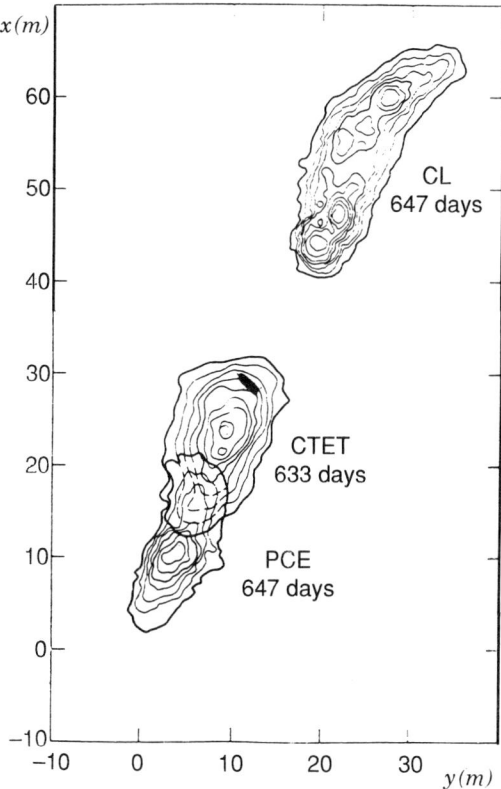

Figure 8.4. Apparent chromatographic separation of organic contaminants in aquifer transport: PCE, tetrachloroethylene; CTET, carbon tetrachloride; CL, chloride. *Source*: Reprinted with permission from Roberts, Goltz, and Mackay, *Water Resources Research*, **22**(13), 2050 (1986). Copyright 1986, American Geophysical Union, Washington, DC.

8.3*b* and *c*, note that the $R = 2$ curve is identical to the $t_2(= \frac{1}{2}t_3)$ curve for $R = 1$. Similarly, when $R = 5$, the peak in concentration has traveled only one-fifth the distance of the $R = 1$ curve. The $R = 5$ curve in Figure 8.3*c* is identical to the $t_1(= 1/5t_3)$ curve in Figure 8.3*b*.

This retardation of the tracer (or contaminant) with increasing R is an important phenomenon in groundwater-transport problems and is the basis for the area of chemical analysis called chromatography. In fact, in environmental transport, this retardation effect is sometimes called the *chromatographic effect*.

For example, Figure 8.4 shows the results of a study of the groundwater transport of several contaminants in a sand aquifer (Roberts et al., 1986). At day zero, the contaminants tetrachloroethylene (PCE) and carbon tetrachloride (CTET), and a chloride tracer (CL) were introduced at the origin of coordinates ($x = y = 0$). The figure shows the location of the contaminant and tracer plumes after nearly 2 years of travel time in the aquifer. Note that the nonadsorbing chloride plume travels the farthest, followed by the carbon tetrachloride and tetrachloroethylene plumes. We imagine here that the PCE has a larger retardation factor than the CTET and, therefore, moves more slowly through the aquifer. This study showed the unambiguous effects of contaminant partitioning and retardation in groundwater flow. In Exercise 8.1, linear partitioning theory is used to calculate the partition coefficients for tetrachloroethylene and carbon tetrachloride in the experiments of Roberts et al.

8.6. ANALYTICAL CHROMATOGRAPHY

One of the most important developments in the analysis of small ("trace") concentrations of organic chemicals was *chromatography*. Chromatography is based on the simple principle that compounds with different adsorptive, partitioning, or size characteristics will travel through separative systems at different rates. Since the rate of travel is specific for specific compounds, the rate of travel through the system can be used to identify the compound. For example, Figure 8.5 shows a chromatographic separation of urea herbicides by a type of chromatography called bonded-phase liquid–liquid chromatography. Notice how the nonretarded compound (called the "solvent") exits the column first, followed in sequence by the more and more strongly retarded urea-herbicides. (The *x* axis in chromatography is normally time, since flow rate of the carrier fluid is usually carefully controlled.) Compare Figure 8.5 with Figure 8.3 and notice the general similarity in response.

According to Karger et al. (1973), there are three basic types of chromatography: liquid–solid chromatography (LSC), liquid–liquid chromatography (LLC), and gas–liquid chromatography (GLC). In LSC, a liquid carrying the compounds to be analyzed is passed through a column containing porous particles saturated with the carrier liquid. The compounds to be analyzed either adsorb to the porous packing, or pass in and out of the porous packing

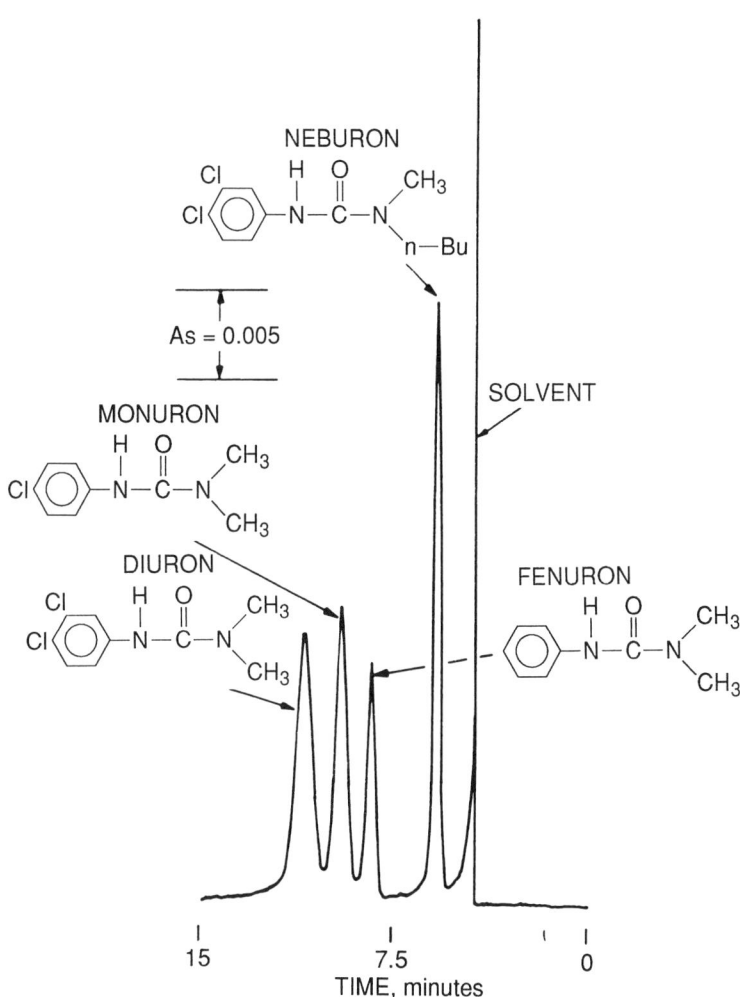

Figure 8.5. Chromatographic analysis of urea-herbicides by liquid–liquid chromatography. *Source*: Reproduced from Snyder and Kirkland, *Introduction to Modern Liquid Chromatography*, © 1979, p. 282. Reprinted by permission of John Wiley & Sons, Inc..

by diffusion (Figure 8.6*a*). Depending on either the adsorptive or diffusional characteristics of the compounds, they will be more or less retarded as they pass through the column. Liquid–solid chromatography is particularly useful for high-boiling-point and thermally unstable compounds, since it is normally carried out at room temperature. Although LSC is not so effective in separating members of a homologous series, it is very sensitive to differences in molecular structure. In LLC, a liquid carrying the compounds to be analyzed passes into a column containing a porous support saturated with a second immiscible liquid. The second liquid is essentially fixed to the porous support and is

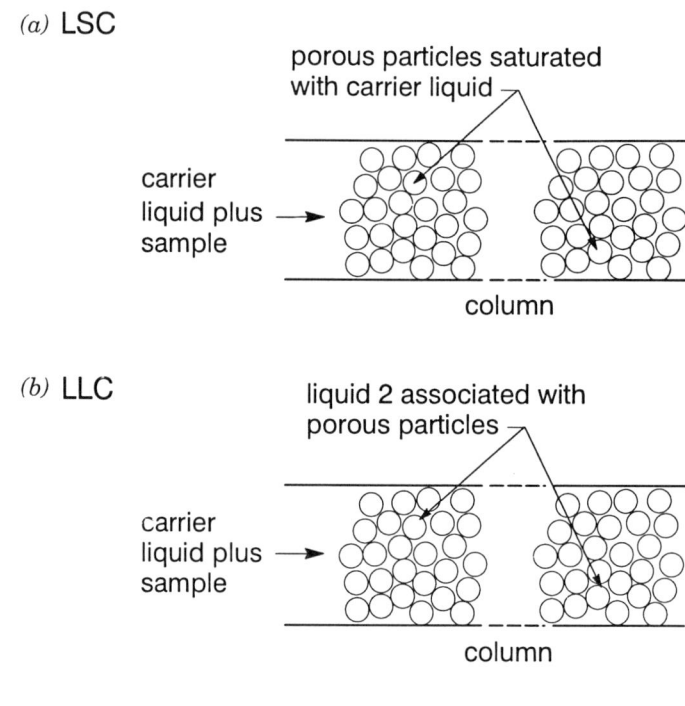

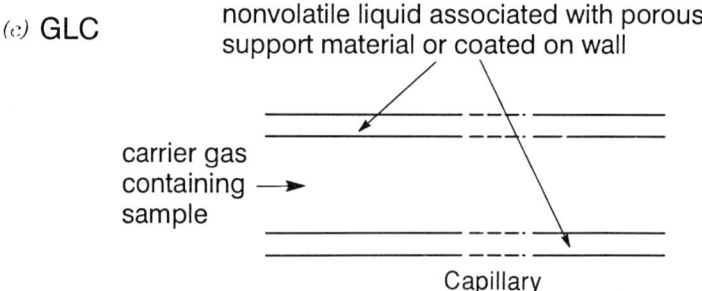

Figure 8.6. Three basic types of chromatography: (*a*) liquid–solid chromatography (LSC); (*b*) liquid–liquid chromatography (LLC); and (*c*) gas–liquid chromatography (GLC).

therefore stationary. Separation occurs when the compounds partition between the first and second liquids (Figure 8.6*b*). Like LSC, LLC is good at separating thermally unstable samples, and larger-molecular-weight compounds. Finally, in GLC, a gas carrying the compounds to be analyzed passes through a capillary column, where the surface of the capillary column is coated with a porous support saturated with a stationary liquid (Figure 8.6*c*). Separation occurs when compounds partition between the stationary liquid coating and the moving carrier gas. Gas–liquid chromatography is the most efficient of separation techniques for low-molecular-weight compounds, that is, it generally has greater separative power than LSC or LLC. In addition, GLC is the most easily automated chromatographic technique. In an important variation of GLC, the column has been connected to a mass spectrometer, which with the aid of information stored in computer files, can quickly identify numerous organic compounds.

The analysis of the first part of this chapter is all we need to understand the main tenets of chromatography. For example, in LSC and LLC, we are essentially dealing with dispersion and partitioning in a one-dimensional porous medium. In GLC (Figure 8.6c), we are not dealing with dispersion in a porous medium. It is more akin to dispersion in laminar tube flow (Section 7.3) with an important difference being that the sample compound can partition into the stationary fluid along the capillary wall. However, the main equations of chromatography pertain to GLC as well.

For example, consider the *chromatogram* shown in Figure 8.7. The chromatogram is essentially an instrument readout that indicates the time that compounds exited the column (usually a paper chart from a strip-chart recorder). Figure 8.7 shows a chromatogram for two compounds, *A* and *B*. Throughout this section, we assume that concentration profiles are always Gaussian. Gaussian peaks are usually found when concentrations are maintained below certain threshold levels, and when the adsorption (or partitioning) is linear and the adsorption–desorption kinetics are fast relative to the movement of the sample through the column. This is referred to as *linear*

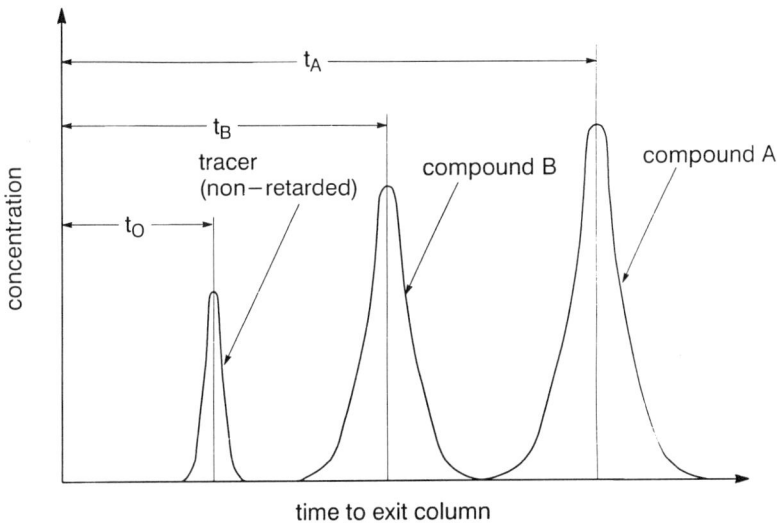

Figure 8.7. Nomenclature for the development of the main equations of chromatography.

elution chromatography. We define the *retention time* for a compound as

$$t_R = \frac{V_R}{Q} \tag{8.37}$$

where V_R is called the *retention volume*, the volume of fluid exiting the column corresponding to the passage of the concentration peak out of the column (i.e., t_A or t_B in Figure 8.7). For nonadsorbing and nonpartitioning solutes, we define their exit time as

$$t_0 = \frac{V_m}{Q} \tag{8.38}$$

where V_m is called the volume of the *mobile phase*; it is the volume of the column corresponding to the void space (for LSC or LLC) or the gas passageway (for GLC). The time t_0 can also be considered the mean travel time of the carrier fluid through the column. The total active volume of the column is defined as

$$V_T = V_m + V_s \tag{8.39}$$

where V_s is the volume of the stationary phase (e.g., the volume of the column inaccessible to the carrier fluid). For a column of fixed length L, we can define the average migration velocity of an adsorbing or partitioning compound as

$$v_s = \frac{L}{t_R} \tag{8.40}$$

and the velocity of the nonadsorbing compound (or mean velocity of carrier fluid) as

$$v_0 = \frac{L}{t_0} \tag{8.41}$$

We defined a relation between v_s and v_0 in Section 8.5 as

$$\frac{v_0}{v_s} = R \tag{8.42}$$

where R is the retardation coefficient. In chromatography, scientists have traditionally defined the *retention* as

$$\mathscr{R} = \frac{1}{R} = \frac{v_s}{v_0} \tag{8.43}$$

In other words, the retention is the inverse of the retardation coefficient. The retention has a value between 0 and 1. When the solute does not move at all (i.e, $v_s = 0$), $\mathscr{R} = 0$; when the compound moves at the maximum speed (i.e., $v_s = v_0$), $\mathscr{R} = 1$. There are two ways in which chromatography can be used in analysis: for compound identification and concentration determination. For identification, the column can be "calibrated" with known compounds. When chromatograms for unknown compounds (or usually mixtures of compounds) are determined, the unknown compounds can be identified by comparing their retention values with the known compounds. To determine compound concentration, recall Eq. 8.33, noting that the total mass of the compound is proportional to the peak height. By performing a calibration with different concentrations of the same compound, it is possible to construct a "calibration curve" relating peak height to concentration. In other applications, determination of the area under the concentration–time curve is most accurate for determination of concentration.

Another important chromatographic relation is between the retention and the number of moles of solute in the mobile and stationary phases. Actually, although the relation was derived by scientists interested in analytical chromatography, the relation has importance in the general area of porous-media transport as well. We return to Eq. 6.119 in Section 8.4, and note that the total moles of solute or compound in a differential element of porous media is the sum of the number of moles in the mobile n_m and stationary n_s phases:

$$c_A \epsilon \Delta V + S_A \rho_b \Delta V = n_m + n_s \tag{8.44}$$

where we have switched to molar concentrations and where ΔV is the differential volume, $\Delta x \Delta y \Delta z$. Adapting a linear isotherm relation:

$$S_A = k_d c_A \tag{6.120}$$

Substituting Eq. 6.120 into Eq. 8.44, and forming the ratio of n_m to the sum $n_s + n_m$, we find:

$$\frac{n_m}{n_m + n_s} = \frac{c_A \epsilon \Delta V}{c_A \epsilon \Delta V + c_A k_d \rho_b \Delta V} = \frac{1}{1 + \dfrac{k_d \rho_b}{\epsilon}} = \frac{1}{R} = \mathscr{R} \tag{8.45}$$

Hence, the fraction of moles of solute or compound in the mobile phase is equal to the retention (or inverse of the retardation).

8.7. FILTRATION

When fluids containing particles pass through porous media, the particles may collide with and stick to the surface of the porous-media grains or fibers. This process of transport through and attachment to the porous media is called *filtration*. Filtration occurs naturally in the environment and is also a classic and important unit process in the treatment of contaminated gas and liquid streams. For example, we have generally felt that drinking water taken from wells is more pure than water taken from lakes, rivers, or streams (although this is not always true). We know that soils act as filters, removing contaminant particles (e.g., bacteria) transported in the groundwater. In waste treatment, filtration is specially designed and optimized to remove particulate contaminants from gas streams (e.g., fiber filters for removal of dust and pollen in air-recirculation systems) and water streams (e.g., granular filters for removal of sediments and bacteria from drinking-water supplies). In this section, we first develop the classical explanations for filtration efficiency and then review some improvements in these classical approaches, which take into consideration forces active near the porous-media surfaces as well as the effect of adjacent filter grains or fibers on flow near the surface of a particular grain or fiber. The head loss a fluid experiences when passing through an engineered filtration system is a critical design consideration; therefore, we end by summarizing the progress made in modeling head loss in clean and dirty filters.

8.7.1. Classic Capture Mechanisms for Isolated Collectors

In this section, we assume that the flow around the porous-media grains or fibers is laminar, and that the collector under examination is isolated from all other collectors (grains or fibers) in the bed or mat. The central grain or fiber under consideration is called the *collector* (Figure 8.8). When a particle strikes a collector, it is assumed to be captured. This is quite similar to our conceptualization of fast coagulation (Section 3.7.1), where we assumed that once two particles collide, they are irreversibly stuck. Since most gas-filtration systems are composed of fibrous media, we assume that the collectors in these systems are composed of cylinders normal to the main flow direction. Since most water-filtration systems have granular media, we simplify these systems by considering the collectors to be spheres.

In the filtration of aerosols and hydrosols, four basic mechanisms are thought to be responsible for the capture of particles by the grains or fibers of the porous media (Figure 8.8). In *direct interception*, a particle approaches the collector along a streamline. Because of its size, or the fact that the particle begins its trip around the collector on a streamline close to the collector, the particle collides with the collector. In the classical approach of analyzing direct interception, we assume the particle follows the streamline perfectly. As we know, when particles are very small, they can experience Brownian motion. The *convective-diffusion* capture mechanism is shown in Figure 8.8. As the particle approaches the collector, it undergoes a jagged and random motion. By chance, it wiggles close enough to the collector surface to be captured. We have already discussed convective-diffusive transport to a spherical particle in Chapter 7; this analysis can help us to understand the convective-diffusion capture mechanism. The *inertial-impaction* mechanism has already been discussed in Sections 2.6 and 2.7. When the density difference between a particle

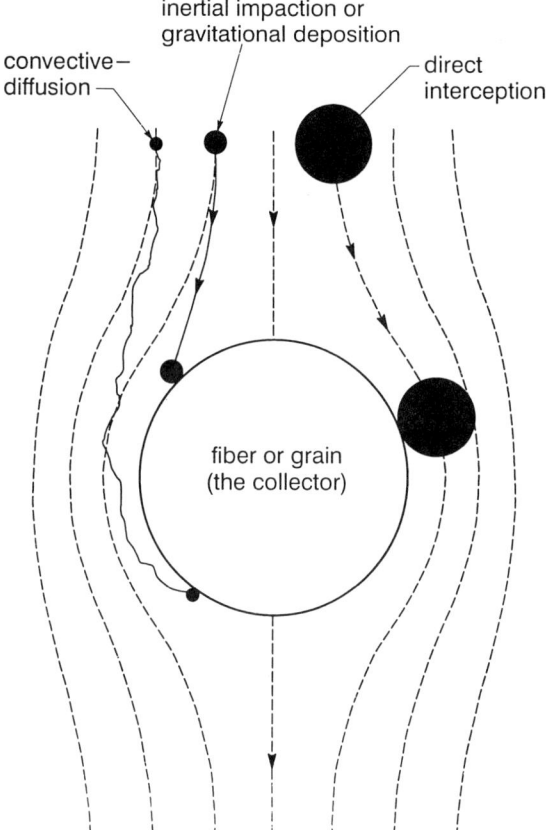

Figure 8.8. Classical particle-capture mechanisms in the filtration of hydrosols and aerosols.

and the fluid is great enough (or when the viscosity of the fluid is small enough), the particle inertia may be great enough (or the drag force small enough) that the particle does not follow the streamline around the collector; hence, the particle diverges from the streamline and strikes the collector (Figure 8.8). The inertial-impaction mechanism is primarily of importance in filtration of particles from gas streams. The particle–fluid density difference in water systems is generally not great enough to make inertial impaction a significant capture mechanism. Finally, in some aqueous systems, the settling velocity of a particle relative to the velocity past the collector can be great enough that the particle settles onto the collector surface (Figure 8.8). This is called the *gravitational deposition*.

We now develop the classical analyses of the four main collection mechanisms. For the case of *direct interception*, we assume that the particle experiences no Brownian motion. This is equivalent to a large Peclet number, or

$$\text{Pe} = \frac{U d_p}{D_{AB}} \to \infty \tag{8.46}$$

Here d_p is the particle diameter, U is the free-stream velocity of the undisturbed flow, and D_{AB} is interpreted as the diffusion coefficient of the particle in the fluid. We have assumed that once the particle strikes the collector surface, it is captured; we can presume then that the particle concentration must change from whatever its value is in the main stream to zero at the collector surface. It is now useful to recall our discussion of concentration boundary layers in Chapter 7. There we learned that in the general convective-diffusion situation, there is a smooth match between the concentration in the main stream and the zero concentration at the particle surface. We can assume that a concentration boundary layer exists in the direct-interception case, but since the Peclet number is very large, we can also assume that the boundary layer is infinitesimally thin. This means that we should be able to use simplified expressions for fluid velocity around the collector, since we only need accurate expressions near the collector surface. This simplification was also used in our approach to the development of convective-diffusive transport to a particle surface in Chapter 7. Hence, for a cylindrical collector (Figure 8.9a), we can imagine a flattened collector surface just like that shown in Figure 7.13. The streamlines are sketched schematically in Figure 8.9b. The expressions for the velocity components in the straightened coordinate system are given by Friedlander (1977):

$$u = 4 A_F U \frac{y}{a} \sin\left(\frac{x}{a}\right) \tag{8.47}$$

and

$$v = -2 A_F U \left(\frac{y}{a}\right)^2 \cos\left(\frac{x}{a}\right) \tag{8.48}$$

Here A_F is a special constant given by

$$A_F = [2(2 - \ln \text{Re})]^{-1} \tag{8.49}$$

where Re is the Reynolds number based on the collector diameter:

$$\text{Re} = \frac{2 a U}{v} \tag{8.50}$$

With the appropriate values for velocity in the region of the collector surface, the computation of the rate of collision of particles with the collector

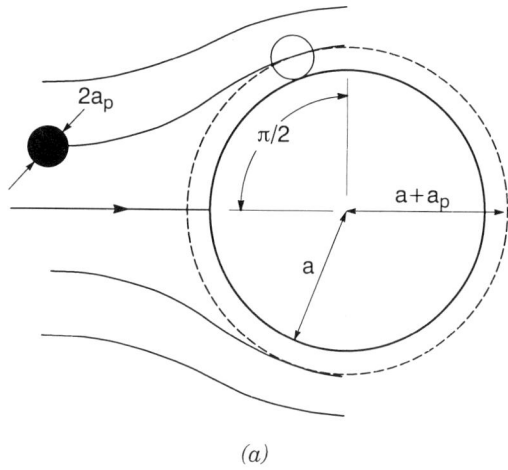

(a)

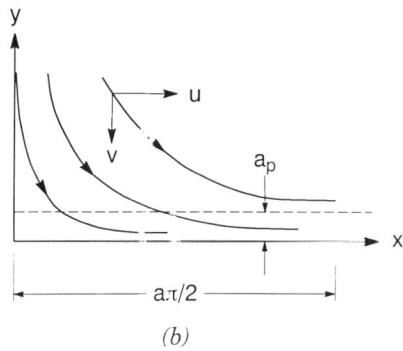

(b)

Figure 8.9. Particle capture by direct interception: (a) flow around a cylindrical collector, and (b) flattened northwest quadrant of the collector with local streamlines.

is quite straightforward: it is simply the mass transfer of particles to the flattened surface shown in Figure 8.9b. However, we must be a little careful in our calculation. Note that since the colliding particle has a finite size (not a mathematical point), the center of mass of the particle can only reach a distance of one particle radius (a_p) from the collector surface. Hence, the correct mass transfer is that of particles across the control volume indicated by the dashed line in Figure 8.9b. In the normal mass-transfer computation, we would form the dot product of the local-velocity vector at the dashed surface with the unit-normal vector. However, we know that this is simply the local normal component of velocity, and we can just integrate the velocity v (Eq. 8.48) over the dashed surface in Figure 8.9b. Hence, the unit width (into page) mass transfer of particles is given by

$$\left\{\begin{array}{l}\text{unit width mass transfer}\\ \text{of particles across}\\ \text{control volume}\end{array}\right\} = W' = 2n_\infty \int_0^{(\pi/2)a} v_{y=a_p}\, dx = 4n_\infty A_F R^2 aU \quad (8.51)$$

where n_∞ is the concentration of particles in the fluid outside of the control volume. In the integral shown above, we use the factor 2 because we have the same flux in the northwest and southwest quadrants of Figure 8.9a. The upper limit in the integral is the actual physical distance along the northwest (or southwest) quadrant of the collector, corresponding to $\pi/2$ rad. Here R is a special parameter called the *interception parameter*:

$$R = \frac{a_p}{a} \quad (8.52)$$

We would now like to calculate the fraction of particles collected by a single collector, which is called the *single-collector efficiency*. Consider the isolated cylindrical collector immersed in a laminar flow shown in Figure 8.10. The (unit width) mass transfer of particles through a projected (unit) area $2a$ is simply $2aUn_\infty$. The single-collector efficiency for the direct interception mechanism for a cylindrical collector (fiber) is defined as the ratio of the mass transfer given by Eq. 8.51 to the quantity $2aUn_\infty$,

$$\eta_{F,\text{di}} = \frac{4n_\infty A_F R^2 aU}{2n_\infty aU} = 2A_F R^2 \quad (8.53)$$

where the subscript di stands for direct interception. The development of the single-collector efficiency for a spherical collector for direct interception follows similar lines, resulting in the following (Yao et al., 1971):

$$\eta_{S,\text{di}} = \frac{3}{2} R^2 \quad (8.54)$$

where R is defined similarly to Eq. 8.52, with a being the spherical collector radius.

As mentioned above, we have already done the necessary calculation in Chapter 7 to determine the collection efficiency for *convective diffusion* at a single spherical collector. In Section 7.7.3 we found that the convective-diffusive mass transfer to a sphere was given by

$$W = 7.98a^{4/3} D_{AB}^{2/3} U^{1/3} n_\infty \quad (7.85)$$

To calculate the collection efficiency for the spherical collector, we follow the

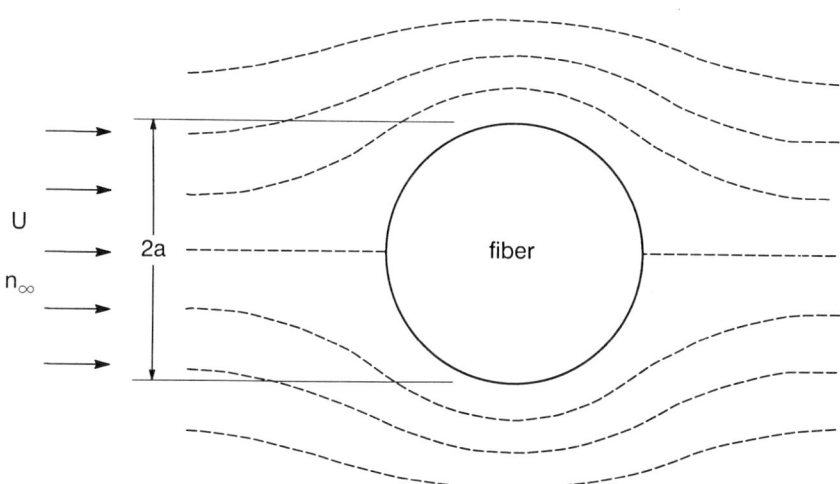

Figure 8.10. Notation for computation of single-collector efficiency for cylindrical collector.

same procedure as above and divide the mass transfer given by Eq. 7.85 to the undisturbed mass transfer through the projected area of the sphere, $\pi a^2 U n_\infty$

$$\eta_{S,\,cd} = \frac{7.98 a^{4/3} D_{AB}^{2/3} U^{1/3} n_\infty}{\pi a^2 n_\infty U} = \frac{7.98}{\pi} \frac{D_{AB}^{2/3}}{a^{2/3} U^{2/3}} = 4.0\,\mathrm{Pe}^{-2/3} \qquad (8.55)$$

where the Peclet number for the collector is defined by Eq. 7.117.

The reader should recall the assumptions involved in the derivations in Section 7.7.3, namely, low Reynolds number flow (Stokes flow), thin-concentration boundary layer (large Schmidt number), and isolation from any other sphere. As before (Eq. 8.46), D_{AB} is the diffusion coefficient of the particle in the fluid and is normally estimated with the Stokes–Einstein formula (Section 6.4). Another less-obvious assumption is that the colliding particle has no physical size itself (note that d_p does not show up in Eq. 8.55). This is the so-called "point particle" assumption (Friedlander, 1977). For a cylindrical collector, the collection efficiency is calculated in a similar fashion (e.g., Friedlander, 1977):

$$\eta_{F,\,cd} = 3.68 A_F^{1/3} \mathrm{Pe}^{-2/3} \qquad (8.56)$$

The derivation of the *gravitational deposition* mechanism is quite straightforward if the reader recalls the development of the differential sedimentation-coagulation mechanism from Section 3.7.2. From Eq. 3.43, we reason that the mass transfer of particles onto a single (stationary) spherical collector by sedimentation is

$$W = \pi(a_p + a)^2 v_t c_\infty \tag{8.57}$$

where v_t can be considered the terminal Stokes settling velocity of the approaching particle (Section 2.2). We form the collection efficiency for gravitational deposition as above:

$$\eta_{S,\text{gd}} = \frac{\pi(a_p + a)^2 v_t c_\infty}{\pi a^2 U c_\infty} \simeq \frac{v_t}{U} \tag{8.58}$$

where the simplified form obtains for small a_p (i.e., $a + a_p \simeq a$).

The last capture mechanism is for *inertial impaction*, which is most important in aerosol-filtration systems. It has been more difficult to derive analytical expressions for collection efficiency during inertial impaction; therefore, numerical solutions of the governing equations have been required. Nevertheless, a good fundamental understanding of inertial impaction can be gained from dimensional analysis. Consider once again a particle approaching a cylindrical collector, as shown in Figure 8.11. In Section 2.9 we learned that the Stokes number is an important dimensionless parameter in particulate systems when particle inertia is important. Now we undertake a full dimensional analysis to establish other important dimensionless parameters for collection efficiency by

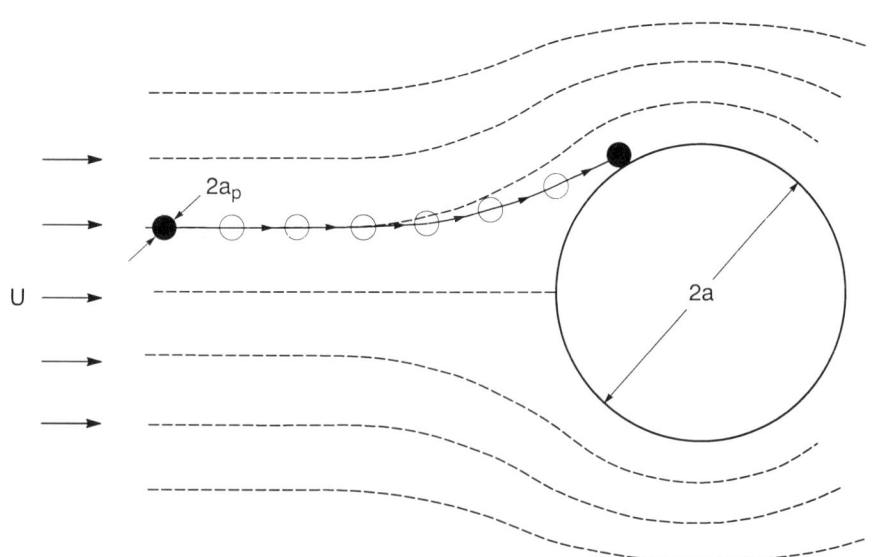

Figure 8.11. Inertial impaction of particle on a cylindrical collector.

inertial impaction. We consider the following physical and flow parameters to be relevant to the situation shown in Figure 8.11:

d_c = collector diameter = $2a$ (L)
d_p = particle diameter = $2a_p$ (L)
U = undisturbed "free-stream" velocity (L/T)
ρ = fluid density (M/L^3)
μ = fluid viscosity $(M\text{-}L^{-1}\text{-}T^{-1})$
ρ_p = particle density (M/L^3)

An analysis called the *Buckingham Pi theorem* states that in a physical problem including n parameters described with m dimensions, there will be $n - m$ independent dimensionless groups describing the system. The n parameters are the six parameters identified above. The number of dimensions is known from examination of the units in the six parameters, which we quickly surmise is three; hence, the number of independent dimensionless groups will be $6 - 3 = 3$. There are a number of ways to derive the important dimensionless groups relevant to this problem. A good discussion of *dimensional analysis* techniques is given in Brodkey and Hershey (1988). We use an approach sometimes called *Raleigh's* method and hypothesize that the following product of n physical parameters will always remain a dimensionless constant:

$$d_c^a d_p^b U^c \rho^d \mu^e \rho_p^f = \text{dimensionless constant} \tag{8.59}$$

The coefficients a–f are rational numbers (often whole numbers). Equation 8.59 simply states the hypothesis that when the parameters of the problem are raised to the correct power, their product will be dimensionless. Exercise 8.3 shows that other models derived in this chapter could have first been studied with dimensional analysis. Now, if the right-hand side is in fact dimensionless, we know that the powers of all fundamental dimensions in the problem (M, L, and T here) must cancel. For example, we can place the dimensions of each parameter into the corresponding term of Eq. 8.59:

$$L^a L^b \left(\frac{L}{T}\right)^c \left(\frac{M}{L^3}\right)^d \left(\frac{M}{L\text{-}T}\right)^e \left(\frac{M}{L^3}\right)^f \tag{8.60}$$

Considering each dimension L, M, and T separately, we conclude that the following three algebraic equations must hold:

$$L: a + b + c - 3d - e - 3f = 0$$
$$M: d + e + f = 0 \tag{8.61}$$
$$T: -c - e = 0$$

From the second and third equations, we find that $d = -e - f$ and $c = -e$. Substituting these into the first equation yields $a = -b - e$. Substituting these into Eq. 8.59 we find:

$$d_c^{-b-e} d_p^b U^{-e} \rho^{-e-f} \mu^e \rho_p^f = \text{dimensionless constant} \qquad (8.62)$$

Now grouping parameters with the same exponents, we quickly find the following:

$$\left(\frac{d_p}{d_c}\right)^b \left(\frac{\mu}{d_c U \rho}\right)^e \left(\frac{\rho_p}{\rho}\right)^f = \text{dimensionless constant} \qquad (8.63)$$

We have thus isolated the three significant dimensionless parameters. The first we recognize as the interception parameter (Eq. 8.52), which was isolated in the analysis of direct interception. The second is just the inverse of the Reynolds number (Eq. 8.50); since the flow field is determined by the Reynolds number, it is natural to expect dependency on the Reynolds number. The third group is the ratio of the density of the particle and fluid. We conclude then that the efficiency of inertial impaction will be a function of these three dimensionless groups.

Earlier we noted that the Stokes number is often an important dimensionless group in inertial impaction. Although the Stokes number did not "fall out" of our first crack at dimensional analysis, it is an easy matter to show that it can. Recall that the "stopping distance" derived in Section 2.8 can be expressed as

$$S = \frac{\rho_p d_p^2 C_c U}{18\mu} \qquad (8.64)$$

where C_c is the slip correction (dimensionless), while all the other parameters in S are already defined above. Since S includes the particle density ρ_p (and several other parameters already considered), we can replace ρ_p in Eq. 8.59 with the stopping distance:

$$d_c^a d_p^b U^c \rho^d \mu^e S^f = \text{dimensionless constant} \qquad (8.65)$$

Since the stopping distance has the units of length, the new form of Eq. 8.60 is

$$L^a L^b \left(\frac{L}{T}\right)^c \left(\frac{M}{L^3}\right)^d \left(\frac{M}{L \cdot T}\right)^e L^f \qquad (8.66)$$

From an analysis similar to that performed above, we find the following new version of Eq. 8.63:

$$\left(\frac{d_p}{d_c}\right)^b \left(\frac{\mu}{d_c U \rho}\right)^e \left(\frac{S}{d_c}\right)^f = \text{dimensionless constant} \qquad (8.67)$$

The first two dimensionless groups are of course the same as those uncovered in the first analysis; the third parameter has been identified in Section 2.9 as the *Stokes* number:

$$\text{St} = \frac{S}{d_c} = \frac{\rho_p d_p^2 C_c U}{18 \mu d_c} \qquad (8.68)$$

Therefore, we can consider that the interception parameter, Reynolds number, and Stokes number will be the significant dimensionless groups describing collection efficiency of a single collector by inertial impaction, or generally that

$$\eta_{F,\text{ii}} = f(R, \text{Re}, \text{St}) \qquad (8.69)$$

where $\eta_{F,\text{ii}}$ is the collection efficiency for inertial impaction on a fiber.

The reader may want to know the values of b, e, and f in Eq. 8.67. Unfortunately, dimensional analysis does not provide these values. All we can say is that inertial impaction should be related to the three dimensionless groups. We need further physical modeling, simulation, and/or experimentation to determine exact dependencies.

Figure 8.12 shows some experimental data (points) and numerical modeling (solid line) for the collection efficiency of sulfuric-acid aerosol particles (0.56–1.3 μm diameter) on 76-μm-diameter platinum wires. The x axis is the Stokes number. As the Stokes number increases, the collection efficiency increases. This makes a lot of sense after consideration of Section 2.9 and the discussion above. The Stokes number can be considered the ratio of the stopping distance (a measure of particle inertia) to the characteristic diameter of the obstacle (the collector in this case). Hence, when the Stokes number increases, the particle inertia increases, and there is greater impaction of the aerosol particles on the collectors. The reader may ask how R and Re come into these experiments, since they were identified above as possible significant dimensionless groups. First, the aerosol particles used have diameters that are small compared with the collector diameter, that is, $R = d_p/d_c = 0.0074–0.017$, and have a fairly small size variation. Therefore, the authors apparently felt that their results were insensitive to the small variations in R. In the experiments represented in Figure 8.12, the authors held the Reynolds number between values of 100 and 330. By presenting their results as independent of the Reynolds number, the authors assume the Reynolds number is effectively constant. Hence, there might well be plots like Figure 8.12 for different ranges of Reynolds number.

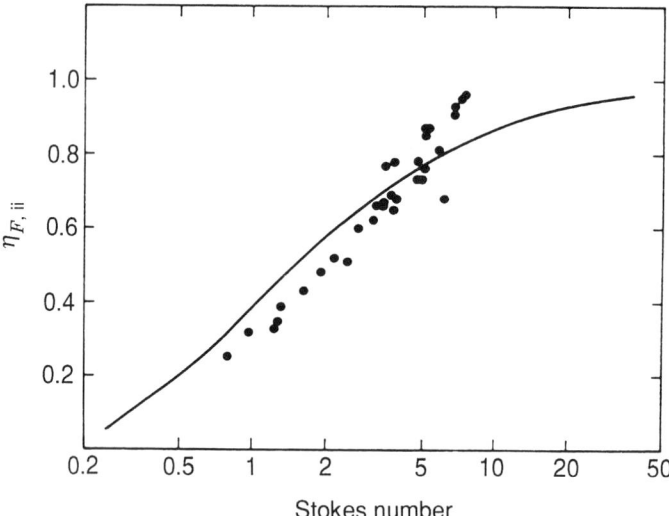

Figure 8.12. Inertial impaction of particles on a cylindrical collector. The dots represent experimental results and the solid line represents theoretical results. *Source*: Reproduced from Friedlander, *Smoke, Dust and Haze*, John Wiley & Sons, New York, 1977. Used with permission of S. K. Friedlander.

The reader may have noted by now that three of the major mechanisms of collection on a single collector are quite similar to three of the major coagulation mechanisms studied in Section 3.7.2. A comparison between analogous collision and coagulation mechanisms is provided in Table 8.2. As with coagulation, we have the problem of how to model situations in which more than one fundamental collection mechanism is significant. Unfortunately, we are not on completely solid theoretical ground if we assume that the different mechanisms are independent. For example, the convective-diffusion collection mechanism was developed above for so-called "point particles" where the particle is assumed to be a mathematical point (i.e., $R \to 0$). In

Table 8.2. Comparison of Analogous Filtration and Coagulation Mechanisms

Coagulation	Filtration
Laminar shear	Direct interception
Brownian diffusion	Convective-diffusion
Differential sedimentation	Gravitational deposition
Isotropic turbulent mixing	—
—	Inertial impaction

certain intermediate particle size ranges, Brownian motion alone may not be able to bring a particle close enough to the collector surface to cause a contact with the surface. However, for these intermediate (or larger) particles, even though the forces of Brownian or convective diffusion may not be large enough to cause a direct contact according to the earlier analysis, the motion may be sufficient to orient a particle such that it impacts the collector through the direct interception mechanism. Nevertheless, we will simply assume that when two (or more) mechanisms are significant at the same time, their effects are additive. Fortunately, even if this is not a good assumption, it will only apply to limited ranges of interaction. Hence, as with coagulation, we assume additivity (and independence) of collection mechanisms. Advanced numerical simulations of collection during filtration allow to investigate the implications of this assumption.

For example, Figure 8.13 shows numerical simulations of the combined effects of convective diffusion, direct interception, and inertial impaction on the filtration of aerosols. Recall from Eq. 8.53 that the collection efficiency for

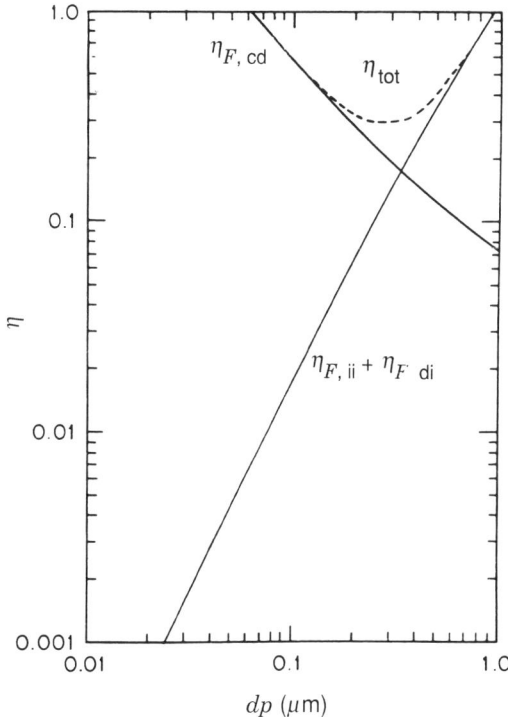

Figure 8.13. Comparison of convective diffusion, direct interception, and inertial impaction collection mechanisms for aerosol filtration. $s = 0.1$, $U = 1$ cm/s, and $d_c = 1.0\ \mu$m. *Source*: Flagan and Seinfeld, *Fundamentals of Air Pollution Engineering*, p. 453, © 1988. Reprinted by permission of Prentice Hall, Upper Saddle River, New Jersey.

direct interception increases with the value of the interception parameter $R = d_p/d_c$. Hence, for a fixed-flow field, this interception efficiency should increase with increasing particle size relative to the fiber diameter. We also just noted that in the case of inertial impaction, the efficiency should also increase with increasing particle size (increasing Stokes number). Therefore, it is clear that collection due to the sum of inertial impaction and interception increases with increasing particle size. However, for convective diffusion, note from Eq. 8.56 that collision efficiency decreases as the $-2/3$ power of the Peclet number. Considering the form of the particle diffusion coefficient (the Stokes–Einstein relation—Eq. 6.25), we conclude that $\eta_{F,cd}$ decreases as the $-2/3$ power of the particle diameter. As shown in Figure 8.13, the form of the different laws suggests a minimum in the total collection efficiency between 0.1 and 1.0 μm. An analogous phenomenon was observed in Figure 3.13, where the minimum coagulation frequency was observed around 1 μm. As we stated earlier in this section, scientists have noted many similarities between coagulation and filtration; the existence of a particle size where minimums in mass transfer exist is another of these similarities.

8.7.2. Effect of Adjacent Collectors on Local Flow Field

In the previous section, we assumed that the flow field around a specific collector was unaffected by the proximity of other collectors. However, as the porosity of a filter decreases, the effects of adjacent collectors on the flow field around any other collector must become and more important. This is a particular concern in the filtration of liquids, where the porosity of the filter can be very low (and the collectors are very close together). As shown in Figure 5.10, the laminar flow around a spherical collector in a porous medium can be very complicated; owing to the randomness of packing of most porous media, the flow field around different grains and fibers of the porous media can also be expected to be different. These problems have in fact been one of the major challenges to modelers of filtration and mass transfer in porous media. Several clever approaches to surmounting these problems are described in a book by Tien (1989). One of the models addressing the more complicated flow phenomena is called the *cell model*, and is usually attributed to Happel (1958). This model imagines the porous medium as an array of identical cells composed of the collector (either a sphere or cylinder) surrounded by a fluid envelope (of either spherical or cylindrical shape). In the cell model, the ratio of the volume of the encircling fluid to the volume of the collector is chosen so that the macroscopic porosity of the medium is matched. For cylindrical collectors, the correction for the effect of crowding by adjacent collectors is given simply by a modified form of the factor A_F of Eq. 8.49. Two different correcting formulas are available, as discussed by Spielman (1977). In Happel's model, the flow field at the surface of the particle is required to conform to the no-slip condition. In addition, the outer boundary of the fluid envelope surrounding the central collector is assumed to be a free surface (Chapter 5); hence, at the outer fluid boundary there is no tangential stress. For Happel's (1959) model of a cylindrical collector, the modified A_F is

$$A_F = \left[-\ln s - 1 + \frac{s^2}{(1+s)^2} \right]^{-1} \tag{8.70}$$

where s is the solid fraction defined in Section 8.2. Kuwabara (1959) imagined another cell-model flow field in which the vorticity (Chapter 5), rather than a free surface at the cell extremity, is assumed to go to zero at this boundary. This yields a slightly different expression for A_F:

$$A_F = \left(\ln s - \frac{3}{2} + 2s - \frac{s^2}{2} \right)^{-1} \tag{8.71}$$

According to Spielman, both expressions for A_F are acceptable, especially because they yield similar results for the velocity field near the cylindrical collector surface.

For the spherical collector similar modifications are available. For example, Eqs. 8.54, 8.55, and 8.58 can be rewritten with a correction factor A_s. Thus, for *direct interception*,

$$\eta_{S,\text{di}} = \frac{3}{2} A_s R^2 \tag{8.72}$$

and for *convective diffusion*,

$$\eta_{S,\text{cd}} = 4.0 A_s^{1/3} \text{Pe}^{-2/3} \tag{8.73}$$

whereas for *gravitational deposition*,

$$\eta_{S,\text{gd}} = A_s \frac{v_t}{U} \tag{8.74}$$

The correction factor A_s is given by Happel's (1958) model as

$$A_s = \frac{2(1 - s^{5/3})}{2 - 3s^{1/3} + 3s^{5/3} - 2s^2} \tag{8.75}$$

8.7.3. Hydrodynamic Retardation

When a particle approaches a collector, at some point we imagine that fluid separating the particle and collector must be squeezed out. This is generally only a significant problem for the direct-interception and gravitational-deposition mechanisms in liquid-filtration systems. O'Melia and Tiller (1993) provide the following corrections based on work by Rajagopalan and Tien (1976). For direct interception, the correction takes the form

$$\eta_{S,\mathrm{di}} = A_s \mathrm{Lo}^{1/8} R^{15/8} \tag{8.76}$$

Here Lo is a dimensionless attraction number called the *van der Waals force number*,

$$\mathrm{Lo} = \frac{4A}{(9\pi\mu d_p^2 U)} \tag{8.77}$$

where A is the Hamaker constant discussed in Chapter 3. For gravitational deposition, the correction takes the form

$$\eta_{S,\mathrm{gd}} = 3.38 \times 10^{-3} A_s G_v^{1.2} R^{-0.4} \tag{8.78}$$

Here G_v is a dimensionless number called the *gravitational parameter*:

$$G_v = \frac{(\rho_p - \rho)g d_p^2}{18\mu U} \tag{8.79}$$

which the reader will note is the ratio of the Stokes law sedimentation velocity to U.

8.7.4. Collection Efficiencies for Fibrous and Granular Filters

The previous sections outlined the classical development of single-collector efficiency in filtration. We now develop the classical relationship between the single-collector efficiency and the removal efficiency of a filter containing many fibers or grains. Consider the differential element shown in Figure 8.14.

We perform a mass balance around the differential element shown in the figure, focussing on the removal of particles by filtration across the element. As usual, the accumulation of particles in the differential element is the difference between the mass transfer at the upstream and downstream faces, or

$$\left\{ \begin{array}{c} \text{rate of mass} \\ \text{accumulation in} \\ \text{the element } (M/T) \end{array} \right\} = U_{\mathrm{ap}} N_l A - U_{\mathrm{ap}} N_{l+\Delta l} A \tag{8.80}$$

where N is the number concentration of particles. The single-fiber efficiency is given in this chapter as the ratio of the mass transfer to a single fiber divided by the upstream mass transfer across the projected surface of the fiber

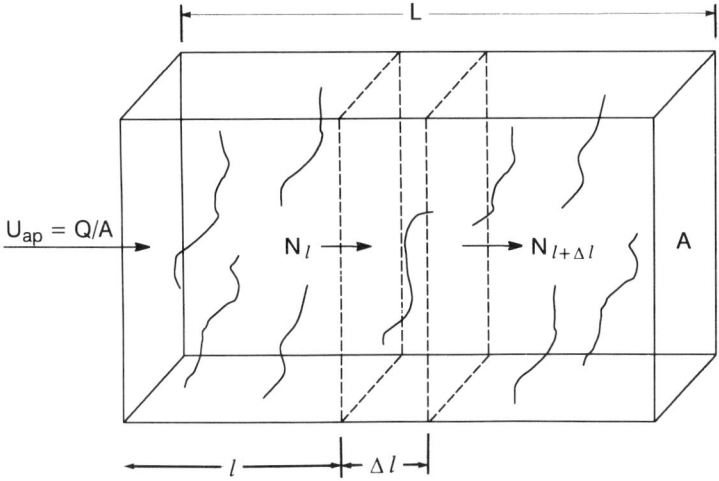

Figure 8.14. Differential element for analysis of overall filter efficiency.

(Probstein, 1989)

$$\eta = \frac{W'}{U N_l 2a} \tag{8.81}$$

where η is the general single collector efficiency (any of the mechanisms discussed in Section 8.7.1), v is the average linear velocity defined in Section 8.2, a is the fiber radius, and W' is the mass transfer to a unit length of a single fiber ($M\text{-}T^{-1}\text{-}L^{-1}$); hence, W' can be simply expressed by rearranging the last equation:

$$W' = 2U_{ap} N_l a \eta \tag{8.82}$$

To obtain the mass transfer to all fibers in the differential element, we must multiply W' by the length of a fiber and the number of fibers in the differential element. The number of fibers in the differential element is simply the total solid volume in the differential element divided by the volume of a single fiber, or

$$\begin{Bmatrix} \text{number of fibers in} \\ \text{differential element} \end{Bmatrix} = \frac{sA\Delta l}{\pi a^2 L_f} \tag{8.83}$$

Therefore, the rate of accumulation of mass in the differential element is

$$U_{ap} N_l A - U_{ap} N_{l+\Delta l} A = 2a U_{ap} N_l \eta L_f \left(\frac{sA\Delta l}{\pi a^2 L_f} \right) \tag{8.84}$$

Rearranging and taking the limit as $\Delta l \to 0$, we find the differential equation appropriate for the differential element of Figure 8.14:*

$$\frac{dN}{dl} = -\frac{2s\eta}{\pi a} N \tag{8.85}$$

A very similar approach can be used to find the differential equation describing particle removal in a bed of spherical collectors (see Exercise 8.5):

$$\frac{dN}{dl} = -\frac{3s\eta}{4a} N \tag{8.86}$$

8.7.5. Attachment Probability

We have assumed that if a particle is transported to the collector surface, it will stick to the surface. We learned in Chapter 3 that in coagulation, even though a collision between two particles is predicted, they do not necessarily contact each other and stick. The same is true in filtration. Interactions between the approaching particle and the collector can either prevent or favor a collision. The effects causing these types of interactions usually occur close to the collector surface, and are considered of colloidal chemical origin. For example, in liquid systems, the van der Waals force can enhance attachment probability. Electrostatic interactions can be either favorable or unfavorable, as in the interaction of charged particles in coagulation. In effect, the collector and particle can both carry a double layer of charged ions.

Unfortunately, there are no widely accepted theories to account generally for the effects of interfacial and colloidal phenomena on attachment probability. Currently, we lump all effects into an unknown parameter called α_d (O'Melia and Tiller, 1993):

$$\alpha_d = \frac{\text{rate at which particles attach to porous media}}{\text{rate at which particles collide with porous media}} \tag{8.87}$$

Just like the alpha value for coagulation, α_d varies from unity (no barrier to attachment) to zero (complete barrier to attachment). Typically, our previous equation for removal efficiency by a bed of spherical particles would be

---

*It also seems reasonable to write Eq. 8.82 in an alternate way, $W' = 2vN_1 a\eta$, where $v = U_{ap}/\epsilon$ (Eq. 8.4). Then following through the steps indicated in Eqs. 8.83–8.85 yields an alternative equation for Eq. 8.85 (Flagan and Seinfeld, 1988)

$$\frac{dN}{dl} = -\frac{2}{\pi}\left(\frac{s}{1-s}\right)\frac{\eta}{a} N \tag{8.85a}$$

Similarly, by a procedure parallel to that leading to Eq. 8.86,

$$\frac{dN}{dl} = -\frac{3}{4}\left(\frac{s}{1-s}\right)\frac{\eta}{a} N \tag{8.86a}$$

modified simply by multiplying the single particle collection efficiency by α_d;

$$\frac{dN}{dl} = -\frac{3s\alpha_d\eta}{4a}N \tag{8.88}$$

Assuming a known boundary condition that at $l = 0$, $N = N_0$, Eq. 8.88 integrates to

$$\ln\frac{N_L}{N_0} = -\frac{3s\alpha_d\eta}{4a}L \tag{8.89}$$

Hence, if the porosity, single-collector efficiency, collector size, and N_L/N_0 are known, Eq. 8.89 can be integrated and the α_d value can be determined from real experiments (see Exercise 8.7).

8.7.6. Head Loss Through Filter Beds and Mats

As shown in Chapter 5, the Poiseuille equation can be used to develop a formula to relate the head loss of a liquid through a granular bed:

$$\frac{\Delta p}{L} = \frac{K\mu S_0^2(1-\epsilon)^2 U_{ap}}{\epsilon^3} \tag{5.71}$$

where ϵ is the bed porosity, μ is the viscosity, K is a constant with a value around 5, U_{ap} is the approach velocity, and S_0 is the *specific surface* of the granular material (Bear, 1988):

$$S_0 = \frac{\text{porous medium surface area}}{\text{unit volume of porous medium}} \tag{5.67}$$

Equation 5.71 is known as the *Carman–Kozeny* equation. Recall that the Poiseuille equation (and therefore the Carman–Kozeny equation) was developed with the assumption of a laminar flow. The appropriate Reynolds number is calculated based on the diameter of the porous media grains:

$$\text{Re} = \frac{\rho U_{ap} d_c}{\mu} \tag{8.90}$$

Above about $\text{Re} = 6$, the flow can no longer be assumed laminar. The *Ergun* equation can be used to compute head loss (Pontius, 1990):

$$\frac{\Delta P}{L} = \frac{4.17\mu S_1^2(1-\epsilon)^2 U_{ap}}{\epsilon^3} + k_e\rho\frac{(1-\epsilon)}{\epsilon^3}S_1 U_{ap}^2 \tag{8.91}$$

where S_1 is the grain surface area per unit of grain volume and k_e has a value of 0.48. Note that this equation is the sum of two terms; the first term is similar to the Carman–Kozeny equation, whereas the second term is similar in structure to the dynamic pressure in the Bernoulli equation (Chapter 5).

In fiber filters used in gas filtration, the head loss for laminar flow conditions is (Hinds, 1982)

$$\frac{\Delta P}{L} = \frac{\mu U_{ap} f(s)}{d_c^2} \tag{8.92}$$

where f is a function of the filter solid fraction:

$$f(s) = 64s^{1.5}(1 + 56s^3) \quad \text{for } 0.006 < s < 0.3 \tag{8.93}$$

8.8. SUMMARY

In this chapter, we demonstrated that the ideas of continuum mechanics could be applied to a porous medium, and that a convective-diffusion equation could be developed for a porous medium. This in turn gave us the ability to model dispersion in porous environments. We found that in some flows, mass transfer could be affected by an important phenomenon referred to as retardation. Because of retardation, dissolved or transported compounds were slowed down relative to nonretarded components. In subsurface environments, retardation is important in determining the overall movement of contaminants and other substances. In the laboratory, retardation is the basis of powerful methods of characterizing unknown organic compounds. These techniques are called chromatography.

In the second half of the chapter, we developed the main theories of single-collector efficiency in filtration. These theories were in turn improved to take into consideration the effects of other collectors and hydrodynamic retardation. These expressions were then incorporated into formulas for the overall removal efficiency of a fibrous or granular filter. Finally, expressions were developed to predict the head loss that a liquid or gas experiences when passing through a filter.

The reader will find the following exercises important for solidifying and extending the principles developed in this chapter.

EXERCISES

8.1. **(a)** Estimate the center of mass of the contaminant and chloride plumes in Figure 8.4, and thereby determine the average velocities of the plumes. From these estimates, determine the apparent retardation factors for PCE and CTET.

   (b) According to linear partitioning theory, the retardation coefficient in a porous medium is given by Eq. 6.122:

$$R = 1 + \frac{k_d \rho_b}{\epsilon} \tag{6.122}$$

   Using the results of part **a**, compute the apparent partition coefficient

k_d for tetrachloroethylene and carbon tetrachloride if the aquifer porosity is 0.33 and the bulk density is 1.8 g/m3.

(c) According to linear chromatographic theory, what is the ratio of moles of solute in the mobile phase to the total number of moles of solute (calculate for both tetrachloroethylene and carbon tetrachloride).

8.2. Compare collision efficiency of direct interception on a fiber when $R = 0.01$ to 0.1, $s = 0.05$ to 0.5, and the Reynolds number is 0.1. Communicate your results using a graph or figure.

8.3. Use dimensional analysis to show that the efficiency of the convective-diffusion collision mechanism should be related to the Peclet number when the significant parameters are the diffusion coefficient, the collector diameter, and the free-stream velocity.

8.4. Yao et al. (1971) studied filtration of latex particles by a bed of spherical glass beads in water. The bed was precoated with a cationic polymer to increase the attachment probability, which can be assumed to be unity. The figure below shows experimental results for the single-collector efficiency versus the filtration rate. Compare these experimental data with the predictions of the two models for convective diffusion (Eqs. 8.55 and 8.73) by constructing lines corresponding to the model prediction. The relevant experimental parameters are shown in the figure.

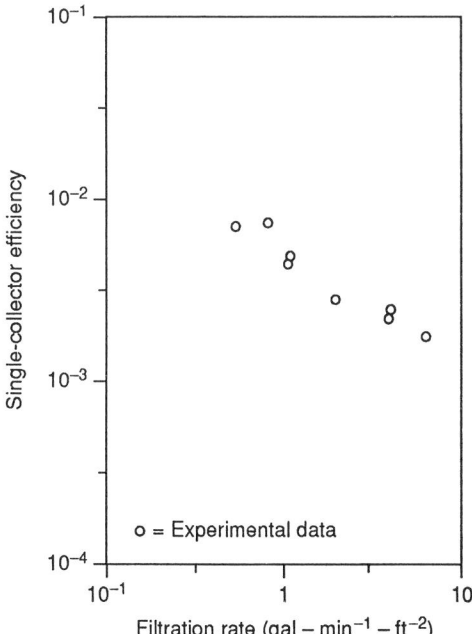

Exercise 8.4. Experimental data for the filtration of latex particles through a bed of spherical glass collectors. $T = 25°C$, $a_p = 0.046\,\mu m$, $a = 0.20\,mm$, and $\epsilon = 0.36$. *Source*: Reprinted with permission from Yao, Habbian, and O'Melia, *Environ. Sci. Technol.*, **5**(11), 1110 (1971). Copyright 1971, American Chemical Society.

8.5. For mass transfer to a single spherical collector show that

$$W = \eta \pi a^2 N_l U_{\mathrm{ap}}$$

After deriving an expression analogous to Eq. 8.83 for the number of spherical collectors in a differential element of a granular filter, derive Eq. 8.86 in the text.

8.6. An aerosol filter has a solids fraction $s = 0.006$ with fiber diameters of $5 \, \mu\mathrm{m}$. Compare the magnitude of the direct interception and convective-diffusion mechanisms for $0.10\text{-}\mu\mathrm{m}$-diameter particles traveling at an approach velocity of $15 \, \mathrm{cm/s}$. $T = 25°\mathrm{C}$.

8.7. The following data were collected by Elimelech and O'Melia (1990) in a study of the filtration of $0.753\text{-}\mu\mathrm{m}$-diameter latex particles on a bed of

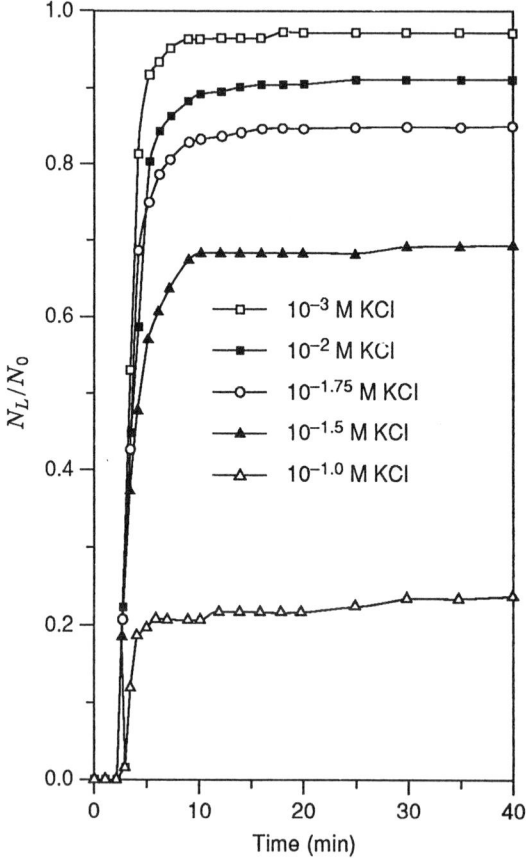

Exercise 8.7. Removal of latex particles in a filter composed of spherical glass collectors. *Source:* Reprinted with permission from Elimelech and O'Melia, *Environ. Sci. Technol.*, **24**, 1534 (1990). Copyright 1990, American Chemical Society.

0.2-mm-diameter glass beads. The curves shown are the normalized exit-particle concentration, N_L/N_0. The different curves are for different KCl concentrations. Removal of particles increases as the KCl concentration increases. The particles have a net negative charge, and when the salt concentration increases, the double layers surrounding the particles are compressed; hence, particle attachment to the collector surface is easier. The removal efficiency increases to approximately steady values over the time shown. The initial period corresponds to flushing of particle-free water from the bed.

(a) What is the most significant single-particle collector efficiency mechanism for an approach velocity of 1.36×10^{-3} m/s? The bed porosity was 0.4. Assume a particle density of 1.05 g/cm^3. Temperature = 24°C.

(b) Using the single-particle collection efficiency identified in part a, determine the α_d value for each of the experiments represented in the figure. The bed depth was 20 cm.

8.8. A glass fiber mat is being used to filter 0.20-μm-diameter particles from an air flow. The air velocity is 100 cm/s, and the glass fibers have a diameter of 4 μm. If the fraction of glass fiber solids is $s = 0.05$, estimate the filter mat thickness required to decrease the aerosol particle concentration by 50%. Consider only the convective-diffusion and direct-interception collection mechanisms. $T = 25$°C.

8.9. The Carman–Kozeny equation provides a prediction of the head loss in a clean gravity filter. Although this has been a very important equation, we have found that in many practical liquid filters, the head loss increases over time as the filter continues to collect particles (becomes "dirty"). Unfortunately, our modeling of head loss in dirty filters has not progressed as far as we would like. Your problem is to try to model head loss in a dirty filter by modifying the Carman–Kozeny equation. You may make the following assumptions if you find them to be helpful: (a) filtration of a monodispersion of small colloids that evenly coat the filter grains and that are evenly filtered throughout the depth of the filter, (b) constant approach velocity, and (c) spherical filter grains. However, since we are breaking new ground here, make any assumptions that you think are reasonable. Your answer should predict an increase in head loss with filter run time. Try to have some fun with this problem and show your modeling muscles!

REFERENCES

Bear, J., *Hydraulics of Groundwater Flow*, McGraw-Hill, New York, 1979.

Bear, J., *Dynamics of Fluids in Porous Media*, Dover Publications, New York, 1988.

Brodkey, R. S. and Hershey, H. C., *Transport Phenomena: A Unified Approach*, McGraw-Hill, New York, 1988.

Elimelech, M. and O'Melia, C. R., "Kinetics of Deposition of Colloidal Particles in Porous Media," *Env. Sci. Technol.*, **24**, 10:1528–1536 (1990).

Flagan, R. C. and Seinfeld, J. H., *Fundamentals of Air Pollution Engineering*, Prentice-Hall, Englewood Cliffs, NJ, 1988.

Freeze, R. A. and Cherry, J. A., *Groundwater*, Prentice-Hall, Englewood Cliffs, NJ, 1979.

Friedlander, S. K., *Smoke, Dust, and Haze*, Wiley, New York, 1977.

Happel, J., "Viscous Flow in Multiparticle Systems: Slow Motion of Fluids Relative to Beds of Spherical Particles," *A.I.Ch.E. J*, **4**, 197–201 (1958).

Happel, J., "Viscous Flow Relative to Arrays of Cylinders," *A.I.Ch.E. J*, **5**, 174–177 (1959).

Hinds, W. C., *Aerosol Technology: Properties, Behavior, and Measurement of Airborne Particles*, Wiley-Interscience, New York, 1982.

Karger, B. L., Snyder, L. R., and Horvath, C., *An Introduction to Separation Science*, Wiley, New York, 1973.

Kuwabara, S., *J. Phys. Soc. Jpn*, **14**, 4:527 (1959).

O'Melia, C. R. and Tiller, C. L., "Physicochemical Aggregation and Deposition in Aquatic Environments," in *Environmental Particles*, J. Buffle and H. P. van Leeuwen, Eds., Lewis Publishers, Boca Raton, FL, 1993.

Rajagopalan, R. and Tien, C., "Trajectory Analysis of Deep-Bed Filtration with the Sphere-in-Cell Porous Media Model," *Am. Inst. Chem. Eng. J.*, **22**, 523–533 (1976).

Roberts, P. V., Goltz, M. N., and MacKay, D. M., "A Natural Gradient Experiment on Solute Transport in a Sand Aquifer: 3. Retardation Estimates and Mass Balances for Organic Solutes," *J. Water Resour. Res.*, **22**, 13:2047–2058 (1986).

Snyder, L. R. and Kirkland, J. J., *Introduction to Modern Liquid Chromatography*, Wiley, New York, 1979.

Spielman, L. A. and Fitzpatrick, J. A., "Theory of Particle Collection under London and Gravity Forces," *J. Colloid Interface Sci.*, **42**, 3:607–623 (1973).

Tien, C., *Granular Filtration of Aerosols and Hydrosols*, Butterworths, Boston, 1989.

Yao, K. M., Habibian, M. T., and O'Melia, C. R., "Water and Waste Water Filtration: Concepts and Applications," *Env. Sci. Technol.*, **5**, 11:1105–1112 (1971).

BIBLIOGRAPHY

Fried, J. J. and Combarnous, M. A., "Dispersion in Porous Media," in *Advances in Hydroscience*, V. T. Chow Ed., Academic, New York, pp. 169–282, 1971.

MacKay, D. M., Freyberg, D. L., and Roberts, P. V., "A Natural Gradient Experiment on Solute Transport in a Sand Aquifer: 1. Approach and Overview of Plume Movement," *J. Water Resour. Res.*, **22**, 13:2017–2029 (1986).

Pinder, G. F., "Groundwater Contaminant Transport Modeling," *Env. Sci. Technol.*, **18**, 4:108A–114A (1984).

Probstein, R. F., *Physicochemical Hydrodynamics: An Introduction*, Butterworths, Boston, MA, 1989.

Spielman, L. A., "Particle Capture from Low-Speed Laminar Flows," *Ann. Rev. Fluid Mech.*, **9**, pp. 297–319 (1977).

9

REACTION KINETICS

9.1. INTRODUCTION

Reactions concern transformation of mass constituents and are extremely important in governing the concentration and occurrence of chemicals and particulate matter in the environment. Reactions usually result in the loss of a reactant and the formation of a product. Schematically, we might write this as

$$\text{reactants} \rightarrow \text{products} \tag{9.1}$$

This book concerns environmental modeling, and we have already encountered a few examples of simple reactions: We studied the oxidation of carbon on the surface of a burning coal particle, the consumption of O_2 and radioactive decay of radon in sediments, the reaction of NO_2 with surfaces in a house, and the consumption of Cl_2 in water and biofilms. Our most extensive consideration of kinetics was in Chapter 3, where we developed expressions for the rate of coagulation of colloids and aerosol particles. These too are considered mathematically as reactions. In these examples, we were interested not only in how much product was formed from reactants, called the stoichiometry, but also the rate at which the process occurred, that is, the kinetics. The purpose of this chapter is to discuss the rates of several generic reaction types and to develop methods for analyzing rate data. These are critical tools for the environmental scientist and engineer.

We point out here that kinetic studies are an important tool in understanding reaction mechanisms. An in-depth discussion of mechanism is outside the scope of this chapter, but the reader can follow up on this subject in some of the excellent chemistry and physical chemistry references at the end of the chapter.

9.2. FIRST-ORDER REACTIONS

Consider that we find the following relationship between reactant A and product B, and that there are no other reactants or products:

$$A \rightarrow B \tag{9.2}$$

Equation 9.2 represents the *stoichiometry* of an irreversible reaction: 1 mol of
A is converted to 1 mol of B. The *reaction rate* is defined as the rate of change
in the concentration of a reactant or product. For computational purposes, we
will need to quantify the reaction rate as a differential equation. We make the
assumption that the rate of change in the concentration of A and B is
proportional to the concentration of A,

$$r_A = \frac{d[A]}{dt} = -k[A] \qquad (9.3)$$

and

$$r_B = \frac{d[B]}{dt} = k[A] \qquad (9.4)$$

Here r_A and r_B stand for the reaction rates of A and B, k is the reaction
rate or kinetic constant, and [A] and [B] represent the appropriate concentrations of A and B. The concentration units could be molar, mass, or colloidal
number concentration in water, or partial pressure or aerosol number concentration in air. In molar concentration units, the correct units for the r values
are mol-L^{-3}-T^{-1}, and the correct units for k are T^{-1}. From Eqs. 9.2 and 9.3,
we conclude:

$$r_A = \frac{d[A]}{dt} = -\frac{d[B]}{dt} = -r_B \qquad (9.5)$$

In the present case, we have used the stoichiometry (Eq. 9.2) to guess at the
kinetics law (Eqs. 9.3 and 9.4). Reactions in which the kinetics are determined
by the stoichiometry are called *elementary* reactions (see Section 9.6).

Reactions 9.3 and 9.4 are called *first-order reactions*. First-order reactions
are called linear because Eqs. 9.3 and 9.4 are linear differential equations.
Integrating Eq. 9.3 with the initial condition that $A = [A]_0$ at $t = 0$, we find:

$$[A] = [A]_0 \exp(-kt) \qquad (9.6)$$

It is also pretty straightforward to solve for the concentration of B. From the
stoichiometry of Eq. 9.2, note that if at time $= 0$ the concentration of B is zero,
then at any other time the concentration of B must be $[B] = [A]_0 - [A]$.
Combining this with Eq. 9.6 yields, for [B]:

$$[B] = [A]_0[1 - \exp(-kt)] \qquad (9.7)$$

Figure 9.1 is a plot of the concentration of A and B as a function of time as
predicted by Eqs. 9.6 and 9.7. As usual, we display the plots in dimensionless
form.

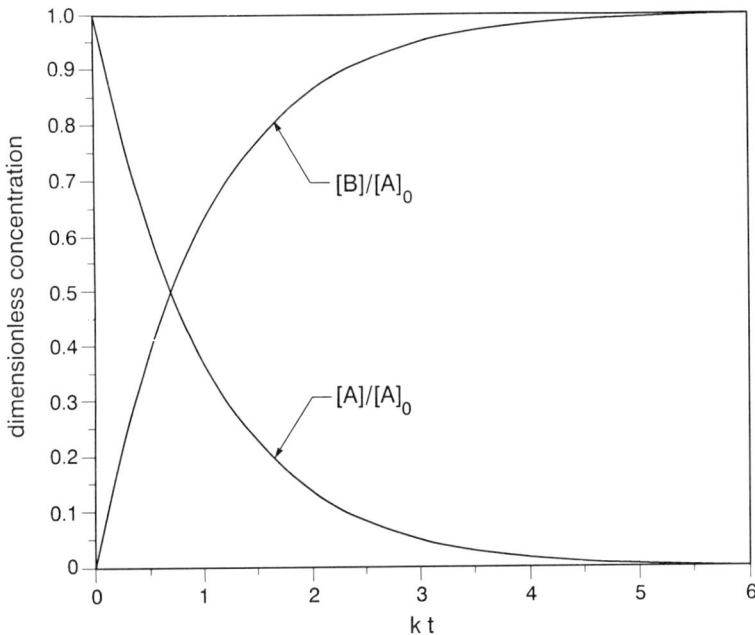

Figure 9.1. Plot of dimensionless concentration of A and B as a function of dimensionless time kt.

As expected, the two plots are mirror images of each other, where the gradient in the concentration of A is the negative of the gradient in concentration of B (Eq. 9.5).

The first-order rate expression says that the rate of change (loss) of A is proportional to the amount of A present, a very elegant and simple hypothesis. But why would reactants, or anything for that matter, decay in this manner? It turns out that examples are all around us:

1. *Death and birth.* Consider two communities, North Gotham and South Gotham. North Gotham just happens to have twice as many citizens as South Gotham. Other than the population difference, all other demographic, socio-logical, and environmental factors in the two communities are the same. The average rate of recorded deaths (e.g., number of deaths per day) in North Gotham would be expected to be exactly twice as high as in South Gotham. So the death rate is proportional to the population — a classic first-order "reaction." By the same token, we would guess the number of births per day in North Gotham to be exactly twice as high as the number of births in South Gotham — another first-order reaction. We can extend this reasoning to other environmental population dynamics, such as the growth and death of microbes.

2. *Disinfection.* Consider a water-filled beaker containing the pathogenic organism *Giardia*. We add a large dose of chlorine to the water. What is the rate of disinfection (i.e., killing) of the organisms? If we assume there is a great excess of disinfectant, and that its concentration is not substantially lowered during the disinfection, the rate of death or disinfection is proportional to the number of organisms still alive. This results in a famous environmental law called the simplified Chick's law for disinfection:

$$\frac{dN}{dt} = -kN \tag{9.8}$$

where N stands for the concentration of live microorganism, and k is the first-order Chick's law constant.*

3. *Radioactive decay.* A given radioactive atom has a fixed probability of decaying during a given interval of time. Now considering a collection of such atoms, the proportion of atoms decaying during a given time interval is the same as the probability of the decay of a single atom. Hence, the actual number of radioactive atoms decaying during a given time interval is proportional to the number of undecayed atoms. This results in a classic linear differential equation identical to Eq. 9.8 above, where N now stands for the number or mass of undecayed atoms.

4. *Light-catalyzed reactions.* Consider a reaction in which A decomposes to B as a result of the addition of radiant energy to the system from the light source (Figure 9.2). The light adds enough energy so that a given molecule of A has a fixed chance of decaying during any given time interval. A stronger light yields a higher probability of decay, and a weaker light yields a lower probability of decay. For certain ideal conditions, the proportion of all molecules of A decaying during a given time interval is the same as the probability of a single molecule decaying during the time interval. Hence, the change in concentration of A is given by a differential equation identical to Eq. 9.8, where we substitute $[A]$ for N.

5. *Diffusion-controlled heterogeneous reactions.* As discussed in Section 6.7.3, when the rate of mass transfer of a substance to a reacting surface is diffusion controlled, the overall mass transfer to the surface of a compound A reacting at the surface is proportional to the concentration of A in the external medium.

---

*The reader might find our kinetic argument objectionable. For example, for a microbe to be "killed," it seems obvious that the disinfectant molecule (e.g., chlorine or ozone) must first diffuse through the cell wall and destroy either the cell wall itself or some important biochemical mechanism within the cell. Diffusion takes time, so we might well expect a lag time in the killing action of the disinfectant. During the lag time, the rate of killing would be less than suggested by first-order kinetics, and therefore, our kinetic law would not fit the kill data over the entire time of the disinfection experiment. This is a reasonable hypothesis, and this type of induction time can be observed in real disinfection studies.

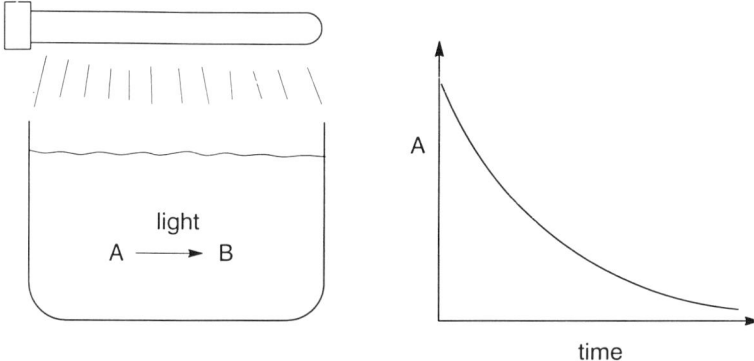

Figure 9.2. Light-catalyzed decomposition of A.

This is true regardless of the actual kinetics of the reaction of A at the surface. In this case, an observer in the continuous medium would record a first-order decay in the concentration of A, yielding an equation identical to Eq. 9.8 where N is replaced by A. These kinds of reactions are fairly common in the environment and industry. For example, where spherical catalyst particles are used in decomposing A, if the reaction at the surface (or in) the catalyst is fast enough that diffusion limitation obtains, first-order kinetics can be observed.

Quite a number of chemical reactions are known to be first order. Table 9.1 lists several first-order reactions of environmental importance. The reader will

Table 9.1. Some Liquid-Phase First-Order Reactions of Environmental Interest

$$MnSO_4(aq) \rightarrow Mn^{2+} + SO_4^{2-}$$
$$Cr^{3+} + H_2O \rightarrow CrOH^{2+} + H^+$$
$$Al^{3+} + H_2O \rightarrow AlOH^{2+} + H^+$$
$$FeOH^{2+} + H_2O \rightarrow Fe(OH)_2^+ + H^+$$
$$H_2S \rightarrow H^+ + HS^-$$
$$H_2CO_3 \rightarrow CO_2 + H_2O$$
$$CO_2 + H_2O \rightarrow H_2CO_3$$
$$HCO_3^- \rightarrow CO_2 + OH^-$$
$$P_3O_{10}^{5-} + H_2O \rightarrow P_2O_7^{4-} + PO_4^{3-}$$
$$P_2O_7^{4-} + H_2O \rightarrow 2PO_4^{3-}$$

Source: Adapted from Pankow and Morgan (1981).

note immediately that in several cases, the stoichiometry of Eq. 9.2 seems to be violated; some of the reactions seem to include a second reactant — water. However, this appearance is indicative of the way in which the equations are written. For example, the third reaction in the table represents the first step in the hydrolysis of AI(III). Chemists have learned that in aqueous systems, the Al^{3+} ion is simply not floating around in a sea of H_2O molecules. Rather, the Al^{3+} ions are thought to have $6 H_2O$ molecules surrounding them in a coordination sphere; this is written, $Al(H_2O)_6^{3+}$. The water molecules are not permanently attached and can be exchanged for other water molecules in solution. In reality, the reaction of Table 9.1 is thought to involve the deprotonation of one of the coordinated H_2O molecules:

$$Al(H_2O)_6^{3+} \rightarrow Al(H_2O)_5OH^+ + H^+ \tag{9.9}$$

Understood in this way, Eq. 9.9 now seems consistent with first-order kinetics.

Often we are presented with experimental data for reaction kinetics. These could be measurements of either product or reactant concentrations over time. If we suspect that the kinetics are first order, we would like to (1) test this hypothesis and (2) if appropriate, determine (fit) the first-order kinetic constant. There are several approaches to answering the question above. The method we follow here (and throughout this chapter) is called the *integral method*. It involves integrating the governing differential equation (which we have already done), and linearizing the equation to test the data fit and determine the rate constant. For example, let's investigate this approach for a reaction that conforms to the stoichiometry of Eq. 9.2. The integrated first-order rate law for the reactant A is given by Eq. 9.6:

$$[A] = [A]_0 \exp(-kt) \tag{9.6}$$

To linearize Eq. 9.6, simply take the natural logarithm of both sides:

$$\ln[A] = \ln[A]_0 - kt \tag{9.10}$$

Recall that $[A]_0$ is the initial reactant concentration. Equation 9.10 describes the equation of a straight line, $y = b + mx$ ($y = \ln[A]$ = dependent variable, $b = \ln[A]_0 = y$ intercept, $x = t$ = independent variable, and $m = -k$ = slope). If a plot of $\ln[A]$ versus time yields a straight line, we have strong evidence to support first- or pseudo-first-order kinetics. An example of the fitting of data to such a first-order kinetic model is found in Exercise 3.5, which concerns the coagulation of river sediments. The following numerical example also performs this analysis on some experimental kinetics data.

Example 9.1. The reaction shown in Eq. 9.2 was studied in the laboratory and the following data were collected describing the conversion of A over time:

Time (min)	$[A]$(mol/L)	$\ln[A]$
0	89.8	4.45
1	83.0	4.42
2	74.0	4.30
4	58.0	4.06
7	44.0	3.78
10	32.0	3.47
20	12.5	2.53
40	2	0.693
70	0.06	−2.81
100	0.0045	−5.40

SOLUTION. We now want to plot the data as described in the linearization discussion above. The first column lists the time values, which provide our x-axis values; the third column is the natural log of the concentration values, and this provides our y-axis values. These are then plotted in the following figure, and a straight line is fit through the data. The straight-line fit seems reasonable, with a correlation coefficient (r^2) value of 0.998. From the slope of the fitted line (−0.10), we estimate $k = 0.10 \, \text{min}^{-1}$. So we have accomplished two goals: we have seen the data fit the first-order kinetics, and we have determined the apparent kinetic constant.

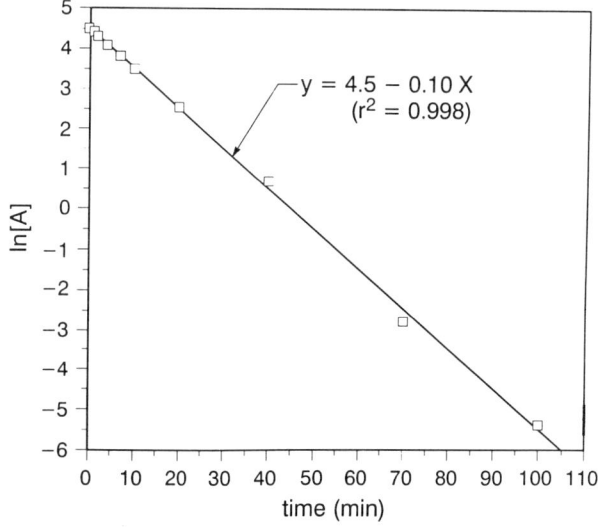

Example 9.1. Test of first-order kinetics.

9.3. SECOND-ORDER REACTIONS

A reaction scheme that has widespread relevance in environmental chemistry is the second-order reaction. Table 9.2 lists a number of reactions that are considered second order.

Consider reactive molecules of type A and B. We might observe over time that, in solution, the following reaction occurs:

$$A + B = P_1 + P_2 \tag{9.11}$$

Here P_1 and P_2 are the products of the reaction between A and B. Equation 9.11 represents the *stoichiometry* of the reaction: 1 mol of A combines with 1 mol of B to form 1 mol of P_1 and 1 mol of P_2. But what are the kinetics of the reaction between A and B? We proceed with the following argument. Consider one molecule of A surrounded by an abundance of B. All molecules are moving around because of their thermal motion. The probability of this molecule of A reacting with any of the B molecules is proportional to the concentration of the B molecules. The larger the concentration of B, the greater the probability of a collision between A and B having enough energy to result in the formation of a product molecule. But now consider that there is actually more than one molecule of A. Each individual molecule can be argued to have the same probability of reaction as derived above. But since there is now a collection of A molecules, the overall probability of reaction between any A molecule and any B molecule is proportional to the product of the concentrations of A and B. This is a famous argument resulting in the idea of *bimolecular*

Table 9.2. Some Liquid-Phase Second-Order Reactions of Environmental Interest

$$H^+HS^- \rightarrow H_2S$$
$$H^+FeOH^{2+} \rightarrow H_2O + Fe^{3+}$$
$$CrOH^+ + H^+ \rightarrow Cr^{3+} + H_2O$$
$$Mn^+ + SO_4^- \rightarrow MnSO_4(aq)$$
$$Ni^{2+} + HP_3O_{10}^{4-} \rightarrow NiHP_3O_{10}^{2-}$$
$$CO_2 + OH^- \rightarrow HCO_3^-$$
$$Fe_{2+} + O_2 \rightarrow Fe(III)$$
$$Fe^{3+} + F^- \rightarrow FeF^{2+}$$
$$FeOH^{2+} + FeOH^{2+} \rightarrow Fe_2(OH)_2^{4+}$$
$$S(\text{-II}) + O_2 \rightarrow S_2O_3^{2-}, SO_3^{2-}, SO_4^{2-}$$
$$Mn^{2+} + O_2 \rightarrow Mn(IV)$$

Source: Adapted from Pankow and Morgan (1981) and Schneider and Schwyn (1990).

kinetics. Applied to Eq. 9.11, bimolecular or second-order kinetics says that

$$r_A = \frac{d[A]}{dt} = -k[B][A] = \frac{d[B]}{dt} = r_B \tag{9.12}$$

By the same token, the kinetic expression for product P is

$$\frac{d[P_1]}{dt} = \frac{d[P_2]}{dt} = k[A][B] \tag{9.13}$$

Note that the units of the kinetic constant k must be different than for the first-order reaction discussed in the last section. The reader should demonstrate that in the molar-unit system, those units are $L^3\text{-mol}^{-1}T^{-1}$. As in Section 9.2, Eqs. 9.12 and 9.13 describe an elementary reaction, one in which the kinetics are suggested by the stoichiometry (Section 9.6).

The easiest way to solve these equations is to define a new variable x which is the amount of A or B reacted, or the amount of P formed. In this case, we let $[A] = [A]_0 - x$ and $[B] = [B]_0 - x$. Then we might rewrite the part of Eq. 9.12 governing the kinetics of A as

$$\frac{dx}{([B]_0 - x)([A]_0 - x)} = kdt \tag{9.14}$$

Given the initial condition that $x = 0$ at $t = 0$, it is fairly easy to show that the solution of Eq. 9.14 is (Exercise 9.2)

$$\frac{1}{[A]_0 - [B]_0} \ln \left\{ \frac{[B]_0([A]_0 - x)}{[A]_0([B]_0 - x)} \right\} = kt \qquad [A]_0 \neq [B]_0 \text{ only} \tag{9.15}$$

This equation can be rearranged further to yield the following (see also Exercise 9.2):

$$[A] = \frac{[A]_0^2 - [A]_0[B]_0}{[A]_0 - [B]_0 \exp[-([A]_0 - [B]_0)kt]} \qquad [A]_0 \neq [B]_0 \text{ only} \tag{9.16}$$

It is interesting to examine the response of Eq. 9.16 over time for different ratios of $[B]_0/[A]_0$. As usual, we nondimensionalize the equation to see its more universal form. Dividing both sides of Eq. 9.16 by $[A]_0$ and rearranging slightly we find:

$$\frac{[A]}{[A]_0} = \frac{1 - \dfrac{[B]_0}{[A]_0}}{1 - \dfrac{[B]_0}{[A]_0} \{ \exp[-([A]_0 - [B]_0)kt] \}} \qquad [A]_0 \neq [B]_0 \text{ only} \tag{9.17}$$

In Figure 9.3, we have plotted Eq. 9.17 against the absolute values of the

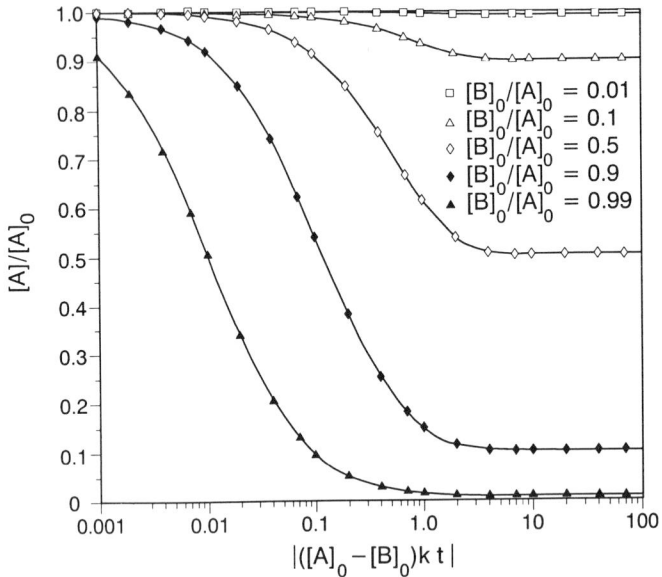

Figure 9.3. Change of concentration of A in second-order reaction between A and B. $[A]_0 > [B]_0$.

argument of the exponential, $([A]_0 - [B]_0)kt$, and for different values of the ratio $[B]_0/[A]_0 < 1$. Note that for $[B]_0/[A]_0 = 0.5$, the concentration of A approaches asymptotically the value $[A] = 0.5\,[A]_0$. This makes great sense: If the initial concentration of B is one half as great as A, and if 1 mol of A combines with 1 mol of B, the ultimate amount of A reacted can only equal one half the amount initially present. Note that all curves approach a different asymptote, defined by the value $1 - [B]_0/[A]_0$.

In Figure 9.4, we have also plotted Eq. 9.17 against values of the argument of the exponential. But here the ratio $[B]_0/[A]_0$ has values greater than 1. Here, of course, the concentration of A must eventually go to zero because the starting concentration of A is less than the starting concentration of B. It is interesting to note that as the ratio $[B]_0/[A]_0$ approaches large values, the curves get very close together. In fact, it can be shown that as $[B]_0/[A]_0$ becomes large, an asymptotic curve is approached. We find in the next section that, surprisingly enough, when $[B]_0/[A]_0 \gg 1$, the kinetics take on the characteristics of a *first-order* reaction.

As in Section 9.2, we might like to see whether Eq. 9.17 will fit data suspected of arising from second-order kinetics. As shown in Exercise 9.3, Eq. 9.17 can be rearranged to

$$\ln\left(\frac{[A]_0[B]}{[B]_0[A]}\right) = ([B]_0 - [A]_0)kt \qquad (9.18)$$

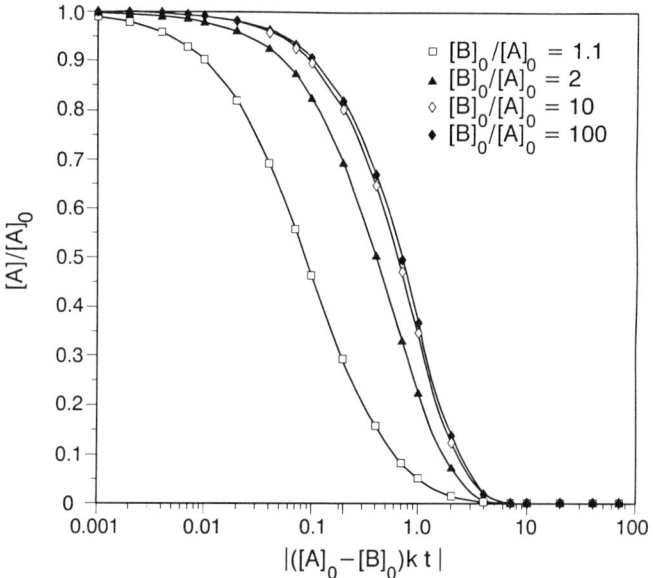

Figure 9.4. Change of concentration of A in second-order reaction between A and B. $[B]_0 > [A]_0$.

This is the equation of a straight line if the left-hand side is the dependent variable, t is the independent variable, $([B]_0 - [A]_0)k$ is the slope, and the y intercept is zero.

When $[B]_0 = [A]_0$, the second-order reaction kinetics are even simpler than those above. From the stoichiometry of Eq. 9.11, if $[B]_0 = [A]_0$, then $[A] = [B]$ at all times. Therefore, the kinetics expression Eq. 9.12 reduces to

$$\frac{d[A]}{dt} = -k[A]^2 \tag{9.19}$$

This equation separates easily and the solution is

$$\frac{[A]}{[A]_0} = \frac{1}{1 + [A]_0 kt} \tag{9.20}$$

To fit Eq. 9.20 to kinetic data, it is easy to see that it can be linearized to

$$\frac{1}{[A]} = \frac{1}{[A]_0} + kt \tag{9.21}$$

Equation 9.21 is the equation of a straight line if $1/[A]$ is the dependent variable, t is the independent variable, k is the slope, and $1/[A]_0$ is the y

intercept. Experimentally, we determine the concentration of A at a series of times and fit Eq. 9.21 to the data. The kinetic constant is measured as the slope of the fitted line.

Finally, we have focused in this section on the change in concentration of A only. We could just as easily have developed analogous equations for the change in concentration of B.

9.4. PSEUDO-FIRST-ORDER REACTIONS

It is interesting to think a bit more about what happens to the second-order reaction Eq. 9.17 if the concentration of B is much greater than the concentration of A. If $[B]_0 \gg [A]_0$, then $[B]_0/[A]_0 \gg 1$. Therefore, in the numerator of Eq. 9.17, we can ignore 1 in comparison to $[B]_0/[A]_0$. In the denominator, note that when $[B]_0 > [A]_0$, the argument of the exponential function is always positive and, therefore, the exponential term is greater than 1. So when $[B]_0/[A]_0 \gg 1$, we can ignore 1 in comparison to $[B]_0/[A]_0 \exp[-([A]_0-[B]_0)kt]$. Finally, the $[B]_0/[A]_0$ in the numerator and denominator cancel, leaving

$$\frac{[A]}{[A]_0} \simeq \exp[([A]_0 - [B]_0)kt] \simeq \exp(-[B]_0kt) \qquad [B]_0 \gg [A]_0 \text{ only} \quad (9.22)$$

Note that this asymptotic form of the governing equation (Eq. 9.17) no longer has the parameter $[B]_0/[A]_0$. Therefore, the asymptotic form is not dependent on the exact values of parameter $[B]_0/[A]_0$ so long as these values are very large. In fact, as $[B]_0/[A]_0$ grows larger, the curves in Figure 9.4 would be indistinguishable from the curve defined by Eq. 9.22.

But let's recall the reaction (Eq. 9.6). The equation for the first-order decomposition of A was given as

$$\frac{[A]}{[A]_0} = \exp(-kt) \qquad (9.6)$$

It is apparent that there is very little mathematical difference between the last two equations. In the arguments of the exponential we find the independent variable time (t) and other (constant) parameters like k and $[B]_0$. We call the reaction represented by Eq. 9.22 a *pseudo-first-order reaction*. In second-order reactions like Eq. 9.11, if the initial concentration of one reactant is much greater than the other reactant (i.e., when the concentration of one reactant is in great excess) the rate of change of the reactant in great deficit appears to be first order or pseudo-first order.

It is a little strange that two equations govern the kinetics of the decomposition of A when $[B]_0/[A]_0 \gg 1$ (Eqs. 9.17 and 9.22). Equation 9.17 derived from a detailed analysis of second-order kinetics and Eq. 9.22 derived from examining a limiting form of the second-order kinetics. In the limit, for $[B]_0/[A]_0 \gg 1$, second-order kinetics is indistinguishable from first-order kinetics. This is not a paradox. It is a mathematical fact.

9.5. ZERO-ORDER REACTIONS

Sometimes when the reaction, Eq. 9.2, is observed in solution, the rate of change in concentration of A is not governed by first-order kinetics. Sometimes the rate of consumption of A is independent of the concentration of A, or

$$r_A = \frac{d[A]}{dt} = -k \tag{9.23}$$

Note that the units of k in this equation are different than in previous sections. The reader should demonstrate that in the molar-concentration system, the units of k in Eq. 9.23 are mol-L^{-3}-T^{-1}.

It seems that the zero-order reaction is controlled by something other than the concentration of A. In fact, zero-order kinetics are very common in the case of catalyzed reactions (see Section 9.6). The catalyst molecule or particle surface may have a fixed number of sites where A can be converted to B. The attachment of A to the reaction site and the subsequent conversion to B take a finite amount of time, which is independent of the abundance of A in the solution. You can think of the catalyst as a gatekeeper, who allows only so many A molecules per unit time to be converted to B.

9.6. ELEMENTARY AND NONELEMENTARY REACTIONS

In Sections 9.2 and 9.3, we examined single, elementary reactions of first and second order. As mentioned above, an *elementary* reaction is one in which the stoichiometry dictates the kinetics. We can make this a little more formal with the following general equation of stoichiometry:

$$aA + bB \rightarrow pP \tag{9.24}$$

The lowercase letters are known as the *stoichiometric coefficients*. In a general reaction involving A and B, we can write a generic reaction-rate expression as

$$\frac{d[A]}{dt} = \frac{d[B]}{dt} = -k[A]^x[B]^y \tag{9.25}$$

The order of a reaction is defined as the sum $x + y$. It is also customary to say that reaction 9.25 is "x order in A" and "y order in B." In an elementary reaction, $x = a$ and $y = b$; hence, the kinetics are dictated by the stoichiometry.

The reader should go back and verify that our definition of an elementary reaction here is consistent with those in Sections 9.2 and 9.3. The reader will also see why the zero-order reaction in Section 9.5 is not considered an elementary reaction — because the kinetics has no relation to the stoichiometry.

Many kinetics studies uncover *nonelementary* reactions, or reactions in which the kinetic law is not dictated by the stoichiometry. For example, the biological utilization of substrates (food) by microorganisms often requires an enzyme catalyst. The following stoichiometry is observed:

$$E + S \rightarrow E + P \tag{9.26}$$

Here E stands for enzyme, S stands for substrate, and P stands for product. Equation 9.26 is typical of many enzyme-catalyzed reactions where the enzyme is not actually consumed, but rather facilitates or lowers the activation energy for another reaction. Scientists long ago postulated that another reaction actually represents what is going on during enzyme-catalyzed utilization. It involves the appearance of an *enzyme–substrate complex*:

$$E + S \underset{k_2}{\overset{k_1}{\rightleftharpoons}} ES \overset{k_3}{\rightarrow} E + P \tag{9.27}$$

Here ES is the enzyme–substrate complex; it is considered a temporary association between the enzyme and substrate molecules. In the last step of Eq. 9.27, the enzyme is released for possible reactions with other substrate molecules. Note also that in the first step of Eq. 9.27 we encounter the first *reversible* reaction in this chapter. A reversible reaction is simply a reaction that can proceed forward and backward. A mathematically derived kinetic law for the overall reaction of Eq. 9.27 has been matched with experimental kinetic data. The mathematical argument begins by writing the kinetics for the ES as

$$\frac{d[ES]}{dt} = k_1[E][S] - k_2[ES] - k_3[ES] \tag{9.28}$$

The reader notes that the reversible part of the reaction is simply handled as a second-order forward reaction (k_1 term) and a first-order backward reaction (k_2 term). Next we assume that after some transient induction period (which can be observed experimentally), the reaction reaches a stage in which the formation and decomposition of ES are equal, or $d[ES]/dt = 0$. Hence, the left-hand side of Eq. 9.28 can be set to zero and, with some rearrangement, we find:

$$[ES] = \frac{k_1[E][S]}{k_2 + k_3} \tag{9.29}$$

Now we imagine that initially (before any reactions), there is some quantity of free enzyme present in solution, $[E]_0$; hence, at any other time, the amount of free (uncomplexed) enzyme is $[E] = [E]_0 - [ES]$. Substituting this into Eq. 9.29, we find, after some rearrangement, that

$$[ES]\left(1 + \frac{k_1[S]}{k_2 + k_3}\right) = \frac{k_1[S][E]_0}{k_2 + k_3} \tag{9.30}$$

However, note that the following relationship must hold:

$$\left(1 + \frac{k_1[S]}{k_2 + k_3}\right) = \frac{k_1[S] + k_2 + k_3}{k_2 + k_3} \tag{9.31}$$

Substituting this into Eq. 9.30 yields

$$[ES] = \frac{k_1[S][E]_0}{k_1[S] + k_2 + k_3} \tag{9.32}$$

Note that during the steady-state period (i.e., following the induction period), the rate of formation of P is equal to the rate of decomposition of S, or

$$\frac{d[P]}{dt} = k_3[ES] = -\frac{d[S]}{dt} \tag{9.33}$$

Therefore, we can write $d[S]/dt$ as

$$\frac{d[S]}{dt} = -\frac{k_1 k_3[S][E]_0}{k_1[S] + k_2 + k_3} \tag{9.34}$$

Finally, dividing numerator and denominator by k_1, we find:

$$\frac{d[S]}{dt} = -\frac{k_3[S][E]_0}{[S] + k_m} \tag{9.35}$$

where k_m is given by

$$k_m = \frac{k_2 + k_3}{k_1} \tag{9.36}$$

Note that Eq. 9.35 is neither first nor second order. The enzyme-catalyzed substrate utilization is called *nonelementary* because it would have been impossible to derive the kinetic law equation (Eq. 9.35) based only on consideration of the overall stoichiometry of Eq. 9.26.

Equation 9.35 has some useful mathematical characteristics that can be used to advantage in determining the values of the kinetic constants. For example, when there is a large excess of substrate present, the denominator is approximated by $[S]$. This cancels with the $[S]$ in the numerator, which indicates that $d[S]/dt$ will simply be proportional to the product $k_3[E]_0$. Therefore, the utilization of substrate is zero order in substrate concentration.* The opposite limit occurs when the concentration of substrate is very small. In this case, $[S]$ is negligible compared with k_m in the denominator; therefore, $d[S]/dt$ will simply equal $k_3/k_m[S][E]_0$. In this limit, the consummation of S is first order in $[S]$. If k_3 is known from investigation at the large S limit, the small S limiting condition might be used to determine k_m. Hence, a clever experimental protocol might be used to determine the kinetic constants of this enzyme-catalyzed reaction.

Finally, note that in our analysis of this nonelementary enzyme-catalyzed reaction, we have essentially proposed a series of two elementary reactions, that is, Eq. 9.27. We have therefore progressed along the trail of understanding the reaction *mechanism*. As stated in the introduction, this too is a critical task of kinetic studies. Although an exhaustive discussion of mechanism is outside the scope of this chapter, many excellent texts go deeper into this subject. The reader is referred to the references at the end of the chapter.

9.7. SIMPLE SERIES AND PARALLEL REACTIONS

As we have already seen in Section 9.6, reactions composed of more than one step are possible. One of the simplest is the two-step *series reaction*:

$$A \xrightarrow{k_1} B \xrightarrow{k_2} C \tag{9.37}$$

Following the procedure of the previous sections, we assume elementary reactions. Hence, the rate of change of A is proposed to be

$$\frac{d[A]}{dt} = -k_1[A] \tag{9.38}$$

---

*Exercise 6.12 developed an equation for the concentration of substrate in a biofilm where the consumption of substrate was zero order in the substrate concentration.

Following Section 9.2 and letting $[A] = [A]_0$ at $t = 0$, we find Eq. 9.6:

$$[A] = [A]_0 \exp(-k_1 t) \tag{9.6}$$

For the intermediate compound B, we note from Eq. 9.37 that it is formed by decomposition of A and lost through its own decomposition to C. Again assuming elementary kinetics, we propose:

$$\frac{d[B]}{dt} = k_1[A] - k_2[B] \tag{9.39}$$

The first step in solving this is to substitute Eq. 9.6 for $[A]$:

$$\frac{d[B]}{dt} = k_1[A]_0 \exp(-k_1 t) - k_2[B] \tag{9.40}$$

This can easily be solved by the integrating-factor method (Exercise 9.5) to yield, for $[B]_0 = 0$:

$$[B] = \frac{k_1[A]_0}{k_2 - k_1} [\exp(-k_1 t) - \exp(-k_2 t)] \qquad \text{for } k_2 \neq k_1 \tag{9.41}$$

We can get $[C]$ simply by performing a mass balance. When $[C]_0 = [B]_0 = 0$, note that $[A] + [B] + [C] = [A]_0$; therefore, $[C] = [A]_0 - [A] - [B]$. Substituting the solutions above for $[A]$ and $[B]$, we quickly get

$$[C] = [A]_0 - [A]_0 \exp(-k_1 t) - \frac{k_1[A]_0}{k_2 - k_1} [\exp(-k_1 t) - \exp(-k_2 t)] \tag{9.42}$$

As usual we try to nondimensionalize these solutions as much as possible for efficient presentation of results. Equation 9.6 is rearranged slightly to

$$\frac{[A]}{[A]_0} = \exp(-k_1 t) \tag{9.6}$$

Next, $[B]$ can be rearranged to

$$\frac{[B]}{[A]_0} = \frac{1}{(k_2/k_1) - 1} \left\{ \exp(-k_1 t) - \exp\left[-\left(\frac{k_2}{k_1}\right) k_1 t \right] \right\} \tag{9.43}$$

Finally, $[C]$ can be expressed nondimensionally as

$$\frac{[C]}{[A]_0} = 1 - \exp(-k_1 t) - \frac{1}{(k_2/k_1) - 1} \left\{ \exp(-k_1 t) - \exp\left[-\left(\frac{k_2}{k_1}\right) k_1 t \right] \right\}$$

$$\tag{9.44}$$

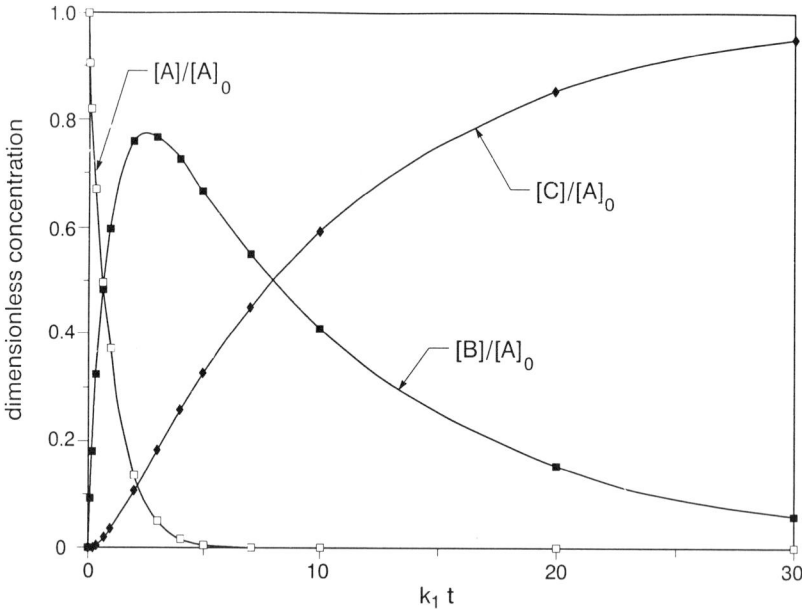

Figure 9.5. Reactant concentration for the simple series reaction where $k_2/k_1 = 0.1$.

Note that in this form, Eqs. 9.6, 9.43, and 9.44 have a single nondimensional independent variable, k_1t, and a single nondimensional parameter, k_2/k_1. Figure 9.5 shows the dimensionless concentration of A, B, and C for a value of $k_2/k_1 = 0.1$. Note in the figure that since k_1 is much greater than k_2, A decays relatively rapidly and B increases relatively rapidly. After enough B is formed, the reaction forming C kicks in and the curves for B and C mirror each other after about $k_1t = 5$. It is easy to see from a mass balance why the curves for B and C mirror each other after a certain point. Note that since $[B]_0 = [C]_0 = 0$, $[A] + [B] + [C] = [A]_0$. However, as noted above, after about $k_1t = 5$, the concentration of A is close to zero. Therefore, $[B] + [C] \approx [A]_0$ or $[C] \approx [A]_0 - [B]$. Then the following must also hold:

$$\frac{[C]}{[A]_0} \approx 1 - \frac{[B]}{[A]_0} \tag{9.45}$$

This confirms that $[B]/[A]_0$ and $[C]/[A]_0$ will be mirror images in Figure 9.5.

In Figure 9.6 we show the concentrations of A, B, and C when $k_2/k_1 = 10$. Here we see a much different response. First, since k_2 is relatively large, any B formed is quickly reacted; therefore, the concentration of B never becomes very large (especially true for $k_1t > 2$). Hence, our mass balance yields $[A]_0 =$

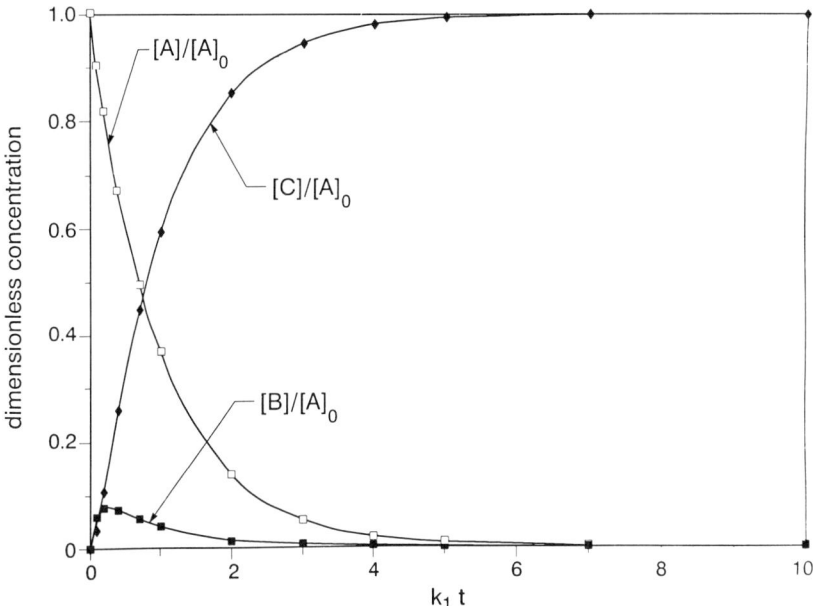

Figure 9.6. Reactant concentration for the simple series reaction where $k_2/k_1 = 10$. Note that the dimensionless time base is shorter than in Figure 9.5.

$[A] + [B] + [C] \approx [A] + [C]$, and $[C] \approx [A]_0 - [A]$, or

$$\frac{[C]}{[A]_0} \approx 1 - \frac{[A]}{[A]_0} \qquad (9.46)$$

This shows that $[A]/[A]_0$ and $[C]/[A]_0$ should be approximately mirror images of each other after a certain time has elapsed; this is demonstrated graphically in Figure 9.6.

Parallel reactions are also very common in chemistry. Consider the following reaction stoichiometry:

$$A + B \xrightarrow{k_1} P_1$$
$$\qquad (9.47)$$
$$C + B \xrightarrow{k_2} P_2$$

Assuming elementary kinetics, the first reaction can be considered a second-order reaction between A and B producing product P_1. By the same token, the second reaction may be considered a second-order reaction between B and C producing a second product, P_2. Note that there is a common reactant B in the two reactions; therefore, the equations are coupled by their competition for reactant B. We can write the differential equation for B as

$$\frac{d[B]}{dt} = -k_1[A][B] - k_2[C][B] \tag{9.48}$$

for A as

$$\frac{d[A]}{dt} = -k_1[A][B] \tag{9.49}$$

and for C as

$$\frac{d[C]}{dt} = -k_2[C][B] \tag{9.50}$$

Given appropriate initial conditions on the concentrations of A, B, and C, Eqs. 9.48–9.50 can be solved numerically. However, a simplified treatment of the equations is possible if we assume that reactants A and C are in great excess of reactant B. In that case, Eqs. 9.48–9.50 can be approximated with pseudo-first-order reactions (Exercise 9.6). When $k_2/k_1 = 5$ and $[A]_0 = [C]_0$, the

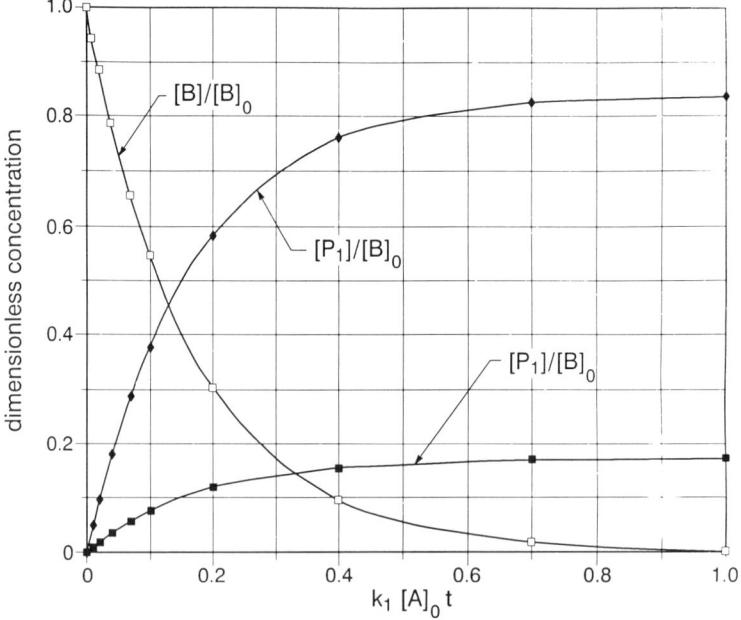

Figure 9.7. Simple parallel reaction with $k_2/k_1 = 5$ and $[A]_0 = [C]_0$. $[A]$ and $[C]$ are in great excess of $[B]$.

profiles of $[B]$, $[P_1]$, $[P_2]$ are shown in Figure 9.7. As expected, the concentration of the common reactant B must go to zero. The concentrations of products P_1 and P_2 approach asymptotic values. Exercise 9.6 demonstrates that the asymptotic ratio of concentrations of P_2 and P_1 is simply given by the values of the two parameters in this problem:

$$\frac{[P_2]}{[P_1]} = \frac{k_2[C]_0}{k_1[A]_0} \tag{9.51}$$

Since $[C]_0/[A]_0 = 1$ and $k_2/k_1 = 5$, the asymptotic concentration ratio in Figure 9.7 is 5.

9.8. REVERSIBLE REACTIONS

We have already encountered a reversible reaction in the nonelementary reaction studied in Section 9.6. We take a closer look at such reactions in this section. Consider first the simple reversible reaction

$$A \underset{k_2}{\overset{k_1}{\rightleftarrows}} B \tag{9.52}$$

As usual, we assume that the forward and backward reactions are elementary; hence, the forward reaction rate is assumed to be

$$r_f = k_1[A] \tag{9.53}$$

and the backward reaction rate is also assumed to be

$$r_b = k_2[B] \tag{9.54}$$

Before progressing more into the overall kinetics of reaction 9.52, let's ask what are the forward and backward reaction rates at equilibrium? At equilibrium, the concentrations of A and B are constant; hence, the forward and reverse reaction rates must balance each other overall. Therefore, at equilibrium, we assume that $r_f = r_b$ and we conclude that

$$\frac{[B]_{eq}}{[A]_{eq}} = \frac{k_1}{k_2} = K_{eq} \tag{9.55}$$

where K_{eq} is a new parameter called the *equilibrium constant*, which is very important in chemistry. If we have the equilibrium constant, we can use charge and mass balances to calculate the equilibrium distributions of products and reactants in a great number of reactions. There are many excellent books on

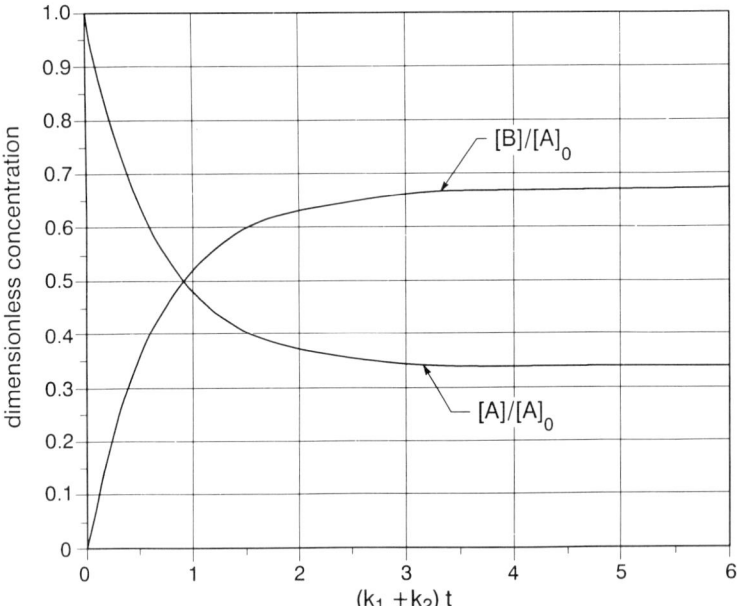

Figure 9.8. Reversible reaction $A \rightleftarrows B$: Plot of dimensionless concentration of A and B as function of dimensionless time for $k_1 = 1$ and $k_2 = 0.5$.

the subject of equilibrium calculations, and the reader is referred to Stumm and Morgan (1981) for an introduction in the environmental context. We now return to the kinetic question by writing the equation for the rate of change in the concentration of A. Continuing with the assumption of elementary reactions, we write

$$\frac{d[A]}{dt} = -k_1[A] + k_2[B] \tag{9.56}$$

We can set this equation up for direct integration if we can get rid of the dependence on the concentration of B. This can be done by noting from the stoichiometry of Eq. 9.52 that $[B] = [B]_0 + ([A]_0 - [A])$. Substituting into Eq. 9.56 we get

$$\frac{d[A]}{(k_1 + k_2)[A] - k_2([B]_0 + [A]_0)} = -dt \tag{9.57}$$

The only variable on the left-hand side is $[A]$ and the only variable on the right-hand side is t. We integrate using the initial condition that $[A] = [A]_0$

at $t = 0$ and find:

$$[A] = [A]_0 \exp(-k_* t) + \frac{k_2([B]_0 + [A]_0)}{k_*} [1 - \exp(-k_* t)] \qquad (9.58)$$

where $k_* = k_1 + k_2$. Now note that as $t \to \infty$, the exponential terms approach zero, and we conclude that the equilibrium concentration of A is given by

$$[A]_{eq} = \frac{k_2([B]_0 + [A]_0)}{k_*} \qquad (9.59)$$

Therefore, Eq. 9.58 can be reexpressed as

$$[A] = [A]_{eq} + [[A]_0 - [A]_{eq}] \exp(-k_* t) \qquad (9.60)$$

The reader is encouraged to show that the limits of this equation are correct—that $[A] = [A]_0$ at $t = 0$ and $[A] = [A]_{eq}$ as $t \to \infty$. An equation analogous to Eq. 9.58 can be derived for the concentration of B:

$$[B] = [B]_0 \exp(-k_* t) + \frac{k_1([B]_0 + [A]_0)}{k_*} [1 - \exp(-k_* t)] \qquad (9.61)$$

It is interesting to compare the response of the reversible reaction (Eq. 9.52) with that of the irreversible first-order reaction (Eq. 9.2) (recall Figure 9.1). For $[B]_0 = 0$, $k_1 = 1$, and $k_2 = 0.5$, Eqs. 9.58 and 9.61 are plotted in Figure 9.8. Note that in contrast to the irreversible reaction in Figure 9.1, the concentrations of both A and B in Figure 9.8 approach nonzero plateau or asymptotic values. The ratio of the plateau concentration values is defined in Eq. 9.55:

$$\frac{[B]_{eq}}{[A]_{eq}} = \frac{k_1}{k_2} = K_{eq} = \frac{1}{0.5} = 2 \qquad (9.55)$$

Indeed, examining the figure, we see that the final concentration of B is twice the final concentration of A. The actual final concentrations of A and B in terms of the initial concentration of A can easily be determined using the equations in this section (Exercise 9.8).

Reversible second-order reactions are also quite common. For example, consider the following reaction:

$$A + B \underset{k_2}{\overset{k_1}{\rightleftarrows}} P \qquad (9.62)$$

Assuming that the forward reaction is second order and the reverse reaction is first order (i.e., elementary kinetics), we can write the differential equation for the product P as

$$\frac{d[P]}{dt} = k_1[A][B] - k_2[P] \tag{9.63}$$

We could approach solution of this equation as we have throughout this chapter, but we will use a simpler approach instead. Consider that for some reason the equilibrium condition has been disturbed. We will define new ultimate equilibrium concentrations to be $[a]$, $[b]$, and $[p]$. The different concentrations can be related in the following way:

$$[A] = [a] + \Delta[A]$$
$$[B] = [b] + \Delta[B] \tag{9.64}$$
$$[P] = [p] + \Delta[P]$$

We assume here that the delta terms are small in relation to the final concentrations, that is, $|\Delta[A]| \ll [a]$, $|\Delta[B]| \ll [b]$, and $|\Delta[C]| \ll [c]$. There are several reasons why the hypothesized nonequilibrium condition might exist. First, the temperature or pressure of the system could be changed. Temperature changes will alter equilibria in liquid and gas systems (see Section 9.10); pressure changes will also change equilibria, mainly in gaseous systems. Second, if any of the concentrations in the system are adjusted or modified, new equilibrium concentrations of all reactants and products will result. Finally, in the dynamic environment, reactants and products may often find themselves (at least temporarily) in a nonequilibrium situation. For example, consider the reaction of SO_2 from the atmosphere in a newly formed raindrop (see Example 9.3).

In any event, we can analyze the nonequilibrium condition by substituting the expressions in Eq. 9.64 into Eq. 9.63. This yields

$$\frac{d}{dt}(p + \Delta[P]) = k_1([a] + \Delta[A])([b] + \Delta[B]) - k_2([p] + \Delta[P]) \tag{9.65}$$

On the left-hand side, the derivative of the constant $[p]$ is zero. Expanding the middle terms in parentheses, we find:

$$\frac{d\Delta[P]}{dt} = k_1([a][b] + [a]\Delta[B] + \Delta[A][b] + \Delta[A]\Delta[B]) - k_2([p] + \Delta[P]) \tag{9.66}$$

Note that the equilibrium terms $k_1[a][b]$ and $k_2[p]$ must cancel each other because of Eq. 9.63. Also, note from the stoichiometry of Eq. 9.62 that

$\Delta[P] = -\Delta[A] = -\Delta[B]$. Substituting into Eq. 9.66 we find:

$$\frac{d\Delta[P]}{dt} = k_1([a]\Delta[P] + [b]\Delta[P] + \Delta[P]^2) - k_2\Delta[P] \qquad (9.67)$$

However, given the conditions stated above on the magnitudes of the delta terms, we can ignore the squared delta term in relation to the other terms; hence, the $\Delta[P]^2$ term is dropped, and rearranging we find:

$$\frac{d\Delta[P]}{dt} = -\frac{\Delta[P]}{\tau} \qquad (9.68)$$

where τ is a constant called the *relaxation time*:

$$\tau = \frac{1}{k_1([a] + [b]) + k_2} \qquad (9.69)$$

Equation 9.68 is an easily separable first-order differential equation, with which we have considerable experience in this book. Consider here that the initial condition is that $\Delta p = \Delta p_{max}$ at time $= 0$ (i.e., the maximum divergence from equilibrium is at the initial time). Integrating we find:

$$\Delta[P] = \Delta[P]_{max} \exp\left(-\frac{t}{\tau}\right) \qquad (9.70)$$

Consider why τ is called a relaxation time: As time progresses, the reaction "relaxes" from its initial disequilibrium state to its final equilibrium value. The kinetics of second-order reactions, especially fast second-order reactions, are often studied with temperature-, pressure-, or electrical field-jump relaxation techniques. In this method, the temperature, pressure, or electric field of the system is changed very quickly. As explained above, this temporarily throws the system out of equilibrium. The concentrations of products (or reactants) are then monitored as the reaction relaxes toward a new equilibrium state. For example, the $\Delta[P]$ data could then be fit to an equation like Eq. 9.70, following linearization such as that performed in Section 9.2. The slope of the fitted line will yield a single value for τ.

Even if we know a and b in Eq. 9.69, we still do not have enough equations to determine the fundamental kinetic parameters k_1 and k_2. Therefore, we might adopt the following approach. Note that if we invert Eq. 9.69, we obtain:

$$\frac{1}{\tau} = ([a] + [b])k_1 + k_2 \qquad (9.71)$$

This is the equation of a straight line, if we consider $1/\tau$ to be y, $([a] + [b])$ to be x, k_1 to be the slope, and k_2 to be the y intercept. We simply repeat the relaxation experiment with new initial concentrations of A and B. The new τ values determined from this experiment and a number of similar experiments can then be used with the appropriate $a + b$ values to fit a straight line to Eq. 9.71.

Example 9.2. Holmes et al. (1968) used electric-field jump methods in a study of the first step in aluminum hydrolysis at 25°C,

$$Al^{3+} + H_2O \underset{k_2}{\overset{k_1}{\rightleftharpoons}} H^+ + AlOH^{2+}$$

The following data for relaxation time were recorded and plotted. Determine k_1, k_2, and the equilibrium constant.

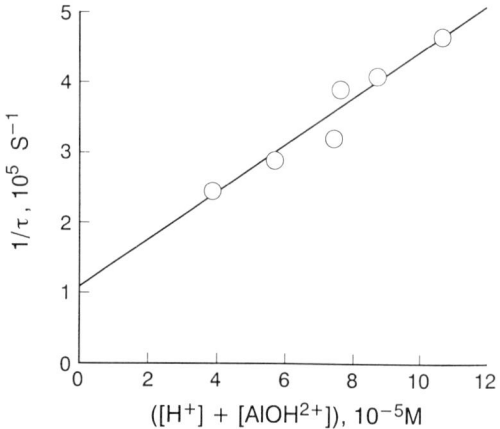

Example 9.2. *Source*: Reprinted with permission from Holmes, Cole, and Eyring, *J. Phys. Chem.*, **72**(1), 302 (1968). Copyright 1968, American Chemical Society.

SOLUTION. First note that water is the solvent in this problem; in dilute aluminum solutions, its concentration is not impacted by the hydrolysis reaction. Hence, the water concentration is constant and the general form of the hydrolysis reaction can be written as

$$A \underset{k_2}{\overset{k_1}{\rightleftharpoons}} B + C$$

But written this way, there is really no difference between it and the general reaction analyzed in this section, that is, Eq. 9.62:

$$A + B \underset{k_2}{\overset{k_1}{\rightleftharpoons}} P \tag{9.62}$$

Therefore, we can borrow the relaxation time estimate shown above (Eq. 9.69) with the provision that we substitute k_1 for k_2 and k_2 for k_1. Then substituting $[H^+]_{eq}$ for $[a]$ and $[AlOH^{2+}]_{eq}$ for b we find:

$$\tau = \frac{1}{k_2([H^+]_{eq} + [AlOH^{2+}]_{eq}) + k_1}$$

Inversion of this equation yields the plot shown above.

From the experimental plot we can measure a slope of 3.4×10^9 M^{-1}-s^{-1} and a y intercept of 1.09×10^5 s^{-1}. Therefore, our analysis tells us that $k_2 = 3.4 \times 10^9$ M^{-1}-s^{-1} and $k_1 = 1.09 \times 10^5$ s^{-1}. Finally, the equilibrium constant at 25°C is given by

$$K_{eq} = \frac{k_1}{k_2} = \frac{1.09 \times 10^5 \ s^{-1}}{3.4 \times 10^9 \ M^{-1} s^{-1}} = 3.2 \times 10^{-5} \ M$$

9.9. CHARACTERISTIC REACTION TIMES

Throughout this book, we have computed characteristic time scales for a large number of physical processes like sedimentation, coagulation, diffusion, and convective diffusion. When a combination of physical and chemical processes controls the occurrence and transport of a given material, an exact mathematical analysis sometimes applies. A number of these cases were presented in Chapters 6 and 7. However, as a prelude to exact analysis or to experimental design, we may simply be satisfied with estimates of characteristic time scales of different physical and chemical processes. We saw in Chapters 6 and 7 that most of the dimensionless parameters describing physical–chemical processes can be expressed as the ratio of time scales. Therefore, it is useful in this chapter to compute characteristic times of different reaction schemes. A characteristic time characterizes the kinetics or speed of a reaction. A fast or slow reaction may impact the overall transport process in different ways.

Most of the reaction time scales in this chapter are fairly obvious from the integrated rate expressions. For the first-order, irreversible reaction, the integrated rate expression is given by Eq. 9.6:

$$\frac{[A]}{[A]_0} = \exp(-kt) \tag{9.6}$$

Here the characteristic time is normally considered the inverse of the reaction rate constant k:

$$\tau = \frac{1}{k} \tag{9.72}$$

Note for this selection of the characteristic time that when $t = \tau$, $\exp(-k\tau) = \exp(-1) = 0.368$ or $[A]/[A]_0 = 0.368$. As we have stated at other junctures in this book, there is nothing special about this selection of the characteristic time (there are others); it is simply mathematically convenient. A similar characteristic time is found for the irreversible, pseudo-first-order reaction discussed in Section 9.4:

$$\tau = \frac{1}{[B]_0 k} \tag{9.73}$$

Time scales for irreversible second-order reactions are handled a little differently. When $[A]_0 = [B]_0$, the integrated rate equation in Section 9.3 was

$$\frac{[A]}{[A]_0} = \frac{1}{1 + [A]_0 kt} \tag{9.20}$$

It is more convenient here to ask at what time A has decreased to one half of its initial concentration. This is called the *half-time*. This corresponds to $[A]/[A]_0 = 0.5$. Substituting into Eq. 9.20, we find that

$$\tau = \frac{1}{[A]_0 k} \tag{9.74}$$

We also approach estimation of characteristic times for reversible reactions by examining the integrated rate equations. For the first-order reversible reaction of Section 9.8, the integrated rate equation is given by Eq. 9.60:

$$[A] = [A]_{eq} + [[A]_0 - [A]_{eq}] \exp(-k_* t) \tag{9.60}$$

As in many other equations in this chapter, the equilibrium concentration of A is approached exponentially; hence, we estimate the characteristic time as the time corresponding to $\exp(-k_* t) = \exp(-1)$, or

$$\tau = \frac{1}{k_*} = \frac{1}{(k_1 + k_2)} \tag{9.75}$$

Finally, for the second-order reversible reaction Eq. 9.62, we consider the characteristic time to be the relaxation time:

$$\tau = \frac{1}{k_1([a] + [b]) + k_2} \tag{9.69}$$

Example 9.3. Schwartz and Freiberg (1981) studied mass transfer and oxidation of SO_2 in aqueous aerosol drops. Aqueous phase SO_2 [called $SO_{2(aq)}$] hydrolyzes to HSO_3^-. The extent of hydrolysis is dependent on pH, as indicated by the following equilibrium reaction:

$$SO_{2(aq)} + H_2O \underset{k_2}{\overset{k_1}{\rightleftharpoons}} H^+ + HSO_3^-$$

The forward and reverse rate constants are established as $k_1 = 3.4 \times 10^6 \, s^{-1}$ and $k_2 = 2.0 \times 10^8 \, L\text{-}mol^{-1}\text{-}s^{-1}$. What is the characteristic time ("relaxation time") of the hydrolysis of SO_2?

SOLUTION. In aqueous chemistry, the solvent water is present in such high concentration that its concentration is assumed to be constant. Therefore if we assume elementary reactions, the forward reaction rate can be expressed as

$$r_f = k_1[SO_{2(aq)}]$$

and the reverse reaction is

$$r_r = k_2[H^+][HSO_3^-]$$

Recall reaction 9.62:

$$A + B \underset{k_2}{\overset{k_1}{\rightleftharpoons}} P \qquad (9.62)$$

Our SO_2 reaction is essentially the same as this, if we write the equation backward (i.e., $A + B \rightleftharpoons P$ is no different than $P \rightleftharpoons A + B$). The relaxation time for Eq. 9.62 was given as Eq. 9.69:

$$\tau = \frac{1}{k_1([a] + [b]) + k_2} \qquad (9.69)$$

Hence in Eq. 9.69, $k_1 = 2.0 \times 10^8 \, L\text{-}mol^{-1}\text{-}s^{-1}$ (i.e., the old k_2) and $k_2 = 3.4 \times 10^6 \, s^{-1}$ (i.e., the old k_1). Here $[a]$ is associated with the equilibrium H^+ concentration and $[b]$ is associated with the equilibrium HSO_3^- concentration; hence, our estimate of the relaxation time of SO_2 hydrolysis is

$$\tau_{SO_2} = \frac{1}{2.0 \times 10^8([a] + [b]) + 3.4 \times 10^6}$$

9.10. ARRHENIUS' LAW AND THE EFFECT OF TEMPERATURE ON REACTION RATE

The rates of most chemical reactions increase with temperature. We have also stated in Section 9.8 that equilibrium constants change with temperature. An old rule of thumb is that the reaction rate doubles for each increase in temperature of 10°C. The understanding of a chemical reaction is not considered complete until the sensitivity of the reaction to temperature is understood. Therefore, many chemical studies include experiments at different temperatures. The effect of temperature on reaction rate can often be modeled with *Arrhenius law*:

$$k = A_f \exp\left(-\frac{E_a}{RT}\right) \tag{9.76}$$

Here R is the universal gas constant, T is the absolute temperature, A_f is called the preexponential factor or the *Arrhenius frequency factor*, and E_a is called the *activation energy*. The activation energy is thought of as the energy barrier that must be surmounted for the reaction to proceed. The activation energy can be considered with the plot shown in Figure 9.9.

The different reacting molecules will have a distribution of energies at any given instant. Only that fraction of molecules that can surmount the energy

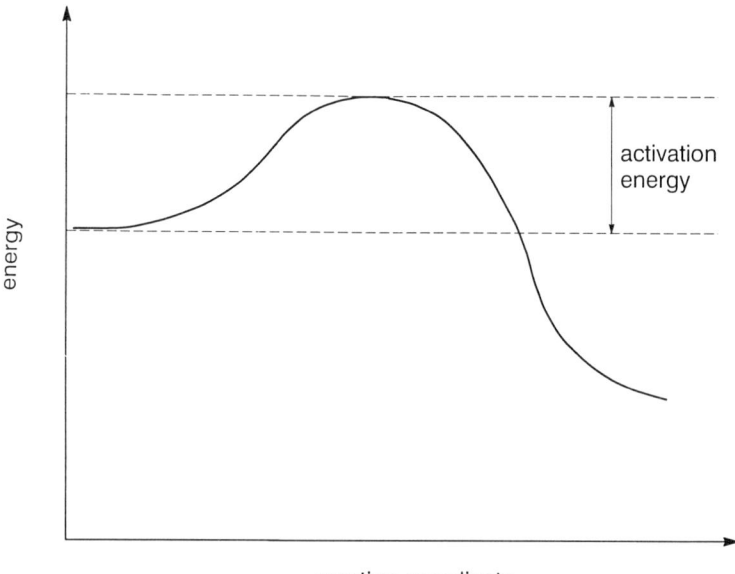

Figure 9.9. Illustration of activation energy in a chemical reaction.

barrier will react. Therefore, the rate of a reaction is then related to E_a; an increase in E_a indicates a slowing down of the reaction (and vice versa). The activation energy of a particular elementary reaction can be determined in the following way. Take the natural log of both sides of Eq. 9.76:

$$\ln k = \ln A_f - \frac{E_a}{R}\left(\frac{1}{T}\right) \tag{9.77}$$

This is the equation of a straight line if $\ln k$ is taken as the y coordinate, $1/T$ is taken as x, $\ln A_f$ is taken as the y intercept, and E_a/R is taken as the slope. Therefore, the activation energy might be determined by measuring reaction rate at a series of different temperatures (Exercise 9.10).

9.11. SUMMARY

Chapter 9 investigated zero-, first, second-, and pseudo-first-order reactions and showed how to integrate the rate laws to predict the course of reactions. Series and parallel reactions were also discussed. In the case of an enzyme-catalyzed reaction, we showed that nonelementary kineics would result from a series of elementary reversible and irreversible steps. The integrated rate laws were used to fit experimental data to the kinetic models. Finally, we showed that most of the models had characteristic times that can be used as another method of characterizing reactions and fitting experimental data.

EXERCISES

9.1. When ammonia and chlorine are contacted in water, nitrogen trichloride can be formed:

$$NH_3 + 3\,HOCl = NCl_3 + 3\,H_2O$$

However, NCl_3 also decomposes in water. The following reaction pathway for NCl_3 decomposition has been suggested:

$$NCl_3 + OH^- \rightarrow NHCl_2 + OCl^-$$

Assuming elementary kinetics, we can hypothesize the following second-order rate law for NCl_3 decomposition:

$$\frac{d[NCl_3]}{dt} = -k[NCl_3][OH^-]$$

However, if $[OH^-]$ is held constant by addition of a pH buffer, then $[OH^-] = $ constant and the kinetics can be considered pseudo-first order in $[NCl_3]$:

$$\frac{d[NCl_3]}{dt} = -k'[NCl_3]$$

where $k' = k[OH^-]$. Saguinsin and Morris (1975) studied the decomposition of NCl_3 at different constant pH values and at 20°C. The kinetic data are shown on the following figure. The solutions containing NCl_3 were monitored with UV spectroscopy at a wavelength of 220 nm. The absorbance (abs) values at 220 nm are proportional to the NCl_3 concentration, so $[NCl_3] = $ const. × abs (where const. is a proportionality constant).

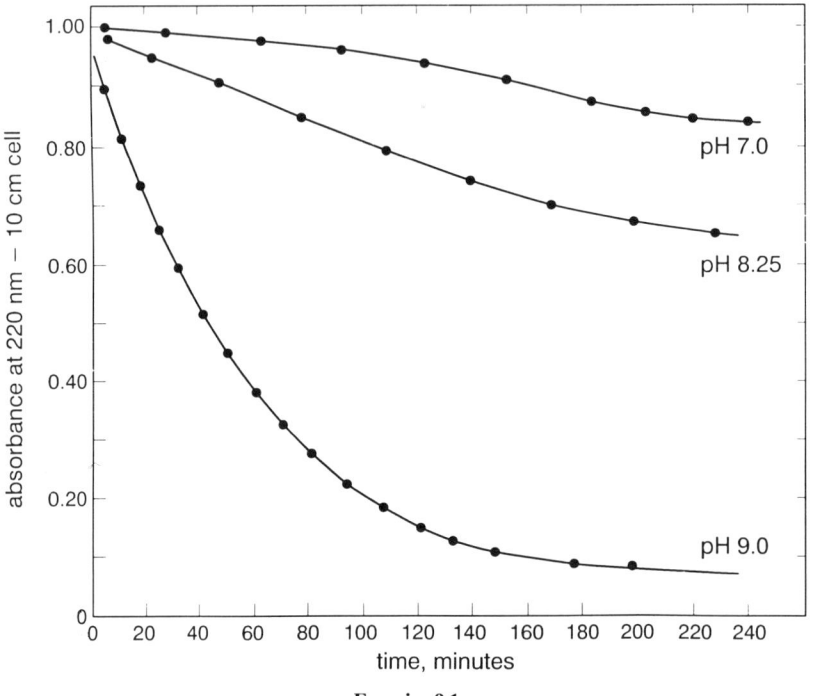

Exercise 9.1

(a) Considering the data for pH $= 9$, use the approach of Sections 9.2 and 9.4 to fit the pseudo-first-order rate constant k' and the apparent second-order rate constant k.

(b) Explain why you did not need to know the constant in $[NCl_3] = $ const. × abs to solve part **a**.

9.2. **(a)** In deriving Eq. 9.15, first show that the following transformation of Eq. 9.14 can be made:

$$\frac{1}{[B]_0 - [A]_0} \left(\frac{1}{[A]_0 - x} - \frac{1}{[B]_0 - x} \right) dx = k dt$$

Next show that Eq. 9.15 follows from direct integration with the boundary condition $x = 0$ at $t = 0$.

(b) To derive Eq. 9.16, first take the exponential of each side of Eq. 9.15 and solve for x. Equation 9.16 then results by defining $[A] = [A]_0 - x$.

9.3. The derivation of Eq. 9.18 is all algebra; however, one step that might not be obvious is the following. Considering the stoichiometry of Eq. 9.11, it must be true that $[B] = [B]_0 - ([A]_0 - [A])$. Therefore, a possible substitution is $[A]_0 - [B]_0 = [A] - [B]$. Derive Eq. 9.18.

9.4. The following scheme has been found to apply to certain chemical reactions:

$$A + B \underset{k_2}{\overset{k_1}{\rightleftarrows}} (AB)^* \overset{k_3}{\rightarrow} P$$

The compound $(AB)^*$ is an intermediate compound sometimes called an activated complex. It may or may not actually be measurable in the solution. Nevertheless, the hypothesis of this intermediate has enabled the modeling of some experimental kinetics data, and given indirect evidence for the existence of the intermediate. Show that the appropriate kinetic law can be given by

$$\frac{d[P]}{dt} = \frac{k_3 k_1 [A][B]}{k_2 + k_3}$$

Hint: As in the case of the enzyme-catalyzed reaction of Section 9.6, you need to assume that the formation of $(AB)^*$ reaches steady state after a short period of time, that is,

$$\frac{d[(AB)^*]}{dt} = 0$$

9.5. Derivation of Eq. 9.41 requires straightforward use of the integrating-factor method.

9.6. Assume that Eq. 9.48 is a pseudo-first order reaction (i.e., $[A]_0 = [C]_0 \gg [B]_0$).
(a) Show that the differential equation for $[B]$ can be written as

$$\frac{d[B]}{dt} = -(k_1 [A]_0 + k_2 [C]_0)[B]$$

Now, for the initial condition $[B] = [B]_0$ at $t = 0$, show that this integrates to

$$\frac{[B]}{[B]_0} = \exp\left[-\left(1 + \frac{k_2}{k_1}\frac{[C]_0}{[A]_0}\right)k_1[A]_0 t\right]$$

(b) For P_1, show that

$$\frac{d[P]_1}{dt} = (k_1[A]_0)[B]$$

Next substitute for $[B]$, integrate for $[P_1]_0 = 0$ at $t = 0$, rearrange, and show that

$$\frac{[P]_1}{[B]_0} = \frac{1}{1 + (k_2/k_1)([C]_0/[A]_0)}\left\{1 - \exp\left[\left(1 + \frac{k_2}{k_1}\frac{[C]_0}{[A]_0}\right)k_1[A]_0 t\right]\right\}$$

Follow a similar procedure to determine the expression for $[P_2]/[B]_0$.

(c) What is the asymptotic ratio of $[P_2]$ and $[P_1]$?

9.7. Water-treatment processes that generate hydroxyl radicals are often used to remove organic chemicals that contaminate groundwater. The hydroxy radical ($\cdot OH$) is a very reactive free-radical species that can be thought of as a piece of a molecule. In an attempt to again become a whole molecule, $\cdot OH$ will attack almost any organic molecule, either adding to a double bond or pulling off a hydrogen atom to become water. Because of this property, the hydroxyl radical can be used to degrade organic contaminants completely to carbon dioxide and water, providing a remediation tool that leaves no harmful by-products.

In some applications, a small amount of toxic material like trichloroethylene (TCE) must be destroyed in the presence of a much larger quantity of "nontarget" material such as the natural organic material that results from the decay of plants, or, in the case of industrial wastewater, some organic material that is either environmentally benign or can subsequently be treated more cheaply by biological methods. This material, usually characterized instrumentally by the measurement of dissolved organic carbon (DOC), may compete for hydroxy radical with the contaminant to the extent that the economic feasibility of the process may depend on the outcome of this competition.

Consider the removal of TCE. The rate equations for TCE and the hydroxyl radical are

$$\frac{d[\text{TCE}]}{dt} = -k_{\cdot\text{OT}}[\cdot \text{OH}][\text{TCE}]$$

and

$$\frac{d[\cdot OH]}{dt} = -k_{\cdot OT}[\cdot OH][TCE] - k_{OD}[\cdot OH][DOC] + S_{\cdot OH}$$

where it is assumed that the treatment produces the $\cdot OH$ radical at a constant rate $S_{\cdot OH}$. In this problem, DOC is assumed to be present in sufficient excess that its concentration does not change significantly during treatment. Thus, the experimentally determined DOC may not change. If the rate constant of the resulting material also does not change much, from the point of view of the effect of the material on the system kinetics, there appears to be no change in the material. The hydroxyl radical reacts essentially as fast as it is produced, so that $d[\cdot OH]/dt = 0$.

(a) Show that the steady-state concentration of $\cdot OH$ is given by

$$[\cdot OH) = \frac{S_{\cdot OH}}{k_{\cdot OT}[TCE] + k_{OD}[DOC]}$$

Substitution of this expression for $[\cdot OH]$ into the TCE rate equation yields

$$\frac{d[TCE]}{dt} = -\left(\frac{k_{\cdot OT}[TCE]}{k_{\cdot OT}[TCE] + k_{OD}[DOC]}\right) \times S_{\cdot OH}$$

(b) To describe the disappearance of TCE, the equation above must be integrated. Show that rearrangement gives

$$\left(1 + \frac{k_{OD}[DOC]}{k_{\cdot OT}[TCE]}\right) d[TCE] = -S_{\cdot OH} dt$$

which integrates to

$$[TCE] = [TCE]_0 - \frac{k_{OD}[DOC]}{k_{\cdot OT}} \ln \frac{[TCE]}{[TCE]_0} - S_{\cdot OH} t$$

where $[TCE]_0$ is the initial concentration of TCE.

(c) For $k_{OD} = 2 \times 10^4$ mg^{-1}-L-s^{-1}, $k_{\cdot OT} = 4.2 \times 10^9$ mol^{-1}-L-s^{-1}, $[DOC] = 2$ mg/L, $[TCE]_0 = 0.05$ mM, and $S_{\cdot OH} = 0.16 \times 10^{-3}$ mol/min. Solve for $[TCE]$ vs. time. Plot your results.

(d) Can you make any statement regarding the apparent reaction order at very small and very large times?

9.8. Determine simple formulas for the final concentrations of A and B in the reversible reaction presented in Figure 9.8. Express answers in the dimensionless terms $[A]_{eq}/[A]_0$ and $[B]_{eq}/[A]_0$.

9.9. Example 9.3 concerned the hydrolysis of $SO_{2(aq)}$ in liquid-aerosol droplets. The second ionization of sulfurous acid is given by the equilibrium relation

$$HSO_3^- \underset{k_2}{\overset{k_1}{\rightleftharpoons}} H^+ + SO_3^{2-}$$

Determine the relaxation time of this reaction if $k_1 \approx 1.9 \times 10^3 \, s^{-1}$ and $k_2 \approx 3 \times 10^{10} \, L\text{-mol}^{-1}\text{-s}^{-1}$.

9.10. The reaction of FO_2 and NO in the atmosphere is of interest in studies of the destruction of atmospheric ozone. The reaction is thought to be irreversible and second-order, where

$$FO_2 + NO \xrightarrow{k_{11}} \text{products}$$

or,

$$\frac{d[FO_2]}{dt} = -k_{11}[FO_2][NO]$$

Li et al. (1995) studied this reaction in a fast discharge mixing device with spectrophotometer detector. They studied the decomposition of FO_2 when NO was held in great excess, and where the above reaction could be assumed to be pseudo-first-order, i.e.,

$$\frac{d[FO_2]}{dt} = -k'[FO_2]$$

where

$$k' = k_{11}[NO]$$

The following data are a plot of k' versus the constant NO concentration at 5 different temperatures. (Note that the abscissa has units of 10^{13} molecules-cm^{-3}, not our normal molar concentration units.)

(a) From these data and the equations above, determine the values of k_{11} for the five different temperatures.

(b) Determine the constants in the Arrhenius equation using the results from part **a**.

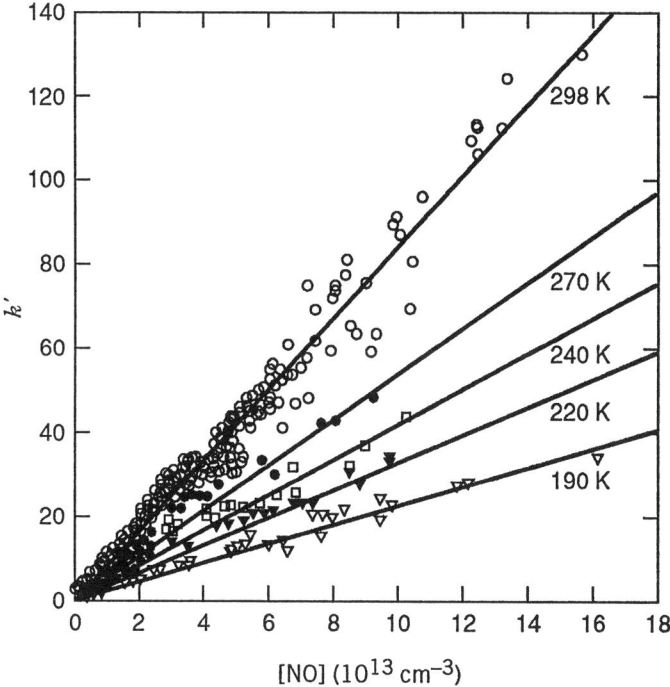

Exercise 9.10. *Source*: Reprinted with permission from Li, Friedl, and Sander, *J. Phys. Chem.*, **99**(36), 13448 (1955). Copyright 1995, American Chemical Society.

REFERENCES

Holmes, L. P., Cole, D. I., and Eyring, E. M., "Kinetics of Aluminum Hydrolysis in Dilute Solutions," *J. Phys. Chem.*, **72**, 1:301–304 (1968).

Li, Z., Friedl, R. R., Sander, S. P., "Kinetics of FO_2 with NO, NO_2, O_3, CH_4, and C_2H_6," *J. Phys. Chem.*, **99**, 36:13445–13451 (1995).

Saguinsin, J. L. S. and Morris, J. C., "The Chemistry of Aqueous Nitrogen Trichloride," *Disinfect. Water Wastewater*, Ann Arbor Science Publishers, Ann Arbor, MI, 1975.

Schwartz, S. E. and Freiberg, J. E., "Mass Transport Limitation to the Rate of Reaction of Gases in Liquid Droplets: Application to Oxidation of SO_2 in Aqueous Solutions," *Atmos. Env.*, **15**, 7:1129–1144 (1981).

Stumm, W. and Morgan, J. J., *Aquatic Chemistry: An Introduction Emphasizing Chemical Equilibria in Natural Waters*, 2nd ed., Wiley-Interscience, New York, 1981.

BIBLIOGRAPHY

Atkins, P. W., *Physical Chemistry*, W. H. Freeman, San Francisco, CA, 1978.

Brezonik, P. L., *Chemical Kinetics and Process Dynamics in Aquatic Systems*, Lewis Publishers, Boca Raton, FL, 1994.

Clark, M. M., Srivastava, R., and David, R., "Mixing and Aluminum Precipitation," *Env. Sci. Technol.*, **27**, 10:2181–2189 (1993).

Freifelder, D., *Physical Chemistry*, Jones and Bartlett, Boston, MA, 1985.

Levenspiel, O., *Chemical Reaction Engineering*, 2nd ed., Wiley, New York, 1972.

Pankow, J. F. and Morgan, J. J., "Kinetics for the Aquatic Environment," *Env. Sci. Technol.*, **15**, 10:1155–1164 (1981).

Schneider, W. and Schwyn, B., In *Aquatic Surface Chemistry*, W. Stumm, Ed., Wiley, New York, 1990.

Snoeyink, V. L. and Jenkins, D., *Water Chemistry*, Wiley, New York, 1980.

Stone, A. T. and Morgan, J. J., "Kinetics of Chemical Transformations in the Environment," in *Aquatic Chemical Kinetics*, W. Stumm, Ed., Wiley-Interscience, New York, 1990.

Weber, W. W., *Physicochemical Processes for Water Quality Control*, Wiley-Interscience, New York, 1972.

10

MIXING AND REACTOR MODELING

10.1. INTRODUCTION

In previous chapters, we found that although some processes and reactions occur in static environments, many others occur in systems in which there is some type of flow. For example, sedimentation and coagulation often occur in flowing systems (Chapters 2 and 3). The extent or degree of sedimentation or coagulation is different in two types of ideal-flow sysems — the ideal plug flow and perfectly mixed reactors. In Chapter 4, we found that partitioning in flowing systems led to interesting separation properties. The analysis was shown to have relevance in laboratory separation processes and groundwater transport of pollutants. Chapter 7 derived detailed equations governing the transport and reaction of materials in both laminar and turbulent-flow systems. There the effects of velocity were parameterized with the Reynolds and Peclet numbers. In Chapter 8, we found that the flow entered our calculation of the transport and reaction of environmental contaminants in porous media.

This chapter develops new procedures for understanding reactions in flowing systems, especially those systems that cannot be described by simple flow models. We begin with an analysis of residence times. The idea here is that a reacting species will be in the reacting environment for only a finite amount of time. Each flowing system is usually comprised of a continuous distribution of residence times and, as a result, we can make good use of concepts from probability and statistics in analyzing the distribution of residence time. From residence times, we move on to develop equations for predicting the extent of chemical reactions in real and ideal-flow systems with different distributions of residence time. The tools developed in this chapter will be used in analyzing a surprising number of real environmental systems and pollution-control devices.

10.2. SIMPLE CLOSED-REACTOR AND RESIDENCE-TIME DISTRIBUTIONS

We first consider the flowing system shown in Figure 10.1a. The system is called a *simple system. A simple system or reactor has a single inlet and a single outlet.* We further specify that it is a *closed system. In a closed system, a given*

(a)

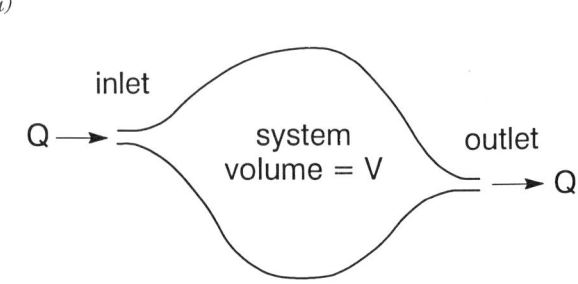

(b)

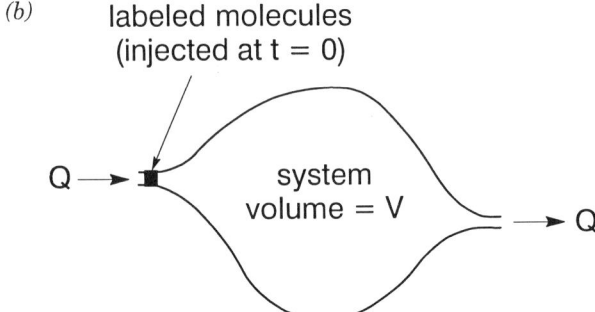

(c)

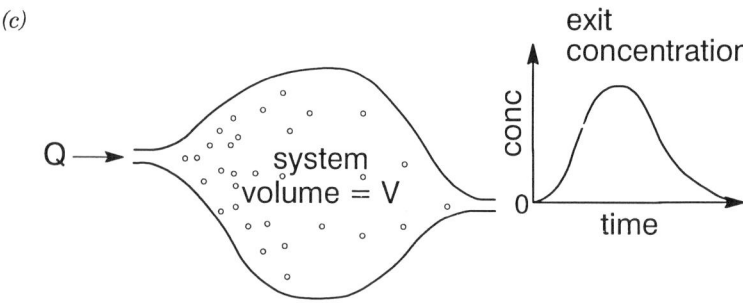

Figure 10.1. (*a*) Definition of a simple, closed-flow system; (*b*) injection of tagged molecules at entrance of system; (*c*) measurement of exit concentration of tagged molecules.

element of fluid can only enter the system once and leave the system once. This assumption may seem a little strange, but in a closed system, we need not consider any molecules in the system which, for some reason, decide to go out the inlet or, once passed through the system, decide to go back in the outlet (or inlet). We do not make any particular assumption about the turbulence and mixing within the system, only that the mixing could be of any type. We do assume the flow is steady, that is, Q = constant.*

Now consider a group of entering molecules that are a different color, have a different light absorbency or conductivity, or are tagged with a radioactive marker. These molecules are assumed to be identical to all other molecules in the system, except that they can be differentiated by the tagging property (e.g., color, light absorbance, radioactivity). These are called the *tracer* molecules. In Figure 10.1b, the tracer molecules are shown by the shaded region in the entrance. Finally, we have at our disposal a quickly responding sensor located at the exit of the system. As shown in Figure 10.1c, as the tracer molecules flow out of the system, the sensor measures the concentration of the tracer molecules leaving the system. The x axis is time, so the $x-y$ plot shown is a measure of the exit concentration versus time.

Now consider Figure 10.2a. Here the concentration versus time plot is divided into a number of segments over equal time intervals $\Delta t = t_{i+1} - t_i$. We define the incremental amount of mass leaving the system over any time interval as

$$\Delta m_{i,i+1} = Q \int_{t_i}^{t_{i+1}} c(t)dt \qquad (10.1)$$

The total mass of tracer injected is simply equal to the sum of all the $\Delta m_{i,i+1}$:

$$M_T = \sum_i \Delta m_{i,i+1} \qquad (10.2)$$

In Figure 10.2b we have plotted the distribution of the $\Delta m_{i,i+1}$ normalized with $M_T \Delta t$. It should be clear from our definitions that the sum of all the $\Delta m_{i,i+1}/M_T \Delta t$ times Δt must equal unity, that is,

$$\sum_i \frac{\Delta m_{i,i+1}}{M_T \Delta t} \Delta t = \frac{1}{M_T} \sum_i \Delta m_{i,i+1} = 1 \qquad (10.3)$$

Therefore, the histogram in Figure 10.b shows the fraction of the total tracer mass exiting the system over time. We can also interpret this as the distribution of residence time in the system. Now imagine what Figure 10.2b looks like as we divide the time range into smaller time increments Δt: The boxes become narrower and the tops of the boxes begin to define the coordinates of a smooth curve. In the limit of $\Delta t \to 0$, the histogram does define a smooth curve, which is shown in Figure 10.2c. This new curve is the *residence-time density function*, $f(t)$.

---

*For the analyses to follow, we also generally assume the flow has been constant for a long time, such that the fluid-flow profiles within the system are well established. The length of time required must be long relative to the mean residence time in the system.

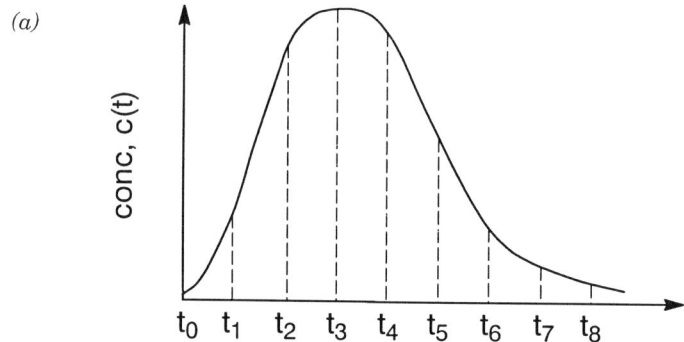

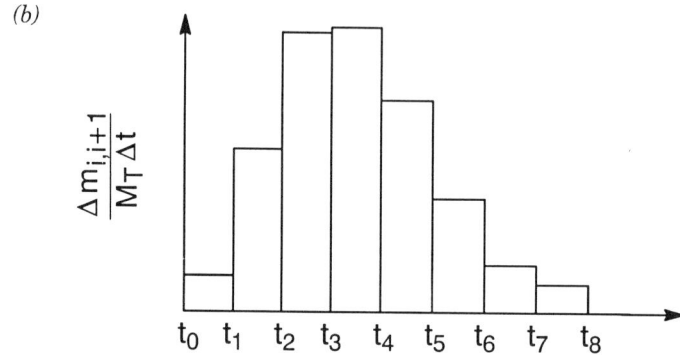

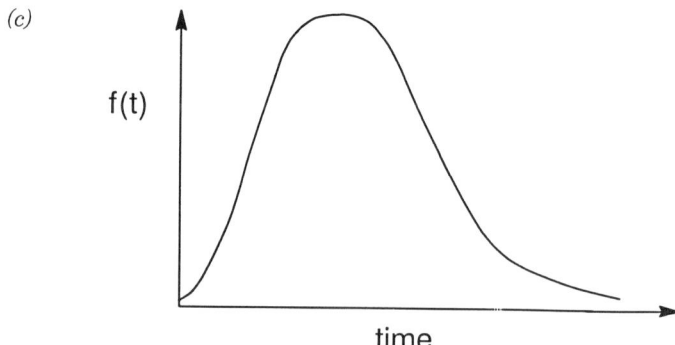

Figure 10.2. (*a*) Concentration vs. time plot at reactor exit; (*b*) fractional tracer mass leaving the system vs. time; (*c*) the residence-time density function, $f(t)$.

The residence-time density function is exactly the same as the *probability density function* of statistics. There are certain exact requirements of probability density functions and, therefore, of residence-time density functions. First, as suggested in the figure, all function values must be greater than or equal to zero, that is, $f(t) \geq 0$. Second, the area under the curve must exactly equal unity:

$$\int_0^\infty f(t)dt = 1 \tag{10.4}$$

Note from Eq. 10.4 that the units of the residence-time density function must be T^{-1} for the integral to be unitless. The next property is related to the calculation of the probability that any given tagged molecule passes out of the system between times t_1 and t_2. From probability theory, this is simply

$$\text{prob}[t_1 \leq t \leq t_2] = \int_{t_1}^{t_2} f(t)dt \tag{10.5}$$

This calculation is shown schematically in Figure 10.3a. It appears that by measuring the residence time of tagged or tracer molecules in a system, we can calculate a continuous function $f(t)$ which describes the residence time of all the tracer molecules flowing through a system at steady state. Further, if there are enough molecules in the group of tagged molecules injected in the system at $t = 0$, we can extend our hypothesis and say that the measured $f(t)$ is the residence-time distribution function for *any molecules* entering the system, or for a subset of molecules that may be subject to reaction within the system. For reactive molecules, $f(t)$ tells us how long the molecules are in the reacting system. Using this information, we may be able to predict the overall extent of reaction within the system. This has been a very powerful idea in chemical and environmental engineering.

The reader will also recall from statistics that the *mean residence time* is easily computed as

$$\bar{t}_{\text{RTD}} = \int_0^\infty f(t)t \, dt \tag{10.6}$$

In residence-time studies, it is often important to compare $\bar{t}_{\text{RTD}}$ to the *theoretical mean residence time*:

$$\bar{t} = \frac{V}{Q} \tag{10.7}$$

where V is the system volume and Q is the steady flow rate through the system (Figure 10.1).

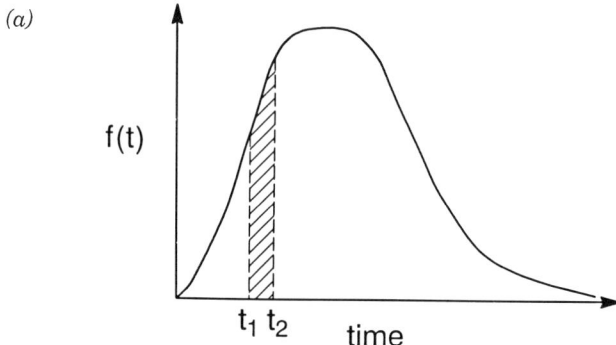

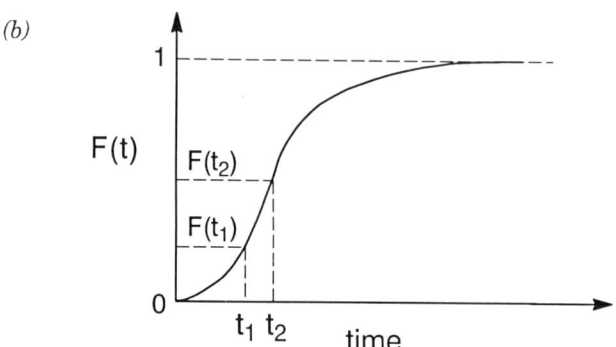

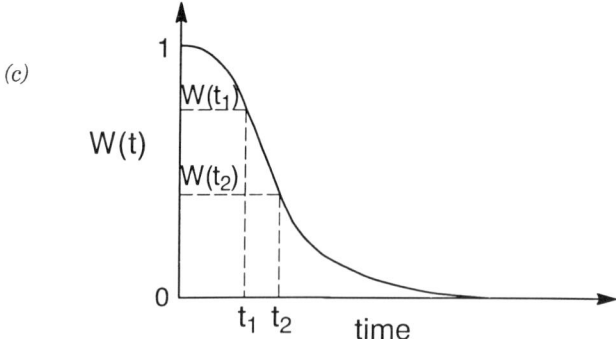

Figure 10.3. (*a*) Residence-time density function showing the fraction of tagged molecules with residence time between t_1 and t_2; (*b*) cumulative residence-time distribution; (*c*) washout function.

Another common way to display and analyze residence times is with the *cumulative residence-time distribution, F(t)*. Just like the cumulative probability distribution of statistics, we define the cumulative residence-time distribution as

$$F(t) = \int_0^t f(t)dt \tag{10.8}$$

The cumulative residence-time distribution is shown in Figure 10.3b. Note from Eqs. 10.4 and 10.8 that $F(t)$ must start from 0 and approach 1 as $t \to \infty$. If presented with the cumulative residence-time distribution $F(t)$, how does one calculate the fraction of molecules with residence time between t_1 and t_2? Note the following from our definition of $F(t)$ and Figure 10.3b:

$$F(t_2) - F(t_1) = \int_0^{t_2} f(t)dt - \int_0^{t_1} f(t)dt = \int_{t_1}^{t_2} f(t)dt = \text{prob}[t_1 \leqslant t \leqslant t_2] \tag{10.9}$$

Thus, the probability of a molecule staying in the system for a time between t_1 and t_2 is simply $F(t_2) - F(t_1)$. The $F(t)$ function is interpreted to be the fraction of molecules in the system with residence time less than or equal to t; a fraction $1 - F(t)$ has residence time greater than or equal to t. For example, in water-treatment technology, regulations for chlorine contactors may stipulate that the contactor must be designed such that 90% of the fluid molecules have a certain minimal residence time. If the F function for the contactor is known, it is simple to check whether the requirements are being met. One simply finds the value of time corrresponding to $F(t) = 0.10$. This value of time is sometimes called the "T_{10}" value for the reactor.

The reader may recall from statistics that there is an exact relationship between $F(t)$ and $f(t)$. Note the following:

$$\frac{dF}{dt} = \lim_{\Delta t \to 0} \frac{\Delta F}{\Delta t} = \lim_{\Delta t \to 0} \frac{F(t + \Delta t) - F(t)}{\Delta t} = \lim_{\Delta t \to 0} \frac{\int_0^{t+\Delta t} f(t)dt - \int_0^t f(t)dt}{\Delta t}$$

$$= \lim_{\Delta t \to 0} \frac{\int_t^{t+\Delta t} f(t)dt}{\Delta t} = \frac{f(t)dt}{dt} = f(t) \tag{10.10}$$

Or summarizing from Eq. 10.10, $f(t)$ is equal to the slope of $F(t)$:

$$f(t) = \frac{dF}{dt} \tag{10.11}$$

The third useful characterization of residence time is called the *washout function*. The washout function is defined by:

$$W(t) = \int_t^\infty f(t)dt \qquad (10.12)$$

The washout function is shown in Figure 10.3c. Note from the definitions that the limits of $W(t)$ must be between 1 and 0. The washout function gives the fraction of molecules in the system with residence time greater than or equal to t or, we might say, it is the fraction of molecules at any time t that has not yet "washed out" of the system.

There are several further connections between the three residence-time functions. First, note that

$$F(t) + W(t) = \int_0^t f(t)dt + \int_t^\infty f(t)dt = \int_0^\infty f(t)dt = 1 \qquad (10.13)$$

Taking the derivative of Eq. 10.13, we find the following relationship:

$$\frac{dF}{dt} + \frac{dW}{dt} = 0 \qquad (10.14)$$

However, we have already proved that $f = dF/dt$, so Eq. 10.14 implies that

$$\frac{dF}{dt} = f = -\frac{dW}{dt} \qquad (10.15)$$

Therefore, at any point in time, the f function is the slope of the F function and the negative slope of the W function. The functions f, F, and W are referred to generically as *residence-time distributions*.

10.3. MEASUREMENT OF RESIDENCE-TIME DISTRIBUTIONS

Suppose you are interested in measuring the residence-time distribution of a chemical reactor, pollution-control device, or even a natural system. There are three main methods in use for simple, closed systems: the *impulse method*, the *negative-step method*, and the *positive-step method*. In fact, we have already developed the essentials of the impulse method. Consider the impulse experiment in Figure 10.4.

As shown in the figure, at some time t_0, we quickly inject a concentrated tracer to simulate a mathematical function called the *Dirac-δ function*. The reader will recall from calculus that the Dirac function has very interesting properties. First, the function is only nonzero at a single point, t_0. Further, the

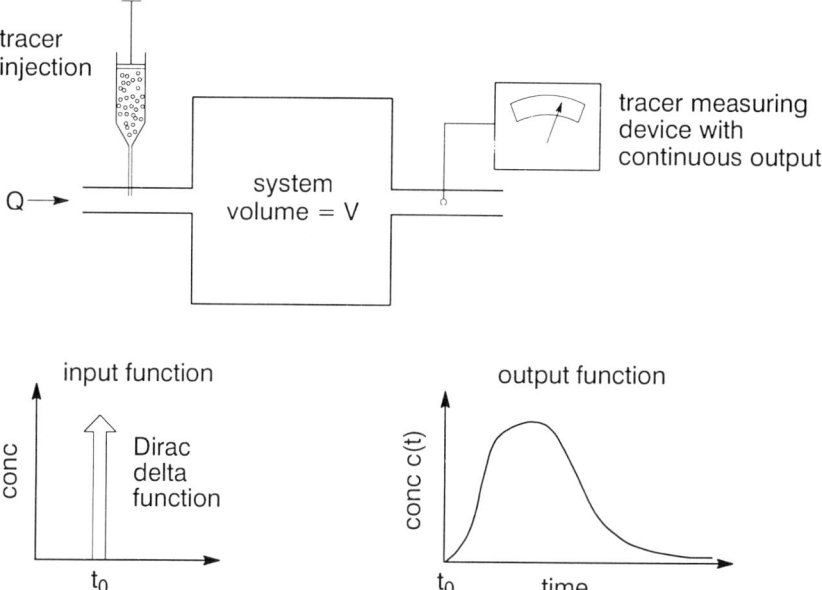

Figure 10.4. Schematic of impulse method of determining residence-time distribution and hypothetical input and output functions.

integral of the Dirac function over the entire time domain is unity:

$$\int_{-\infty}^{+\infty} \delta(t - t_0)dt = 1 \tag{10.16}$$

Since the function only has a nonzero value at t_0, Eq. 10.16 implies that the function must approach infinity at $t = t_0$. Finally, one of the more useful properties of the Dirac delta function is that when it is multiplied by another function and integrated, it returns the value of the other function at t_0:

$$\int_{-\infty}^{+\infty} \delta(t - t_0)f(t)dt = f(t_0) \tag{10.17}$$

As shown in Figure 10.4, the Dirac function is usually represented by an upward pointing arrow.

After injection of the tracer, a measuring device records the concentration of tracer leaving the system. We have shown a skewed distribution of outlet concentration as an example. The concentration output curve, called the $c(t)$ curve, is not the residence-time density function, $f(t)$; however, it is a simple matter to get $f(t)$ from $c(t)$:

$$f(t) = \frac{c(t)}{\displaystyle\int_0^\infty c(t)dt} \qquad (10.18)$$

So for the impulse-tracer experiment, the $f(t)$ function is obtained by normalizing the $c(t)$ function with the area under the $c(t)$ curve. This operation is analogous to the transformation performed in Section 10.2, where we converted tracer mass distribution to an $f(t)$ curve.

The positive-step experiment is outlined in Figure 10.5. In this experiment, the valve of the tracer-containing reservoir is suddenly opened at $t = t_0$. As tracer begins to flow into the system, the output sensor starts to record an increasing concentration at the exit. The concentration increases toward the limit of the steady-input concentration, C_0. By performing a mass balance around the inlet, we can calculate C_0 as

$$C_0 = \frac{C_r Q_r}{Q + Q_r} \qquad (10.19)$$

Here C_r is the tracer concentration in the reservoir, and Q_r is the flow rate out of the reservoir. The $c(t)$ output curve should remind the reader of the

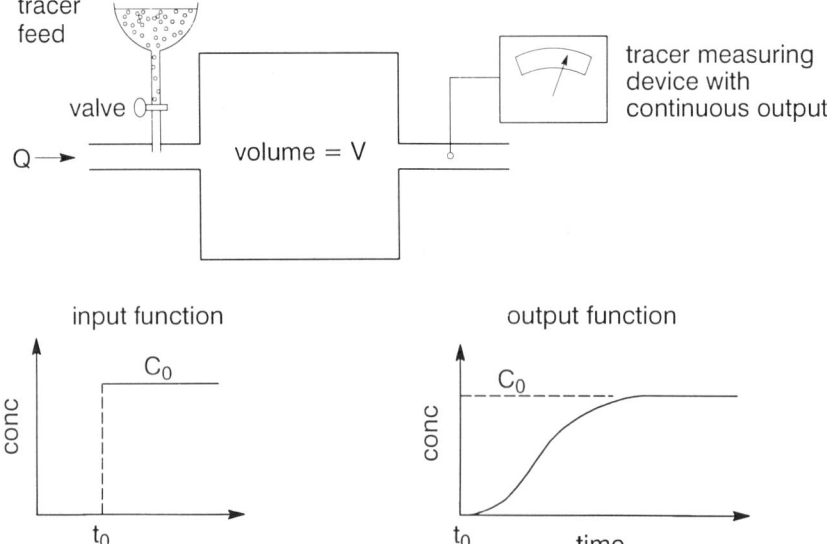

Figure 10.5. Schematic of positive-step method of determining residence-time distribution, and hypothetical input and output functions.

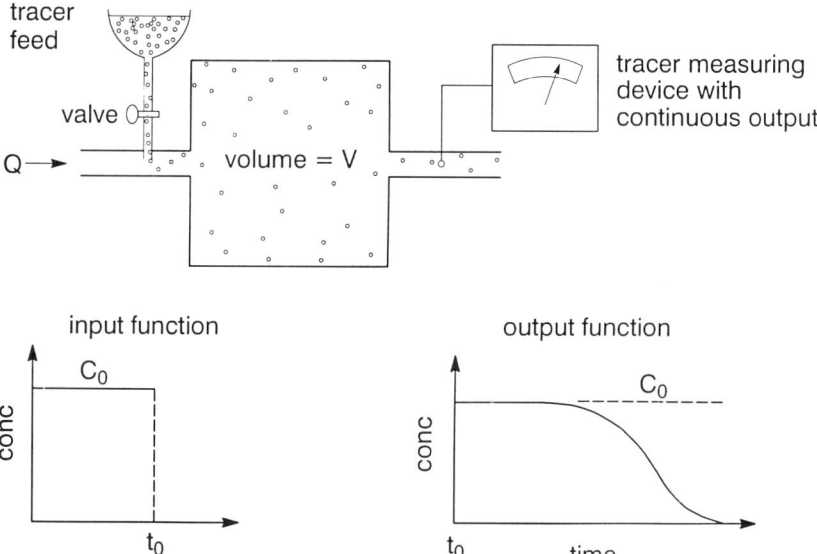

Figure 10.6. Schematic of negative-step method of determining residence-time distribution, and hypothetical input and output functions.

cumulative residence-time distribution function, $F(t)$, curve of Figure 10.3b. In fact, there is a simple transformation between the two:

$$F(t) = \frac{c(t)}{C_0} \qquad (10.20)$$

Thus, for the positive-step experiment, the cumulative distribution function, $F(t)$, is obtained by normalizing the output concentration with C_0. How does one then determine the residence-time density, $f(t)$, function? It is simply the derivative of $F(t)$, as shown by Eq. 10.15.

Finally, the negative-step experiment is shown in Figure 10.6. In this experiment the valve to the tracer-containing reservoir has been open for a long time, such that the exit concentration from the reactor has reached a steady-state value, C_0, which is calculated with Eq. 10.19. At $t = t_0$, the value is suddenly closed. The output concentration from the system decreases with time to a value of zero. The $c(t)$ curve should remind the reader of the washout, $W(t)$, function from the last section. There is a simple transformation between the $c(t)$ and $W(t)$ functions:

$$W(t) = \frac{c(t)}{C_0} \qquad (10.21)$$

Hence, for the negative-step experiment, the washout function, $W(t)$, is obtained by normalizing the output concentration with C_0. The residence time density, $f(t)$, can then be determined in accordance with Eq. 10.15.

10.4. RESIDENCE-TIME DISTRIBUTIONS FROM DISCRETE DATA

Up to this point, we have imagined that a sufficiently fast-responding sensor continuously monitors and records the concentration of tracer leaving the system. However, in most cases, we may not be lucky enough to have such a system. For example, if a radioactive tracer or marker is used, an appropriate continuous concentration monitor may not exist. Other tracers may have to be measured by gas chromatography, and a quickly responding measurement system may not exist. In these and several other cases, discrete samples may be transported off-site for later analysis. These are sometimes called "grab samples." In any event, we discuss in this section how to apply the preceding formulas in the case of discrete samples.

Suppose that samples are taken at a series of times t_i (e.g., t_0, t_1, t_2,..., $t_{i=imax}$); there is also a discrete series of effluent concentration values, $c(t_i)$ [e.g., $c(t_0)$, $c(t_1)$, $c(t_2) \ldots c(t_{i=imax})$], corresponding to the time values. The *discrete residence-time density function*, $f(t_i)$, is easily determined with discrete data. The appropriate equation, analogous to Eq. 10.18, is

$$f(t_i) = \frac{c(t_i)}{\text{area}} \qquad (10.22)$$

In this equation "area" is the area under a curve constructed from the discrete grab samples:

$$\text{area} = \sum_{i=0}^{imax-1} \left[\frac{c(t_{i+1}) + c(t_i)}{2} \right] (t_{i+1} - t_i) \qquad (10.23)$$

We say that Eq. 10.23 is "the discrete approximation" of the integral in the numerator of Eq. 10.18. With these values of $f(t_i)$, the *discrete cumulative residence-time distribution* is calculated with the discrete approximation of Eq. 10.8. At some time $t_{i=I}$, this is

$$F(t_{i=I}) = \sum_{i=1}^{I} \left[\frac{f(t_{i-1}) + f(t_i)}{2} \right] (t_i - t_{i-1}) \qquad (10.24)$$

where $F(t_{i=0})$ is always set to a value of zero.*

*For purposes of plotting F and W, it is slightly more accurate to plot values at the average time values $(t_i + t_{i-1})/2$.

Finally, it is straightforward to calculate the *discrete washout function*, $W(t_{i=I})$, as

$$W(t_{i=I}) = 1 - F(t_{i=I}) \tag{10.25}$$

Two other important calculations should be made when analyzing discrete data. The first is the discrete estimate of the mean residence time. This is simply the discrete approximation to Eq. 10.6:

$$\bar{t}_{RTD} = \sum_{i=0}^{imax-1} \left[\frac{(t_i + t_{i+1})}{2} \right] \left[\frac{f(t_i) + f(t_{i+1})}{2} \right] (t_{i+1} - t_i) \tag{10.26}$$

This value of the mean residence time should be compared with the theoretical mean residence time estimate (Eq. 10.7). However, imagine a situation where it is impractical to measure a system volume (e.g., a lake), or where it is impossible to measure a system flow rate (e.g., a very corrosive or toxic flow stream). If just the system flow rate or volume is unknown, Eq. 10.26 can be used to estimate the mean residence time. This can be substituted into Eq. 10.7 and the unknown V or Q can be calculated.*

The last calculation is to determine the mass of tracer exiting the system during an impulse tracer experiment. In terms of the continuous function $c(t)$, this is simply

$$\{\text{outlet tracer mass}\} = Q \int_0^\infty c(t)dt \tag{10.27}$$

But note that the integral is just the area under the curve, which can be calculated for discrete data with Eq. 10.23. So it is a simple matter to calculate the mass of tracer leaving the system. This value should always be compared with the mass of tracer injected into the system. Sometimes the injected mass is known exactly, but other times it may actually be an unknown in which we have great interest (e.g., the unknown mass of contaminant entering a system). The practical calculation of these quantities is demonstrated in the following numerical example.

Example 10.1. An impulse-tracer experiment was performed on a small laboratory reaction chamber (which is also a simple, closed system). The first two columns of the following spreadsheet give the measurement time and the measured concentrations exiting the system. The flow rate through the system is known to be $2 \text{ m}^3/\text{hr}$, but the system volume is unknown. Calculate the residence-time density function, the residence-time cumulative distribution function, the washout function, the mean residence time, the system volume, and the mass of tracer injected into the systems. Also plot $f(t)$, $F(t)$, and $W(t)$ versus time.

---

*The strategy outlined in this paragraph assumes there are no extreme flow pathologies such as dead zones, which are highly stagnant regions that are effectively cutoff from possible contact with the main flow. When dead zones exist, the effective volume of the system could be less than V.

SOLUTION. In the third column of the spreadsheet, Eq. 10.23 is used to compute the incremental contributions to the area under the $c(t)$ versus time curve. The sum of all these incremental areas is shown as 126.3 mg-min-L^{-1}. The discrete $f(t)$ values are calculated with Eq. 10.22, and these are provided in column 4. The incremental contributions to the discrete estimate of the mean residence time are calculated with Eq. 10.26, and these are shown in column 5; the sum of these values is the mean residence time, 7.95 min. The discrete values of the cumulative residence-time distribution and washout function are calculated with Eqs. 10.24 and 10.25, and are shown in columns 6 and 7. The values for $f(t)$, $F(t)$, and $W(t)$ are plotted in the accompanying figures.

Knowing the mean residence time and the system flow rate, we can easily calculate the system volume from Eq. 10.7 if we estimate $\bar{t} = \bar{t}_{RTD}$:

$$V = Q\bar{t}_{RTD} = 2\,\frac{m^3}{hr} \times \frac{hr}{60\,min} \times 7.95\,min = 0.265\,m^3$$

The mass of tracer is calculated with Eq. 10.27:

$$\{\text{outlet tracer mass}\} = Q \int_0^\infty c(t)dt = 2\,\frac{m^3}{hr} \times \frac{hr}{60\,min} \times 126.3\,\frac{mg\text{-}min}{L} \times \frac{1000\,L}{m^3}$$

$$= 4210\,mg = 4.21\,g$$

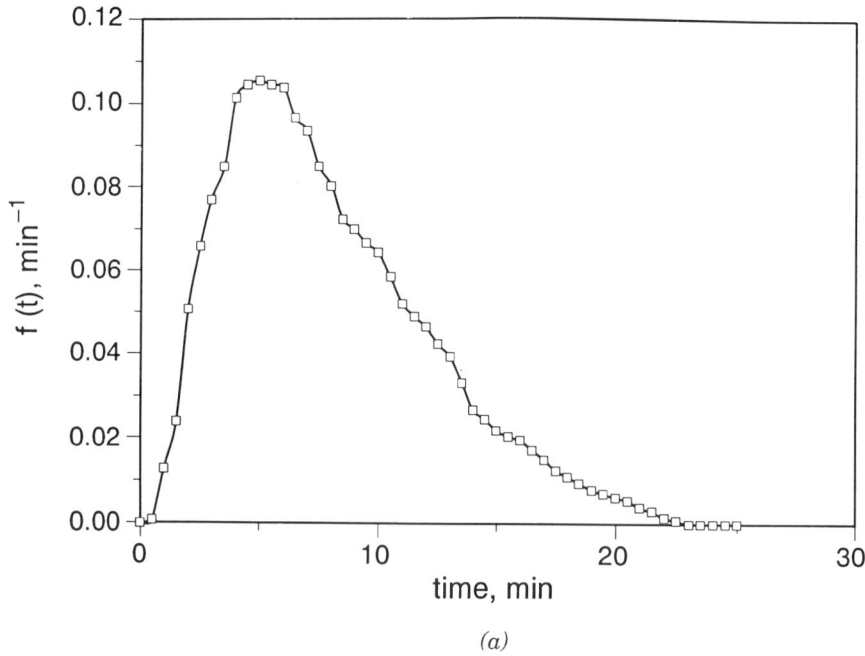

(a)

Example 10.1(a)

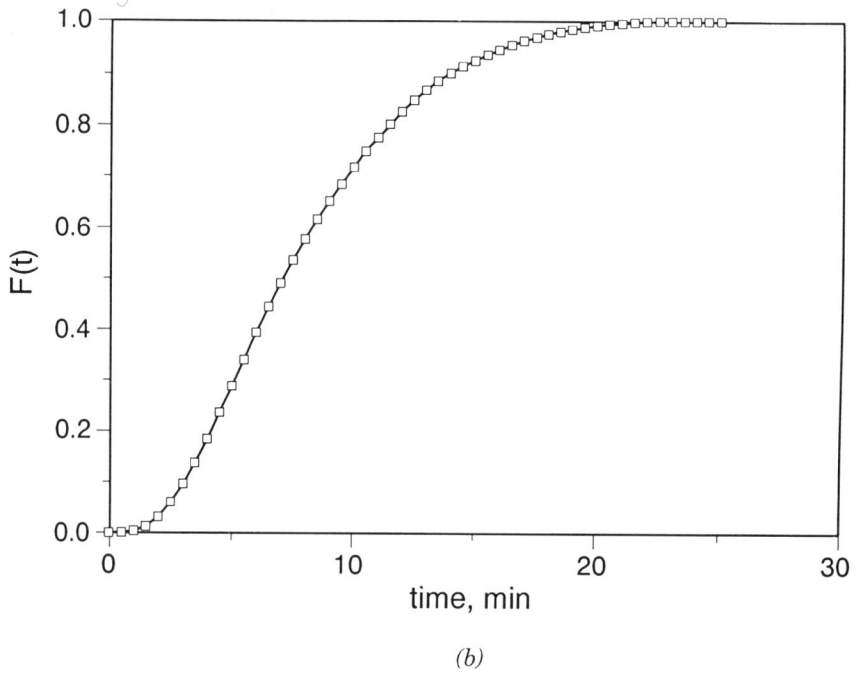

(b)

Example 10.1(b)

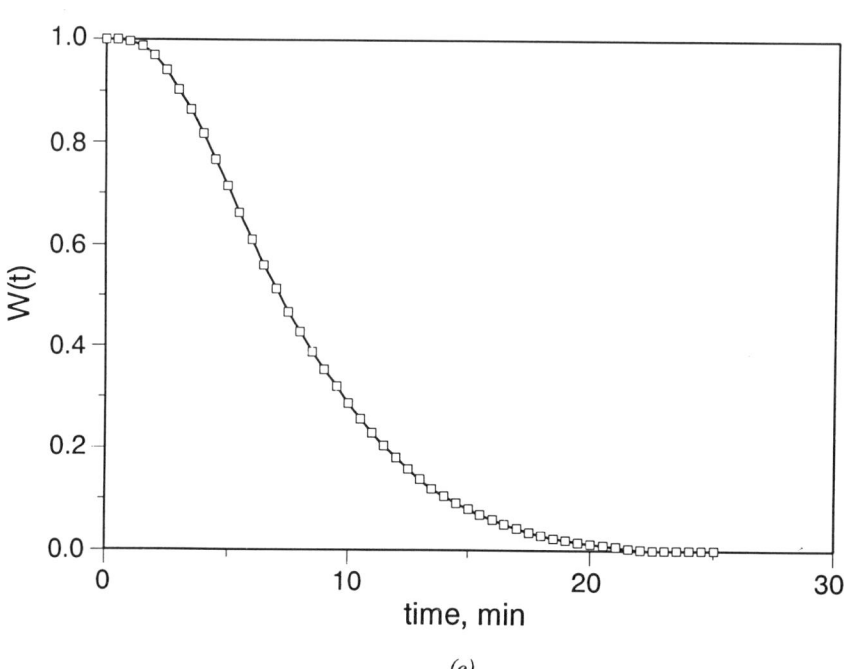

(c)

Example 10.1(c)

Time (min)	Concentration (mg/L)	Area i^a	$f(t)$	$t_i{}^b$	$F(t)$	$W(t)$
0	0	0.025	0.00E+00	4.95E−05	0.00E+00	1.00E+00
0.5	0.1	0.425	7.92E−04	2.52E−03	1.98E−04	1.00E+00
1	1.6	1.15	1.27E−02	1.14E−02	3.56E−03	9.96E−01
1.5	3	2.35	2.28E−02	3.26E−02	1.27E−02	9.87E−01
2	6.4	3.675	5.07E−02	6.55E−02	3.13E−02	9.69E−01
2.5	8.3	4.5	6.57E−02	9.80E−02	6.04E−02	9.40E−01
3	9.7	5.1	7.68E−02	1.31E−01	9.60E−02	9.04E−01
3.5	10.7	5.875	8.47E−02	1.74E−01	1.36E−01	8.64E−01
4	12.8	6.5	1.01E−01	2.19E−01	1.83E−01	8.17E−01
4.5	13.2	6.625	1.05E−01	2.49E−01	2.34E−01	7.66E−01
5	13.3	6.625	1.05E−01	2.75E−01	2.87E−01	7.13E−01
5.5	13.2	6.575	1.05E−01	2.99E−01	3.39E−01	6.61E−01
6	13.1	6.325	1.04E−01	3.13E−01	3.91E−01	6.09E−01
6.5	12.2	6	9.66E−02	3.21E−01	4.41E−01	5.59E−01
7	11.8	5.625	9.34E−02	3.23E−01	4.89E−01	5.11E−01
7.5	10.7	5.2	8.47E−02	3.19E−01	5.33E−01	4.67E−01
8	10.1	4.8	8.00E−02	3.14E−01	5.75E−01	4.25E−01
8.5	9.1	4.475	7.21E−02	3.10E−01	6.13E−01	3.87E−01
9	8.8	4.3	6.97E−02	3.15E−01	6.48E−01	3.52E−01
9.5	8.4	4.125	6.65E−02	3.18E−01	6.82E−01	3.18E−01
10	8.1	3.875	6.41E−02	3.14E−01	7.15E−01	2.85E−01
10.5	7.4	3.5	5.86E−02	2.98E−01	7.45E−01	2.55E−01
11	6.6	3.2	5.23E−02	2.85E−01	7.73E−01	2.27E−01
11.5	6.2	3.025	4.91E−02	2.81E−01	7.98E−01	2.02E−01
12	5.9	2.825	4.67E−02	2.74E−01	8.22E−01	1.78E−01
12.5	5.4	2.6	4.28E−02	2.62E−01	8.45E−01	1.55E−01
13	5	2.3	3.96E−02	2.41E−01	8.65E−01	1.35E−01
13.5	4.2	1.9	3.33E−02	2.07E−01	8.84E−01	1.16E−01
14	3.4	1.625	2.69E−02	1.83E−01	8.99E−01	1.01E−01
14.5	3.1	1.475	2.45E−02	1.72E−01	9.12E−01	8.85E−02
15	2.8	1.35	2.22E−02	1.63E−01	9.23E−01	7.68E−02
15.5	2.6	1.275	2.06E−02	1.59E−01	9.34E−01	6.61E−02
16	2.5	1.175	1.98E−02	1.51E−01	9.44E−01	5.60E−02
16.5	2.2	1.025	1.74E−02	1.36E−01	9.53E−01	4.67E−02
17	1.9	0.875	1.50E−02	1.20E−01	9.61E−01	3.86E−02
17.5	1.6	0.75	1.27E−02	1.05E−01	9.68E−01	3.17E−02
18	1.4	0.65	1.11E−02	9.39E−02	9.74E−01	2.57E−02
18.5	1.2	0.55	9.50E−03	8.17E−02	9.79E−01	2.06E−02
19	1	0.475	7.92E−03	7.24E−02	9.84E−01	1.62E−02
19.5	0.9	0.425	7.13E−03	6.65E−02	9.88E−01	1.25E−02
20	0.8	0.375	6.33E−03	6.01E−02	9.91E−01	9.11E−03
20.5	0.7	0.3	5.54E−03	4.93E−02	9.94E−01	6.14E−03
21	0.5	0.225	3.96E−03	3.79E−02	9.96E−01	3.76E−03
21.5	0.4	0.15	3.17E−03	2.58E−02	9.98E−01	1.98E−03
22	0.2	0.075	1.58E−03	1.32E−02	9.99E−01	7.92E−04
22.5	0.1	0.025	7.92E−04	4.50E−03	1.00E+00	1.98E−04
23	0	0	0.00E+00	0.00E+00	1.00E+00	0.00E+00
23.5	0	0	0.00E+00	0.00E+00	1.00E+00	0.00E+00
24	0	0	0.00E+00	0.00E+00	1.00E+00	0.00E+00
24.5	0	0	0.00E+00	0.00E+00	1.00E+00	0.00E+00
25	0	0	0.00E+00	0.00E+00	1.00E+00	0.00E+00

^{a}Area = sum of column 3 = 126.3.
$^b\bar{t}$ = sum of column 5 = 7.95.

10.5. PERFECT MIXING AND IDEAL PLUG FLOW

We now consider the two most important ideal residence-time distributions: *perfect mixing and ideal plug flow*. The concept of perfect mixing has been encountered at several junctures in this book, although we have waited until now to formally derive the residence-time distribution functions. To derive the residence-time distributions for perfect mixing, consider Figure 10.7.

This figure is only slightly altered from Figure 10.6. It depicts a washout or negative-step change experiment, as described in Section 10.3. We have added some new nomenclature, which is consistent with the systems approach we have used throughout this book (e.g., recall Section 1.2). The steady inlet concentration is called c_{in}, the concentration within the system is called c_{sys}, and the outlet concentration is called c_{out}. Just as in Section 1.2, we write a mass balance on the tracer:

$$V \frac{dc_{sys}}{dt} = Qc_{in} - Qc_{out} \tag{10.28}$$

Dividing through each side by V, we find:

$$\frac{dc_{sys}}{dt} = \frac{c_{in}}{\bar{t}} - \frac{c_{out}}{\bar{t}} \tag{10.29}$$

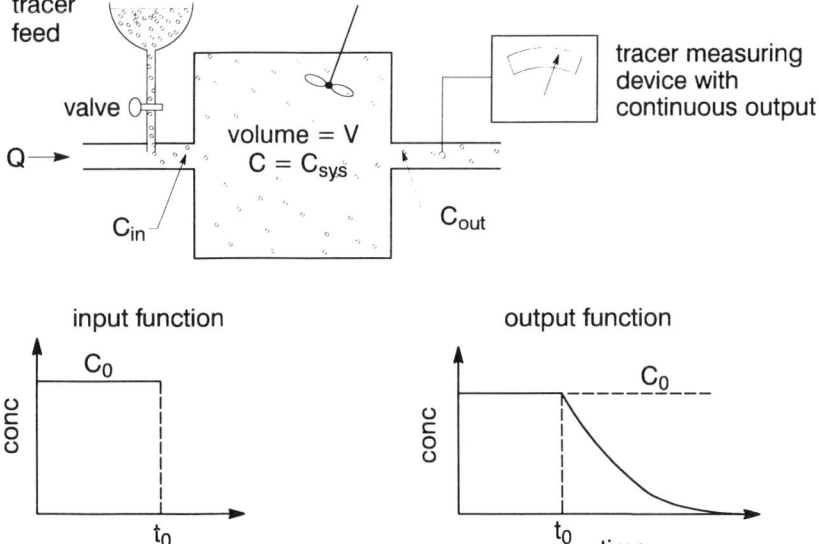

Figure 10.7. A washout experiment in a system with perfect mixing.

where $\bar{t}$ is the theoretical mean residence time defined by Eq. 10.7. *In perfect mixing, which is indicated by the small mixing impeller in Figure 10.7, the mixing is so good or complete, that the tracer concentration within the system is uniform.* The reader should note that this is a pretty extreme and probably unrealistic assumption. It indicates that any tracer entering the system is *instantaneously* mixed with the rest of the tracer already in the system. In perfect mixing, the contents of the system are homogeneous and, therefore, $c_{out} = c_{sys}$ always. In the washout experiment, the inlet, outlet, and interior concentrations are the same prior to $t = t_0$; the inlet concentration goes to zero at $t = t_0$. Therefore, if we consider the system response for $t \geq t_0$ only, we can say that $c_{in} = 0$. Making these changes to Eq. 10.29, we find:

$$\frac{dc_{out}}{dt} = -\frac{1}{\bar{t}} c_{out} \qquad (10.30)$$

Although the inlet concentration is zero after $t = t_0$, we can fix the initial condition in the system as $c_{sys} = c_{out} = C_0$ at $t = t_0$. Rearranging and integrating Eq. 10.30, we find:

$$\frac{c_{out}}{C_0} = \exp\left(-\frac{t}{\bar{t}}\right) \qquad (10.31)$$

However, since this was a properly conducted washout experiment, we know from Eq. 10.21 that

$$W(t) = \frac{c_{out}}{C_0} = \exp\left(-\frac{t}{\bar{t}}\right) \qquad (10.32)$$

Hence, we have just derived the *washout function for perfect mixing*. The *residence-time density function for perfect mixing* is derived immediately from Eqs. 10.15 and 10.32:*

$$f(t) = -\frac{dW(t)}{dt} = \frac{1}{\bar{t}} \exp\left(-\frac{t}{\bar{t}}\right) \qquad (10.33)$$

Ideal plug flow is another critical concept for environmental engineers, which we have also encountered previously at several junctures in this book. Consider the ideal plug-flow system shown in Figure 10.8. The critical assumptions of ideal plug flow are that velocity throughout the plug-flow

---

*The term continuous stirred tank reactor, or CSTR, is often used to describe the ideal reactor we call the perfect mixer. But the terminology of this book is preferred over CSTR since our terminology more accurately describes the ideal state of mixing assumed.

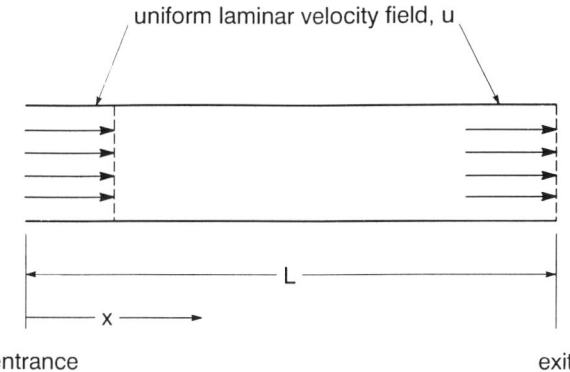

Figure 10.8. A system with ideal plug flow.

system is steady, uniform (constant across the flow section), and laminar, and that there is no molecular diffusion of the tracer or other transported material. Although laminar flow in systems like that in Figure 10.8 is not unusual for low Reynolds numbers (recall Chapter 5), we have learned that a uniform (constant), one-dimensional, velocity field can only be expected at the entrance of such a system; momentum boundary layers always cause a rounding off of the velocity profile to match the no-slip condition of all real fluids at flow-field boundaries (such as the walls shown in Figure 10.8). Since the assumptions made here are somewhat unrealistic, we refer to the model as *ideal* plug flow.

To derive the residence-time density function for the ideal plug flow shown in Figure 10.8, we can consider the positive-step change experiment shown in Figure 10.9. Here we have replaced the mixed vessel depicted in previous experiments with the ideal plug-flow reactor. As explained above, the velocity field is assumed to be a uniform (constant), one-dimensional laminar flow with no molecular diffusion or dispersion. We must also make a subtle assumption about the tracer injection. We must assume that at the injection point (the reactor "entrance"), the tracer is uniformly distributed across the flow section. Everything else in the figure is identical to the positive-step change experiment outlined previously in Figure 10.5. The general transport equation, the convective-diffusion equation, was developed in Sections 7.2 and 7.7. For the case of a one-dimensional, steady, laminar flow with no molecular diffusion, Eq. 7.83 can be simplified to

$$-u \frac{\partial c}{\partial x} = \frac{\partial c}{\partial t} \qquad (10.34)$$

We consider the system to be "relaxed" prior to the step change in inlet concentration at $t = t_0$. Taking the Laplace transform of Eq. 10.34, we find

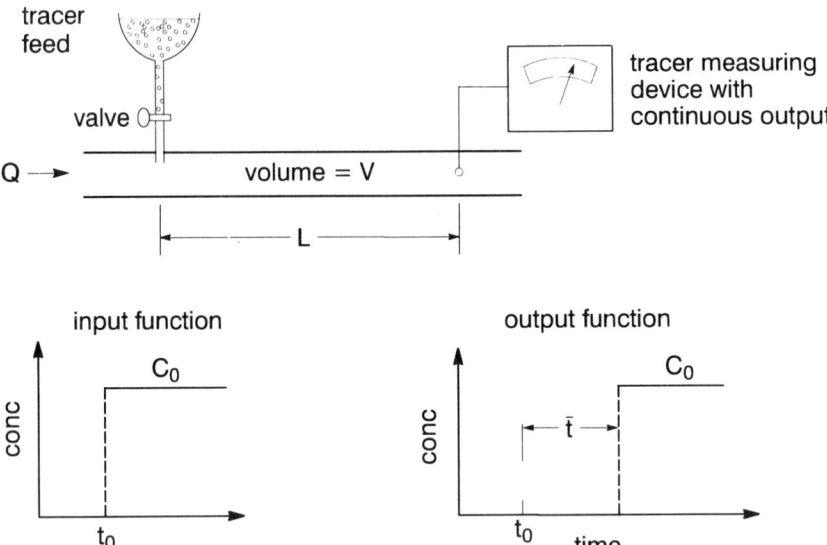

Figure 10.9. A positive-step change experiment for determining cumulative residence-time distribution for ideal plug flow.

(Nauman and Buffham, 1983):

$$-u\,\frac{d\bar{c}(s)}{dx} = s\bar{c}(s) \tag{10.35}$$

where the overbar indicates Laplace transformation.

Hence, Laplace transformation turns a partial differential equation into an ordinary differential equation. In Laplace transformation notation, our variables and boundary conditions can be expressed as $\bar{c}(s) = \bar{c}(x, s)$, $\bar{c}_{in}(s) = \bar{c}(0, s)$, and $\bar{c}_{out}(s) = \bar{c}(L, s)$. For the initially relaxed system, we apply the positive-step change function shown in Figure 10.9; therefore, for $t \geqslant t_0$, $c_{in} = C_0$. The Laplace transform of C_0 for $t \geqslant t_0$ is found from a table of Laplace transforms:

$$\bar{c}_{in}(s) = C_0 s^{-1} \tag{10.36}$$

which is essentially our initial condition. We can now separate Eq. 10.35, which yields:

$$\int_{\bar{c}_{in}(s)}^{\bar{c}_{out}(s)} \frac{d\bar{c}(s)}{\bar{c}(s)} = -\frac{s}{u} \int_0^L dx \tag{10.37}$$

Integrating, we find:

$$\frac{\bar{c}_{out}(s)}{\bar{c}_{in}(s)} = \exp\left(-\frac{s}{u}L\right) \tag{10.38}$$

Now we substitute Eq. 10.36 for $\bar{c}_{in}(s)$, which yields:

$$\frac{\bar{c}_{out}(s)}{C_0} = \frac{1}{s}\exp\left(-\frac{s}{u}L\right) \tag{10.39}$$

From a table of inverse Laplace transforms we find:

$$\frac{c_{out}}{C_0} = H\left(t - \frac{L}{u}\right) \tag{10.40}$$

where H is the Heaviside unit-step function:

$$H[(t - t_0) - a] = \begin{cases} 0 & (t - t_0) < a \\ 1 & (t - t_0) > a \end{cases} \tag{10.41}$$

However, since the left-hand side of Eq. 10.40 is the definition of the cumulative residence-time distribution function, $F(t)$, we conclude from Eqs. 10.40 and 10.41 that for ideal plug flow,

$$F(t) = \begin{cases} 0 & t < \bar{t} \\ 1 & t > \bar{t} \end{cases} \tag{10.42}$$

where we have set $t_0 = 0$ and have noted that for the ideal plug flow reactor, $\bar{t} = L/u$.

To derive the residence-time density function, $f(t)$, for ideal plug flow, we take the derivative of $F(t)$, which yields:

$$f(t) = \delta(t - \bar{t}) \tag{10.43}$$

Therefore, the residence-time density function for ideal plug flow is simply a Dirac delta function centered on $t = \bar{t}$. Recall that the input for the properly performed impulse experiment is a delta function. From Eq. 10.43, we note that the f function for ideal plug flow is also a delta function, only delayed by an amount of time equal to the mean residence time. Equation 10.43 indicates then that in ideal plug flow, all fluid molecules have the identical residence time, $\bar{t}$. Therefore, in ideal plug flow, a delta-function input results in a delayed delta-function output.

It will be of considerable use to display the various residence-time densities in dimensionless form. For example, the *dimensionless residence-time density* for

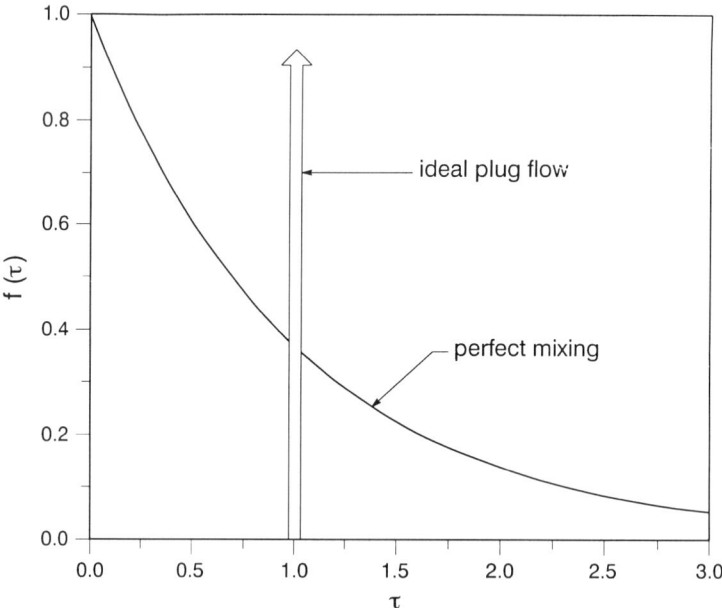

Figure 10.10. Comparison of dimensionless residence time density functions for perfect mixing and ideal plug flow.

perfect mixing is obtained simply by multiplying $f(t)$ by $\bar{t}$:

$$\bar{t}f(t) = \exp\left(-\frac{t}{\bar{t}}\right) = \exp(-\tau) = f(\tau) \tag{10.44}$$

where we have replaced $t/\bar{t}$ with the dimensionless time τ:

$$\tau = \frac{t}{\bar{t}} \tag{10.45}$$

In Figure 10.10, we plot the dimensionless residence-time density functions for perfect mixing and ideal plug flow versus τ. The upward-pointing arrow is the standard representation of the Dirac function.

After what we have said about the assumptions of these two models, the reader might wonder what the use is for two such unlikely residence-time models. The answer is that even though mixing in real systems can never achieve the assumption of these models, we find that almost all residence-time distributions, either empirical or measured, are in a sense bounded by the perfect mixer and ideal plug-flow reactor. This is a rather important theoretical and practical point, whose value will be realized when we consider the effect of residence time on reactions within a system.

10.6. MOMENTS OF RESIDENCE-TIME DISTRIBUTIONS

We have indicated that residence-time distributions are solidly based in statistical theory. Therefore, any residence-time distribution can theoretically be described by its moments. Knowledge of moments can be used in fitting residence-time data to residence-time models (e.g., the perfect mixier, ideal plug-flow reactor, and other models to be discussed) and is currently an important issue in groundwater transport of contaminants.

The *moments about the origin* can be calculated with two formulas. The first is

$$\mu_n = \int_0^\infty t^n f(t)dt \tag{10.46}$$

We have already discussed the first moment about the origin in Section 10.2, the so-called *mean residence time*:

$$\mu_1 = \bar{t}_{RTD} = \int_0^\infty f(t)dt \tag{10.6}$$

An alternative approach uses the washout function:

$$\mu_n = n \int_0^\infty t^{n-1} W(t)dt \tag{10.47}$$

One advantage of Eq. 10.47 over 10.46 is that because of the lower power of t in the integrand, the moments estimated from Eq. 10.47 are less sensitive to inaccuracies in the extreme values ("tail") of the residence-time distribution. Recall that experimentally measured residence times must use some analytical technique to measure tracer concentrations. For long times, the tail of the distribution must approach zero concentration. All analytical techniques become less accurate as the measured concentration approaches zero. In Eq. 10.47, the tail concentration values are less heavily weighted because of the lower power of t involved in the calculation.

Another calculation frequently used is the *moment about the mean*. This is defined as

$$\mu'_n = \int_0^\infty (t - \bar{t})^n f(t)dt \tag{10.48}$$

By far the most important moment about the mean is the second moment, the *variance*:

$$\sigma^2 = \mu'_2 = \int_0^\infty (t - \bar{t})^2 f(t)dt \tag{10.49}$$

The moments about the origin and the moments about the mean are mathematically related. By expanding under the integral in Eq. 10.48 and directly integrating, it can be shown that (Exercise 10.2)

$$\mu_0' = \mu_0 = 1 \tag{10.50}$$

$$\mu_1' = 0 \tag{10.51}$$

$$\mu_2' = \mu_2 - (\bar{t})^2 \tag{10.52}$$

$$\mu_3' = \mu_3 - 3\bar{t}\mu_2 + 2(\bar{t})^3 \tag{10.53}$$

$$\mu_4' = \mu_4 - 4\mu_3\bar{t} + 6\mu_2(\bar{t})^2 - 3(\bar{t})^4 \tag{10.54}$$

The third moment is related to the skewness, and the fourth moment is related to the kurtosis. It is also possible to derive the inverse relationships (Exercise 10.3) between the moments:

$$\mu_2 = \mu_2' + (\bar{t})^2 \tag{10.55}$$

$$\mu_3 = \mu_3' + 3\mu_2'\bar{t} + (\bar{t})^3 \tag{10.56}$$

$$\mu_4 = \mu_4' + 4\mu_3'\bar{t} + 6\mu_2'(\bar{t})^2 + (\bar{t})^4 \tag{10.57}$$

Example 10.2. For the impulse-tracer experiment performed in Example 10.1, determine the second, third, and fourth moments about the mean.

SOLUTION. Since the data recorded in Example 10.1 are discrete data, we first need to develop the formula for the discrete approximation of Eq. 10.48. This follows very easily from Eq. 10.26:

$$\mu_n' = \sum_{i=0}^{imax-1} \left[\frac{(t_i + t_{i+1})}{2} - \bar{t} \right]^n \left[\frac{f(t_i) + f(t_{i+1})}{2} \right] (t_{i+1} - t_i) \tag{10.58}$$

In the following spreadsheet, Eq. 10.58 is used to calculate the three higher moments about the origin. The results are $\mu_2' = 18.36$, $\mu_3' = 60.5$, and $\mu_4' = 1050$.

It is also possible to calculate the moments of the two empirical residence-time distributions, perfect mixing and ideal plug flow. Applying Eq. 10.48 to the perfect mixer (Eq. 10.33), we compute:

$$\sigma^2 = \frac{1}{\bar{t}} \int_0^\infty \exp\left(-\frac{t}{\bar{t}}\right)(t - \bar{t})^2 dt \tag{10.59}$$

Expanding under the integral and integrating (see Exercise 10.4), we quickly find for the perfect mixer that

$$\sigma^2 = (\bar{t})^2 \tag{10.60}$$

Time (min)	Concen-tration (mg/L)	Area i^a	$f(t)$	First Moment about Origin[b]	Second Moment about Mean[c]	Third Moment about Mean[d]	Fourth Moment about Mean[e]
0	0	0.025	0.00E+00	4.95E−05	1.17E−02	−9.03E−02	6.96E−01
0.5	0.1	0.425	7.92E−04	2.52E−03	1.74E−01	−1.26E+00	9.04E+00
1	1.6	1.15	1.27E−02	1.14E−02	4.09E−01	−2.74E+00	1.83E+01
1.5	3	2.35	2.28E−02	3.26E−02	7.15E−01	−4.43E+00	2.75E+01
2	6.4	3.675	5.07E−02	6.55E−02	9.45E−01	−5.39E+00	3.07E+01
2.5	8.3	4.5	6.57E−02	9.80E−02	9.63E−01	−5.01E+00	2.60E+01
3	9.7	5.1	7.68E−02	1.31E−01	8.92E−01	−4.19E+00	1.97E+01
3.5	10.7	5.875	8.47E−02	1.74E−01	8.20E−01	−3.44E+00	1.45E+01
4	12.8	6.5	1.01E−01	2.19E−01	7.04E−01	−2.61E+00	9.64E+00
4.5	13.2	6.625	1.05E−01	2.49E−01	5.37E−01	−1.72E+00	5.49E+00
5	13.3	6.625	1.05E−01	2.75E−01	3.82E−01	−1.03E+00	2.78E+00
5.5	13.2	6.575	1.05E−01	2.99E−01	2.52E−01	−5.54E−01	1.22E+00
6	13.1	6.325	1.04E−01	3.13E−01	1.45E−01	−2.46E−01	4.17E−01
6.5	12.2	6	9.66E−02	3.21E−01	6.83E−02	−8.19E−02	9.82E−02
7	11.8	5.625	9.34E−02	3.23E−01	2.18E−02	−1.52E−02	1.06E−02
7.5	10.7	5.2	8.47E−02	3.19E−01	1.63E−03	−3.25E−04	6.47E−05
8	10.1	4.8	8.00E−02	3.14E−01	3.44E−03	1.04E−03	3.11E−04
8.5	9.1	4.475	7.21E−02	3.10E−01	2.27E−02	1.82E−02	1.46E−02
9	8.8	4.3	6.97E−02	3.15E−01	5.76E−02	7.49E−02	9.75E−02
9.5	8.4	4.125	6.65E−02	3.18E−01	1.06E−01	1.91E−01	3.44E−01
10	8.1	3.875	6.41E−02	3.14E−01	1.62E−01	3.74E−01	8.60E−01
10.5	7.4	3.5	5.86E−02	2.98E−01	2.17E−01	6.09E−01	1.71E+00
11	6.6	3.2	5.23E−02	2.85E−01	2.76E−01	9.11E−01	3.01E+00
11.5	6.2	3.025	4.91E−02	2.81E−01	3.46E−01	1.32E+00	5.00E+00
12	5.9	2.825	4.67E−02	2.74E−01	4.14E−01	1.78E+00	7.65E+00
12.5	5.4	2.6	4.28E−02	2.62E−01	4.74E−01	2.28E+00	1.09E+01
13	5	2.3	3.96E−02	2.41E−01	5.12E−01	2.71E+00	1.44E+01
13.5	4.2	1.9	3.33E−02	2.07E−01	5.06E−01	2.94E+00	1.70E+01
14	3.4	1.625	2.69E−02	1.83E−01	5.11E−01	3.22E+00	2.03E+01
14.5	3.1	1.475	2.45E−02	1.72E−01	5.40E−01	3.67E+00	2.50E+01
15	2.8	1.35	2.22E−02	1.63E−01	5.70E−01	4.16E+00	3.04E+01
15.5	2.6	1.275	2.06E−02	1.59E−01	6.14E−01	4.79E+00	3.74E+01
16	2.5	1.175	1.98E−02	1.51E−01	6.41E−01	5.32E+00	4.42E+01
16.5	2.2	1.025	1.74E−02	1.36E−01	6.29E−01	5.53E+00	4.87E+01
17	1.9	0.875	1.50E−02	1.20E−01	5.99E−01	5.57E+00	5.18E+01
17.5	1.6	0.75	1.27E−02	1.05E−01	5.70E−01	5.59E+00	5.48E+01
18	1.4	0.65	1.11E−02	9.39E−02	5.46E−01	5.63E+00	5.79E+01
18.5	1.2	0.55	9.50E−03	8.17E−02	5.08E−01	5.49E+00	5.93E+01
19	1	0.475	7.92E−03	7.24E−02	4.80E−01	5.43E+00	6.13E+01
19.5	0.9	0.425	7.13E−03	6.65E−02	4.69E−01	5.53E+00	6.53E+01
20	0.8	0.375	6.33E−03	6.01E−02	4.49E−01	5.53E+00	6.80E+01
20.5	0.7	0.3	5.54E−03	4.93E−02	3.89E−01	4.98E+00	6.38E+01
21	0.5	0.225	3.96E−03	3.79E−02	3.15E−01	4.19E+00	5.58E+01
21.5	0.4	0.15	3.17E−03	2.58E−02	2.26E−01	3.12E+00	4.31E+01
22	0.2	0.075	1.58E−03	1.32E−02	1.21E−01	1.74E+00	2.48E+01
22.5	0.1	0.025	7.92E−04	4.50E−03	4.34E−02	6.42E−01	9.50E+00
23	0	0	0.00E+00	0.00E+00	0.00E+00	0.00E+00	0.00E+00
23.5	0	0	0.00E+00	0.00E+00	0.00E+00	0.00E+00	0.00E+00
24	0	0	0.00E+00	0.00E+00	0.00E+00	0.00E+00	0.00E+00
24.5	0	0	0.00E+00	0.00E+00	0.00E+00	0.00E+00	0.00E+00
25	0	0	0.00E+00	0.00E+00	0.00E+00	0.00E+00	0.00E+00

[a]Area = sum of column 3 = 126.3.
[b]$\bar{t}$ = sum of column 5 = 7.95 min.
[c]Second moment = sum of column 6 = 18.36 min^2.
[d]Third moment = sum of column 7 = 6.05E+01 min^3.
[e]Fourth moment = sum of column 8 = 1.05E+03 min^4.

Applying Eq. 10.48 to the ideal plug-flow mixer (Eq. 10.43) with $t_0 = 0$, we compute:

$$\sigma^2 = \int_0^\infty \delta(t - \bar{t})(t - \bar{t})^2 dt \tag{10.61}$$

However, in Section 10.3 we noted that a useful property of the Dirac delta function is that when it is integrated with another function [like $(t - \bar{t})^2$ in Eq. 10.61], it returns the value of that function at $t = \bar{t}$; hence, for the ideal plug flow reactor,

$$\sigma^2 = 0 \tag{10.62}$$

Shortly it will be convenient to characterize the dimnsionless moments, particularly the *dimensionless variance*, of different residence-time models and measurements. The dimensionless variance is simply defined as

$$\sigma_\tau^2 = \frac{\sigma^2}{(\bar{t})^2} \tag{10.63}$$

Thus, the dimensionless variance for the perfect mixer is 1, and the dimensionless variance for the ideal plug-flow reactor remains 0. In the formula above tau stands for dimensionless time (Eq. 10.45). Recall that at the end of Section 10.5, we noted that the perfect mixer and ideal plug-flow reactor will bound the performance of most ideal and real reactor systems. We will find, then, that for the vast majority of residence-times distributions:

$$0 \leqslant \sigma_\tau^2 \leqslant 1 \tag{10.64}$$

For example, for the residence-time distribution measured in Example 2, $\sigma_\tau^2 = 18.36/(7.95)^2 = 0.29$.

10.7. OTHER RESIDENCE-TIME MODELS

Several other models have been useful in characterizing residence-time distributions. The models discussed in this section have one or more empirical parameters. The two models discussed previously, the perfect mixer and the ideal plug-flow reactor, have one parameter, that is, the mean residence time. However, $\bar{t}$ is often known independently (e.g., from a tracer study or from $\bar{t} = V/Q$), and is therefore usually not considered an "empirical parameter." Hence, the perfect mixer and the ideal plug-flow reactor are usually referred to as zero-parameter models.

10.7.1.   Tanks In Series

A useful model for residence time is shown in Figure 10.11. As indicated in the figure, the tanks-in-series model imagines a number of perfectly mixed tanks of equal size arranged in series. The number of tanks can have any integral value from 1 to ∞. Generally, the sum of the volumes of all the tanks is equal to the volume of the real system being modeled, that is,

$$V = V + V_2 + V_3 + \cdots + V_n \tag{10.65}$$

Using transfer functions and linear systems modeling, it can be shown that the residence-time density function for the tanks-in-series model is (Nauman and Buffham, 1983)

$$f(t) = \frac{N^N t^{N-1}}{(\bar{t})^N (N-1)!} \exp\left(-\frac{Nt}{\bar{t}}\right) \tag{10.66}$$

where N, the number of tanks in the series, is the parameter of this one-parameter model. As in Section 10.5, the dimensionless residence-time density is obtained simply by multiplying $f(t)$ by the mean residence time $\bar{t}$ and replacing $t/\bar{t}$ with τ. This modifies Eq. 10.66 slightly:

$$\bar{t}f(t) = \frac{N^N t^{N-1}}{(\bar{t})^{N-1}(N-1)!} \exp\left(-\frac{Nt}{\bar{t}}\right) = \frac{N^N \tau^{N-1}}{(N-1)!} \exp(-N\tau) = f(\tau) \tag{10.67}$$

In Figure 10.12, we have plotted the dimensionless residence-time density versus the dimensionless time τ. Note that as the number of tanks increases, the residence-time density moves from the exponential distribution of the single perfectly mixed tank ($N = 1$) to a distribution that increasingly seems to be centered at $\tau = 1$. In fact, as the number of tanks approaches ∞, the tanks-in-series model approaches the Dirac function centered at $\tau = 1$. Therefore, as N approaches ∞, the residence-time density for the tanks-in-series model approaches the residence-time density for ideal plug flow. We see that by varying the parameter N, the tanks-in-series model characterizes mixing varying between the perfect mixer and the ideal plug-flow mixer. This is one of the chief strengths of the tanks-in-series model.

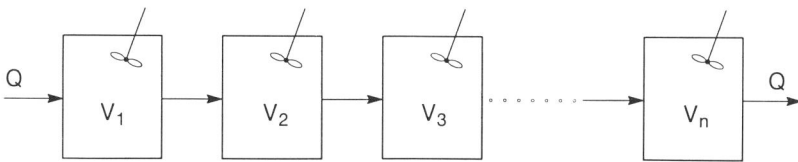

Figure 10.11. The tanks-in-series model.

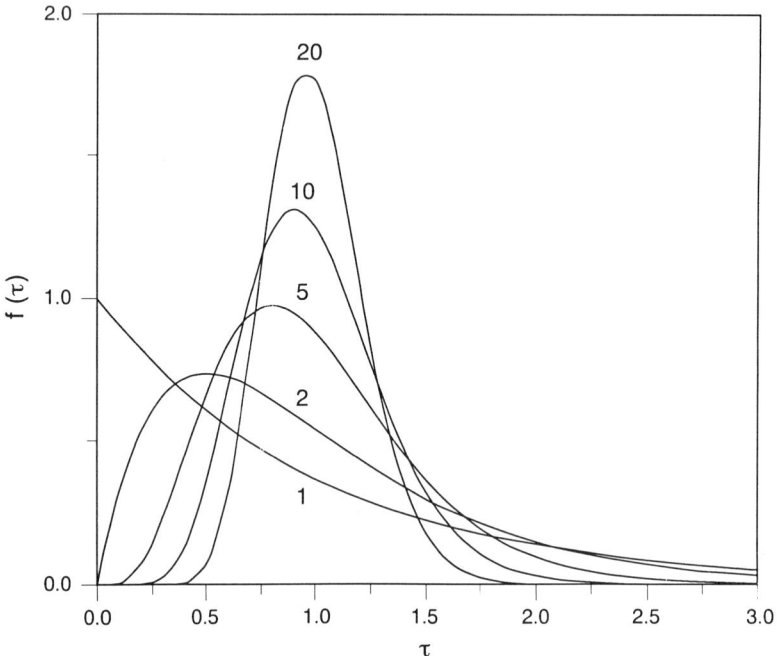

Figure 10.12. Dimensionless residence-time density for tanks-in-series model. The numbers next to the curves are the number of tanks in the series.

The dimensionless variance for the tanks-in-series model is quite simple (Nauman and Buffham, 1983):

$$\sigma_\tau^2 = \frac{1}{N} \tag{10.68}$$

Note that this formula is consistent with our earlier description of the dimensionless variances of the ideal plug flow and perfect mixers: For $N = 1$, $\sigma_\tau^2 = 1$ (i.e., perfect mixing), and for $N \to \infty$, $\sigma_\tau^2 = 0$ (i.e., plug flow).

10.7.2. Gamma-Function Extension to Tanks-In-Series Model

One notices that the shape of the residence-time density function for the tanks-in-series model is much more sensitive to changes in N when N is small. For example, in Figure 10.12, note the fairly large changes in $f(t)$ as N changes from 1 to 2, and the much smaller changes in $f(t)$ that would result if N changes from 10 to 11. One might wonder then, whether N can take on fractional values, especially in the low-N range. In other words, can N be considered a continuous rather than a discrete variable? It is possible to have nonintegral values of N, and the resulting residence-time density function is (Nauman and Buffham, 1981)

$$f(t) = \frac{N^N t^{N-1}}{(\bar{t})^N \Gamma(N)} \exp\left(-\frac{Nt}{\bar{t}}\right) \tag{10.69}$$

Here $\Gamma(N)$ is the gamma function:

$$\Gamma(N) = \int_0^\infty \exp(-x) x^{N-1} dx \tag{10.70}$$

The gamma function is tabulated in many reference books and is available on many spreadsheet and equation-solver programs. With Eq. 10.69, it is possible to have much more freedom in modeling residence-time distributions, especially in the low-N range. Note that, as opposed to all the models considered previously, there is no rigorous interpretation of N when it takes on fractional values.

The dimensionless variance for the gamma-function extension of the tanks-in-series model is also given by Eq. 10.68. We have said previously that the vast majority of residence-time distributions had dimensionless variances between 0 and 1. When N in the gamma-function extension takes on fractional values greater than 1, we find dimensionless variance values between 0 and 1, just as in the original tanks-in-series model. But can we let N take on a fractional value less than 1? In Figure 10.13, we present the dimensionless form of the gamma-function extension for some fractional values of N greater than and less than 1.

First, note the expected behavior for $N = 1$, which is the same as the perfect mixer of Figures 10.10 and 10.12. The curve for $N = 1.5$ seems to split the difference between the curves for $N = 1$ and 2. However, the curve for $N = 0.5$ demonstrates behavior not yet seen in any of the residence-time densities examined in this chapter. The early values ($\tau < 0.2$) appear much higher than the curve for $N = 1$, whereas most of the later values fall below the curve for $N = 1$. The curve for $N = 0.5$ seems to suggest that a high proportion of molecules have very short residence times; in fact, it suggests that more molecules have short residence times than in perfect mixing (i.e., $N = 1$). Nauman and Buffham (1983) call this phenomenon *bypassing*. This means that some fraction of the fluid molecules pass too quickly out of the system—they essentially bypass the system. Although this type of reactor performance has been discussed in the chemical engineering literature, it has not often been documented in the environmental engineering literature, and we will not further consider the issue of bypassing. In summary, the gamma-function extension to the tanks-in-series model can be thought of as an empirical

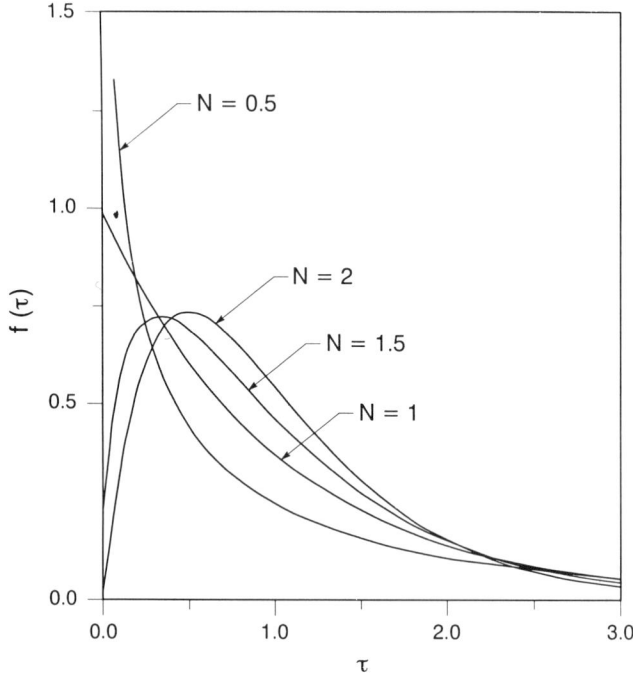

Figure 10.13. Gamma-function extension to the tanks-in-series model.

residence-time distribution, with a single parameter N, which can be considered a continuous variable. It can be useful for fitting residence-time distributions which do not correspond well with small integral values of N in the tanks-in-series model.

10.7.3.   Fractional Tubularity

Often when residence-time density functions are measured, the density function is zero-valued for a finite amount of time (generally some fraction of the mean residence time), quickly peaks, and is then followed by an exponential decay. In other words, the residence-time density resembles that for a delayed perfect mixer (Figure 10.15). A model that is sometimes useful in simulating this behavior is shown in Figure 10.14. It is called a fractional-tubularity model because the flow first passes through an ideal plug-flow reactor (PFR), which is called a tubular- or piston-flow reactor in some books. After passing through the ideal PFR, the flow then enters a perfect mixer. It is a fairly simple matter to guess the residence-time density function for this model. First note that any ideal PFR delays the exit of the tracer for a time period equal to its mean residence time. Thus, if there is a pulse injection of tracer at the entrance of the

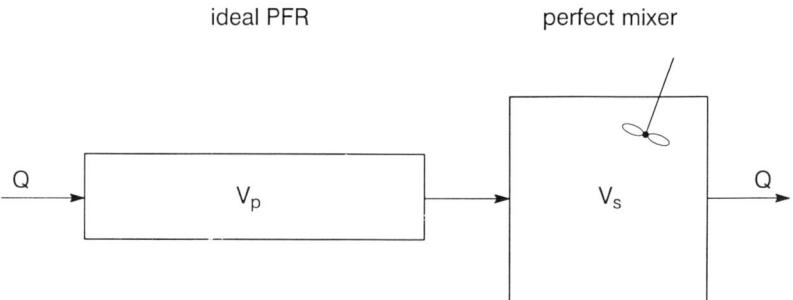

Figure 10.14. Reactor configuration for fractional-tubularity model.

PFR at $t = 0$, the tracer exits the PFR as a perfect impulse (Dirac function) at a time $t = \bar{t}_p$ later, where $\bar{t}_p$ is the mean residence time of the plug-flow reactor:

$$\bar{t}_p = \frac{V_p}{Q} \tag{10.71}$$

At $t = \bar{t}_p$, all the tracer enters the perfect mixer just as if we were performing an impulse experiment. Therefore, we conclude that the following must represent the residence-time density for the fractional-tubularity model:

$$f(t) = \begin{cases} 0 & t < \bar{t}_p \\ \dfrac{1}{\bar{t}_s} \exp\left(-\dfrac{t - \bar{t}_p}{\bar{t}_s}\right) & t > \bar{t}_p \end{cases} \tag{10.72}$$

Here $\bar{t}_s$ is the mean residence time in the perfect mixer:

$$\bar{t}_s = \frac{V_s}{Q} \tag{10.73}$$

In the fractional-tubularity model, we usually require that

$$\bar{t}_p + \bar{t}_s = \bar{t} \tag{10.74}$$

or that the sum of the mean residence times of the two subsystems must equal the overall mean residence time of the system. The dimensionless variance of the fractional-tubularity model is

$$\sigma_\tau^2 = (1 - \tau_p)^2 = \tau_s^2 \tag{10.75}$$

where τ_p and τ_s are the dimensionless residence time of the PFR and perfect mixers:

$$\tau_p = \frac{\bar{t}_p}{\bar{t}} \tag{10.76}$$

and

$$\tau_s = \frac{\bar{t}_s}{\bar{t}} \tag{10.77}$$

Although the fractional-tubularity model seems to have two parameters, $\bar{t}_p$ and $\bar{t}_s$ (or τ_p and τ_s), it is actually a one-parameter model because Eq. 10.74 shows that $\bar{t}_p$ and $\bar{t}_s$ are not independent (and from Eqs. 10.74, 10.76, and 10.77, $\tau_p + \tau_s = 1$). To see what this density function looks like, note that, using the definitions given above, the dimensionless residence-time density for $\tau > \tau_p$ can be reduced to

$$\bar{t}_s f(t) = \exp\left[-\frac{1}{\tau_s}(\tau - \tau_p)\right] = f(\tau) \tag{10.78}$$

The dimensionless residence-time density for the fractional-tubularity model is plotted in Figure 10.15 for $\tau_p = 0.2$ and $\tau_s = 0.8$. Note that the distribution is

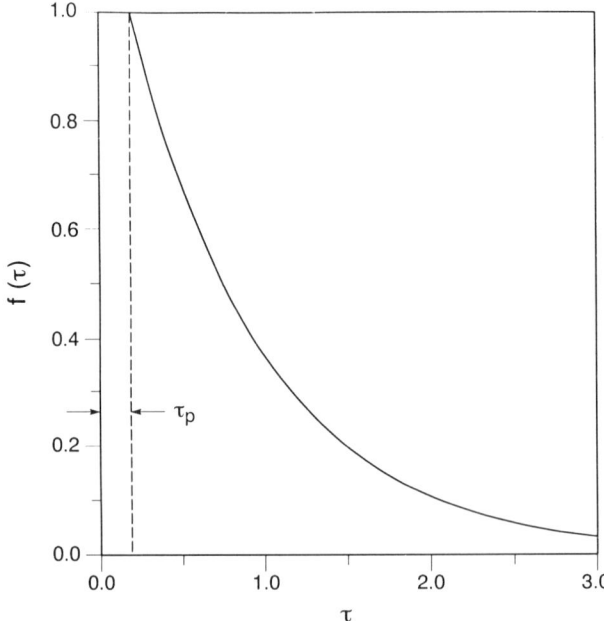

Figure 10.15. Residence-time density function for the fractional-tubularity model; $\tau_p = 0.2$ and $\tau_s = 0.8$.

characterized by a sharp discontinuity at $\tau = \tau_p$ (called the first-appearance time).

One might also wonder if the fractional tubularity model can be extended to consider an ideal plug-flow reactor followed by an N tanks-in-series configuration. This has been done, and has been found to be useful in residence-time modeling. It is pursued further in Exercise 10.5.

10.7.4. Crossflow and Stagnancy

In certain mixing systems, some of the fluid may be considered relatively stagnant. For example, injected tracer might move through part of the system relatively quickly, while being retained for longer periods in the stagnant regions. Consider air moving through a house between two open windows on opposite ends of the house. As the air moves through the entrance window, out the first room, down the hall past rooms without windows, and finally out of the house, there can be an exchange of air parcels and molecules with numerous relatively stagnant zones (e.g., closets and other rooms). Tracer entering relatively stagnant regions of the house will definitely be slowed down, and will exit the house at a much later time. One model developed to simulate this effect is called the *crossflow model*, shown in Figure 10.16. Note that tracer cannot directly exit the second, stagnant subsystem. This leads to the stagnancy

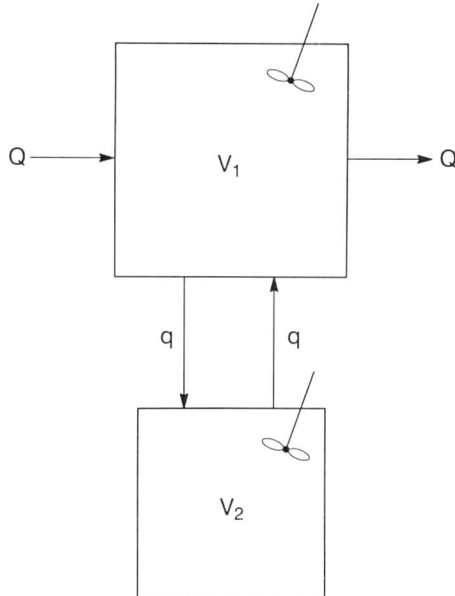

Figure 10.16. A crossflow model.

property, which is usually indicated by a long tail in the residence-time density function. The residence-time density function for the stagnancy model is given as (Nauman and Buffham, 1983)

$$f(t) = \frac{(1 - a_1\bar{t}_2)\exp(-a_1 t) - (1 - a_2\bar{t}_2)\exp(-a_2 t)}{(a_2 - a_1)(1 + q/Q)\bar{t}_1\bar{t}_2} \qquad (10.79)$$

Here $\bar{t}_1 = V_1/(Q + q)$, $\bar{t}_2 = V_2/q$, and

$$a_1, a_2 = \frac{1}{2}\left[\frac{\bar{t}_1 + \bar{t}_2}{\bar{t}_1\bar{t}_2} \pm \sqrt{\frac{(\bar{t}_1 + \bar{t}_2)^2}{\bar{t}_1^2\bar{t}_2^2} \frac{4}{(1 + q/Q)\bar{t}_1\bar{t}_2}}\right] \qquad (10.80)$$

The stagnancy model effectively has two parameters, q/Q and V_1/V_2. Note that because of the two exponential terms, the model also has two time constants, a_1^{-1} and a_2^{-1}. This allows a fair amount of flexibility in modeling long tails in residence-time densities.

A stagnancy property exists in other models we have studied. For example, recall from Chapters 5 and 7 that in laminar pipe flow, the central section of the flow travels much faster than the fluid near the pipe walls. Tracer that diffuses to the near-wall fluid is delayed to a certain extent relative to tracer in the faster-moving fluid, as if it had entered a stagnant region. The net result of this radial diffusion or mixing was that the tracer exiting the system was spread out over time. Exercise 10.6 investigates the use of a partitioning model from Chapter 4 to explore the stagnancy property.

10.7.5. Recycle Models

Many types of recycle reactors are found in environmental-engineering processes, especially biological processes. In the generic-recycle reactor, a fixed fraction of material leaving the reactor is recycled to the reactor inlet. This does not necessarily violate the assumption of a closed system, as we always define the system to include the recycle connections. Several recycle reactor models are shown in Figure 10.17, where the dotted lines define the system.

The recycle reactor is important in biological processes for a number of reasons. First, the recycle flow helps maintain high concentrations of microorganisms in suspended-growth systems. Also, we will see that as the recycle flow is increased, the reactor takes on the characteristics of a perfect mixer. In the perfect mixer, any incoming material is instantaneously diluted. This is a useful characteristic when compounds toxic to the biological process enter the system. In fixed-film biological processes, the recycle flow increases the fluid velocity past the microbiological films, and is thought to thereby decrease the thickness of concentration boundary layers and increase mass transfer. Finally, in fluidized-bed reactors with recycle, the recycle flow is actually critical for keeping the bed fluidized.

(a)

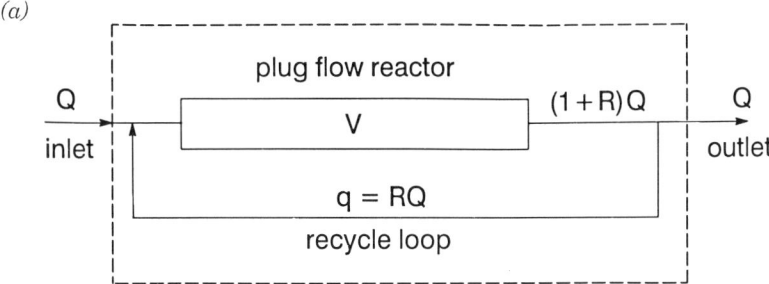

(b)

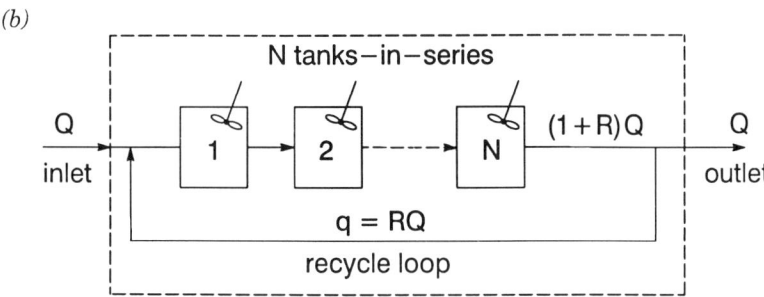

(c)

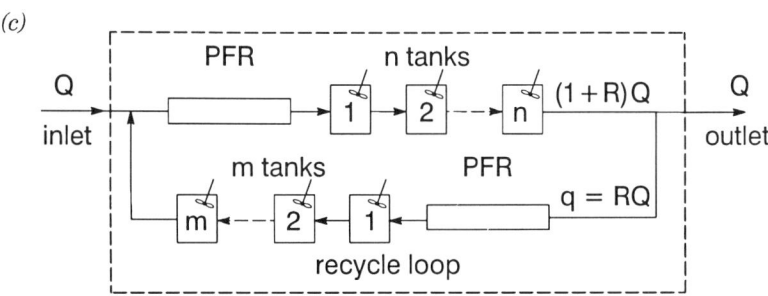

Figure 10.17. Three versions of the recycle reactor: (a) plug-flow plus recycle; (b) N tanks in series plus recycle; (c) plug flow plus tanks in series in main and recycle sections.

Note that in the recycle reactor, the recycle flow q is defined as RQ, where R is the *recycle ratio*:

$$R = \frac{q}{Q} \tag{10.81}$$

Because of the recycle, the inlet and outlet to the main reactor section must be $(1 + R)Q$. Note that in the first two models in Figure 10.17, the recycle line is assumed to have an infinitesimally small volume itself; therefore, the residence time in the recycle loop itself is zero. Fluid entering the recycle loop is immediately transported to the head of the reactor.

The reactor in Figure 10.17a is a plug-flow reactor with recycle. Consider an impulse experiment at the inlet of this recycle reactor. The time for the impulse of tracer to pass through the plug-flow section and appear first at the system outlet must be

$$\text{first appearance time} = \frac{V}{(1 + R)Q} = \frac{\bar{t}}{1 + R} \tag{10.82}$$

The fraction of tracer leaving the system at the first appearance time must be given by the ratio of the exit flow and $(1 + R)Q$, that is,

$$\left\{ \begin{matrix} \text{fraction tracer} \\ \text{leaving system} \end{matrix} \right\} = \frac{Q}{(1 + R)Q} = \frac{1}{1 + R} \tag{10.83}$$

This tracer appears in the exit stream as a Dirac function:

$$\delta\left(t - \frac{\bar{t}}{1 + R} \right) \tag{10.84}$$

However, at this same instant, a certain fraction of tracer is recycled to the front of the system, and this fraction is defined by the ratio of the recycle flow to $(1 + R)Q$:

$$\left\{ \begin{matrix} \text{fraction tracer} \\ \text{recycled} \end{matrix} \right\} = \frac{RQ}{(1 + R)Q} = \frac{R}{1 + R} \tag{10.85}$$

This recycled tracer reaches the system outlet at twice the first appearance time. On this second passage, the fraction leaving the system is

$$\left\{ \begin{matrix} \text{fraction of tracer exiting} \\ \text{system at second passage} \end{matrix} \right\} = \left(\frac{R}{1 + R} \right)\left(\frac{1}{1 + R} \right) \tag{10.86}$$

This material leaves the system as a second Dirac impulse. Continuing this thought experiment we conclude that the residence-time density function for the plug-flow with recycle model must be an infinite series of delta functions (Rippin, 1967):

$$f(t) = \sum_{i=1}^{\infty} \left(\frac{1}{1 + R} \right)\left(\frac{R}{1 + R} \right)^{i-1} \delta\left(t - \frac{i\bar{t}}{1 + R} \right) \tag{10.87}$$

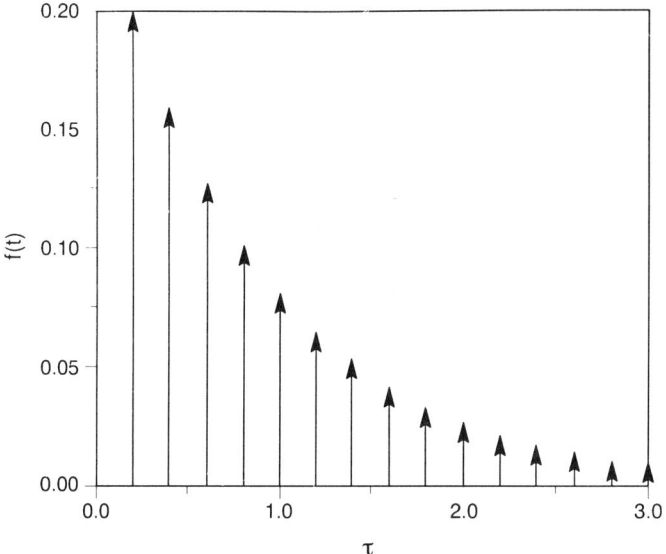

Figure 10.18. Plot of the plug-flow plus recycle model for $R = 4$.

A plot of Eq. 10.87 for $R = 4$ is shown in Figure 10.18. Note how the recycle reactor seems to mimic an exponential distribution of residence time. We note from Eq. 10.87 that as the recycle ratio increases, the density of Dirac functions increases; in the limit of $R \to \infty$, we would find the residence time density for perfect mixing.

The mean residence time of the plug-flow with recycle model is given by Eq. 10.7, and the dimensionless variance is given as (Fu et al., 1971)

$$\sigma_\tau^2 = \frac{R}{1 + R} \tag{10.88}$$

In Figure 10.17b, we show an N tanks-in-series model with recycle. The residence time density for this model is given by Fu et al. (1971) as

$$f(t) = \frac{1}{R} \sum_{i=1}^{\infty} \left(\frac{R}{1+R} \right)^i \frac{1}{\bar{T}_N^{iN}} \frac{t^{iN-1}}{(iN-1)!} \exp\left(-\frac{t}{\bar{T}_N} \right) \tag{10.89}$$

Here $\bar{T}_N$ is defined by

$$\bar{T}_N = \frac{V_n}{Q + q} = \frac{V_n}{(1 + R)Q} \tag{10.90}$$

where V_n is the volume of one of the tanks in the N tanks-in-series model (Figure 10.11). The mean residence time for this model can be expressed as

$$\bar{t} = N(1 + R)\bar{T}_N \tag{10.91}$$

and the dimensionless variance is (Fu et al., 1971)

$$\sigma_t^2 = \frac{(1 + NR)}{N(1 + R)} \tag{10.92}$$

Undoubtedly, the most versatile recycle residence-time density model is presented in Figure 10.17c. Here the main reactor section is composed of ideal plug flow plus n tanks in series, and the recycle line is composed of ideal plug flow plus m tanks in series. In the most general model, the two tanks-in-series subsystems can have different numbers of tanks in series (n and m), and the different reactors in the two subsystems can have different volumes. (T_1 and T_2 are the mean residence times of the two tanks-in-series subsystems, and τ_1 and τ_2 are the mean residence times of the two plug-flow subsystems.) The resulting expression for the residence-time density function is fairly complicated (Gibilaro, 1971). A less-complicated expression derives if we assume that $T_1/n = T_2/m$, or if one of the tanks-in-series subsystems has zero volume (i.e., either the main or recycle subsystem has only a plug-flow component). In this case, the residence-time density is given by (Gibilaro, 1971)

$$f(t) = \frac{1}{1 + R} \sum_{i=1}^{\infty} \left(\frac{R}{1 + R}\right)^{i-1} \frac{\gamma^\epsilon t_*^{\epsilon-1} \exp(-\gamma t)}{\Gamma(\epsilon)} H(t_*) \tag{10.93}$$

When H is the unit-step function (recall Eq. 10.41), γ is

$$\gamma = \frac{T_2}{m} \qquad \epsilon = m(i - 1) \text{ for } T_1 = 0 \tag{10.94}$$

$$\gamma = \frac{T_1}{n} \qquad \epsilon = ni \text{ for } T_2 = 0 \tag{10.95}$$

or

$$\gamma = \frac{T_1}{n} \qquad \epsilon = ni + m(i - 1) \text{ for } \frac{T_1}{n} = \frac{T_2}{m} \tag{10.96}$$

and t_* is given by

$$t_* = t - [ni\tau_1 - m(i - 1)\tau_2] \tag{10.97}$$

The use of the gamma function in Eq. 10.93 implies that the number of tanks in series does not have to be an integral value.

10.8. AXIAL-DISPERSION MODEL

The reader will recall from Chapters 7 and 8 that an important model for convective-diffusive transport in laminar and turbulent flows, as well as flow in porous media, was the one-dimensional dispersion model (recall Figure 8.1). The axial-dispersion model is often useful in modeling flows in systems that approximate one-dimensional dispersion problems, for example, in long narrow reactors, packed beds, and fluidized beds. The one-dimensional convective-diffusion equation takes the general form (cf., Eqs. 7.81 and 8.26):

$$\frac{\partial c}{\partial t} + u \frac{\partial c}{\partial x} = D \frac{\partial^2 c}{\partial x^2} \tag{10.98}$$

In this equation, D can be interpreted as the molecular-diffusion coefficient in laminar-flow transport problems (Chapter 7), the one-dimensional dispersion coefficient in turbulent-transport problems (Chapter 7), or the one-dimensional hydrodynamic dispersion coefficient for porous-media transport problems (Chapter 8). Considering the flow situation shown in Figure 10.9, the solution to this equation for a plane source of tracer at the origin (an impulse input at $x = 0$ and $t = 0$) has already been given in Chapter 8 as

$$c(x, t) = \frac{M}{(4\pi t)^{1/2}} \exp\left[-\frac{(x - ut)^2}{4Dt}\right] \tag{10.99}$$

where M is the unit area mass of tracer injected (units $= M/L^2$; recall discussion of Section 6.7.2). We would like to convert $c(x, t)$ into a regular residence-time density function, which is accomplished with Eq. 10.18:

$$f(t) = \frac{c(t)}{\displaystyle\int_0^\infty c(t)dt} \tag{10.18}$$

With regard to the numerator, note that if we carefully consider the mass of tracer leaving the system:

$$\int_0^\infty c(t)u \, dt = u \int_0^\infty c(t)dt = M \tag{10.100}$$

This equation differs slightly from Eq. 10.27 because of the unusual units of M.

Combining Eqs. 10.18, 10.99, and 10.100, we find that

$$f(t) = \frac{u}{2(\pi Dt)^{1/2}} \exp\left[-\frac{(x - ut)^2}{4Dt}\right] \tag{10.101}$$

For the one-dimensional system shown in Figure 10.9, the mean residence time is defined by $\bar{t} = L/u$; letting $x = L$, Eq. 10.101 can be rearranged to

$$f(t) = \frac{1}{\bar{t}} \frac{1}{2(\pi\,\mathrm{Pe}^{-1})^{1/2}} \exp\left[-\frac{(L - ut)^2}{4Dt}\right] \tag{10.102}$$

where Pe is the Peclet number:

$$\mathrm{Pe} = \frac{Lu}{D} \tag{7.8}$$

Further rearrangement of the argument of the exponential yields the final form for the dimensionless residence-time density function:

$$\bar{t}f(t) = \frac{1}{2(\pi\,\mathrm{Pe}^{-1})^{1/2}} \exp\left[-\frac{(1 - \tau)^2}{4\,\mathrm{Pe}^{-1}}\right] = f(\tau) \tag{10.103}$$

The response indicated by Eq. 10.103 is the familiar Gaussian curve shown in Figure 8.1 (i.e., a perfect Gaussian curve convected past the system outlet). This model has a single parameter, the Peclet number. The dimensionless variance is given by Levenspiel (1972) as

$$\sigma_\tau^2 = \frac{2D}{uL} = 2\,\mathrm{Pe}^{-1} \tag{10.104}$$

It turns out that the solution represented by Eq. 10.103 can only be accurate for small amounts of dispersion, or equivalently, for large Peclet numbers. When dispersion is large, the shape of the output concentration curve changes significantly during passage of the tracer out of the system. Hence, Eq. 10.103 cannot apply for small Peclet numbers.

Numerical techniques have been used to solve the governing equation, Eq. 10.98, for closed systems; unlike Eq. 10.103, this solution is not restricted to high Peclet numbers. There has apparently been some development in thought concerning the appropriate boundary conditions for solving Eq. 10.98 in both closed and open systems. The basic idealization of open and closed axial-dispersion systems is shown in Figure 10.19. Note that in the closed system, there is a discontinuity in the diffusion or dispersion coefficient when crossing the entrance or exit boundaries. Essentially, there is no diffusion or dispersion just before and just after the system boundaries, only within the system. The

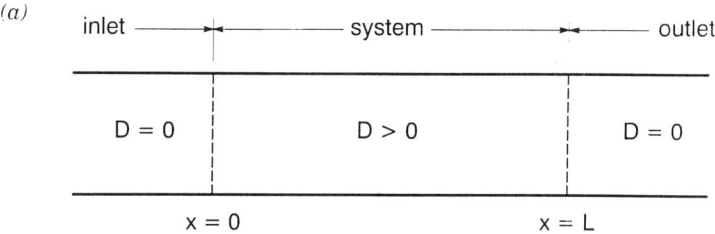

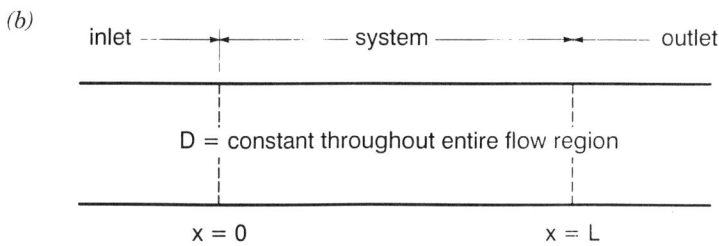

Figure 10.19. Axial-dispersion models for (a) closed and (b) open systems.

closed-system assumption is generally good in systems with high velocities in the inlet and outlet. Nauman and Buffham (1983) have argued that by considering flux across the inlet boundary to be continuous, the correct boundary condition for the closed-system inlet is

$$uc_{in}(t) = uc(t, 0+) - D \left.\frac{\partial c}{\partial x}\right|_{x=0+} \tag{10.105}$$

where $0+$ refers to a location an infinitesimal distance inside the inlet boundary. At the outlet of the closed system, the flow passes from a mixed region to an unmixed region; therefore, there can not be a composition charge moving across the exit. Hence, the appropriate boundary conditions must be

$$\left.\frac{\partial c}{\partial x}\right|_{L-} = 0 \quad \text{for all } t \tag{10.106}$$

where $L-$ refers to a location an infinitesimal distance inside the outlet boundary. The numerical solution of Eq. 10.98 for these boundary conditions is shown in Figure 10.20. This is an extremely interesting plot. Note that for a very small Peclet number, which corresponds to an infinite dispersion or diffusion coefficient (or a very small product uL), the residence-time density function has an exponential shape identical to the perfect mixer. On reflection, this of course makes sense. When diffusion or dispersion is infinite, the tracer

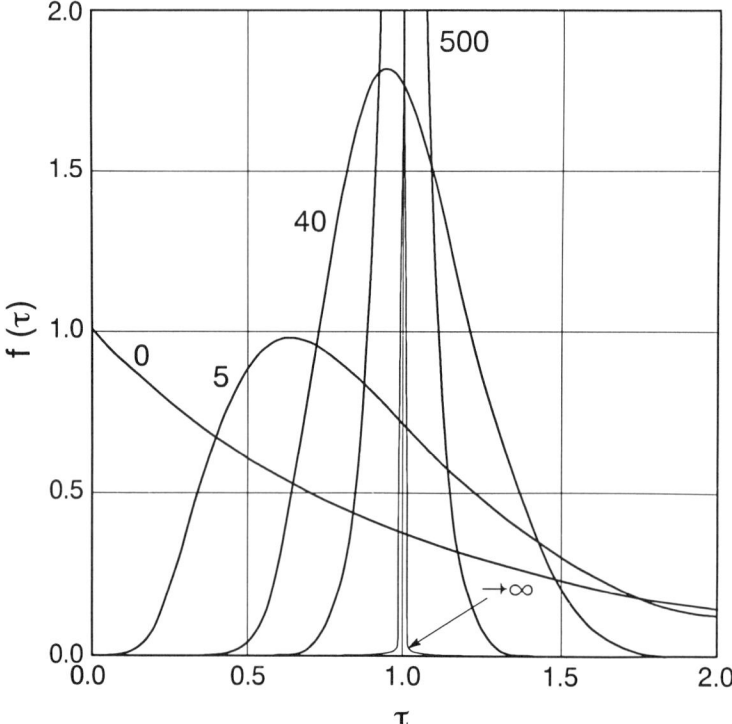

Figure 10.20. Dimensionless residence-time density function for axial dispersion in a closed system. The numbers next to the curves are Peclet numbers. *Source*: Reproduced from Levenspiel, *Chemical Reaction Engineering*, p. 277, © 1972. Reprinted by permission of John Wiley & Sons, Inc.

or transported material is instantaneously mixed within the closed system. As the Peclet number grows larger, however, the residence-time density function approaches the Gaussian shape discussed earlier. As the Peclet number approaches infinity, $f(t)$ approaches the Dirac function indicative of ideal plug flow.

The dimensionless variance for the closed-axial dispersion model is (Nauman and Buffham, 1983)

$$\sigma_\tau^2 = \frac{2}{\text{Pe}} - \frac{2}{\text{Pe}^2}[1 - \exp(-\text{Pe})] \qquad (10.107)$$

The reader should verify that the limits do indeed correspond to 0 (ideal plug flow) and 1 (perfect mixing) by substituting a wide range of Peclet numbers into Eq. 10.107.

As we have stressed, the numerical solution shown in Figure 10.20 is strictly for closed systems; the reader will recall that in a closed system, molecules can only enter and leave the system once. But in diffusive or dispersive systems like the axial-dispersion system, the system may not be truly closed. Recall that dispersive phenomena rely on molecular diffusion, turbulent diffusion, or mechanical dispersion (e.g., in laminar flows, turbulent flows, or porous-media flows, respectively). Although the flow is assumed to be moving to the right in the systems shown in Figure 10.19, at any given instant, diffusion or dispersion can be transporting mass in any direction. It may also be that at some particular point or time, mass is actually moving to the left. Therefore, it is possible for mass to leave for a short time through the entrance, or come back in the exit. Then the system cannot be considered closed in a strict sense. This situation is diagrammed in Figure 10.19b. The likelihood of open-system characteristics increases as the diffusion or dispersion coefficient increases, or as the product uL decreases (i.e., for small values of the Peclet number). For open systems, the appropriate boundary conditions are based on continuity of flux and concentration across the inlet and outlet boundaries (Nauman and Buffham, 1983). The solution to the one-dimensional dispersion equation for the open system is given for $x = L$ as (Levenspiel, 1972)

$$f(t) = \frac{1}{\bar{t}} \frac{1}{2\left[\pi \frac{t}{\bar{t}} Pe^{-1}\right]^{1/2}} \exp\left[-\frac{(L-ut)^2}{4\frac{t}{\bar{t}} Dt}\right] \qquad (10.108)$$

This can also easily be reexpressed in dimensionless terms as a one-parameter model:

$$\bar{t}f(t) = \frac{1}{2(\pi\tau Pe^{-1})^{1/2}} \exp\left[-\frac{(1-\tau)^2}{4\tau Pe^{-1}}\right] = f(\tau) \qquad (10.109)$$

Figure 10.21 shows a plot of Eq. 10.109 for different values of the Peclet number. As discussed above, as the Peclet number grows larger, the prediction of Eq. 10.109 approaches the perfect Gaussian curve predicted by Eq. 10.103. However, at the opposite extreme of a very small Peclet number, we no longer find the exponential curve for perfect mixing in a closed system. This too makes sense. In an open system like Figure 10.19b with extremely high dispersion, the tracer or transported material simply diffuses to infinity in the positive and negative x directions; the tracer concentration approaches zero for all values of space and time. Note the flat function in Figure 10.21 for $Pe = 0.1$. The dimensionless variance for axial dispersion in an open system is (Levenspiel, 1972)

$$\sigma_\tau^2 = \frac{2}{Pe} + \frac{8}{Pe^2} \qquad (10.110)$$

Indeed, as the Peclet number becomes very small, the dimensionless variance grows without limit.

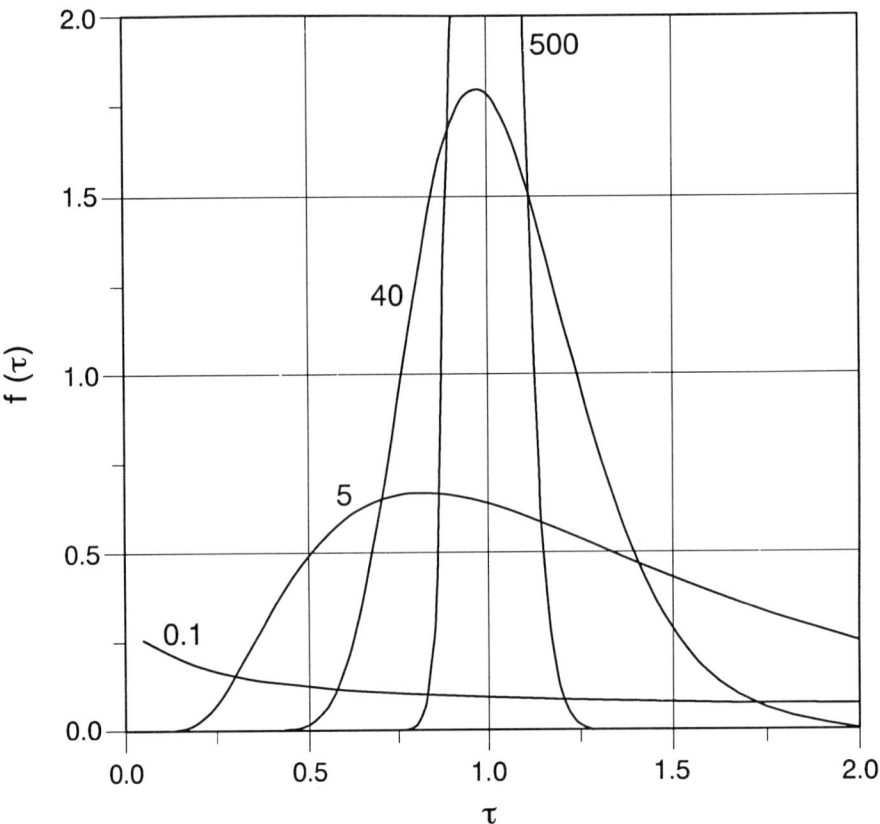

Figure 10.21. Dimensionless residence-time density function for axial dispersion in an open system. The numbers next to the curves are Peclet numbers.

10.9.   FITTING RESIDENCE-TIME DISTRIBUTIONS TO DATA

A student's first impulse when modeling real continuous-flow systems with the theoretical residence-time models developed in this chapter is often to pick the model that most resembles the real system physically. For example, if the system is a long tube with recycle (see Figure 10.17a), the inclination is to use the plug-flow with recycle model developed in Section 10.7.5. However, the student is often shocked to see that the theoretical model prediction does not look much like the real measured residence-time density function. In fact, the actual residence-time data may look much more like that predicted by the N tanks-in-series plus recycle, or even the axial-dispersion model. The student has experienced a real problem with modeling residence-time distributions: there is often no a priori method to select a residence-time model based on the configuration, dimensions, flow, or mixing in the real system. An exception to this rule is for systems that approximate one-dimensional axial dispersion. Researchers have had some success in correlating Reynolds numbers and Peclet numbers for long, narrow, packed beds (Levenspiel, 1972).

But getting back to the problem encountered above, how does one model residence times in real systems? The only alternative is often to fit real residence-time data to one of the theoretical models by searching for the model with the best fit. Two methods of fitting are in common use — the first entails calculating the model parameters from the moment of the real residence-time data (generally the first and second moments), and the second involves selecting the model parameters through minimization of the error between the real data and the model prediction.

For the first method, note that many of the parameters of the theoretical models developed in this chapter are expressed in terms of the mean residence time and variance. Therefore, if the mean residence time and variance can be calculated from an actual tracer experiment, the model parameters can also be calculated. However, the reader will quickly note a problem with this method: numerous models can be fit to the data and some might actually do a very poor job of approximating the real residence-time data. Therefore, the moments method may not be very good at discriminating between different candidate models. The second method seeks to select the model parameter(s) that minimizes the error between the model and the real residence-time data. It is called *time-domain fitting*. Time-domain fitting suffers from one problem of the moments method: numerous models can be fit. However, the time-domain method is usually superior to the methods moment because it can optimize the fit for any given model, and it can be used to discriminate between various candidate models through comparison on an error term developed below.

The time-domain method seeks those parameter values that minimize the sum of the squared errors (SSE) between the data and model, that is, we attempt to minimize (Nauman and Clark, 1991)

$$\text{SSE}(P_1, P_2, \ldots) = \sum_i \left[f_{\text{data}}(t_i) - f_{\text{model}}(t_i) \right]^2 \qquad (10.111)$$

where P_i are the model parameters. Actually, it is very easy to accomplish time-domain fitting with Eq. 10.111 in modern spreadsheats, either by trial-and-error calculation of the SSE for different guesses of the model parameters, or with the mathematical optimization routine included in most new spreadsheet programs. As mentioned above, the time-domain fitting can be used in two ways. For a selected model, the parameters leading to the minimal SSE can be determined. If several candidate models are being investigated, time-domain fitting can be used to discriminate the model with the minimal SSE. The time-domain and the moments methods are investigated in Exercises 10.7 and 10.8.

There are some downsides to the fitting procedures outlined here. One is that the fitting procedure may not tell you much about mixing in the real systems. For example, if the real system is a single-mixed tank or other chamber, and if the N tanks-in-series model is found to fit real tracer data very well, what does the modeling procedure tell you about the flow and mixing? As noted above, this is a problem with traditional residence-time modeling procedures. Second, if a model is found that provides an excellent fit to the data, can this information be used to predict the residence-time distribution for a higher or lower flow rate, or for a larger- or smaller-scale version of the test system? Generally, the answer to these questions is no. This is because we have modeled the residence-time distribution and not the fundamental mixing.

However, new techniques are becoming available to a wider audience of environmental modelers. These techniques and software permit the numerical solution of the Navier–Stokes and convective-diffusion equations (with reactions) for the actual boundary conditions in the real flow system; in other words, we can model the mixing. One such technique was discussed in Section 5.13.5. In the future, these techniques will be used more frequently to understand and model flow, transport, and reaction in many different mixing systems.

10.10. MIXING AND REACTIONS

The previous sections of this chapter focused on characterizing residence-time distributions. The major reason for this effort is to determine the effect of mixing on reactions of environmental interest. The focus of this section is to show how residence-time data can be used to predict the course of reactions and to derive theoretical limits on the extent of reaction in flowing systems.

10.10.1. Reaction Probabilities and Conversion for First-Order Reactions

Recall that in Section 9.2 we derived the rate equation for an elementary, irreversible, first-order reaction of $A \rightarrow B$ in a homogeneous, batch (nonflow) reactor. The integrated rate equation was

$$\frac{[A]}{[A]_0} = \exp(-kt) \tag{9.6}$$

where $[A]_0$ is the initial concentration of A and k has units of T^{-1}. We ask the question, for any given molecule of A, what is the probability that it will react (i.e., be converted to B) during time t? If $t = 0$, no A would react, and the probability of reaction would be zero; if $t \rightarrow \infty$, all A would react during time t, and the probability of reaction would be unity. It follows from Eq. 9.6 that the probability of reaction of any molecule subject to the first-order reaction must be

$$P_r = 1 - \exp(-kt) = 1 - \frac{[A]}{[A]_0} \tag{10.112}$$

Now let's switch our frame of reference to a single, continuous-flow, mixed reactor like any of those outlined earlier in the chapter. If the inlet concentration of A is $[A]_{in}$, what is the probability that any entering A molecule reacts while in the system? This is a more difficult question, because unlike the batch reactor, the A molecules entering the system will all have different residence times (or times in which they can react). Molecules in the system for long-time have a high reaction probability; molecules in the system for a short time have a low reaction probability. We have to weigh the simple (batch) reaction probability derived above with the residence time. The correct method of doing this is

$$\bar{P}_r = \int_0^\infty P_r f(t)dt = \int_0^\infty [1 - \exp(-kt)]\, f(t)dt \tag{10.113}$$

where $\bar{P}_r$ is the reaction probability in the flow system. In analogy to the batch reactor, $\bar{P}_r$ must simply be related to the inlet and outlet concentrations as

$$\bar{P}_r = 1 - \frac{[A]_{out}}{[A]_{in}} = \text{conversion} \tag{10.114}$$

Equation 10.114 is an important equation in the chemical-engineering literature. It is the definition of the *chemical conversion*. For example, consider the limits to conversion: When $[A]_{out}$ is zero, the reactant has been completely converted, or the conversion is unity; when $[A]_{out}$ is equal to $[A]_{in}$, no reaction has taken place and the conversion is zero.

Now we substitute Eq. 10.114 into Eq. 10.113 and evaluate the integral:

$$\text{conversion} = 1 - \frac{[A]_{out}}{[A]_{in}} = \int_0^\infty f(t)dt - \int_0^\infty \exp(-kt) f(t)dt$$

$$= 1 - \int_0^\infty \exp(-kt)\, f(t)dt \tag{10.115}$$

Finally, by slight rearrangement we find:

$$\frac{[A]_{out}}{[A]_{in}} = \int_0^\infty \exp(-kt)\, f(t)dt \tag{10.116}$$

Equation 10.116 is a very powerful equation. It says that *to predict the course of any irreversible first-order reaction in a flow system, all we need to know is the residence-time density function and the reaction rate constant.* This equation has two main functions: (1) it can be used to establish the theoretical bounds on first-order reactions; (2) it can be used with real residence-time data to predict the course of first-order reactions in real flow systems. This second aspect is investigated in the following numerical example.

Example 10.3. For the system examined in Example 10.1, determine $[A]_{out}/[A]_{in}$ if A is subject to an irreversible, first-order reaction. Investigate first-order reaction rate constants of 0.05, 0.1, and 0.2 min^{-1}.

SOLUTION. Since the experiment described in Example 10.1 used discrete data, we will have to develop a discrete approximation of Eq. 10.116. That is simply

$$\frac{[A]_{out}}{[A]_{in}} = \sum_{i=0}^{imax-1} \exp\left[-k\left(\frac{t_i + t_{i+1}}{2}\right)\right]\left[\frac{f(t_i) + f(t_{i+1})}{2}\right](t_{i+1} - t_i) \quad (10.117)$$

This equation is used in the following spreadsheet (opposite page). The answers are found as the sums of columns 5–7, that is, for $k = 0.05\,\text{min}^{-1} = [A]_{out}/[A]_{in} = 0.687$, for $k=0.1\,\text{min}^{-1}=[A]_{out}/[A]_{in}=0.49$, and for $k=0.2\,\text{min}^{-1}=[A]_{out}/[A]_{in} = 0.275$. As expected, the predictions of this model are sensitive to the reaction-rate constant — the larger the kinetic constant, the greater the change in concentration.

The next goal in this section is to compute the chemical conversion for four of the theoretical models examined earlier in the chapter. Three of these are quite easy to derive with Eq. 10.116. For a first-order reaction in a *perfect mixer*, we simply substitute the $f(t)$ for the perfect mixer (Eq. 10.33) into Eq. 10.116:

$$\frac{[A]_{out}}{[A]_{in}} = \int_0^\infty \exp(-kt)\,\frac{1}{\bar{t}}\exp\left(-\frac{1}{\bar{t}}\right)dt = \frac{1}{\bar{t}}\int_0^\infty \exp\left[-\left(k + \frac{1}{\bar{t}}\right)t\right]dt \quad (10.118)$$

which is easily evaluated as

$$\frac{[A]_{out}}{[A]_{in}} = \frac{1}{k\bar{t} + 1} \quad (10.119)$$

The chemical conversion is simply

$$\text{conversion} = 1 - \frac{[A]_{out}}{[A]_{in}} = 1 - \frac{1}{k\bar{t} + 1} = \frac{k\bar{t}}{k\bar{t} + 1} \quad (10.120)$$

Conversion for the ideal *plug-flow reactor* is also easily calculated by substituting the $f(t)$ for this mixer (Eq. 10.43) into Eq. 10.116:

$$\frac{[A]_{out}}{[A]_{in}} = \int_0^\infty \exp(-kt)\delta(t - \bar{t})dt = \exp(-k\bar{t}) \quad (10.121)$$

Time (min)	Concentration (mg/L)	Area i^a	$f(t)$ (min^{-1})	$[A]_{out}/[A]_{in}$		
				$k = 0.05^b$	$k = 0.1^c$	$k = 0.2^d$
0	0	0.025	0.00E+00	1.95E−04	1.93E−04	1.88E−04
0.5	0.1	0.425	7.92E−04	3.24E−03	3.12E−03	2.90E−03
1	1.6	1.15	1.27E−02	8.55E−03	8.04E−03	7.09E−03
1.5	3	2.35	2.38E−02	1.70E−02	1.56E−02	1.31E−02
2	6.4	3.675	5.07E−02	2.60E−02	2.32E−02	1.86E−02
2.5	8.3	4.5	6.57E−02	3.11E−02	2.71E−02	2.06E−02
3	9.7	5.1	7.68E−02	3.43E−02	2.92E−02	2.11E−02
3.5	10.7	5.875	8.47E−02	3.86E−02	3.20E−02	2.20E−02
4	12.8	6.5	1.01E−01	4.16E−02	3.36E−02	2.20E−02
4.5	13.2	6.625	1.05E−01	4.14E−02	3.26E−02	2.03E−02
5	13.3	6.625	1.05E−01	4.03E−02	3.10E−02	1.84E−02
5.5	13.2	6.575	1.05E−01	3.91E−02	2.93E−02	1.65E−02
6	13.1	6.325	1.04E−01	3.66E−02	2.68E−02	1.43E−02
6.5	12.2	6	9.66E−02	3.39E−02	2.42E−02	1.23E−02
7	11.8	5.625	9.34E−02	3.10E−02	2.16E−02	1.04E−02
7.5	10.7	5.2	8.47E−02	2.79E−02	1.90E−02	8.74E−03
8	10.1	4.8	8.00E−02	2.52E−02	1.67E−02	7.30E−03
8.5	9.1	4.475	7.21E−02	2.29E−02	1.48E−02	6.16E−03
9	8.8	4.3	6.97E−02	2.14E−02	1.35E−02	5.35E−03
9.5	8.4	4.125	6.65E−02	2.01E−02	1.23E−02	4.65E−03
10	8.1	3.875	6.41E−02	1.84E−02	1.10E−02	3.95E−03
10.5	7.4	3.5	5.86E−02	1.62E−02	9.46E−03	3.23E−03
11	6.6	3.2	5.23E−02	1.44E−02	8.23E−03	2.67E−03
11.5	6.2	3.025	4.91E−02	1.33E−02	7.40E−03	2.28E−03
12	5.9	2.825	4.67E−02	1.21E−02	6.57E−03	1.93E−03
12.5	5.4	2.6	4.28E−02	1.09E−02	5.75E−03	1.61E−03
13	5	2.3	3.96E−02	9.39E−03	4.84E−03	1.29E−03
13.5	4.2	1.9	3.33E−02	7.56E−03	3.80E−03	9.62E−04
14	3.4	1.625	2.69E−02	6.31E−03	3.09E−03	7.44E−04
14.5	3.1	1.475	2.45E−02	5.59E−03	2.67E−03	6.11E−04
15	2.8	1.35	2.22E−02	4.99E−03	2.33E−03	5.06E−04
15.5	2.6	1.275	2.06E−02	4.59E−03	2.09E−03	4.33E−04
16	2.5	1.175	1.98E−02	4.13E−03	1.83E−03	3.61E−04
16.5	2.2	1.025	1.74E−02	3.51E−03	1.52E−03	2.85E−04
17	1.9	0.875	1.50E−02	2.92E−03	1.23E−03	2.20E−04
17.5	1.6	0.75	1.27E−02	2.44E−03	1.01E−03	1.71E−04
18	1.4	0.65	1.11E−02	2.07E−03	8.30E−04	1.34E−04
18.5	1.2	0.55	9.50E−03	1.71E−03	6.68E−04	1.02E−04
19	1	0.475	7.92E−03	1.44E−03	5.49E−04	8.00E−05
19.5	0.9	0.425	7.13E−03	1.25E−03	4.67E−04	6.48E−05
20	0.8	0.375	6.33E−03	1.08E−03	3.92E−04	5.17E−05
20.5	0.7	0.3	5.54E−03	8.42E−04	2.98E−04	3.74E−05
21	0.5	0.225	3.96E−03	6.16E−04	2.13E−04	2.54E−05
21.5	0.4	0.15	3.17E−03	4.00E−04	1.35E−04	1.53E−05
22	0.2	0.075	1.58E−03	1.95E−04	6.42E−05	6.94E−06
22.5	0.1	0.025	7.92E−04	6.35E−05	2.03E−05	2.09E−06
23	0	0	0.00E+00	0.00E+00	0.00E+00	0.00E+00
23.5	0	0	0.00E+00	0.00E+00	0.00E+00	0.00E+00
24	0	0	0.00E+00	0.00E+00	0.00E+00	0.00E+00
24.5	0	0	0.00E+00	0.00E+00	0.00E+00	0.00E+00
25	0	0	0.00E+00	0.00E+00	0.00E+00	0.00E+00

[a]Area = sum of column 3 = 126.3

[b]Sum of column 5 = 6.87E−01.

[c]Sum of column 6 = 4.90E−01.

[d]Sum of column 7 = 2.75E−01.

(Recall the discussion of the properties of the Dirac function in Sections 10.3 and 10.6.) Note that Eq. 10.121 is almost identical to the integrated equation for the homogeneous batch reactor from Chapter 9:

$$\frac{[A]}{[A]_0} = \exp(-kt) \tag{9.6}$$

Thus, we highlight a simple but important result: $[A]_{out}/[A]_{in}$ for a first-order reaction in an ideal plug-flow reactor with mean residence time $\bar{t}$ is identical to $[A]/[A]_0$ in a homogeneous batch reactor if we let $t = \bar{t}$ in the batch reactor. Therefore, for first-order reactions, we can think of the plug-flow reactor as a batch reactor which is turned off at $t = \bar{t}$.

Finally, conversion in the ideal plug-flow reactor is

$$\text{conversion} = 1 - \frac{[A]_{out}}{[A]_{in}} = 1 - \exp(-k\bar{t}) \tag{10.122}$$

It is now rewarding to compare conversion for a first-order reaction in the perfect mixer and the ideal plug-flow reactor. In Figure 10.22, we plot Eqs. 10.120 and 10.122 versus the dimensionless time $k\bar{t}$.

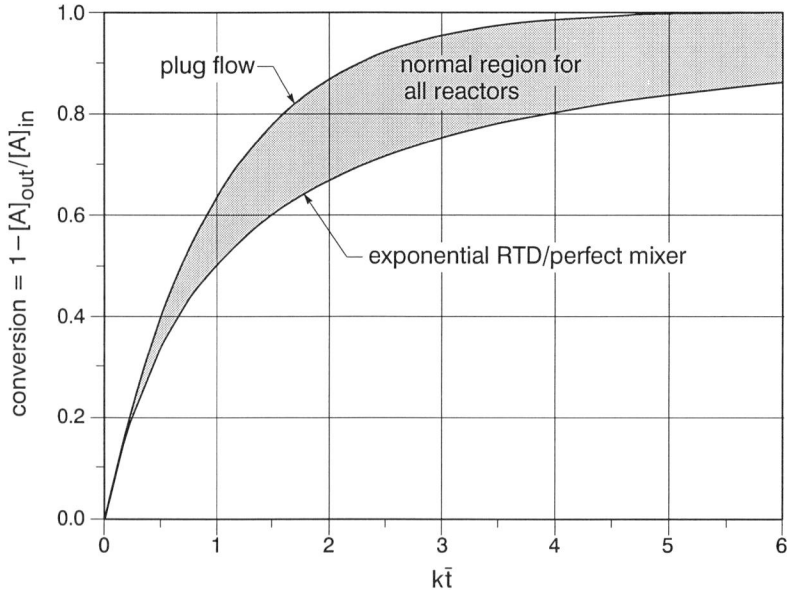

Figure 10.22. Conversion for irreversible first-order reaction according to perfect mixer and ideal plug-flow models.

Several important reactor principles are demonstrated in Figure 10.22. First, and most obvious, as k or $\bar{t}$ (or the product $k\bar{t}$) increase, so does the chemical conversion. Second, for a fixed $k\bar{t}$, the plug-flow reactor always outperforms (has a higher conversion than) the perfect mixer.* Third, for a fixed conversion, the product $k\bar{t}$ is always smaller in the plug-flow reactor. This last point has important consequences in reactor design. For example, consider a process with a fixed flow Q. Because $k\bar{t}$ is smaller in the plug-flow reactor, it can have a smaller $\bar{t}$ to achieve the same conversion. Since $\bar{t} = V/Q$ (Eq. 10.7), the plug-flow reactor can achieve the same conversion as the perfect mixer with a smaller-volume reactor (see Exercise 10.10).

The next model to be considered is for a first-order reaction in an N *tanks-in-series* system. Substituting Eq. 10.66 into Eq. 10.116 and integrating yields (see Exercise 10.12):

$$\frac{[A]_{\text{out}}}{[A]_{\text{in}}} = \frac{N^N}{(N + k\bar{t})^N} = \frac{1}{\left(1 + \dfrac{k\bar{t}}{N}\right)^N} \tag{10.123}$$

As we have stressed in this chapter, the models for perfect mixing and ideal plug flow usually set the bounds for conversion of all other reactors. Therefore, you can imagine that the curves generated with Eq. 10.123 will fall within the "normal region" indicated in Figure 10.22 (Exercise 10.12).

For the *axial-dispersion* model, the techniques of this section have not led to an analytical solution for the conversion of first-order reactions. It is therefore necessary to return to the governing partial differential equation, which for a first-order reaction is (Nauman and Buffham, 1983)

$$\frac{\partial [A]}{\partial t} + u \frac{\partial [A]}{\partial x} = D \frac{\partial^2 [A]}{\partial x^2} - k[A] \tag{10.124}$$

For steady-state and closed-system boundary conditions (Eqs. 10.105 and 10.106), the solution to Eq. 10.124 is

$$\frac{[A]_{\text{out}}}{[A]_{\text{in}}} = \frac{4a \exp\left[\text{Pe}\,\dfrac{(1 - a)}{2}\right]}{(1 + a)^2 - (1 - a)^2 \exp(-a\,\text{Pe})} \tag{10.125}$$

where

$$a = \sqrt{1 + \frac{4k\bar{t}}{\text{Pe}}} \tag{10.126}$$

*The plug-flow reactor outperforms the perfect mixer for all positive reaction orders, that is, for all $r = k[A]^n$ where $n \geqslant 0$ (Levenspiel, 1972).

In our review of the residence-time density function for the closed-system axial-dispersion model in Section 10.8, we demonstrated, using Figure 10.20, that for the limit of Pe $\rightarrow \infty$, the axial-dispersion model approaches the limit of the plug-flow reactor; for Pe $\rightarrow 0$, the axial-dispersion model approaches the limit of the perfect mixer. The same conclusion is made for a first-order reaction in the closed-axial dispersion system. For Pe $\rightarrow \infty$, Eq. 10.125 approaches the equation describing conversion in a plug-flow reactor (Eq. 10.122):

$$\frac{[A]_{\text{out}}}{[A]_{\text{in}}} = \exp(-k\bar{t}) \tag{10.122}$$

whereas for Pe $\rightarrow 0$, Eq. 10.125 approaches the equation for conversion in a perfect mixer (Eq. 10.119):

$$\frac{[A]_{\text{out}}}{[A]_{\text{in}}} = \frac{1}{k\bar{t} + 1} \tag{10.119}$$

10.10.2. Conversion for Second-Order Reactions

To compute conversion for irreversible second-order reactions, we abandon the reaction-probability approach of the last section and return to the familiar theme of the mass balance. For the *perfect mixer*, we adopt the systems approach (box model—Section 1.6) and apply it to a reactor like that in Figure 10.7. Thus, the mass accumulation is equal to the mass in minus the mass out plus the change due to reaction; for steady-state conditions this yields:

$$0 = Q[A]_{\text{in}} - Q[A]_{\text{out}} + Vr_A \tag{10.127}$$

where r_A is the rate of reaction of A. Rearranging, we find Levenspiel's (1972) *performance equation* for the steady-state *perfect mixer*:

$$\frac{V}{Q} = \bar{t} = \frac{[A]_{\text{in}} - [A]_{\text{out}}}{-r_A} \tag{10.128}$$

Now we consider the equation for irreversible second-order kinetics from Section 9.3:

$$r_A = \frac{d[A]}{dt} = -k[A]^2 \tag{9.19}$$

Substituting this equation (with $[A] = [A]_{\text{out}}$—the perfect mixing assumption) into the performance equation and rearranging, we find (Exercise 10.13):

$$\frac{[A]_{\text{out}}}{[A]_{\text{in}}} = \frac{2}{1 + \sqrt{1 + 4k\bar{t}[A]_{\text{in}}}} \tag{10.129}$$

The conversion for the second-order reaction in a perfect mixer is then

$$\text{conversion} = 1 - \frac{[A]_{\text{out}}}{[A]_{\text{in}}} = 1 - \frac{2}{1 + \sqrt{1 + 4k\bar{t}[A]_{\text{in}}}} \tag{10.130}$$

For the *plug-flow reactor*, we will perform a mass balance on a differential element of the reactor (Section 1.4). Consider again the plug-flow reactor shown in Exercise 2.15, in which we replace all C's with $[A]$. Then the accumulation of mass in the differential element is the mass in minus the mass out plus the change due to reaction; for steady-state conditions this yields:

$$0 = Q[A]_l - Q[A]_{l+\Delta l} + \Delta V r_A \tag{10.131}$$

Rearranging we find:

$$\frac{\Delta V}{Q} = \frac{[A]_l - [A]_{l+\Delta l}}{-r_A} \tag{10.132}$$

Next we adopt the trick outlined in Section 1.4, and use a truncated Taylor series to express the concentration of A at $l + \Delta l$ in terms of the concentration at l:

$$[A]_{l+\Delta l} = [A]_l + \frac{d[A]}{dl} \Delta l \tag{10.133}$$

Substituting into Eq. 10.132 and taking the limit as $\Delta l \to 0$, we find:

$$\frac{dV}{Q} = \lim_{\Delta l \to 0} \frac{\Delta V}{Q} = \lim_{\Delta l \to 0} \left[\frac{[A]_l - \left([A]_l + \frac{d[A]}{dl} \Delta l \right)}{-r_A} \right] = \frac{d[A]}{r_A} \tag{10.134}$$

Now we integrate the differential equation over the entire plug-flow reactor:

$$\frac{1}{Q} \int_0^V dV' = \int_{[A]_{\text{in}}}^{[A]_{\text{out}}} \frac{d[A]}{r_A} \tag{10.135}$$

where V' is a dummy integration variable. This yields Levenspiel's (1972) *performance equation* for the steady-state *ideal plug-flow reactor*:

$$\frac{V}{Q} = \bar{t} = \int_{[A]_{\text{in}}}^{[A]_{\text{out}}} \frac{d[A]}{r_A} \tag{10.136}$$

To determine conversion for the irreversible second-order reaction, we substitute Eq. 9.19 into Eq. 10.136:

$$\bar{t} = \int_{[A]_{in}}^{[A]_{out}} \frac{d[A]}{r_A} = \int_{[A]_{in}}^{[A]_{out}} \frac{d[A]}{-k[A]^2} = \frac{1}{k}\left(\frac{1}{[A]_{out}} - \frac{1}{[A]_{in}}\right) \quad (10.137)$$

or rearranging:

$$\frac{[A]_{out}}{[A]_{in}} = \frac{1}{1 + k\bar{t}[A]_{in}} \quad (10.138)$$

Conversion for a second-order reaction in the plug-flow reactor is then

$$\text{conversion} = 1 - \frac{[A]_{out}}{[A]_{in}} = 1 - \frac{1}{1 + k\bar{t}[A]_{in}} \quad (10.139)$$

In Figure 10.23 we compare conversion for a second-order reaction in the perfect mixer and the ideal plug-flow reactors, where the independent variable is the *dimensionless reaction parameter*, $k\bar{t}[A]_{in}$ (Nauman and Buffham, 1983). Most of the comments made on Figure 10.22 also pertain to Figure 10.23. Note

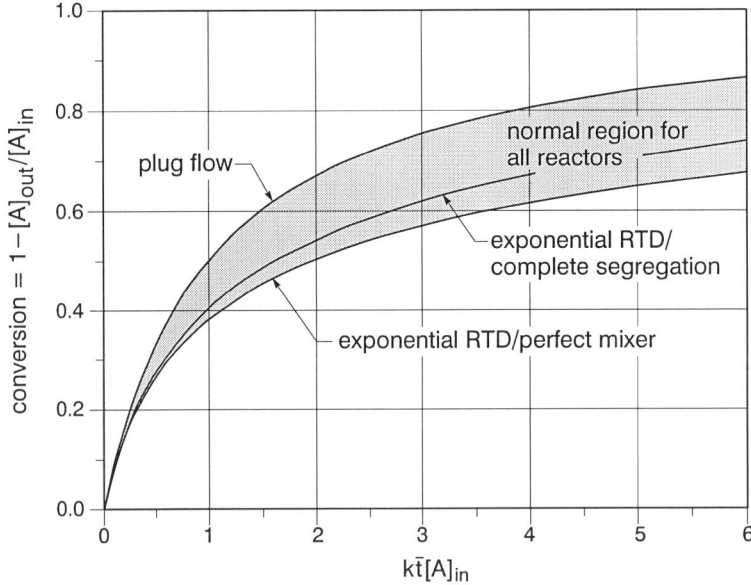

Figure 10.23. Conversion for irreversible second-order reaction according to perfect mixer and ideal plug-flow models. Intermediate curve labeled exponential RTD/complete segregation is discussed in Section 10.10.3.

that because the kinetics are nonlinear (i.e., second order), the initial concentration $[A]_{in}$ is required to determine conversions.

10.10.3. Complete Segregation and Micromixing

In Section 10.9, we pointed out that one problem in our residence-time analyses is that we have basically modeled the residence-time distribution, and not the mixing. Around the middle of this century, chemical engineers became very much occupied with trying to understand more about mixing, generally within the framework of residence-time modeling. One of their first questions was: Is it possible to have two reactors with the same residence-time distribution, but with different mixing on the molecular scale? Consider the two reactors in Figure 10.24.

We are quite familiar by now with the perfect mixer in Figure 10.24a, and know that it gives the familiar exponential-shaped residence-time density function (Eq. 10.33). But consider also the new reactor shown in Figure 10.24b. It is composed of a parallel arrangement of ideal plug-flow reactors. We have designed the new reactor with variable residence times in the different plug-flow sections. We have picked the size of the sections so that the average residence-

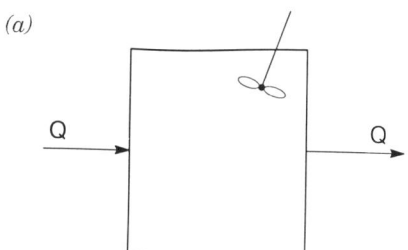

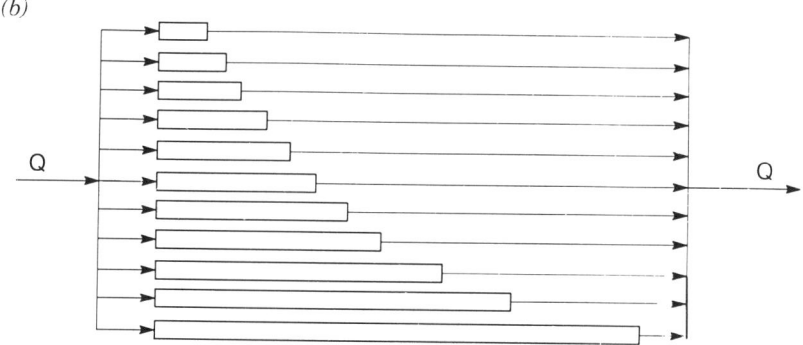

Figure 10.24. (a) Perfect mixer and (b) segregated-flow reactor. Note that residence time in the piping connecting the plug-flow subsystems is zero.

time distribution for the entire reactor mimics an exponential distribution. In fact, if we carefully control the size of the plug-flow elements so that the mean residence time for the entire reactor is the same as in the perfect mixer, and if we let the number of the plug-flow elements go to infinity, we can duplicate exactly the exponential residence-time density for the perfect mixer (Eq. 10.33). So what does this have to do with the issue of molecular scale mixing?

Recall that in the perfect mixer, any molecule entering the system is instantaneously mixed with all other molecules already in the system; molecular-scale mixing is instantaneous. The perfect mixer is therefore referred to as a *maximum mixedness* reactor. On the other hand, in the *ideal* plug-flow reactor, a molecule entering the system does not mix with any other molecules. In the reactor in Figure 10.24b, molecules entering the plug-flow subsystems mix with no other molecules, at least not until the final (infinitesimal) instant when all molecules come together at the exit of the reactor. So in this *completely segregated flow* reactor, all molecules are segregated from each other. But as outlined above, we have carefully designed the reactor to have the exponential residence-time distribution of Eq. 10.33. Thus, this thought experiment has yielded the answer to our first question: theoretical reactors can be imagined in which there are different levels of molecular-scale mixing, yet which will have the same distribution of residence time. Moving to the real world, although no reactors perform just like either of those in Figure 10.24, we do believe that there are different degrees of molecular mixing in real reactors. Imagine, for example, a relatively stagnant pond with a small inflow and outflow, in which mixing is poor. Then it is likely that certain elements of fluid may never come into contact with other elements before exiting the system. Depending on the internal mixing, real reactors and systems will have different kinds of molecular-scale mixing.

We now investigate the implication of the flow segregation on chemical conversion. We noted in Section 10.10.1 that conversion in a plug-flow reactor is identical to a batch reactor, when the residence time in the plug-flow reactor is the same as the operation time in the batch reactor. Therefore, knowing the mean residence time of each of the plug-flow elements in Figure 10.24b, we should be able to calculate the overall conversion. In fact, all we have to do is properly weigh the conversion in each plug-flow subsystem, which is done with the residence-time density function (Nauman and Buffham, 1983),

$$c_{out} = \int_0^\infty c_{batch}(t) f(t) dt \qquad (10.140)$$

where $c_{batch}(t)$ is the concentration in a batch reactor at time t.

First, we investigate conversion for the irreversible second-order reaction, Eq. 9.19. We learned in the last section that in the plug-flow reactor,

$$\frac{[A]_{out}}{[A]_{in}} = \frac{1}{1 + k\bar{t}[A]_{in}} \qquad (10.138)$$

For this reaction, the exit concentration in the batch reactor is found by replacing $\bar{t}$ with t:

$$\frac{[A]_{\text{batch}}(t)}{[A]_{\text{in}}} = \frac{1}{1 + kt[A]_{\text{in}}} \qquad (10.141)$$

Substituting this and Eq. 10.33 into Eq. 10.140 yields

$$\frac{[A]_{\text{out}}}{[A]_{\text{in}}} = \frac{1}{\bar{t}} \int_0^\infty \frac{\exp(-t/\bar{t})}{1 + k\bar{t}[A]_{\text{in}}} \, dt = \frac{\exp(1/k\bar{t}[A]_{\text{in}})}{k\bar{t}[A]_{\text{in}}} \int_{(k\bar{t}[A]_{\text{in}})^{-1}}^\infty \frac{\exp(-z)}{z} \, dz \qquad (10.142)$$

where z is a dummy integration variable. The last integral in Eq. 10.142 is called the exponential integral, and can be found in mathematical handbooks or evaluated numerically with many modern mathematics packages. The conversion corresponding to various reaction parameters, $k\bar{t}[A]_{\text{in}}$, is presented in Figure 10.23 (curve labeled "exponential RTD/complete segregation"). Note that the segregated flow reactor with exponential residence time yields higher conversion than the perfect mixer (since it has some of the superior characteristics of a plug-flow reactor), but has a lower conversion than the single ideal plug-flow reactor (apparently because of the exponential distribution of residence time). Figure 10.23 emphasizes the principle that for the second-order reaction, two reactors can have the same residence-time distribution, but different chemical conversion. Or stated in another way, *residence time* does not uniquely determine conversion for *second-order reactions*. The reason for the difference in conversion in the two reactors is something called *micromixing*, which involves microscopic mixing in systems. The difference between the two lower curves in Figure 10.23 is due to the extremely different micromixing environments in the two reactors—complete segregation and maximum mixedness.

Having established a micromixing effect for the second-order reaction, we might ask if micromixing is a factor in all reactions. The answer is—sometimes. One reaction that is known to be insensitive to micromixing is the first-order reaction studied in this chapter. For the irreversible first-order reaction, the exit concentration in a batch reactor is found by replacing $\bar{t}$ with t in Eq. 10.122:

$$\frac{[A]_{\text{batch}}(t)}{[A]_{\text{in}}} = \exp(-kt) \qquad (10.143)$$

Substituting this and Eq. 10.33 into Eq. 10.140 yields

$$\frac{[A]_{\text{out}}}{[A]_{\text{in}}} = \int_0^\infty \exp(-kt) \frac{1}{\bar{t}} \exp\left(-\frac{t}{\bar{t}}\right) dt = \frac{1}{k\bar{t} + 1} \qquad (10.144)$$

which is the same as Eq. 10.119 for the perfect mixer. So the first-order reaction is insensitive to micromixing; segregation has no impact.

A great deal of work by chemical engineers has been done in determining which reactions and which types of mixing can lead to micromixing sensitivity. The models and tests used in characterizing micromixing have gone far beyond the simple introduction provided here (Villermaux, 1986). We cannot summarize the vast literature on micromixing, but instead will attempt to summarize for the reader those real mixing and chemical reaction problems in which micromixing can be expected to influence the course of a chemical reaction (David and Clark, 1991; Nauman and Clark, 1991):

1. The residence-time distribution must be significantly different from plug flow. As we saw in the examples in this section, conversion of chemical reactions is usually pretty straightforward in a plug-flow reactor. There is a possible exception to this rule. When reactants are introduced to a reactor at different points (so-called non-*premixed* reactants), micromixing can even have an effect in systems that approach ideal plug flow.

2. The actual chemical reactor must have some segregation characteristics. This is most common in reaction systems in which the mixing is poor, and when the reactants are not premixed.

3. Single reactions must be nonlinear. (Recall that we found first-order, i.e., linear, reactions to be insensitive to micromixing.) Also, reactions with higher conversions are more sensitive to micromixing. This can be seen in Figure 10.23 — note that the further we move out along the x axis, the greater the difference between the segregated and perfect-mixing predictions. Thus, for the second-order reaction, systems operating at large values of $k\bar{t}[A]_{in}$ will show higher sensitivity to micromixing.

As we said above, the models of micromixing have become very complex and sophisticated. One issue that has arisen in micromixing modeling is the earliness or lateness of molecular-level mixing. A demonstration of the impact of this phenomenon is provided in Exercise 10.16.

10.11. SUMMARY

This chapter reviewed the statistical basis of residence time in simple closed systems. Three kinds of residence-time distributions were developed: the residence-time density function, the cumulative residence-time distribution function, and the washout function. Three methods of measuring real residence-time distributions were outlined, and example calculations of residence-time distributions and moments were provided. Several theoretical models for residence-time distributions were developed. They include the perfect mixer, the ideal plug-flow reactor, the N tanks-in-series model (and gamma-function extension), the fractional-tubularity model, the stagnancy model, the recyle model, and several versions of the axial-dispersion model. The moments method and time-domain fitting procedures were used to show (1) how theoretical models could be fit to real data and (2) how residence-time models could be discriminated based on minimization of error.

The main motivation for studying residence time is to understand its effect on reactions. In the last sections of the chapter, we investigated how residence time and mixing affected chemical conversion for irreversible first- and second-order reactions. We found that based on the theoretical models, the limits to chemical reaction could be estimated for real-flow systems. Finally, we described a new phenomenon called micromixing, which is a result of variations in molecular-scale mixing.

The exercises attempt to extend the principles developed in this chapter and show how the generic formulas developed can be applied to "nonchemical" reactions like sedimentation and adsorption in flowing systems.

EXERCISES

10.1. The following tracer data was collected during an impulse tracer test on a system of unknown volume. Assume a simple closed system, and determine and plot $f(t)$, $F(t)$, and $W(t)$. Also determine the system volume if the flow rate is $1.25\,\mathrm{m^3/hr}$.

Time (min)	$c(t)$ (mg/L)	Time (min)	$c(t)$ (mg/L)
0	0	18	20
1	0	19	17
2	13	20	14
3	30	21	11
4	54	22	8
5	68	23	6
6	80	24	5
7	85	25	4
8	90	26	3
9	87	27	2
10	82	28	2
11	73	29	1
12	64	30	1
13	56	31	1
14	46	32	0
15	38	33	0
16	33	34	0
17	26		

10.2. Expand the integral in Eq. 10.48 and integrate to prove Eqs. 10.50–10.54.

10.3. Derive Eqs. 10.55–10.57 by rearrangement and successive substitution of Eqs. 10.52–10.54.

10.4. Derive Eq. 10.60 by expanding under the integral and integrating. Hint: also recall Eq. 10.4.

10.5. The purpose of this problem is to derive the residence-time function for an ideal plug-flow system followed by an N tanks-in-series system.

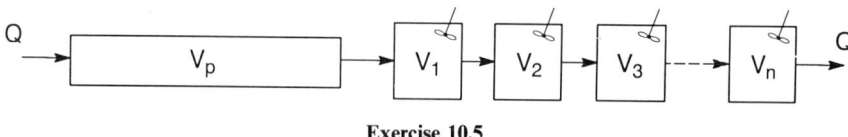

Exercise 10.5

We define V_p as the volume of the plug-flow subsystem and V_s as the total volume $(V_1 + V_2 + V_3 + \cdots + V_n)$ of the N tanks-in-series sub-system; hence, $V = V_p + V_s$. Argue why the residence-time density function should be

$$f(t) = \begin{cases} 0 & t < \bar{t}_p \\ \dfrac{N^N (t - \bar{t}_p)^{N-1}}{\bar{t}_s^N (N - 1)!} \exp\left(-\dfrac{N(t - \bar{t}_p)}{\bar{t}_s}\right) & t > \bar{t}_p \end{cases}$$

where

$$\bar{t}_s = \frac{V_s}{Q}$$

10.6. In Section 4.6, we developed a model for partitioning between flowing and stationary phases. The thought experiment used to develop the governing equation (Eq. 4.49) can now be recognized as a legitimate positive-step tracer experiment; hence, Eq. 4.52 is actually the F function developed in this chapter. Derive the residence-time density function (f) corresponding to the flow and partitioning model (Eq. 4.52).

10.7. If Exercise 10.1 was completed, (a) fit the N tanks-in-series model to the discrete residence-time data using the time-domain fitting procedure, and (b) compute the variance of the residence-time data and compute N for the tanks-in-series model using the moments approach (Eq. 10.68). Hint: For time-domain fitting, simply create columns in the spreadsheet for $f(t_i)$ corresponding to different (integral) values of N in the tanks-in-series model, and then create columns to compute the SSE (Eq. 10.111). The best-fit model is then the one with the minimal SSE. Did parts (a) and (b) yield the same tanks-in-series model?

10.8. If Exercise 10.1 was completed, use time-domain fitting to fit the fractional-tubularity model (see the hint in Exercise 10.7). Does the model seem to fit the data? Which model — N tanks-in-series or fractional tubularity — best fits the data?

10.9. Using the mean residence time determined in Example 10.1, determine the theoretical conversion limits of the perfect mixer and the ideal plug-flow reactor for the reaction-rate constants used in Example 10.3. Compare these conversions with the conversions determined using the actual residence-time data in Example 3. Which theoretical model most closely approximates the real system?

10.10. For a conversion of 0.8 and a first-order reaction-rate constant of $0.2 \, \text{min}^{-1}$, use Eqs. 10.120 and 10.122 (or Figure 10.22) to determine $\bar{t}$ for the perfect mixer and the ideal plug-flow reactor. If the system flow rate is $0.1 \, \text{m}^3/\text{s}$, what are the required reactor sizes?

10.11. Derive Eq. 10.123 following the procedures used in deriving Eq. 10.119. Hints: You will probably need to refer to a definite integrals table. Also, from calculus, $\Gamma(N + 1) = N!$, so $\Gamma(N) = (N - 1)!$ Verify for $N = 2, 4$, and 10 that the predictions of conversion for the N tanks-in-series model are bounded by the two curves in Figure 10.22.

10.12. At several junctures in this book, we have mentioned that environmental "reactions" other than strict chemical reactions can be modeled or cast in terms of reaction terminology. Compare the curve in Figure 2.8 for continuous sedimentation in a perfect mixer with the curve in Figure 10.22 for a first-order reaction in a perfect mixer. (You will find that the curve in Figure 2.8 overlays the curve in Figure 10.22!)

(a) Show that the apparent first-order reaction-rate constant for sedimentation in a perfect mixer is

$$k_s = \frac{v_t}{D}$$

where v_t is the Stokes sedimentation velocity and D is the rectangular sedimentation system depth. Hint: you will need to consider the equations for fractional removal in sedimentation and conversion of first-order reactions.

(b) Verify that for continuous sedimentation is the N tanks-in-series system

$$\frac{C_{out}}{C_{in}} = \left(\frac{1}{1 + \dfrac{v_t}{N V_0}} \right)^N$$

where V_0 is the overflow velocity of Section 2.7.

10.13. Derive Eq. 10.129 following the instructions in the text. Hint: You will need to rearrange your intermediate equation as a quadratic equation in $[A]_{in}/[A]_{out}$. Then solve using the quadratic equation from elementary algebra.

10.14. Schnoor and McAvoy (1981) studied the transport of pesticide in a Midwestern lake. Pesticide entering the lake was considered to have three possible fates: (1) dissolved pesticide could adsorb to suspended solids (e.g., colloidal material) in the lake water; (2) pesticide adsorbed on suspended solids could effectively be lost from the system through sedimentation of the solids to the bottom of the lake; (3) dissolved pesticide could be washed out of the lake. Adsorption to suspended solids was modeled as

$$r_a = k_a M_s C$$

where r_a is the adsorption rate $(M\text{-}L^{-3}\text{-}T^{-1})$, k_a is the adsorption rate constant $(L^3\text{-}T^{-1}\text{-}M^{-1})$, M_s is the constant suspended solids concentration (M/L^3), and C is the dissolved pesticide concentration (M/L^3). Here M_s is assumed to be constant in the lake at steady state, so the equation above can be considered pseudo-first order (Chapter 9). The desorption rate was modeled as

$$r_d = k_d C_p$$

where r_d is the desorption rate $(M\text{-}L^{-3}\text{-}T^{-1})$, k_d is the first-order desorption rate constant (T^{-1}), and C_p is the particulate-associated pesticide concentration, that is, the lake pesticide concentration associated with the suspended solids.

(a) Show that, at equilibrium, the two equations above describe the linear isotherm

$$R = K_p C$$

where $R = C_p/M_s$ is the amount of pesticide adsorbed per unit mass of suspended solids (M/M) and $K_p = k_a/k_d$ is the partition or adsorption coefficient. Hint: At equilibrium, $r_a = r_d$.

For the loss of pesticide associated with suspended solids settled to the bottom of the lake, Schnoor and McAvoy assumed, for a perfectly mixed lake (see Exercise 10.12):

$$r_s = -k_s C_p$$

where r_s is the sedimentation loss rate (T^{-1}) and k_s is the first-order sedimentation loss constant.

(b) Assuming the lake is a steady-state perfect mixer, use Eq. 10.128 to show that

$$C_{in} - C = -\bar{t}(-k_a M_s C + k_d C_p)$$

and

$$C_{p,in} - C_p = -\bar{t}(-k_d C_p + k_a M_s C - k_s C_p)$$

where C_{in} and $C_{p,in}$ are the dissolved- and particulate-associated concentration in the lake inflow and $\bar{t}$ is the mean residence time in the lake. Defining the total pesticide concentrations $C_T = C + C_p$ and $C_{T,in} = C_{in} + C_{p,in}$, derive the following equation:

$$C_{T,in} = C_T + k_s \bar{t} C_p$$

From part **a**, we know that

$$C_p = R M_s = K_p C M_s$$

(c) Using this relationship and $C_T = C + C_p$, show that

$$C = \frac{C_T}{1 + K_p M_s}$$

and, therefore,

$$C_p = \frac{K_p M_s C_T}{1 + K_p M_s}$$

Finally, show that the steady-state total pesticide concentration in the perfectly mixed lake is

$$C_T = \frac{C_{T,in}}{1 + \dfrac{k_s \bar{t} K_p M_s}{1 + K_p M_s}}$$

10.15. This problem concerns adsorption on small particles in a flow system at steady state. Referring to Section 6.7.4, note that by combining Eqs. 6.107–6.109 we find

$$c_{A,t} = c_{A,0} - (c_{A,0} - c_{A,\infty}) \left[1 - \sum_{n=1}^{\infty} \frac{6\alpha(\alpha + 1) \exp\left(\dfrac{q_n^2 t}{\tau}\right)}{9 + 9\alpha + q_n^2 \alpha^2} \right]$$

(a) Identifying $c_{A,\text{batch}}$ with $c_{A,t}$ and c_{in} with $c_{A,0}$, show that substituting the equation above into Eq. 10.140 yields

$$\frac{c_{A,\text{in}} - c_{A,\text{out}}}{c_{A,\text{in}} - c_{A,\infty}} = 1 - \int_0^\infty \sum_{n=1}^\infty \left[\frac{6\alpha(\alpha + 1)\exp\left(\dfrac{q_n^2 t}{\tau}\right)}{9 + 9\alpha + q_n^2\alpha^2} \right] f(t)dt$$

(b) Substitute the $f(t)$ for the N tank-in-series model into the equation above and show the following:

$$\frac{c_{A,\text{in}} - c_{A,\text{out}}}{c_{A,\text{in}} - c_{A,\infty}} = 1 - \frac{6\alpha(\alpha + 1)}{(N - 1)!} \sum_{n=1}^\infty \frac{\Gamma(N)}{9 + 9\alpha + q_n^2\alpha^2 \left(\dfrac{q_n^2}{N}\dfrac{\bar{t}}{\tau} + 1\right)}$$

where $\Gamma(N)$ is the complete gamma function. Plot the equation versus $\bar{t}/\tau$ for $N = 1, 2$, and 4 and $\alpha = 1.0$.

(c) Adham et al. (1991) showed that, for a perfect mixer, the following equation describes adsorption performance:

$$\frac{c_{A,\text{in}} - c_{A,\text{out}}}{c_{A,\text{in}} - c_{A,\infty}} = 1 - \frac{6}{\pi^2} \sum_{n=1}^\infty \frac{1}{n^4\pi^2 \dfrac{\bar{t}}{\tau} + n^2}$$

Also, plot this equation on the plot created in part **b**. Does micromixing have an impact on adsorption performance for the exponential residence-time density?

10.16. In Section 3.7.3 we found a simplified equation for Brownian motion coagulation

$$\frac{dN_\infty}{dt} = -\frac{4hT}{3\mu} N_\infty^2$$

Now consider coagulation in ideal plug flow and perfectly mixed systems. Using the approach of Section 10.10.2, determine equations for $N_{\infty,\text{out}}/N_{\infty,\text{in}}$. How is Figure 10.23 related to these equations?

10.17. In Section 10.7.3, we presented the fractional tubularity model as shown in Figure 10.15, that is, with a plug-flow section preceding a perfect-mixer section. However, exactly the same residence-time density (Eq. 10.72) results if the order of the reactors is reversed, that is, if the perfect mixer precedes the plug-flow element. Although the residence-time distributions are the same, molecular-scale mixing occurs relatively late in Figure 10.15, but instantaneously in the reversed configuration. It turns out that the reversed configuration is the maximum mixedness reactor for the fractional-tubularity model (Nauman and Buffham, 1983). According to our discussion in Section 10.10.3, we have reason to suspect that micromixing may result in a difference in conversion in the two reactor schemes.

(a) Consider the case of $k\bar{t}[A]_{in} = 1$ for the overall fractional-tubularity reactor and $\bar{t}_p = \bar{t}_s$. Use Eqs. 10.129 and 10.138 to show that $[A]_{out}^{(2)}/[A]_{in}^{(1)} = 0.528$ for the original configuration of Figure 10.15 and $[A]_{out}^{(2)}/[A]_{in}^{(1)} = 0.536$ for the reversed configuration. Here the superscripts refer to the first and second reactors.

(b) Rework part **a** for $k\bar{t}[A]_{in} = 4$ for the overall reactor.

(c) Explain the impact of mixing earliness in the fractional-tubularity system.

REFERENCES

Adham, S. S., Snoeyink, V. L., Clark, M. M., and Bersillon, J. L., "Prediction and Verification of the Performance of Powdered Activated Carbon for Removal of Organic Compounds in the PAC/UF Process," *Journal of the American Water Works Association*, **83**, 12:81–91 (1991).

David R. and Clark, M. M., "Micromixing," in *Mixing in Coagulation and Flocculation*, A. Amirtharajah, M. M. Clark, and R. Trussell, Eds., American Water Works Association Research Foundation, Denver, CO, 1993.

Fu, B., Weinstein, H., Berstein, B., and Shaffer, A. B., "Residence Time Distributions of Recycle Systems — Integral Equation Formulation," *Ind. Eng. Process Des. Dev.*, **10**, 4:501–508 (1971).

Gibilaro, L. G., "The Recycle Flow-Mixing Model," *Chem. Eng. Sci.*, **26**, 299–304 (1971).

Levenspiel, O., *Chemical Reaction Engineering*, 2nd ed., Wiley, New York, 1972.

Nauman, E. B. and Buffham, B. A., *Mixing in Continuous Flow Systems*, Wiley-Interscience, New York, 1983.

Nauman, E. B. and Clark, M. M., "Residence Time Distributions," in *Mixing in Coagulation and Flocculation*, American Water Works Association Research Foundation, Denver, CO, 1993.

Rippin, D. W., "The Recycle Reactor as a Model of Incomplete Mixing," *Ind. Eng. Chem. Fundam.*, **6**, 4:488–492 (1967).

Schnoor, J. L. and McAvoy, D. C., "Pesticide Transport and Bioconcentration Model," *J. Env. Eng. Div., Proc. Amer. Soc. Eng.*, **107**, EE6: 1229–1246 (1981).

Villermaux, J., "Micromixing Phenomena in Stirred Reactors," in *Encyclopedia of Fluid Mechanics*, Gulf Publishing, West Orange, NJ, 1986.

BIBLIOGRAPHY

Clark, M. M., Srivastava, R., and David, R., "Mixing and Aluminum Precipitation," *Env. Sci. Technol.*, **27**, 10:2181–2189 (1993).

Doquang, Z., "Etudes Experimentale et Numerique des Performances des Contacteurs de Desinfection de l'Eau par le Chlore," Doctoral Thesis, l'Institut National des Sciences Appliqueés de Toulouse, Toulouse, France, 1993.

APPENDIX I

SI UNITS AND PHYSICAL CONSTANTS

Table I.1. SI Units

Quantity	Name	Symbol	Definition
Base Units			
Length	meter	m	
Mass	kilogram	kg	
Time	second	s	
Electrical current	ampere	A	
Temperature	kelvin	K	
Amount of substance	mole	mol	
Luminous intensity	candela	cd	
Derived Units			
Acceleration	meter/second^2		m-s^{-2}
Angular velocity	radian/second		rad-s^{-1}
Angular acceleration	radian/second^2		rad-s^{-2}
Area	meter^2		m^2
Density	kilogram/meter^3		kg-m^{-3}
Electrical charge	coulomb	C	A-s
Electrical potential	volt	V	W-A^{-1} = J-C^{-1}
Electrical resistance	ohm	Ω	V-A^{-1}
Energy, work, and heat	joule	J	N-m = kg-m^2-s^{-2}
Entropy	joule/kelvin		J-K^{-1}
Force	newton	N	kg-m-s^{-2}
Frequency	hertz	Hz	s^{-1}
Power	watt	W	J-s^{-1} = kg-m^2-s^{-3}
Pressure and stress	pascal	Pa	N-m^{-2} = kg-m^{-1}-s^{-2}
Velocity	meter/second		m-s^{-1}
Viscosity (dynamic)	pascal-second		Pa-s
Viscosity (kinematic)	meter^2/second		m^2-s^{-1}
Volume	meter^3		m^3

Table I.2. SI Prefixes

Factor	Prefixes	Symbol	Factor	Prefixes	Symbol
10^{-12}	pico	p	10	deca	da
10^{-9}	nano	n	10^2	hecto	h
10^{-6}	micro	μ	10^3	kilo	k
10^{-3}	milli	m	10^6	mega	M
10^{-2}	centi	c	10^9	giga	G
10^{-1}	deci	d	10^{12}	tera	T

Table I.3. Physical Constants

Quantity	Symbol	Value	SI Unit
Acceleration of gravity	g	9.807	$m\text{-}s^{-2}$
Avogadro number	L_A	6.022×10^{23}	mol^{-1}
Boltzmann constant	k	1.381×10^{-23}	$J\text{-}K^{-1}$
Proton charge	e	1.602×10^{-19}	C
Electron mass	m	9.1095×10^{-31}	kg
Faraday constant	F	9.648×10^4	$C\text{-}mol^{-1}$
Gas constant	R	8.314	$J\text{-}K^{-1}\text{-}mol^{-1}$
Vacuum permittivity	ϵ_0	8.854×10^{-12}	$J^{-1}\text{-}C^2\text{-}m^{-1}$
Standard atmosphere	p_0	1.013×10^5	Pa
Zero of celsius scale	T_0	273.15	K

BIBLIOGRAPHY

Atkins, P. W., *Physical Chemistry*, W. H. Freeman, San Francisco, CA, 1978.

Probstein, R. F., *Physicochemical Hydrodynamics: An Introduction*, Butterworths, Boston, MA, 1989.

Stumm, W. and Morgan, J. J., *Aquatic Chemistry: An Introduction Emphasizing Chemical Equilibria in Natural Waters*, Wiley, New York, 1981.

APPENDIX II

REVIEW OF VECTORS

A *vector* is specified at a point in space by both magnitude and direction. Some examples of vectors are force, velocity, acceleration, and stress. A *scalar* is specified at a point in space by a single number—its magnitude. Some examples of scalars are temperature, pressure, density, and concentration. At a point in space, a vector can be represented uniquely by three numbers called the *components* of the vector. For example, in a rectangular coordinate system (a *cartesian* space), a velocity vector **u** can be represented as shown in Figure II.1.

In Figure II.1, u_1, u_2, and u_3 are the components of the vector **u**. The magnitude of **u**, indicated by $|\mathbf{u}|$ is

$$u = |\mathbf{u}| = \sqrt{u_1^2 + u_2^2 + u_3^2} \tag{II.1}$$

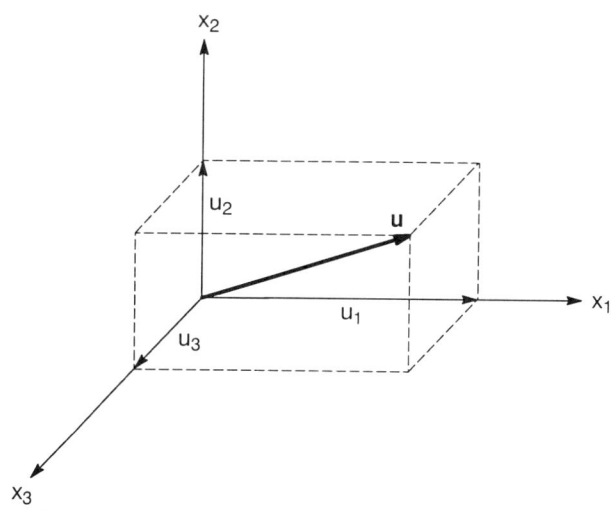

Figure II.1. Coordinate system for representation of vector **u**.

539

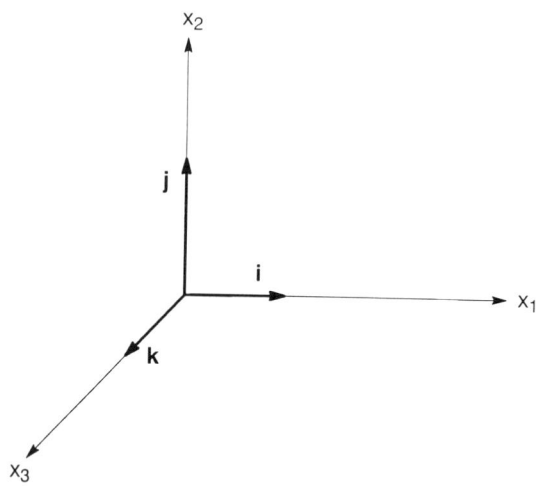

Figure II.2. Illustration of basic unit vectors collinear with coordinate axes.

Unit vectors are special vectors which have a magnitude of 1. One important group of unit vectors points along the three coordinate axes, as indicated in Figure II.2. From our definition of unit vectors,

$$|\mathbf{i}| = |\mathbf{j}| = |\mathbf{k}| = 1 \qquad (\text{II.2})$$

With the help of Figures II.1 and II.2, we see that a mathematical way of expressing the vector **u** in a cartesian coordinate system is

$$\mathbf{u} = u_1\mathbf{i} + u_2\mathbf{j} + u_3\mathbf{k} \qquad (\text{II.3})$$

Note in Eq. II.3 that u_1, u_2, and u_3 can be thought of as scalars multiplying unit vectors along the coordinate axes.

Some basic laws and operations of vector algebra are useful:

1. Vector addition is *commutative* and *associative*. Examples will suffice to define these terms:

$$\text{commutative property: } \mathbf{A} + \mathbf{B} = \mathbf{B} + \mathbf{A} \qquad (\text{II.4})$$

$$\text{associative property: } \mathbf{A} + (\mathbf{B} + \mathbf{C}) = (\mathbf{A} + \mathbf{B}) + \mathbf{C} \qquad (\text{II.5})$$

2. Products of scalars and vectors can be formed. A satisfactory example is given in Eq. II.3.

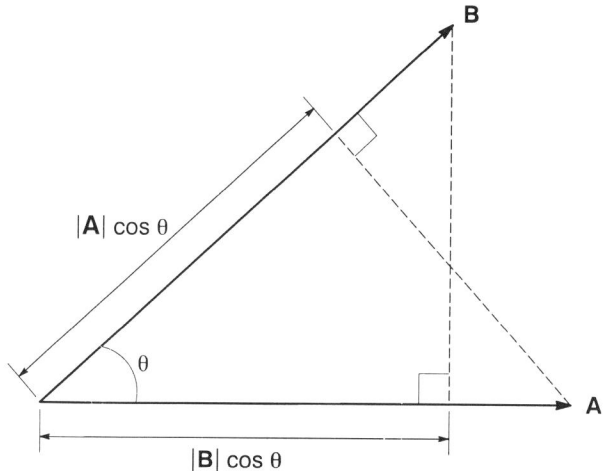

Figure II.3. Geometrical interpretation of a dot product.

3. The *scalar* or *dot* product of two vectors is defined mathematically by

$$\mathbf{A} \cdot \mathbf{B} = |\mathbf{A}| |\mathbf{B}| \cos \theta \tag{II.6}$$

where θ is the angle between the positive directions of $\mathbf{A}$ and $\mathbf{B}$. Figure II.3 shows that, geometrically, the dot product is the product of the magnitude of one vector times the projection of the other upon it. The dot product of two vectors results in a scalar.

An important use of the dot product is in the computation of the magnitude of a vector normal to a surface element. In Figure II.4, $\mathbf{n}$ is the *unit outward normal vector*, which has a magnitude of 1 and is perpendicular to the surface element ds. The dot product of $\mathbf{n}$ and $\mathbf{u}$ is

$$\mathbf{u} \cdot \mathbf{n} = |\mathbf{u}| |\mathbf{n}| \cos \theta = u(1) \cos \theta = u \cos \theta \tag{II.7}$$

Geometrically, $\mathbf{u} \cdot \mathbf{n}$ is seen as the projection of $\mathbf{u}$ on $\mathbf{n}$. Physically, it is the component of $\mathbf{u}$ normal to the surface element *ds*. This is an extremely important relation in this book because the mass flux through a surface element *ds* is equal to $\mathbf{u} \cdot \mathbf{n}$ times the local mass concentration (also see discussion below Eq. 1.15).

Several other properties derive from the definition of the dot product:

a. The dot product of a vector with itself is the magnitude of the vector squared. For example, referring to Figure II.3,

$$\mathbf{A} \cdot \mathbf{A} = |A| |A| \cos(0) = A^2 \tag{II.8}$$

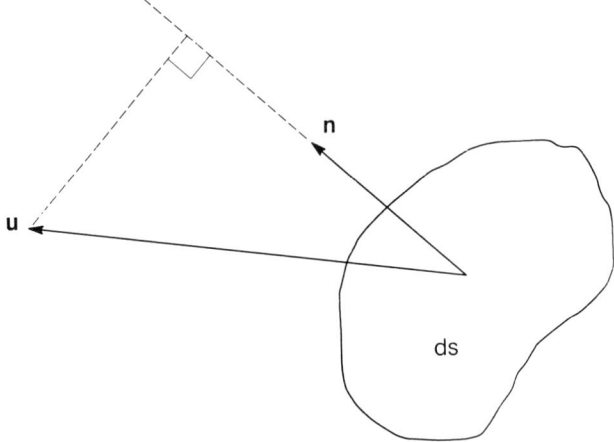

Figure II.4. Illustration of the projection of a velocity vector on a unit-normal vector at surface element *ds*.

The following identities also become evident:

$$\mathbf{i} \cdot \mathbf{i} = \mathbf{j} \cdot \mathbf{j} = \mathbf{k} \cdot \mathbf{k} = 1 \tag{II.9}$$

b. The dot products of different unit vectors $\mathbf{i}$, $\mathbf{j}$, and $\mathbf{k}$ are zero:

$$\mathbf{i} \cdot \mathbf{j} = \mathbf{j} \cdot \mathbf{k} = \mathbf{k} \cdot \mathbf{i} = 0 \tag{II.10}$$

Thus, Eq. II.10 says that the dot product of perpendicular vectors is zero.

c. Dot multiplication is *distributive* over addition; for example:

$$\mathbf{A} \cdot (\mathbf{B} + \mathbf{C}) = \mathbf{A} \cdot \mathbf{B} + \mathbf{A} \cdot \mathbf{C} \tag{II.11}$$

d. Dot multiplication is *commutative*:

$$\mathbf{A} \cdot \mathbf{B} = \mathbf{B} \cdot \mathbf{A} \tag{II.12}$$

e. Finally, since $\mathbf{A}$ and $\mathbf{B}$ in Figure II.3 have three cartesian components,

$$\mathbf{A} \cdot \mathbf{B} = (a_1\mathbf{i} + a_2\mathbf{j} + a_3\mathbf{k}) \cdot (b_1\mathbf{i} + b_2\mathbf{j} + b_3\mathbf{k})$$

Because dot multiplication is distributive over addition, we find:

$$\mathbf{A} \cdot \mathbf{B} = a_1 b_1 + a_2 b_2 + a_3 b_3 \tag{II.13}$$

4. The cross multiplication of vectors $\mathbf{A}$ and $\mathbf{B}$ (also called the *cross product*

or *vector product*) yields a vector whose magnitude is given by

$$\mathbf{A} \times \mathbf{B} = |\mathbf{A}|\,|\mathbf{B}|\sin\theta \tag{II.14}$$

and whose direction is perpendicular to the plane containing $\mathbf{A}$ and $\mathbf{B}$ according to the "right-hand" rule. Several properties follow from the given definitions above:

a. Cross multiplication is not commutative (in general, $\mathbf{A} \times \mathbf{B} \neq \mathbf{B} \times \mathbf{A}$), although it is distributive over addition.

b. The following identities are found among the unit vectors $\mathbf{i}$, $\mathbf{j}$, and $\mathbf{k}$:

$$\mathbf{i} \times \mathbf{i} = \mathbf{j} \times \mathbf{j} = \mathbf{k} \times \mathbf{k} = 0 \tag{II.15}$$

$$\mathbf{i} \times \mathbf{j} = -\mathbf{j} \times \mathbf{i} = \mathbf{k} \tag{II.16}$$

$$\mathbf{j} \times \mathbf{k} = -\mathbf{k} \times \mathbf{j} = \mathbf{i} \tag{II.17}$$

$$\mathbf{k} \times \mathbf{i} = -\mathbf{i} \times \mathbf{k} = \mathbf{j} \tag{II.18}$$

c. Considering the components of $\mathbf{A}$ and $\mathbf{B}$:

$$\mathbf{A} \times \mathbf{B} = (a_1\mathbf{i} + a_2\mathbf{j} + a_3\mathbf{k}) \times (b_1\mathbf{i} + b_2\mathbf{j} + b_3\mathbf{k}) \tag{II.19}$$

and because cross multiplication is distributive over addition:

$$\mathbf{A} \times \mathbf{B} = (a_2b_3 - a_3b_2)\mathbf{i} - (a_1b_3 - a_3b_1)\mathbf{j} + (a_1b_2 - a_2b_1)\mathbf{k} \tag{II.20}$$

Equation II.20 can be represented as a determinant:

$$\mathbf{A} \times \mathbf{B} = \begin{vmatrix} \mathbf{i} & \mathbf{j} & \mathbf{k} \\ a_1 & a_2 & a_3 \\ b_1 & b_2 & b_3 \end{vmatrix} \tag{II.21}$$

From Eq. II.21, it is easy to see why cross multiplication is not commutative: interchanging rows of a determinant changes its sign.

5. The concept of the *gradient of a scalar* is used extensively in physics:

$$\text{grad } \phi = \nabla\phi = \frac{\partial\phi}{\partial x_1}\mathbf{i} + \frac{\partial\phi}{\partial x_2}\mathbf{j} + \frac{\partial\phi}{\partial x_3}\mathbf{k} \tag{II.22}$$

The gradient of ϕ can be considered an operation on ϕ; therefore, we sometimes make use of the gradient or "del" operator:

$$\nabla = \left(\mathbf{i}\frac{\partial}{\partial x_1} + \mathbf{j}\frac{\partial}{\partial x_2} + \mathbf{k}\frac{\partial}{\partial x_3}\right) \tag{II.23}$$

From vector calculus, recall that $\nabla\phi$ points in the direction of the greatest spatial change (gradient) of ϕ. The magnitude of $\nabla\phi$ is equal to this greatest spatial change in ϕ. Other terminology is sometimes used in talking about gradients of scalar fields, that is, if a vector field, say F, is the gradient of a scalar field, say ϕ, then F is called a *conservative vector field* and ϕ is called the *potential* at the point in question. A famous example is the calculation of pressure in hydrostatics:

$$\nabla P = \rho \mathbf{g} \tag{5.8}$$

where ρ is the fluid density and $\mathbf{g}$ is the acceleration of gravity.

6. The *Laplacian* of a scalar also shows up frequently in physics:

$$\nabla^2\phi = \frac{\partial^2\phi}{\partial x_1^2} + \frac{\partial^2\phi}{\partial x_2^2} + \frac{\partial^2\phi}{\partial x_3^2} \tag{II.24}$$

7. The del operator is used in computing the *divergence* of a vector:

$$\nabla \cdot \mathbf{A} = \left(\mathbf{i}\,\frac{\partial}{\partial x_1} + \mathbf{j}\,\frac{\partial}{\partial x_2} + \mathbf{k}\,\frac{\partial}{\partial x_3} \right) \cdot (A_1\mathbf{i} + A_2\mathbf{j} + A_3\mathbf{k})$$
$$\nabla \cdot \mathbf{A} = \frac{\partial A_1}{\partial x_1} + \frac{\partial A_2}{\partial x_2} + \frac{\partial A_3}{\partial x_3} \tag{II.25}$$

The divergence has many important interpretations in physics. In fluid mechanics, it shows up in the continuity equation for an incompressible fluid (Chapter 1):

$$\nabla \cdot \mathbf{u} = 0 \tag{1.24}$$

8. The *curl* of a vector is given by

$$\nabla \times \mathbf{A} = \left(\mathbf{i}\,\frac{\partial}{\partial x_1} + \mathbf{j}\,\frac{\partial}{\partial x_2} + \mathbf{k}\,\frac{\partial}{\partial x_3} \right) \times (A_1\mathbf{i} + A_2\mathbf{j} + A_3\mathbf{k})$$
$$\nabla \times \mathbf{A} = \left(\frac{\partial A_3}{\partial x_2} - \frac{\partial A_2}{\partial x_3} \right)\mathbf{i} - \left(\frac{\partial A_3}{\partial x_1} - \frac{\partial A_1}{\partial x_3} \right)\mathbf{j} + \left(\frac{\partial A_2}{\partial x_1} - \frac{\partial A_1}{\partial x_2} \right)\mathbf{k} \tag{II.26}$$

Note that Eq. II.26 can be represented by a determinant:

$$\begin{vmatrix} \mathbf{i} & \mathbf{j} & \mathbf{k} \\ \dfrac{\partial}{\partial x_1} & \dfrac{\partial}{\partial x_2} & \dfrac{\partial}{\partial x_3} \\ A_1 & A_2 & A_3 \end{vmatrix} \tag{II.27}$$

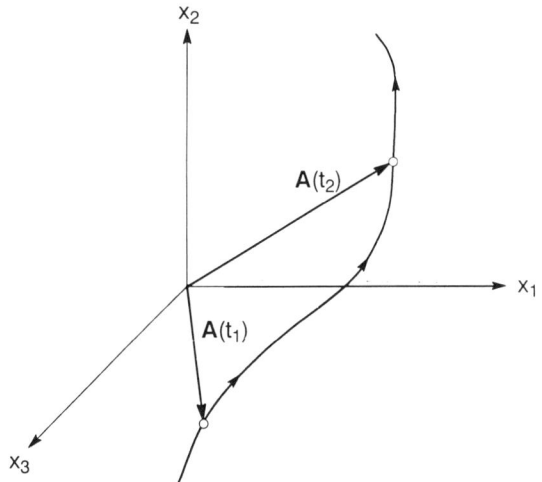

Figure II.5. Particle coordinates as a function of time.

In fluid mechanics, the curl has an important interpretation related to *solid-body rotation* at a point (Chapter 2).

The discussion above indicates that vectors are functions of space variables. They can also be functions of time. For example, if we follow a moving point in space and time, the coordinates of the particle could be described as shown in Figure II.5. For example, at t_1 in the figure, the position vector could be given by

$$\mathbf{A}(t_1) = A_1(t_1)\mathbf{i} + A_2(t_1)\mathbf{j} + A_3(t_1)\mathbf{k} \tag{II.28}$$

We can also discuss a vector derivative. For example, the derivative of the position vector **A** would be

$$d\mathbf{A}(t) = dA_1(t)\mathbf{i} + dA_2(t)\mathbf{j} + dA_3(t)\mathbf{k} \tag{II.29}$$

Finally, the derivative of vector functions can be integrated over properly defined lines, surfaces, and volumes.

There are numerous important *integral theorems* of vector calculus, such as the *divergence theorem*:

$$\iint \mathbf{A} \cdot \mathbf{n} ds - \iiint \nabla \cdot \mathbf{A} dV \tag{II.30}$$

The divergence theorem says that in a region of space bounded by a surface, the *surface integral* of the dot product of the outward unit-normal vector **n** and a vector **A** is equal to the volume integral of the divergence of **A**. **A** and the divergence of **A** must be continuous on the surface and throughout the region bounded by the surface.

BIBLIOGRAPHY

Borisenko, A. I. and Tarapov, I. E., *Vector and Tensor Analysis with Applications*, rev. English ed. (translated by Richard A. Silverman), Dover Publications, New York, 1979.

Wylie, C. R., *Advanced Engineering Mathematics*, McGraw-Hill, New York, 1975.

APPENDIX III

EQUATIONS OF FLUID MECHANICS AND CONVECTIVE-DIFFUSION IN RECTANGULAR, CYLINDRICAL, AND SPHERICAL COORDINATES

Table III.1. The Equation of Continuity

Rectangular coordinates (x, y, z):

$$\frac{\partial \rho}{\partial t} + \frac{\partial}{\partial x}(\rho u_x) + \frac{\partial}{\partial y}(\rho u_y) + \frac{\partial}{\partial z}(\rho u_z) = 0$$

Cylindrical coordinates (r, θ, z):

$$\frac{\partial \rho}{\partial t} + \frac{1}{r}\frac{\partial}{\partial r}(\rho r u_r) + \frac{1}{r}\frac{\partial}{\partial \theta}(\rho u_\theta) + \frac{\partial}{\partial z}(\rho u_z) = 0$$

Spherical coordinates (r, θ, ϕ):

$$\frac{\partial \rho}{\partial t} + \frac{1}{r^2}\frac{\partial}{\partial r}(\rho r^2 u_r) + \frac{1}{r \sin \theta}\frac{\partial}{\partial \theta}(\rho u_\theta \sin \theta) + \frac{1}{r \sin \theta}\frac{\partial}{\partial \phi}(\rho u_\phi) = 0$$

Table III.2. Components of the Stress Tensor for Newtonian Fluids

Rectangular Coordinates (x, y, z)	Cylindrical Coordinates (r, θ, z)	Spherical Coordinates (r, θ, ϕ)
$\tau_{xx} = \mu \left[2 \dfrac{\partial u_x}{\partial x} - \dfrac{2}{3} (\nabla \cdot \mathbf{u}) \right]$	$\tau_{rr} = \mu \left[2 \dfrac{\partial u_r}{\partial r} - \dfrac{2}{3} (\nabla \cdot \mathbf{u}) \right]$	$\tau_{rr} = \mu \left[2 \dfrac{\partial u_r}{\partial r} - \dfrac{2}{3} (\nabla \cdot \mathbf{u}) \right]$
$\tau_{yy} = \mu \left[2 \dfrac{\partial u_y}{\partial y} - \dfrac{2}{3} (\nabla \cdot \mathbf{u}) \right]$	$\tau_{\theta\theta} = \mu \left[2 \left(\dfrac{1}{r} \dfrac{\partial u_\theta}{\partial \theta} + \dfrac{u_r}{r} \right) - \dfrac{2}{3} (\nabla \cdot \mathbf{u}) \right]$	$\tau_{\theta\theta} = \mu \left[2 \left(\dfrac{1}{r} \dfrac{\partial u_\theta}{\partial \theta} + \dfrac{u_r}{r} \right) - \dfrac{2}{3} (\nabla \cdot \mathbf{u}) \right]$
$\tau_{zz} = \mu \left[2 \dfrac{\partial u_z}{\partial z} - \dfrac{2}{3} (\nabla \cdot \mathbf{u}) \right]$	$\tau_{zz} = \mu \left[2 \dfrac{\partial u_z}{\partial z} - \dfrac{2}{3} (\nabla \cdot \mathbf{u}) \right]$	$\tau_{\phi\phi} = \mu \left[2 \left(\dfrac{1}{r \sin \theta} \dfrac{\partial u_\phi}{\partial \phi} + \dfrac{u_r}{r} + \dfrac{u_\theta \cot \theta}{r} \right) - \dfrac{2}{3} (\nabla \cdot \mathbf{u}) \right]$
$\tau_{xy} = \tau_{yx} = \mu \left[\dfrac{\partial u_x}{\partial y} + \dfrac{\partial u_y}{\partial x} \right]$	$\tau_{r\theta} = \tau_{\theta r} = \mu \left[r \dfrac{\partial}{\partial r} \left(\dfrac{u_\theta}{r} \right) + \dfrac{1}{r} \dfrac{\partial u_r}{\partial \theta} \right]$	$\tau_{r\theta} = \tau_{\theta r} = \mu \left[r \dfrac{\partial}{\partial r} \left(\dfrac{u_\theta}{r} \right) + \dfrac{1}{r} \dfrac{\partial u_r}{\partial \theta} \right]$
$\tau_{yz} = \tau_{zy} = \mu \left[\dfrac{\partial u_y}{\partial z} + \dfrac{\partial u_z}{\partial y} \right]$	$\tau_{\theta z} = \tau_{z\theta} = \mu \left[\dfrac{\partial u_\theta}{\partial z} + \dfrac{1}{r} \dfrac{\partial u_z}{\partial \theta} \right]$	$\tau_{\theta\phi} = \tau_{\phi\theta} = \mu \left[\dfrac{\sin \theta}{r} \dfrac{\partial}{\partial \theta} \left(\dfrac{u_\phi}{\sin \theta} \right) + \dfrac{1}{r \sin \theta} \dfrac{\partial u_\theta}{\partial \phi} \right]$
$\tau_{zx} = \tau_{xz} = \mu \left[\dfrac{\partial u_z}{\partial x} + \dfrac{\partial u_x}{\partial z} \right]$	$\tau_{zr} = \tau_{rz} = \mu \left[\dfrac{\partial u_z}{\partial r} + \dfrac{\partial u_r}{\partial z} \right]$	$\tau_{\phi r} = \tau_{r\phi} = \mu \left[\dfrac{1}{r \sin \theta} \dfrac{\partial u_r}{\partial \phi} + r \dfrac{\partial}{\partial r} \left(\dfrac{u_\phi}{r} \right) \right]$
$(\nabla \cdot \mathbf{u}) = \dfrac{\partial u_x}{\partial x} + \dfrac{\partial u_y}{\partial y} + \dfrac{\partial u_z}{\partial z}$	$(\nabla \cdot \mathbf{u}) = \dfrac{1}{r} \dfrac{\partial}{\partial r} (r u_r) + \dfrac{1}{r} \dfrac{\partial u_\theta}{\partial \theta} + \dfrac{\partial u_z}{\partial z}$	$(\nabla \cdot \mathbf{u}) = \dfrac{1}{r^2} \dfrac{\partial}{\partial r} (r^2 u_r) + \dfrac{1}{r \sin \theta} \dfrac{\partial}{\partial \theta} (u_\theta \sin \theta) + \dfrac{1}{r \sin \theta} \dfrac{\partial u_\phi}{\partial \phi}$

Table III.3. Navier–Stokes Equations for a Newtonian Fluid with Constant Density

Rectangular coordinates (x, y, z)

$$\rho\left(\frac{\partial u_x}{\partial t} + u_x\frac{\partial u_x}{\partial x} + y_y\frac{\partial u_x}{\partial y} + u_z\frac{\partial u_x}{\partial z}\right) = \mu\left[\frac{\partial^2 u_x}{\partial x^2} + \frac{\partial^2 u_x}{\partial y^2} + \frac{\partial^2 u_x}{\partial z^2}\right] - \frac{\partial p}{\partial x} + \rho g_x$$

$$\rho\left(\frac{\partial u_y}{\partial t} + u_x\frac{\partial u_y}{\partial x} + u_y\frac{\partial u_y}{\partial y} + u_z\frac{\partial u_y}{\partial z}\right) = \mu\left[\frac{\partial^2 u_y}{\partial x^2} + \frac{\partial^2 u_y}{\partial y^2} + \frac{\partial^2 u_y}{\partial z^2}\right] - \frac{\partial p}{\partial y} + \rho g_y$$

$$\rho\left(\frac{\partial u_z}{\partial t} + u_x\frac{\partial u_z}{\partial x} + u_y\frac{\partial u_z}{\partial y} + u_z\frac{\partial u_z}{\partial z}\right) = \mu\left[\frac{\partial^2 u_z}{\partial x^2} + \frac{\partial^2 u_z}{\partial y^2} + \frac{\partial^2 u_z}{\partial z^2}\right] - \frac{\partial p}{\partial z} + \rho g_z$$

Cylindrical coordinates (r, θ, z):

$$\rho\left(\frac{\partial u_r}{\partial t} + u_r\frac{\partial u_r}{\partial r} + \frac{u_\theta}{r}\frac{\partial u_r}{\partial \theta} - \frac{u_\theta^2}{r} + u_z\frac{\partial u_r}{\partial z}\right)$$
$$= \mu\left[\frac{\partial}{\partial r}\left(\frac{1}{r}\frac{\partial}{\partial r}(ru_r)\right) + \frac{1}{r^2}\frac{\partial^2 u_r}{\partial \theta^2} + \frac{\partial^2 u_r}{\partial z^2} - \frac{2}{r^2}\frac{\partial u_\theta}{\partial \theta}\right] - \frac{\partial p}{\partial r} + \rho g_r$$

$$\rho\left(\frac{\partial u_\theta}{\partial t} + u_r\frac{\partial u_\theta}{\partial r} + \frac{u_\theta}{r}\frac{\partial u_\theta}{\partial \theta} + \frac{u_r u_\theta}{r} + u_z\frac{\partial u_\theta}{\partial z}\right)$$
$$= \mu\left[\frac{\partial}{\partial r}\left(\frac{1}{r}\frac{\partial}{\partial r}(ru_\theta)\right) + \frac{1}{r^2}\frac{\partial^2 u_\theta}{\partial \theta^2} + \frac{\partial^2 u_\theta}{\partial z^2} + \frac{2}{r^2}\frac{\partial u_r}{\partial \theta}\right] - \frac{1}{r}\frac{\partial p}{\partial \theta} + \rho g_\theta$$

$$\rho\left(\frac{\partial u_z}{\partial t} + u_r\frac{\partial u_z}{\partial r} + \frac{u_\theta}{r}\frac{\partial u_z}{\partial \theta} + u_z\frac{\partial u_z}{\partial z}\right) + \mu\left[\frac{1}{r}\frac{\partial}{\partial r}\left(r\frac{\partial u_z}{\partial r}\right) + \frac{1}{r^2}\frac{\partial u_z}{\partial \theta^2} + \frac{\partial^2 u_z}{\partial z^2}\right] - \frac{\partial p}{\partial z} + \rho g_z$$

Spherical coordinates (r, θ, ϕ):

$$\rho\left(\frac{\partial u_r}{\partial t} + u_r\frac{\partial u_r}{\partial r} + \frac{u_\theta}{r}\frac{\partial u_r}{\partial \theta} + \frac{u_\phi}{r\sin\theta}\frac{\partial u_r}{\partial \phi} - \frac{u_\theta^2 + u_\phi^2}{r}\right)$$
$$= \mu\left[\frac{\partial}{\partial r}\left(\frac{1}{r^2}\frac{\partial}{\partial r}(r^2 u_r)\right) + \frac{1}{r^2\sin\theta}\frac{\partial}{\partial \theta}\left(\sin\theta\frac{\partial u_r}{\partial \theta}\right)\right.$$
$$\left. + \frac{1}{r^2\sin^2\theta}\frac{\partial^2 u_r}{\partial \phi^2} - \frac{2}{r^2\sin\theta}\frac{\partial}{\partial \theta}(u_\theta\sin\theta) - \frac{2}{r^2\theta}\frac{\partial u_\phi}{\partial \phi}\right] - \frac{\partial p}{\partial r} + \rho g_r$$

$$\rho\left(\frac{\partial u_\theta}{\partial t} + u_r\frac{\partial u_\theta}{\partial r} + \frac{u_\theta}{r}\frac{\partial u_\theta}{\partial \theta} + \frac{u_\phi}{r\sin\theta}\frac{\partial u_\theta}{\partial \phi} + \frac{u_r u_\theta}{r} - \frac{u_\phi^2\cot\theta}{r}\right)$$
$$= \mu\left[\frac{1}{r^2}\frac{\partial}{\partial r}\left(r^2\frac{\partial u_\theta}{\partial r}\right) + \frac{1}{r^2}\frac{\partial}{\partial \theta}\left(\frac{1}{\sin\theta}\frac{\partial}{\partial \theta}(u_\theta\sin\theta)\right)\right.$$
$$\left. + \frac{1}{r^2\sin^2\theta}\frac{\partial^2 u_\theta}{\partial \phi^2} + \frac{2}{r^2}\frac{\partial u_r}{\partial \theta} - \frac{2\cot\theta}{r^2\sin\theta}\frac{\partial u_\phi}{\partial \phi}\right] - \frac{1}{r}\frac{\partial p}{\partial \theta} + \rho g_\theta$$

Table III.3. *(Continued)*

$$\rho\left(\frac{\partial u_\phi}{\partial t} + u_r\frac{\partial u_\phi}{\partial r} + \frac{u_\theta}{r}\frac{\partial u_\phi}{\partial \theta} + \frac{u_\phi}{r\sin\theta}\frac{\partial u_\phi}{\partial \phi} + \frac{u_\phi u_r}{r} + \frac{u_\theta u_\phi}{r}\cot\theta\right)$$

$$= \mu\left[\frac{1}{r^2}\frac{\partial}{\partial r}\left(r^2\frac{\partial u_\phi}{\partial r}\right) + \frac{1}{r^2}\frac{\partial}{\partial \theta}\left(\frac{1}{\sin\theta}\frac{\partial}{\partial \theta}(u_\phi\sin\theta)\right)\right.$$

$$\left. + \frac{1}{r^2\sin^2\theta}\frac{\partial^2 u_\phi}{\partial \phi^2} + \frac{2}{r^2\sin\theta}\frac{\partial u_r}{\partial \phi} + \frac{2\cot\theta}{r^2\sin\theta}\frac{\partial u_\theta}{\partial \phi}\right] - \frac{1}{r\sin\theta}\frac{\partial p}{\partial \phi} + \rho g_\phi$$

Table III.4. The Convective-Diffusion Equation for a Constant ρ and D_{AB}

Rectangular coordinates (x, y, z):

$$\frac{\partial c_A}{\partial t} + \left(u_x\frac{\partial c_A}{\partial x} + u_y\frac{\partial c_A}{\partial y} + u_z\frac{\partial c_A}{\partial z}\right) + D_{AB}\left(\frac{\partial^2 c_A}{\partial x^2} + \frac{\partial^2 c_A}{\partial y^2} + \frac{\partial^2 c_A}{\partial z^2}\right) + R_A$$

Cylindrical coordinates (r, θ, z):

$$\frac{\partial c_A}{\partial t} + \left(u_r\frac{\partial c_A}{\partial r} + u_\theta\frac{1}{r}\frac{\partial c_A}{\partial \theta} + u_z\frac{\partial c_A}{\partial z}\right) = D_{AB}\left(\frac{1}{r}\frac{\partial}{\partial r}\left(r\frac{\partial c_A}{\partial r}\right) + \frac{1}{r^2}\frac{\partial^2 c_A}{\partial \theta^2} + \frac{\partial^2 c_A}{\partial z^2}\right) + R_A$$

Spherical coordinates (r, θ, ϕ):

$$\frac{\partial c_A}{\partial t} + \left(u_r\frac{\partial c_A}{\partial r} + u_\theta\frac{1}{r}\frac{\partial c_A}{\partial \theta} + u_\phi\frac{1}{r\sin\theta}\frac{\partial c_A}{\partial \phi}\right)$$

$$= D_{AB}\left(\frac{1}{r^2}\frac{\partial}{\partial r}\left(r^2\frac{\partial c_A}{\partial r}\right) + \frac{1}{r^2\sin\theta}\frac{\partial}{\partial \theta}\left(\sin\theta\frac{\partial c_A}{\partial \theta}\right) + \frac{1}{r^2\sin^2\theta}\frac{\partial^2 c_A}{\partial \phi^2}\right) + R_A$$

BIBLIOGRAPHY

Bird, R. V., Stewart, W. D., and Lightfoot, E. N., *Transport Phenomena*, Wiley, New York, 1960.

Panton, R. L., *Incompressible Flow*, Wiley, New York, 1984.

APPENDIX IV

PHYSICAL PROPERTIES OF WATER AND AIR

Table IV.1. Physical Properties of Water (SI Units)

Temperature (°C)	Density, ρ (kg/m^3)	Dynamic Viscosity, μ (kg-m^{-1}-s^{-1})	Kinematic Viscosity, v (m^2/s)
0	999.9	1.787E−03	1.787E−06
5	1000.0	1.519E−03	1.519E−06
10	999.7	1.307E−03	1.307E−06
20	998.2	1.002E−03	1.004E−06
30	995.7	7.975E−04	8.009E−07
40	992.2	6.529E−04	6.580E−07
50	988.1	5.468E−04	5.534E−07
60	983.2	4.665E−04	4.745E−07
70	977.8	4.042E−04	4.134E−07
80	971.8	3.547E−04	3.650E−07
90	965.3	3.147E−04	3.260E−07
100	958.4	2.818E−04	2.940E−07

Table IV.2. Physical Properties of Air at Standard Atmospheric Pressure (SI Units)

Temperature ($^\circ$C)	Density, ρ (kg/m^3)	Dynamic Viscosity, μ (kg-m^{-1}-s^{-1})	Kinematic Viscosity, v (m^2/s)
-40	1.514	1.57E$-$05	1.04E$-$05
-20	1.395	1.63E$-$05	1.17E$-$05
0	1.292	1.71E$-$05	1.32E$-$05
5	1.269	1.73E$-$05	1.36E$-$05
10	1.247	1.76E$-$05	1.41E$-$05
15	1.225	1.80E$-$05	1.47E$-$05
20	1.204	1.82E$-$05	1.51E$-$05
25	1.184	1.85E$-$05	1.56E$-$05
30	1.165	1.86E$-$05	1.60E$-$05
40	1.127	1.87E$-$05	1.66E$-$05
50	1.109	1.95E$-$05	1.76E$-$05
60	1.060	1.97E$-$05	1.86E$-$05
70	1.029	2.03E$-$05	1.97E$-$05
80	0.9996	2.07E$-$05	2.07E$-$05
90	0.9721	2.14E$-$05	2.20E$-$05
100	0.9461	2.17E$-$05	2.29E$-$05
200	0.7461	2.53E$-$05	3.39E$-$05
300	0.6159	2.98E$-$05	4.84E$-$05
400	0.5243	3.32E$-$05	6.34E$-$05
500	0.4565	3.64E$-$05	7.97E$-$05
1000	0.2772	5.04E$-$05	1.82E$-$04

BIBLIOGRAPHY

Munson, B. R., Young, D. F., and Okiishi, T. H., *Fundamentals of Fluid Mechanics*, Wiley, New York, 1990.

INDEX

553

ENVIRONMENTAL SCIENCE AND TECHNOLOGY
List of Titles (*Continued*)